Forum for Interdisciplinary Mathematics

T0171804

The *Forum for Interdisciplinary Mathematics* (FIM) series publishes high-quality monographs and lecture notes in mathematics and interdisciplinary areas where mathematics has a fundamental role, such as statistics, operations research, computer science, financial mathematics, industrial mathematics, and bio-mathematics. It reflects the increasing demand of researchers working at the interface between mathematics and other scientific disciplines.

More information about this series at http://www.springer.com/series/13386

Sujit Kumar Bose

Numerical Methods of Mathematics Implemented in Fortran

Sujit Kumar Bose
S. N. Bose National Centre for Basic
Sciences
Kolkata, West Bengal, India

Additional material to this book can be downloaded from http://extras.springer.com.

ISSN 2364-6748   ISSN 2364-6756   (electronic)
Forum for Interdisciplinary Mathematics
ISBN 978-981-13-7116-5   ISBN 978-981-13-7114-1   (eBook)
https://doi.org/10.1007/978-981-13-7114-1

Library of Congress Control Number: 2019933853

This Springer imprint is published by the registered company Springer Nature Singapore Pte Ltd.
The registered company address is: 152 Beach Road, #21-01/04 Gateway East, Singapore 189721,
Singapore

|| Mangalācaranam ||

Om sahanāvavatu sahanaubhunaktu
sahavíryyaṁkarvāvahai |
Tejasvināvadhítamastu mā vidvíśāvahai ||

|| *Om Śāntiḥ Śāntiḥ Śāntiḥ* ||

{O the Universal Consciousness protect and nourish us (both teacher and student) to work with great vigour to enlighten our intellect, without creation of any jealousy or rancour whatsoever (in that endeavour). Let peace, peace, peace be everywhere.}

Preface

Numerical computation of mathematically formulated problems is universal in sciences and engineering. The mathematical models appearing in the two disciplines essentially boil down to formalisms of analysis and linear algebra, and for obtaining useful results from them, numerical methods were devised from time to time. The numerical methods were laid on a firm mathematical foundation later on, during the course of development of the subject. The invention of computing machines, such as the present-day digital computer, has greatly increased the power of numerically solving difficult scientific problems.

The gamut of scientific computation consists of the development of methods of treating different mathematical models, their mathematical analysis and writing computer codes for obtaining desired results. A course on numerical methods of mathematics is an integral part of the undergraduate science and engineering studies. This study sometimes forms a part of the postgraduate courses as well. The emphasis on methods, analysis and coding for computer implementation varies in different course types. A typical course may consist of the principal methods, and some programming, with little intricacies of analysis. However, a theoretical course may emphasise the methods together with their analyses and possibly some algorithms of the methods developed. This book is designed in such a manner that a course instructor can draw requisite material for any of the courses leaving the remaining parts for reference purposes. For laying emphasis on the methods that form the bulk of the courses, simple examples and exercises are given for hand calculation with the aid of a calculator. For courses emphasising the underlying mathematical theory, the topics are presented in separate sections. At the other end where the reader would like to have complete computer codes, concise computer programs are provided that follow the diverse methods developed in a general manner. In some cases, programming tricks become necessary for obtaining correct and accurately computed results from computer programs. Such tricks are rarely employed in this book. The succinct programs presented as subroutine procedures in the text are applicable under very general settings, so as to be useful even in research investigations. Special packages often popular with students sometimes give disappointingly incorrect results and are hard to debug on account of their opacity. Open source codes, however, are often long-drawn statements, not clearly reflecting the corresponding algorithm followed. The subroutines presented in the book are free from these difficulties on account of their transparency.

For writing the computer codes, I have chosen to use the Fortran programming language. Fortran is a powerful, easy-to-learn and traditional language for scientific computation in which there are a number of inbuilt intrinsic functions. The present version Fortran 2003 is fully backwards-compatible with its earlier popular versions FORTRAN 77 and Fortran 90/95. The language has been upgraded to Fortran 2010, and a future version Fortran 2018 is proposed to be launched during this year. The latest versions are the only ones usable in supercomputers and in modern multiprocessor desktops. There is, therefore, some merit in retaining Fortran for the purpose of scientific computation, especially because of the fact that a large number of source codes mentioned in Chap. 1 are written in that language. At present, to my knowledge, the output in Fortran is alphanumeric in nature. Conversion of output to desirable graphic displays, however, is easily accomplished by using some graphic software.

The book is divided into ten chapters. Chapter 1 gives a brief description of computer software and types of errors encountered in computing due to truncation of mathematical formulae and digitisation of decimal numbers. Elements of Fortran statements used in the book are also described in this chapter. Elementary illustrative examples are given in the chapter for initiation in Fortran programming to beginners. Chapters 2 and 3 treat the fundamental problem of solving a single equation or a system of several equations possessing real and complex roots, respectively. A major part of Chap. 3 deals with the most important topic of solving a system of linear equations. Chapter 4 provides the study of the classic problem of approximation of a function by interpolation over the given data points, deferring the treatment of the more theoretical topics of approximation to Chap. 8. Chapter 4 provides the methods which lead to the topics of numerical differentiation and integration and to the solution of ordinary and partial differential equations numerically. These topics are respectively covered in Chaps. 5–7. Chapters 9 and 10 treat the computationally fascinating, but mathematically difficult subjects of the matrix—eigenvalues and the fast Fourier transform for computing Fourier integrals. Short biographical sketches of the principal discoverers of the weighty material have been given to enliven the reader as one proceeds to uncover the masterly techniques. Tomes have been written by mathematicians and programmers on topics of each of the ten chapters. In a moderately sized book like the present one, the selection of the material arranged in some logical manner, therefore, became a necessity. I have kept the benchmark of choice to understand the basics, keeping in view the utility of application of the routines to the extent of research-level problems. A suite of 38 ready-to-use subroutines is given that directly follow the different methods developed in the book.

The book is an outcome of teaching the subject to undergraduate and postgraduate students of engineering in which the students provided a huge contribution as feedback. No less has been the effect on my own travails in computing for my research problems that have spanned for at least the past five decades. The compilation of this book is a reflection of these experiences. As writing progressed, I had to sift through several books including one of my own that strayed away from the benchmark laid down by myself. The ones that I found useful to varying degrees are thankfully acknowledged at the end of the book. Thanks are also due to S. N. Bose National Centre for Basic Sciences, Kolkata, for providing me the library facility to look at new texts that have appeared in recent years.

Kolkata, India   Sujit Kumar Bose

Contents

About the Author

Sujit Kumar Bose is a distinguished computer educationist and author of several useful books on computational and numerical mathematics. He was a Research Consultant at the Department of Civil Engineering, Indian Institute of Technology Kharagpur, from 2002 to 2014. He was also member of the Guest Faculty at the Department of Computer Science, Jadavpur University, Kolkata, from 1998 to 2006; a Senior Professor of Mathematics at the S. N. Bose National Centre for Basic Sciences, Kolkata, India, from 1993 to 1998; Professor and Head of the Department of Mathematics at the National Institute of Technology, Durgapur, India, from 1975 to 1993; an Assistant Professor at the West Bengal Educational Service, from 1965 to 1975; and a Lecturer at the West Bengal Junior Educational Service, from 1960 to 1965. In addition, he was a postdoctoral fellow at the University of California, Los Angeles, from 1972 to 1974.

Professor Bose is associated with several respected research institutes: Fellow of the Indian Academy of Sciences, India; member of the Indian Society of Theoretical and Applied Mechanics at the Department of Mathematics, Indian Institute of Technology Kharagpur; member of the Calcutta Mathematical Society; and member of the Indian Statistical Institute, Kolkata. He is awarded the Prof. B. Sen Memorial Gold Medal for the Best Research in Elasticity, in 1978, and the Sardar Vallabhbhai Patel Award for the Excellence in Researches in Mathematical Sciences, in 2012. He obtained B.Sc. (Math Hons, 1957) and M.Sc. (Applied Mathematics, 1959) as well as D.Sc. (Applied Mathematics, 1971), all from the University of Calcutta, winning several gold medals. He was a Visiting Scholar at the Computational Mechanics Center, Georgia Institute of Technology, USA, and the Department of Applied Mathematics at the University of Twente, The Netherlands in 1989. Professor Bose has research interests in applied mathematics, solid mechanics, fluid mechanics and sediment transport.

Chapter 1
Computation in FORTRAN

Mathematically formulated problems pervade all branches of science and engineering. To an extent, this trend is discernible in social sciences like economics and business management as well. Lately, medical science is also catching up with this trend. When a mathematical formulation is simple in form, one can manage the calculations for the solution of the problem using paper, pencil and a *digital calculator*. On the other hand, if calculations for the solution are long and complicated in nature, one is compelled to consider the general mathematical form of the problem. **Scientific** or **Numeric Computing** is concerned with restructuring the mathematical formulation into an **algorithm** or a procedure involving simple arithmetic operations, $+$, $-$, $\times$ and $/$, suitable for machine calculations by a digital computer. Algorithms offer answers or **output** for the problem in numeric form that could be presented in visual form for better display if necessary. Initially, an algorithm requires some special information or **input** for producing the output. A **digital computer** works on the concept of *stored program*, which is an algorithm coded in *machine language*, generated by the computer itself, after the algorithm for the solution of the problem in hand is written in some **programming language** like **FORTRAN**. The techniques of FORTRAN programming are briefly elaborated in this chapter by considering simple but useful algorithms. FORTRAN programs for complicated algorithms appear in subsequent chapters that deal with different kinds of generalised mathematical formulations. The current versions of Fortran are very convenient for coding mathematical expressions and procedures.

The term **algorithm** is named after **Abu Jafár Muhammad ibn Musa Al-Khwarizimi** (A.D. 825), Arabic mathematician living in Baghdad, who first wrote a treatise of Arithmetic using Hindu decimal numbers. This book made its way from Baghdad to Spain and translated into Latin by English mathematician Robert of Chester and Spanish by John of Saville. In this way, Hindu decimal numbers were introduced to the west. Al-Khwarizimi also authored a book on algebra and its applications.

© Springer Nature Singapore Pte Ltd. 2019
S. K. Bose, *Numerical Methods of Mathematics Implemented in Fortran*, Forum for Interdisciplinary Mathematics,
https://doi.org/10.1007/978-981-13-7114-1_1

1.1   Calculators and Computers

Calculators and computers are number-crunching digital electronic machines. Numbers or more precisely positive *integers* and zero are known in this country since ancient times in the **decimal system** of representation. The system consists of ten digits 0, 1, 2, $\cdots$, 9 in which ten represented by 10 forms the **base** of the system. One is familiar with the representation of the succeeding numbers using the ten digits with *place value*. Let $d_1, d_2, \cdots , d_n$ be any of the decimal digits, then in the integer $d_1 d_2 \cdots d_n$ (not to be confused with the product of $d_1, d_2, \cdots$, and d_n) , d_n is said to be in *unit* position, d_{n-1} in *ten's* position, d_{n-2} in *hundred's* position and so on. Interestingly, the number can be expressed as a polynomial in 10:

$$d_1 d_2 \cdots d_n = d_1 \times 10^{n-1} + d_2 \times 10^{n-2} + \cdots + d_{n-1} \times 10^1 + d_n = \sum_{i=0}^{n-1} d_{n-i} 10^i$$

$$(1.1)$$

For example, one can write

$$13 = 1 \times 10^1 + 3$$
$$544 = 5 \times 10^2 + 4 \times 10^1 + 4$$

The two 4's in 544 have different place value although the digits are the same!

   Negative integers and *fractional (or rational) numbers* were, respectively, created to meet the necessity of closing the subtraction and division processes for all numbers. One is familiar with the representation of negative numbers by putting a $-$ sign in front of the number. Similarly, the decimal representation of fractional numbers $(\cdot / \cdot)$ is well known. These numbers were known in this country before Al Khwarizimi's time by about 800 A.D. Like an integer a terminating fractional number $d_1 d_2 \cdots d_n . d_{n+1} d_{n+2} \cdots d_{n+m}$ can also be represented as a polynomial in 10:

$$d_1 d_2 \cdots d_n . d_{n+1} d_{n+2} \cdots d_{n+m} = d_1 \times 10^n + d_2 \times 10^{n-1} + \cdots + d_n \times 10^0$$
$$+ d_{n+1} \times 10^{-1} + d_{n+2} \times 10^{-2} + \cdots + d_{n+m} \times 10^{-m}$$
$$= \sum_{i=-m}^{n-1} d_{n-i} 10^i \qquad (1.2)$$

Thus, for example, one can write

$$13.544 = 1 \times 10^1 + 3 \times 10^0 + 5 \times 10^{-1} + 4 \times 10^{-2} + 4 \times 10^{-3}$$

Fractional and irrational numbers together are called *real numbers*. Because of physical limitation of a computer, nonterminating reals like the *irrational numbers* $\sqrt{2} = 1.41421356\cdots$ and $\pi = 3.141592653\cdots$ cannot be handled exactly in practice. Only *sufficiently accurate* representation by terminated the two reals such as 1.414214 and 3.141593 is possible depending on the accuracy required for computation. In such approximations, the digits that are present in the exact representation of the numbers are called **significant digits**. Thus, in the above two representations, there are *six* significant digits, the digits 4 and 3 in the two numbers appearing in the last place having no *significance* because these two digits are obtained by replacing the succeeding digits present in the two numbers. Following the same idea, zeros in front of integers and those at the end of decimal fraction have no significance and therefore not written. For example, 013.5440 is written as 13.544 as the dropped zeros do not contribute to the number and have no significance. The difficulty of exactly calculating with such transcendental numbers by a computer, following some algorithm, is self-evident.

Remark The representations (1) and (2) of integers and reals is suggestive that numbers can be represented in any base B by replacing 10 by B and the digits by the admissible ones. For example, if $B = 2$ with only two digits 0 and 1, a number is called a **binary digit** or **bit**. For $B = 8$ and $B = 16$, the numbers are, respectively, called **octal** and **hexadecimal**. Conversion from one system to another is possible following the polynomial representation.

1.1.1 Computer Driver: Software

A computer needs clear-cut instructions to start up and be ready to process user jobs. A set of instructions to carry out any such function is called a **program** and a group of such programs that are put into a computer to operate and control its activities is called **Software**. A program is stored in a **file** and a software consists of a number of files. In a computer, there are three essential kinds of software, which are stored in hard disk. These are: (i) **operating system**, (ii) **language processors** or **translators** or **compilers** and (iii) user application programs.

The *Operating System* (or OS) manages the resources of a computer and schedules its operation. It comes as a number of **system files** and acts as an interface between the hardware and other kinds of software. Its functions include: (a) monitoring the use of the machine's resources, (b) controlling and coordinating peripheral devices like the I/O devices, (c) helping user to develop programs, (d) helping user programs execute its instructions, and (e) dealing with any faults that may occur in the computer and inform the user. An OS comes with the computer hardware. Popular operating systems are **Windows** and **Linux**.

Application programs following certain algorithms for solving mathematically formulated problems are written by users in some programming language. Such user-written programs are called *Source Code*. **FORTRAN** (standing for FORmula TRANslation) was especially designed by IBM Corporation and launched in the year 1957 for writing of programs for the purpose of scientific computing. At present, vastly improved versions of the language are available both as open source and as commercial software. In these versions, program writing can be done in modular structures that follow the steps of the corresponding mathematical formulation exactly. This speciality makes the task of checking of any Fortran program quite easily. Other later generation well-known programming languages are C developed by Bell Laboratories, and JAVA developed by Sun Microsystems. The latter two languages, however, are of general nature, not necessarily oriented towards scientific computations that have led to rapid growth of *Information Technology*, making the computer *ubiquitous*. MATLAB developed by Math Works Inc., and PYTHON developed by Centrum Wiskunde and Informatica are two other programming tools, popular with the scientific computing community. These two languages work slightly differently, adopting a number of special purpose inbuilt tools leading to directly obtaining visual outputs, if required. In this book, the FORTRAN programming language has been adopted for important reasons. It is best suited for learning higher level programming, extending from scratch to programs for supercomputers. Being the first language developed for scientific computing, it has a vast repository of known source codes for direct use. Diverse codes such as those for numerical weather prediction, finite element analysis of structures, computational fluid dynamics, computational physics, crystallography and computational chemistry are all available in Fortran only. To name a few, Fission Reaction Event Yield Algorithm (FREYA) of Nuclear Physics, Vienna Ab Inition Simulation Package (VASP) and MicrOMEGAs of Materials Science and Numerical Electromagnetic Code (NEC) of electromagnetism belong to this category. The Fortran source codes are transparent and can be modified whenever required. Finally, it is interoperable by *binding* with programs written in other languages, such as C, PYTHON and MATLAB. Being a language oriented towards scientific computations, it handles complex valued functions easily, a facility not available in JAVA. An assessment states that Fortran source codes can be faster in execution by as much as 30%, compared to source codes written in other languages. The saving in execution time is therefore substantial for large programs.

The task of writing Fortran programs following algorithms of mathematical methods is mostly straightforward, save for the fact that errors in computing inevitably occur as described in the next section. In some cases, especially when a huge number of operations are involved during the execution of the programs, spurious output may be returned as output. Great care is therefore needed in writing Fortran codes, for bypassing such traps. In pathologically bad cases, the accurate computation may even not be possible. Such cases are however rare, leaving scope for research in order to overcome the difficulties present in the mathematical method followed.

1.2 Errors in Computing

We suppose that a scientific problem is mathematically formulated, in its exactitude, as best as possible. By this, we mean that even though the formulation is obtained following certain assumptions regarding the actual nature of the problem, the formulation by itself does not violate any mathematical rule. Frequently such formulation, or mathematical model of the actual problem use tools of analysis and its calculi, and thus invoke infinitesimal and infinity of magnitude. Computation, whether done by paper and pencil or by machine, is, on the other hand, always finite state. Any quantity, say x, is thus approximated by another quantity $\tilde{x}$, as closely as possible, depending on the capacity of the machine (or hand computation). Thus, an **error** creeps into the quantity x, which is precisely defined by the relation

$$\text{True Value} = \text{Approximate Value} + \text{Error}$$

or
$$\text{Error} = \text{True Value} - \text{Approximate value} = x - \tilde{x} \qquad (1.3a)$$

This error is sometimes called **Absolute Error**. In contrast, the **Relative Error** is defined as

$$\text{Relative Error} = \frac{\text{Error}}{\text{True Value}} = \frac{\text{Error}}{x} \qquad (1.3b)$$

and similarly one can define

$$\text{Relative Error with respect to } \tilde{x} = \frac{\text{Error}}{\text{Approximate Value}} = \frac{\text{Error}}{\tilde{x}} \qquad (1.3c)$$

For example, let $x = 1/3 = 0.3333 \cdots$ (to infinity), and $\tilde{x} = 0.333$. Then, the error $= 1/3 - 0.333 = 1/3 \times 10^{-3}$; relative error $= 1/3 \times 10^{-3}/(1/3) = 10^{-3}$ and relative error with respect to $\tilde{x} = 1/3 \times 10^{-3}/(0.333) = 1.00\dot{1} \times 10^{-3}$. In the last quantity, 001 recurs infinitely after the decimal.

We are generally interested in the absolute error, but the relative error is a better index when a quantity is very small or very large. For, consider the case of $x = 10^{-9}$ and $\tilde{x} = 10^{-6}$ (both close to 0), then error $= -0.999 \times 10^{-6}$ is still small but the relative error $= -999$ is large. Similarly one can construct examples for very large numbers.

When there are several quantities and each one is approximated suitably, then the errors compound. In computing, the endeavour is to keep the accumulated error firmly in check, so that one obtains results to some desired accuracy. To achieve this goal, one has to diagnose the nature of approximations, which come into play while computing, and trace the nature of the errors. This question is examined in the following subsections.

1.2.1 Truncation Errors

Most analytical techniques use infinite series, differentiation and integration. The occurrence of functions like $\sin(x)$, $\cos(x)$, $\exp(x)$, $\ln(x)$ is commonplace. These are transcendental functions having convergent infinite series representation like

$$\sin(x) = x - \frac{x^3}{3!} + \frac{x^5}{5!} - \cdots + (-1)^{N-1} \frac{x^{2N-1}}{(2N-1)!} + \cdots \infty \qquad (1.4)$$

For computation, the series must be replaced by computable expression. Since the series is rapidly convergent, one can *truncate* the series up to N terms and approximate the function by a polynomial of degree $2N - 1$. Calculators and computers use this technique in a sophisticated manner to 'evaluate' the function with very high accuracy (see Chaps. 4 and 8 for introduction of such techniques). In differentiation, given a function $f(x)$, its derivative is defined by

$$f'(x) = \lim_{h \to 0} \frac{f(x+h) - f(x)}{h} \qquad \text{(fixed } x\text{)} \qquad (1.5)$$

which again is an infinite process with decreasing values of h. So one can approximate it by $[f(x + h) - f(x)]/h$ where h is fixed but small. This procedure can be extended to higher order derivatives as well as to partial derivatives of a function of several variables. Definite integrals are commonly defined by Riemann integration. The integral of a bounded function $f(x)$ on $[a, b]$ is

$$\int_a^b f(x)\, dx = \lim_{h \to 0} \left[h \sum_{i=1}^{(b-a)/h} f[a + (i - 1 + \theta)h] \right] \qquad 0 \le \theta \le 1 \qquad (1.6)$$

where equal subintervals are chosen for subdivision of the interval $[a, b]$. So, an approximation could be the expression within the outer brackets, for a small value of h.

The truncation in the above mentioned infinite processes leads to **truncation error**. It depends on N and h as the case may be. A practical way to estimate the truncation error is to increase N or decrease h and observe the computed results. If the computed results 'settle down' or 'converge', then one might decide that the truncation error is small enough to produce an acceptable approximate result. Such a methodology is called **convergence test** and most programs have one or more of such convergence tests for obtaining approximate results to the desired degree of accuracy. *This does not, however, mean that convergence occurs in all cases whenever a test is passed.* In fact counter-examples can be constructed to prove that *testing for convergence may be a mathematically unsolvable problem.* One such example is given below. Consider the integral

$$A = \int_1^2 \frac{\sin x}{x}\, dx$$

We subdivide $[1, 2]$ into 10 subintervals of length 0.1 each and obtain

$$A = \left[\int_{1.0}^{1.1} + \int_{1.1}^{1.2} + \cdots + \int_{1.9}^{2.0} \right] \left(\frac{\sin x}{x}\, dx \right)$$

Now suppose that an integral in the above expression is approximated by a single subdivision of $h = 0.1$ and the function value by the mean of the values of the integrand at the end points of the subdivision. The approximation of the first integral in the expression for A is then

$$\int_{1.0}^{1.1} \frac{\sin x}{x} \, dx \approx 0.1 \times \frac{1}{2} \left[\frac{\sin(1.0)}{1.0} + \frac{\sin(1.1)}{1.1} \right]$$

Similarly for the other sub-integrals. Adding them, we get

$$A \approx 0.05 \times \left[\frac{\sin(1.0)}{1.0} + 2 \times \left\{ \frac{\sin(1.1)}{1.1} + \frac{\sin(1.2)}{1.2} + \cdots + \frac{\sin(1.9)}{1.9} \right\} + \frac{\sin(2.0)}{2.0} \right]$$
$$= 0.65921$$

This method of approximation is called the Composite Trapezoidal Rule (see Chap. 5, Sect. 2.3). The exact value is however $A = 0.65932 \cdots$ obtained by using an accurate integrator. Evidently, there is agreement up to three decimal places. Now suppose that the approximation passes a convergence test with some prescribed accuracy. We show that an arbitrary function $g(x)$ could also yield the same approximation. For this purpose, define a saw-tooth function $s(x)$ by

$$s(x) = \begin{cases} 0 & \text{for} \quad x = x_i = 1 + 0.1 \times (i - 1) \\ 10(x - x_i) & \text{for} \quad x_i < x \leq x_{i+1} \end{cases}$$

where $i = 1, 2, 3 \cdots, 11$. Now define a function $g(x) = (\sin x)/x + C \times s(x)$. Its integral is

$$G = \int_1^2 \frac{\sin x}{x} + C \int_1^2 s(x) \, dx = A + C \times \frac{1}{2} = A + 0.5 \times C$$

where C is an *arbitrary constant*. In the above equation, each tooth of $s(x)$ is a triangle of base 0.1 and height 1; so the integral of $s(x)$ that equals the total area under $s(x)$ is $(\frac{1}{2} \times 0.1 \times 1) \times 10 = 1/2$. Now if the composite trapezoidal rule is applied to G, its approximation by the rule equals that of A, because $s(x)$ vanishes at all the points of subdivision. Thus, the approximations of G and A are the same even though $g(x)$ and $\sin x/x$ differ arbitrarily and so it may be impossible to compute an integral like G accurately, by adopting some particular numerical integration method.

   In *ordinary circumstances*, one may *assume 'pathological difficulty' of the above nature does not exist*, so that simple convergence test suffice. Sometimes more than one test is employed to guard against difficult cases, yielding accurate results. For example, in the case of evaluation of A, one may (1) decrease h by increasing the number of subintervals, and (2) count the subintervals in which the integrand oscillates (non-monotonic). If the computed value settles to the desired number of decimal places and there are few subintervals of swing in the function, the computed result may be accepted. In practice, additional tests seldom take up excessive computational time.

Whenever truncation is resorted to, it is of paramount importance to know, how fast the error goes to zero as parameters like h and N vary. The answer to such question is best represented by using the Landau 'big O' symbol. It is defined as follows:

Definition A function $f(x)$ is said to be of the **order of the function** $g(x)$ if

$$\lim_{x \to x_0} \left| \frac{f(x)}{g(x)} \right| = C < \infty \tag{1.7}$$

for some constant $C > 0$ and on writes $f(x) = O(g(x))$ or $f(x)$ is big-oh $g(x)$. In this case $|f(x)| \le K|g(x)|$, $K > 0$ for $x \to x_0$. If $C = 0$, then $f(x)$ is said to be of **higher order** than $g(x)$ and one writes $f(x) = o(g(x))$ or $f(x)$ is little-oh of $g(x)$. From the definition, it is easy to check the following properties:

(i) $f(x) = o(g(x)) \Rightarrow f(x) = O(g(x))$
(ii) $f(x) = K \cdot O(g(x)) \Rightarrow f(x) = O(g(x))$, where $K = $ constant
(iii) $f(x) = O(g_1(x))$ and $g_1(x) = O(g_2(x)) \Rightarrow f(x) = O(g_2(x))$
(iv) $f_1(x) = O(g_1(x))$ and $f_2(x) = O(g_2(x)) \Rightarrow f_1(x) \cdot f_2(x) = O(g_1(x) \cdot g_2(x))$
(v) $f(x) = O(g_1(x) \cdot g_2(x)) \Rightarrow f(x) = g_1(x) \cdot O(g_2(x))$

Analogues of properties $(ii) - (v)$ also hold for the symbol 'o'.

When the error in a truncation process is like $O(h^p)$ or $O(1/N^k)$, the convergence is said to be of the order h^p or $1/N^k$, respectively. The term **rate** is often used synonymously to order. We can use the definition to examine the order of convergence of $\sin(x)$ and $f'(x)$ considered in Eqs. (1.4) and (1.5). Denoting by $S_N(x)$ the sum of the first N terms of Eq.(1.4), which approximates $\sin(x)$, we have according to Eq. (1.3a)

$$\text{error} = \sin(x) - S_N(x) = (-1)^{N+1} \frac{x^{2N+1}}{(2N + 1)!} + \cdots$$

Since the series on the right-hand side of the above equation is uniformly convergent, we get

$$\lim_{N \to \infty} \left| \frac{\text{error}}{x^{2N+1}/(2N + 1)!} \right| = 1, \quad \text{or,} \quad \text{error} = O\left(\frac{x^{2N+1}}{(2N + 1)!} \right), \quad \text{as } N \to \infty$$

The process is therefore rapidly convergent, as the sequence $x^{2N+1}/(2N + 1)!$ rapidly converges to 0 for all x. For the approximation of the derivative $f'(x)$ considered in Eq. (1.5), if we use Taylor's theorem

$$f(x + h) = f(x) + hf'(x) + \frac{h^2}{2!} f''(x + \theta h), \quad 0 < \theta < 1$$

The error (1.3a) is therefore given by

$$\text{error} = f'(x) - \frac{f(x + h) - f(x)}{h} = -\frac{h}{2} f''(x + \theta h)$$

So, $\lim_{h \to 0} |\text{error}/h| = \frac{1}{2} f''(x)$ or error $= O(h)$ as $h \to 0$. Thus, the error is of the *first* order. It is a slowly convergent process. The topic of order of convergence of methods for evaluating definite integrals like Eq. (1.6) will be considered in Chap. 5. In most of the chapters, truncation of exact expressions will be the key to develop numerical methods.

1.2.2 Roundoff Errors

Computers have certain limitations. The Central Processing Unit (CPU) is a two-state *binary* machine and can store numbers in bits 0 and 1 only. So a decimal number entered by a user is first encoded in binary bits. In fact, all other information is also encoded in binary bits. Processing of numbers is performed by storing a number in a register of the CPU and the latter has always a given capacity called *word length*. This brings in a limitation on the size of numbers and infinitely large numbers cannot be stored.

The storage in a register is organised in two ways: according to whether a number is an **integer** or **real**. The corresponding arithmetic is also organised accordingly. These are called fixed-point arithmetic and Floating-point arithmetic.

1°. Fixed-point Arithmetic

Consider a positive or negative **integer** x represented by n significant digits $\pm d_1 d_2 \cdots d_n$ in base B. For a computer $B = 2$, in which case $d_i = $ a bit, either 0 or 1; whereas for the user $B = 10$. In both the cases, one can write

$$x = \pm(d_1 B^{n-1} + d_2 B^{n-2} + \cdots + d_{n-1} B + d_n) = \sigma \sum_{i=0}^{n-1} d_{n-i} B^i \tag{1.8}$$

where σ is positive or negative sign, according to the sign of x. The number of digits n is called *mantissa* length of x.

Let $N (\geq n)$ be the word length of the computer, and then σ can be represented by a bit of the register (+ by 0 and − by 1). The biggest number in magnitude that can be stored is $x_m = \pm(11 \cdots 1)$, in which all the remaining $N - 1$ bits are 1. Then,

$$x_m = \sigma(1 \times 2^{N-2} + 1 \times 2^{N-3} + \cdots + 1 \times 2 + 1) = \pm(2^{N-1} - 1)$$

where $B = 2$. For example, in 32-bit processor, $N = 32$ and we obtain

$$x_m = \pm(2^{31} - 1) = \pm 2147483647$$

in decimal digits. The number is ten digits long and bigger numbers cannot be stored.

The arithmetic of integers is carried out in a special way by the processor and is called **fixed-point arithmetic**.

$2°$. Floating-point Arithmetic

Next consider a (nonterminating) **real** x represented by $\pm d_1 d_2 \cdots d_n \cdot d_{n+1} d_{n+2} \cdots$ in base B (2 or 10). We can write

$$x = \pm(d_1 B^{n-1} + d_2 B^{n-2} + \cdots d_{n-1} B + d_n + d_{n+1} B^{-1} + d_{n+2} B^{-2} + \cdots)$$

$$= \sigma B^n \sum_{i=1}^{\infty} d_i B^{-i} \tag{1.9}$$

where as before σ stands for $+$ or $-$, n represents the **exponent** and the digits $d_1, d_2, \cdots$ make up the mantissa. When x is a *rational* number in base 10, the infinite series terminates or else it is a recurring decimal. In the former case the *mantissa* length is finite and in the latter case, it is infinite. As examples, we have the representations

$$\frac{365}{2} = 182.5 = 0.1825 \times 10^3$$

$$\frac{22}{7} = 3.14285\dot{7} = 0.314285\dot{7} \times 10^1$$

The equivalent right-hand side representations are in **normal exponent form**. Irrational numbers, on the other hand, are nonrecurring, nonterminating decimals like

$$\sqrt{2} = 0.141421356237309 \cdots \times 10^1$$
$$e = 0.271828182845904 \cdots \times 10^1$$
$$\pi = 0.314159265358979 \cdots \times 10^1$$
$$1/\pi = 0.318309886183790 \cdots \times 10^{-2}$$

with infinite mantissa length. Trivially, integers can also be expressed in the exponent form, like

$$13 = 13.0 = 0.14 \times 10^2, \quad 544 = 544.0 = 0.54 \times 10^3$$

It is not practical to do arithmetic exactly with infinite strings of digits representing floating-point numbers. In a machine, the word length is finite, consisting of N bits in a number base $B = 2$. So a word is organised in the following manner. Leaving one bit for σ, the rest are divided into two groups. One of the two groups is reserved for the exponent n and the rest of the bits are meant for the mantissa. The size for each group depends upon the make of the processor. This organisation determines the maximum real number in magnitude that the memory can hold. For example, in a 32-bit processor, 8 bits may be reserved for the exponent. The maximum binary exponent range is then $11111111 = 2^8 - 1 = 255$ (decimal). This range can be arranged from -128 to 127, which means that the exponent could range from 2^{-128} to 2^{127} or 10^{-39} to 10^{38}. Any decimal exponent higher than 38 would not be processed issuing **overflow** message. On the other hand, the decimal exponents < -39 are considered negligible and *set to* 0. Regarding the mantissa, the available number of bits is $= 38 - 1 - 8 = 23$ and the maximum possible mantissa becomes $2^{23} - 1 = 8388607$ with seven decimal digits. In conclusion, numbers between $\pm 0.8388607 \times 10^{-39}$ and $\pm 0.8388607 \times 10^{38}$ can be stored for processing in such a computer.

Suppose in general that the exponent lies between n_- and n_+ and the mantissa length of a word is m. To accommodate a number x in the memory, 'rounding' is necessary for the last available (mth)digit. One is acquainted with the rounding rule which states that *if the first discarded digit is greater than or equal to B/2 'round up' by 1 the last retained significant digit, otherwise keep the last retained digit unchanged.* When $B = 10$ one is familiar with the fact that that the last retained digit is increased by 1, if the first discarded digit is ≥ 5, while in the contrary case it is left unchanged. According to this rule if $Rd(x)$ denotes the rounded value of x, then from Eq. (1.9)

$$Rd(x) = \begin{cases} \sigma B^n (\sum_{i=1}^m d_i B^{-i} + B^{-m}), & d_{m+1} \geq B/2 \\ \sigma B^n \sum_{i=1}^m d_i B^{-i}, & d_{m+1} < B/2 \end{cases} \qquad (1.10)$$

where $d_i \in \{0, 1, \cdots, B - 1\}$, $(1 \leq i \leq m)$ and $d_1 \neq 0$ if $x \neq 0$. If one attempts to approximate x by $Rd(x)$, the error occurs. The relationship of $Rd(x)$ to x, through the relative error is given by the following theorem.

Theorem 1.1 *If x is approximated by Rd(x), then*

$$Rd(x) = x(1 + \epsilon) = \frac{x}{1 + \eta} \qquad (1.11a)$$

where the relative error $-\epsilon$ and the relative error η with respect to $Rd(x)$ satisfy

$$|\epsilon|, |\eta| \leq \frac{1}{2} B^{-m+1} \qquad (1.11b)$$

Proof By definition $-\epsilon = (x - Rd(x))/x$ and $-\eta = (x - Rd(x))/Rd(x)$. For finding the upper bound on $|\epsilon|$ and $|\eta|$, consider the expression of $Rd(x) - x$. For $d_{m+1} < B/2 \Rightarrow d_{m+1} \leq B/2 - 1$, we have from Eqs. (1.9) and (1.10)

$$-\sigma[Rd(x) - x] = B^n \sum_{i=m+1}^{\infty} d_i B^{-i} \leq B^n d_{m+1} B^{-m-1} + B^n d_{m+2} B^{-m-2}$$

$$\leq B^{n-m-1} \left(\frac{B}{2} - 1\right) + B^n \cdot B \cdot B^{-m-2} = \frac{1}{2} B^{n-m}$$

where $d_{m+2} < B$ has been used. On the other hand, if $d_{m+1} \geq B/2$,

$$\sigma[Rd(x) - x] = B^n \left(\sum_{i=1}^{m} d_i B^{-i} + B^{-m} \right) - B^n \sum_{i=1}^{\infty} d_i B^{-i}$$

$$= B^{n-m} - B^n d_{m+1} B^{-m-1} - B^n \sum_{i=m+2}^{\infty} d_i B^{-i}$$

$$\leq B^{n-m} - B^n \cdot \frac{1}{2} \cdot B^{-m-1} = \frac{1}{2} B^{-m+1}$$

Hence in both the cases of (1.10), $|Rd(x) - x| \leq \frac{1}{2} B^{n-m}$

Now, we have $d_1 \geq 1$ and so from Eq. (1.9)

$$\sigma x = B^n \sum_{i=1}^{\infty} d_i B^{-i} \geq B^n d_1 B^{-1} \geq B^{n-1}$$

Using the two inequalities, we obtain

$$|\epsilon| = \left| \frac{Rd(x) - x}{x} \right| \leq \frac{1}{2} B^{n-m}/B^{n-1} = \frac{1}{2} B^{-m+1}$$

Again, the roundoff rule implies

$$\sigma Rd(x) \geq B^n d_1 B^{-1} \geq B^{n-1}$$

and we obtain the bound for $|\eta|$, as in the case of $|\epsilon|$.   □

We now take up the question of arithmetic operation on two rounded floating-point numbers x and y with m digit mantissa in base B. Denote any of the arithmetic operations $+$, $-$, $\times$, $(\cdot/\cdot)$ by o and obtain x o y. Evidently, the mantissa length of x o y will be greater than m. For instance, if $m = 4$ and $x = 0.1234 \times 10^5$, $y = 0.5678 \times 10^{-3}$, then $x + y = 0.123400005678 \times 10^5$. Therefore, such longish operation is carried out by first storing the computed x o y in a long word length register and then roundoff the number to m digit mantissa. The procedure is called **floating-point operation**. If it is denoted by Fl, then $Fl(x \text{ o } y)$ results in $Rd(x \text{ o } y)$, i.e. $Fl(x \text{ o } y) = Rd(x \text{ o } y)$.

The organisation of arithmetic in a processor is special. We consider here the principles adopted for addition, viz. $Fl(x + y)$ with mantissa length m. Suppose $x = \sigma_1 10^{n_1} \times .d_1 d_2 \cdots$ and $y = \sigma_2 10^{n_2} \times .d_1' d_2' \cdots$ where $n_1 \geq n_2$. Writing the numbers x and y as $2m$-digit floating-point numbers having the same exponent n_1 (i.e. as **double precision** numbers) and normalising the sum to m-digit mantissa by rounding, we obtain the requisite floating-point sum $Fl(x + y)$. When $n_1 - n_2 > m$, the method always yields $Fl(x + y) = x$. For example with $m = 4$ in the previous example $n_1 - n_2 = 5 - (-3) = 8 > 4$ and $Fl(x + y) = 0.1234 \times 10^5 = x$. On the other hand when $n_1 - n_2 \leq m$, there are different possibilities, illustrated by the following examples:

Examples 1. Let $B = 10$, $m = 4$

(*i*) Let $x = 0.1234 \times 10^2$ and $y = 0.5678 \times 10^0$; then

$$
\begin{array}{r}
.12340000 \times 10^2 \\
+ .00567800 \times 10^2 \\
\hline
.12907800 \times 10^2
\end{array} \Rightarrow Fl(x+y) = 0.1291 \times 10^2
$$

(*ii*) Let $x = 0.4321 \times 10^{-2}$ and $y = 0.8764 \times 10^{-2}$; then

$$
\begin{array}{r}
.43210000 \times 10^{-2} \\
+ .87640000 \times 10^{-2} \\
\hline
1.30850000 \times 10^{-2}
\end{array} \Rightarrow Fl(x+y) = 0.1309 \times 10^{-1}
$$

(*iii*) Let $x = 0.1000 \times 10^1$ and $y = -0.9987 \times 10^0$; then

$$
\begin{array}{r}
.10000000 \times 10^1 \\
- .09987000 \times 10^1 \\
\hline
.00013000 \times 10^1
\end{array} \Rightarrow Fl(x+y) = 0.1300 \times 10^{-2} \qquad \square
$$

1.2.2.1 Floating-Point Error Propagation

Once roundoff error is committed, it contaminates subsequent results. We examine case (*iii*) of Example 1 considered above in more detail regarding error propagation. The two numbers x and y are nearly equal in magnitude and have opposite sign. Suppose to exactly *five* digit mantissa $x = 0.99955 \times 10^0$ and $y = -0.99873 \times 10^0$. Rounding to *four* digits $Fl(x) = 0.1000 \times 10^0$ and $Fl(y) = -0.9987 \times 10^0$, we obtain the two summands of case (*iii*). The relative errors $-\epsilon_x$ and $-\epsilon_y$ given by $Fl(x) = x(1 + \epsilon_x)$ and $Fl(y) = y(1 + \epsilon_y)$ yield $\epsilon_x = 0.45020 \times 10^{-3}$, $-\epsilon_y = 0.30038 \times 10^{-4}$. From the treatment of case (*iii*) $Rd[Fl(x) + Fl(y)] = 0.1300 \times 10^{-2} = 0.0013000 \times 10^0(1 + \epsilon)$ and so $\epsilon = 0$. But,

$$
\begin{aligned}
Fl[Fl(x) + Fl(y)] &= [Fl(x) + Fl(y)](1 + \epsilon) \\
&= x(1 + \epsilon_x) + y(1 + \epsilon_y)](1 + \epsilon) = (x + y)(1 + \rho)
\end{aligned}
$$

where

$$
\begin{aligned}
\rho &= \frac{x[\epsilon + \epsilon_x(1 + \epsilon)] + y[\epsilon + \epsilon_y(1 + \epsilon)]}{x + y} \\
&= \frac{0.99955 \times 0.45020 \times 10^{-3} - (0.99873) \times (-0.30038 \times 10^{-4})}{0.99955 - 0.99873} \\
&= 0.5854 = 58.54\%
\end{aligned}
$$

Thus, the absolute value of the relative error in the last rounding operation in adding x and y is 58.54%. This value is manifold larger in magnitude than the values of ϵ_x and ϵ_y at -0.045% and -0.003%, respectively. Clearly, the reason for amplification in the error is due to *cancellation of digits* in the sum $x + y$, in which x and y are nearly of the same magnitude but opposite in sign.

The propagation of large errors when two nearly equal numbers are subtracted is generally true. The propagation of error can be judged qualitatively in a general manner by considering the **condition number** of a problem. Let y be a quantity obtained from n data x_1, $x_2, \cdots$, x_n by arithmetic operations, so that

$$y = \phi(x_1, x_2, \cdots, x_n) \tag{1.12}$$

We study how errors in x_1, $x_2, \cdots$, x_n affect the result y when an error δx_i is committed in x_i. Thus, the first order approximation to the relative error becomes

$$\frac{\delta y}{y} = \sum_{i=1}^{n} \left(\frac{x_i}{\phi(x_1, x_2, \cdots, x_n)} \frac{\partial \phi}{\partial x_i} \right) \cdot \frac{\delta x_i}{x_i} \tag{1.13}$$

The numbers $\dfrac{x_i}{\phi} \dfrac{\partial \phi}{\partial x_i}$, $i = 1, 2, \cdots$, n are called **condition numbers** of the problem (1.12). When the absolute values of the condition numbers are less than or equal to 1, then the problem is said to be **well conditioned**; otherwise, it is called **poorly conditioned**. When the problem is well conditioned

$$\left| \frac{\delta y}{y} \right| \leq \sum_{i=1}^{n} \left| \frac{\delta x_i}{x_i} \right| = \text{sum of absolute relative errors in } x_1, x_2, \cdots, x_n$$

Evidently, in this case, the relative error in y remains reasonably bounded by relative errors in x_1, $x_2, \cdots$, x_n.

We use the above analysis to investigate the error in the computation of $y = x_1 \circ x_2$, where as before o stands for $+$, $-$, $\times$, $(\cdot/\cdot)$. The absolute condition numbers are

$$\left| \frac{x_i}{x_1 \pm x_2} \right|, \quad i = 1, 2$$

for $\pm$ operations. In the cases $\times$ and $\delta - \dot{-} -$, the numbers are always equal to 1. Hence we have an important conclusion that

(a) $\times$ and $(\cdot/\cdot)$ *are well-conditioned operations.*

(b) $+$ *and* $-$ *are well conditioned as long as the summands have, respectively, the same or opposite signs.* And

(c) $+$ *and* $-$ *are poorly conditioned whenever the two summands are of approximately the same absolute value, but have, respectively, opposite or the same signs.*

Example 1(*iii*) illustrates (c) for poor condition of subtracting nearly equal positive quantities.

1.2.2.2 Numerical Stable Evaluation of Formulae

Numerical evaluation of mathematical formulae is central to scientific computing. We tend to evaluate them in a straightforward manner. A formula evaluation always reduces to performing a sequence of elementary arithmetic operations. If a formula contains subtraction of nearly equal positive quantities, then there is a possibility of growing roundoff errors, rendering the evaluation **unstable**. To assure the stability of the overall process, each individual step must be made **numerically stable**. We illustrate the stabilisation procedure in the case of two examples.

Suppose it is required to solve the quadratic equation

$$ax^2 + bx + c = 0$$

The two roots of the equation are given by the **Sridhara (about 870–930 A.D.)** solution

$$x = \frac{-b \pm \sqrt{b^2 - 4ac}}{2a}$$

The two roots can be written as

$$x_1 = \frac{1}{2a}[-b - \text{sgn}(b)\sqrt{b^2 - 4ac}], \quad x_2 = \frac{1}{2a}[-b + \text{sgn}(b)\sqrt{b^2 - 4ac}]$$

If $b^2 > 4ac$, both the roots are real. If moreover $b^2 >> 4ac$ (the symbol $>>$ meaning that b^2 is large compared to $4ac$), then $\sqrt{b^2 - 4ac}$ nearly equals $|b|$. Hence subtraction of nearly equal positive quantities is involved in x_2, rendering its evaluation unstable. The difficulty can be circumvented by noting that $x_1 x_2 = c/a$ or,

$$x_2 = \frac{c}{ax_1}$$

The nature of stability of the two formulations can also be examined from the stand point of condition number. For this purpose, we simplify by setting $a = 1$, $b = 2p$ and $c = q$, so that the equation is taken as $x^2 + 2px + q = 0$, without loss of generality. Assume $p > 0$ where $p^2 >> |q|$, so that

$$x_2 = -p + \sqrt{p^2 - q}$$

The two methods of computation are thus:

Method 1 Compute $r = p^2$, $s = r - q$, $u = \sqrt{s}$,,

$$x_2 := \phi_1(u) = -p + u$$

and

Method 2 Compute r, s, u as in Method 1 and $x_1 = -p - u$,

$$x_2 := \phi_2(u) = \frac{q}{x_1} = -\frac{q}{p+u}$$

For *Method 1*, $\partial\phi_1/\partial u = 1$ and so

$$\text{Cond. No. of Method 1} = \frac{u}{\phi_1(u)} \frac{\partial\phi_1}{\partial u} = \frac{u}{-p+u} \cdot 1 = \frac{\sqrt{p^2 - q}}{-p + \sqrt{p^2 - q}}$$

$$= -\frac{1}{q}\sqrt{p^2 - q}\left(p + \sqrt{p^2 - q}\right)$$

$$> \begin{cases} -2p^2/q, & \text{if } q < 0 \\ 2(p^2/q - 1), & \text{if } q > 0 \end{cases}$$

Since $p^2 >> |q|$, the condition number is always greater than 1 and the method is poorly conditioned.

For *Method 2*, $\partial\phi_2/\partial u = q/(p+u)^2$ and so

$$\text{Cond. No. of Method 2} = \frac{u}{\phi_2(u)} \frac{\partial\phi_2}{\partial u} = -\frac{u(p+u)}{q} \times \frac{q}{(p+u)^2}$$

$$= -\sqrt{p^2 - q}/\left(p + \sqrt{p^2 - q}\right)$$

Since the absolute value of the right-hand side is always less than 1, the method is well conditioned.

In the second example, we illustrate how roundoff errors can play havoc with computation. Consider the problem of solving the difference equation

$$x_{n+1} = (p+1)x_n - q, \quad p > 0, \ q > 0$$

whose solution is simply $x_n = q/p$ (check). Suppose p to be somewhat larger than q, say $p = 3$ and $q = 1$ and we iterate the equation starting with $x_1 = q/p$. Writing a computer program to perform the iteration (see Example 5, Sect. 1.4.13) and executing it, reveals that the results for higher values are grossly absurd. The absurdity stems from the fact that whatever roundoff error is committed in computing x_1 is multiplied by a factor $p + 1 = 4$ at each step of iteration, vitiating later stage results. The remedy lies in performing the recursion in the descending order

$$x_n = (x_{n+1} + q)/(p+1)$$

Starting with an arbitrary value of x_n for sufficiently large n. Arbitrarily choosing this value of n to be 30, and $x_{30} = 10$, the results of recursion are given in the following table:

	$x_{n+1} = 4x_n - 1$	$x_n = \frac{1}{4}(x_{n+1} + q)$
n	$x_1 = 1/3$	$x_{30} = 10$
1	0.33333	0.33333
$\vdots$	$\vdots$	$\vdots$
6	0.33334	0.33333
7	0.33337	0.33333
8	0.33350	0.33333
9	0.33398	0.33333
10	0.33594	0.33333
11	0.34375	0.33333
12	0.37500	0.33333
13	0.50000	0.33333
14	1.00000	0.33333
15	3.00000	0.33333

It is remarkable to note the accuracy of the results of the third column, even though the starting value is grossly in error. This happens because at every iteration, the enormous quantity is multiplied by a small factor 1/4.

Recurrence relations are important in the context of solving partial differential equations, where properly solving them in a stable manner becomes an important question (see Chap. 7). In general, when a computational problem has a large number of arithmetic operations, a systematic analysis of error propagation is a difficult task. One could try to account for error with each arithmetic operation as the computation proceeds, but such forward error analysis is generally very complicated, whenever the formulae in question have nonlinear structure. As stated before, close watch on the acceptability of output generated, during execution of a program becomes a necessity.

1.3 Complexity of Algorithms

An algorithm is a procedure for implementing the computation of a mathematical method. Starting from an input it yields the output following some unique instructions, constituted of mostly arithmetic operations $+, \, - \times, \, (\cdot/\cdot)$ along with comparison operations $<, \, \leq, \, >, \, \geq, \, =, \, \neq$ and replacement operation $\leftarrow$. Many a time, to simplify the description, one uses the operations $\sqrt{\cdot}, \, | \cdot |$, sin, cos exp, and ln, that are actually subalgorithms. Sometimes, other subalgorithm written elsewhere are used. An algorithm must have the following properties. (*i*)**Definiteness**: Every step in an algorithm must be unambiguously defined. Every possible situation must be accounted for. (*ii*) **Finiteness**: The process must stop after a finite number of steps.

And (iii) **General Applicability**: The algorithm should work on an entire class of problems, so that the solution of a specific problem in its class is obtained by merely changing the input. The requirement of *definiteness* in an algorithm has led to the development of precise algorithmic language, but in this text, we use an informal form called *pseudocode*, sufficient for the development of a *computer program*. A program is merely a computer code of an algorithm, written in a programming language like FORTRAN. The structure of an algorithm and its corresponding program are so closely related that except in complicated cases, one may write the program trivially from its algorithm. In simple cases, therefore, one directly writes a program based on the method, for execution on a computer.

In general, there may be several alternative algorithms that accomplish the same task. This raises the important question of **complexity** of an algorithm and methods of reducing it for economic computing. There are two types of complexity: *static* and *dynamic*. Static complexity can be in the form of length of algorithm and the number instructions in it. Dynamic complexity, on the other hand, may be in terms of **run time** and **memory size requirement** of a computer. These measures depend on the amount of input and hence are of much more practical significance. Run time complexity is amenable to mathematical treatment when the input size grows without bound. Here the complexity depends on the growth of the basic arithmetic operations with an increase in the number of input data. Thus, *complexity* is defined as the *function that maps the input data to the number of basic arithmetic operations carried out in the algorithm*. If n be the number of input data, then the complexity function is a function of n. Since one is interested in the limit $n \to \infty$, the Landau symbol 'O' is sufficient for describing the function.

We now give some examples of algorithms together with their complexity. Such algorithms find application in programming.

1°. Naive evaluation of a polynomial:

$$P(x) := a_0 x^n + a_1 x^{n-1} + \cdots + a_{n-1} x + a_n = \sum_{i=0}^{n} a_{n-i} x^i$$

The pseudocode of the algorithm reads:

Input: Integer n, real $(a_0, a_1, \cdots, a_n)$, real x

Output: Real P

Computational Steps:

(i) $P \leftarrow a_n$

(ii) For $i \leftarrow 1, 2, \cdots, n$;

$\qquad P \leftarrow P + a_{n-i} \cdot x^i$

End

Here the first line notes the input: n, a_0, a_1, $\cdots$, a_n and x. Their types are also noted, as the computer stores them differently. The second line notes the output variable P, which is real. Lastly, the computational steps for obtaining P are written down. The sum is generated by recursion, taking into account terms containing increasing powers of x one by one. The recursion is started in step (i) with P set equal to or *assigned* the value a_n. In step (ii), $P \leftarrow P + a_{n-i}x^i$ means that assignment to P is taking place such that new value of P = old value of $P + a_{n-i}x^i$. In this way, augmentation in P is done for incorporating the different terms of the series. Thus, for $i \leftarrow 1$, value of $P \leftarrow a_n + a_{n-1}x$; for $i \leftarrow 2$, value of $P \leftarrow (a_n + a_{n-1}x) + a_{n-2}x^2$ and so on. Finally for $i \leftarrow n$, we get the value of the polynomial P.

There are only two basic arithmetic operations in the algorithm: addition $(+)$ and multiplication $(\cdot)$. Noting that x^i is i times the product of x involving $i - 1$ multiplications, the complexity becomes

$$\sum_{i=1}^{n}(1 + 1 + i - 1) = n + \frac{n(n+1)}{2} = \frac{1}{2}n(n+3) = O(n^2) \quad \text{for} \to \infty$$

$2°$. Evaluation of a polynomial by Horner's scheme.

For this scheme, we rewrite $P(x)$ as

$$P(x) = a_n + x(a_{n-1} + x(a_{n-2} + \cdots + x(a_1 + a_0x)))$$

This is equivalent to the recursion

$$P_1(x) = a_1 + a_0x$$
$$P_2(x) = a_2 + xP_1(x) = a_2 + x(a_1 + a_0x)$$
$$\cdots\cdots\cdots\cdots\cdots$$
$$P_n(x) = a_n + xP_{n-1}(x) = P(x)$$

The pseudocode now becomes

Input: Integer n, real $(a_0, a_1, \cdots, a_n)$, real x

Output: Real P

Computational steps:

(i) $P := a_0$

(ii) For $i = 1, 2, \cdots, n$;

$\quad$ $P = a_i + xP$

$\quad$ End

Satisfy yourselves that the algorithm generates the said recursion. The complexity of this algorithm is (since there is one $+$ and one $\cdot$ in step (ii))

$$\sum_{i=1}^{n}(1+1) = 2n = O(n) \quad \text{for} \quad n \to \infty$$

Obviously, compared to the naive algorithm, there is huge economy in computation for large values of n.

3°. Find the maximum of a set of numbers $a_1, a_2, \cdots, a_n$.

Suppose b is the biggest among the given set of numbers. For $i = 1, 2, \cdots, n$ compare b with a_i. If $b < a_i$, replace b by a_i. This process will yield b. The algorithm will thus be

> Input: Integer n, real $(a_1, a_2 \cdots, a_n)$
>
> Output: Real b
>
> Computational steps:
>
> (i) $b \leftarrow -0.8388607 \times 10^{-39}$ (smallest number that can be stored);
>
> (ii) For $i \leftarrow 1, 2, \cdots, n$
>
>   If $b < a_i$ then $b \leftarrow a_i$
>
> End

Counting the number of comparison operations, the complexity is $\sum_{i=1}^{n}(1) = n$.

4°. Given a set of numbers $a_1, a_2, \cdots, a_n$ $(n \geq 2)$, sort the numbers in ascending order of magnitude.

Consider two elements a_i and a_j $(i < j)$. If $a_i > a_j$, swap a_i and a_j. The operation is accomplished by considering a temporary element t such that a_i is moved to t, a_j is moved to a_i and finally t is moved to a_j. The pseudocode of the algorithm thus becomes

> Input: Integer n, real $(a_1, a_2, \cdots, a_n)$
>
> Output: Real $(a_1, a_2, \cdots, a_n)$
>
> Computational steps:
>
> For $i \leftarrow 1, 2, \cdots, n$; $j = i+1, \cdots, n$
> if $(a_i > a_j)$ then
> $t \leftarrow a_i$
> $a_i \leftarrow a_j$
> $a_j \leftarrow t$
>
> End

Counting the number of comparisons $\geq$, complexity $\sum_{i=1}^{n}(n-i) = n \cdot n - \frac{1}{2}n(n+1)$

$= \frac{1}{2}n(n-1) = O(n^2)$ for $n \to \infty$.

In all of the above examples, the complexity is of the form $O(n^p)$. The complexity is then simply called NP. For many problems, however, no NP complexity of algorithm has been discovered, and computing for such problems in reasonable run time is a challenge.

Exercises

1. Given $x = 0.8726 \times 10^2$ and $y = 0.3142 \times 10^1$ in a four-digit mantissa decimal computer with roundoff, determine the results of operations

$$x + y, \quad x - y, \quad x \times y, \quad \text{and} \quad x\delta(\cdot/\cdot)y$$

Also, find the absolute error in each case.
[$0.9040 \times 10^2, 0.8412 \times 10^2, 0.2742 \times 10^3$ and 0.2777×10^1].

2. In the four-digit computer stated in Ex. 1, given

$$x = 0.9678 \times 10^4, \quad y = 0.6780 \times 10^3, \quad z = 0.7800 \times 10^2$$

show that

$(i) \quad (x + y) + z \neq x + (y + z)$
$(ii) \quad (x \times y) \times z = x \times (y \times z)$
$(iii) \ z \times (x + y) \neq z \times x + z \times y$

Does the equality in (ii) hold in the case of a five-digit computer? (This example shows that in general, the associative and distributive properties of floating-point addition and multiplication do not exactly hold).
[(i) $(x + y) + z = 0.1044 \times 10^4$, $x + (y + z) = 0.1043 \times 10^4$ (ii) $(x \times y) \times z = 0.51184 \times 10^9$, $x \times (y \times z) = 0.51177 \times 10^9$, (iii) L.H.S$=0.8081 \times 10^6$, R.H.S$= 0.8078 \times 10^6$].

3. Let x and y be two real numbers in base B. Show that in an m bit mantissa computer, the relative error ρ in a floating point operation satisfies

$$|\rho| \leq \left| \frac{x}{x \pm y} \right| |\epsilon_x| + \left| \frac{y}{x \pm y} \right| |\epsilon_y| \quad \text{for addition or subtraction } x \pm y$$

and $$|\rho| \leq |\epsilon_x| + |\epsilon_y|, \text{ to first order for multiplication } xy \text{ or division } x/y$$

where $|\epsilon_x|, |\epsilon_y| \leq \frac{1}{2} B^{-m+1}$.

Hence estimate the relative error in computing (i) $(x + y) \times z$ and (ii) $x^2 + y/z$, for $x = 1.2$, $y = 45.6$ and $z = 3.4$, when $B = 10$ and $m = 4$.
[Use Theorem 1.1 for the first part or, alternatively you may use Eq. (1.13) of Sect. 1.2.2.1. (i) 2.2×10^{-3}, (ii) 10^{-3}].

4. Find the expression for the relative error estimate in the computation of

$$(i)\ \sqrt{x^2 + y^2} \qquad (ii)\ e^{ax} \sin by$$

$[(i)\ \epsilon,\ (ii)\ \epsilon[|ax| + |by \cot by|],$ where $\epsilon = \frac{1}{2} \cdot 10^{-m}].$

5. Show that in the computation of:

(i) $x^2 + 2x$ is better procedure than $x(x + 2)$

(ii) $(y + z) + x$ is better procedure than $(x + y) + z$, when $|x| > |y| > |z|$.

6. Find the condition number and condition of the following functions:

$$(i)\ \sin x, \quad (ii)\ \sqrt{x}, \quad (iii)\ x^3, \quad (iv)\ 1/(x - 1), \quad (v)\ e^x$$

$[(i)\ x/\tan x < 1,\ (ii)\ 1/2,\ (iii)\ 3,\ (iv)\ -x/(x - 1),\ (v)\ x].$

7. Rearrange the following expressions to avoid loss of accuracy due to subtraction cancellation:

(i) $1 - \cos x,$ near $x = 0$ (ii) $\dfrac{x - \sin x}{\tan x},$ near $x = 0$

(iii) $(x + a)^4 - a^4,$ near $x = 0$ (iv) $\sqrt{x^2 + 1} - x,$ near $x = \infty$

$[(i)\ \sin^2 x/(1 + \cos x),\ (ii)\ (x + \sin x)/\tan x - 2\cos x,$
$(iii)\ x^4 + 4x^3 a + 6x^2 a^2 + 4xa^3,\ (iv)\ 1/(\sqrt{x^2 + 1} + x)].$

8. Demonstrate dramatic instability in the computation of e^x for $x = -12$, using the partial sum of its Taylor series expansion:

$$S_n = \sum_{i=0}^{n} \frac{x^i}{i!}, \quad n = 0,\ 1,\ 2,\ 3,\ \cdots\cdots$$

Use the technique of Example 8, Sect. 1.4.12 for writing a FORTRAN program.

9. Show that

(i) $n^2,\quad 10,000 + .00001n^2,\quad n^2 + 2/n - 5n,\quad 10n^2 + \ln n - \sin(n)$ are $O(n^2)$ as $n \to \infty$.

(ii) $1/n^2,\quad 1/(n \ln n),\quad 2/n^2 + 3/n^3$ are $o(1/n)$ as $n \to \infty$

(iii) If sequence $\{x_n\} \to x$ as $n \to \infty$ then $x_n = x + o(1)$

(iv) $\sin h = h + O(h^3) = h - \dfrac{h^3}{6} + o(h^3)$ as $h \to 0$

1.4 Elements of FORTRAN

The name FORTRAN is an acronym for FORmula TRANslation. This high-level language was devised at IBM Corporation in 1957, that has been updated from time to time. The currently available versions are Fortran 90/95, and Fortran/2013. Fortran/2018 is scheduled to be launched in the year 2018. All of the new versions are backward compatible with FORTRAN77 which was in wide use in the earlier years. The thrust of FORTRAN from the beginning has been to tackle mathematically formulated problems, for which the computer was developed.

A FORTRAN program consists of a sequential arrangement of a set of FORTRAN *statements*. A FORTRAN statement is written in a line and a long statement can be continued in the next line by typing the character and at the end of the line. If two or more small length statements can be accommodated in a line, they must be separated by the character ';'. A statement can optionally be given a **label** by writing numerical digits not starting with 0, before beginning the statement. For convenience, the group of digits is written so as to look like a number, e.g. 10 or 100. A FORTRAN statement is written using the following FORTRAN characters:

(*i*) **Alphabets**: A, B, C, $\cdots$, X, Y, Z

or/and

a, b, c,$\cdots\cdots$, x, y, z

in which 'or' applies for Windows operating system, while 'and' is for Linux. Windows does not distinguish between capital and lower case letters, whereas they are distinct in Linux. It may be added that originally FORTRAN admitted only the capital letters, but this restriction has now been removed.

(*ii*) **Numerals**: 0,1,2, $\cdots$, 9 and

(*iii*) **Special characters**: These are

Blank	,	Comma	<	Less than
= Equals	.	Decimal point	>	Greater than
+ Plus	'	Single quote	$	Dollar
− Minus	:	Colon	%	Percent
* Asterisk	!	Exclamation point	"	Quote
/ Slash	&	Ampersand	?	Question mark
(Left parethesis	)	Right parethesis	;	Semi colon

The role of the characters is on expected lines with the exception that = has a special meaning. * and / play multiple roles, depending on the context of use. The second special character in the second row is the decimal point and not a stop or period sign. The first special character 'blank' is used to give a good textual presentation to a statement. This fact implies that the absence or presence of some blanks does not make a statement incorrect.

In a program, constant numbers are written as **Fortran constants**. They are of *three* types: **Integer, Real and Complex**, depending on how the numbers are stored in the memory of the computer.

(*i*) **Integer Constants.** Using the ten numerals, a constant is written as in elementary arithmetic, with or without a + or − sign. No decimal point is present in such constants. The storage is done as in fixed-point arithmetic discussed in Sect. 1.2.2.

Examples 1. 0, 1, +5, −13, −5108.

± 0 is equivalent to 0.

(*ii*) **Real Constants.** These are constants with terminating fractional parts and are stored in the way discussed in Floating-point arithmetic of Sect. 1.2.2. The decimal point must be present in such constants. Real constants can be written in two forms:

(*a*) **Fractional Form.** This form is the same as that in arithmetic, with generic form *m.n* with or without + or − sign. *m* is the integer part and *n* the fractional part of the number.

Examples 2. 0.0, 2.5, +31.4798, −378.605.

Fractional numbers are internally converted to the form as under:

(*b*) **Exponent Form.** The generic (normal) form of such constants is .mEn, with or without + or − sign. Here *m* is the mantissa and *n* the exponent, E replacing the base 10 of the number system.

Examples 3. 0.0E0, 0.258E2, −0.15E4, −0.44693E−3

One can also shift the exponent, making appropriate changes in the mantissa part into a fractional form.

Examples 4. For the numbers of Example 2, one can write

$$0.0E0, 2.58E1, -1500.0E0, -0.044693E-2$$

Such forms are internally converted into normal form.

(*iii*) **Complex Constants.** Following the algebraic definition of complex numbers, complex constants are written as a pair of real constants, within parentheses (and), separated by a comma.

Examples 5. (0., 0.), (0.,1.), (2.3, −4.691), (−1.234E2, .56978E−3)

Note that (0., 1.) is the purely imaginary number *i* which is sometimes also written as *j*.

Irrational numbers such as $\sqrt{2}$ and π must be approximated by roundoff (see floating-point arithmetic, Sect. 1.2.2). In a 32-bit machine, seven significant digits in the mantissa can be retained. Thus, $\sqrt{2}$ and π could, respectively, be approximated by 1.414214 and 3.141593. The same procedure should be applied to nonterminating rational numbers. For higher accuracy in computation, **double precision** constants written by D instead of E in the exponent form are used. About 15 significant digits can be retained in the mantissa in this way. In this case, two 'words' combine to store the number. In double precision, $\sqrt{2}$ and π can thus be written as 1.41421356237310D0 and 3.14159265358979D0. Double precision complex constants can be written as a pair of double precision real constants. But some of the compilers may restrict the real and imaginary parts to **single precision** constants of the first type only, in which case a constant is stored in 32 bits of memory, or in **4 bytes** (1 byte = 8 bits) of a 32-bit machine. In contrast, a double precision real constant occupies 8 bytes, and consequently, a double precision complex constant occupies 16 bytes.

Certain Fortran constants may appear several times in a Fortran program. Such constants can be given short symbolic names for convenience and used in the program instead of their value. Such names are called **Named Constants** and must be declared by the **PARAMETER** statement with the *syntax*

PARAMETER (named_constant = value list)

where the different named_constants and their respective values must be separated by commas. A more elaborate PARAMETER statement, explicitly declaring the type of constants is a *type declaration* (see Sect. 1.4.2) of the form

REAL, PARAMETER :: named_constant = value list

where the statement part 'REAL' indicates that the corresponding FORTRAN value is a REAL constant. Similarly, INTEGER and COMPLEX named constants can be declared by writing INTEGER or COMPLEX instead of REAL. It is obvious that such declarations should be made in the beginning, before statements of execution occur in the program.

Examples 1. π = 3.141593, e = 2.718282, and i = $\sqrt{-1}$ can be declared as

PARAMETER(pi = 3.141593, e = 2.718282, ci = (0.0, 1.0))

or as,

REAL, PARAMETER :: pi = 3.141593, e = 2.718282
COMPLEX, PARAMETER :: ci = (0.0, 1.0)

The complete Fortran language suit has a long list of admissible statements and features that help program development [32]. Restricting our objective of generic numeric program development, we proceed to describe the most useful ones. Capital alphabets are used for emphasis.

1.4.1   The PROGRAM, END and COMMENT Statements

A program for solving a problem in hand can be given a name. Its syntax is

PROGRAM   program name

It should be the first statement of the program. The statement is optional with default program name MAIN.

The mandatory last statement of a program is

END PROGRAM

or simply END. This statement marks the physical end of the program; any other statement following it is not considered part of the program.

A long program is usually written with a modular structure, where specialised jobs are assigned to other function types or subprograms (see function and subroutine subprograms, Sect. 1.4.10). The statement END is used to end such programs also. Other structures are also ended with the statement END, followed by name of the structure.

COMMENT statements are used to append explanations in a program (main or sub). Its syntax is

!   comment (to end of line)

In practice, it (i.e. !) is placed to the right of a Fortran statement, providing an explanation of the statement. For providing explanation of groups of statements, comments occupying whole lines may be put preceding the statements.

1.4.2   FORTRAN Variables and Type Declarations

Unlike a single letter symbol in mathematics, a Fortran variable can be a chain of alphabets and numerals, starting always with an alphabet. One is allowed to join words by __ to form a variable name.

Examples 1. X, A1, Z1BYZ2, Frequency, square_root, inverse_of

Variables are of three types: **integer, real** and **complex** depending on the data types (constants), which they assume for values. The types in single precision are specified by type declaration statements INTEGER, REAL and COMPLEX with syntaxes:

INTEGER :: integer variable list (separated by commas)

REAL :: real variable list (separated by commas)

COMPLEX :: complex variable list (separated by commas)

In double precision, the syntaxes are one of the following three ways:

DOUBLE PRECISION :: real variable list (separated by commas)

DOUBLE COMPLEX :: complex variable list (separated by commas)

or,

REAL*8 :: real variable list (separated by commas)

COMPLEX*16 :: complex variable list (separated by commas)

or,

REAL(KIND=8) :: real variable list (separated by commas)

COMPLEX(KIND=8) :: complex variable list (separated by commas)

The portion KIND= is optional in the last two statements and one can simply write REAL(8) and COMPLEX(8). In single precision KIND=4 and is optional. The KIND attribute in fact has a much wider use [32].

Variable lists in single precision can be shortened by the default option that integer variables starting with the alphabets I, J, K, L, M, N may be omitted from an integer variables list. Similarly, real variables starting with A–H and O–Z may be omitted from real variables list. *For portability, however, it is a good practice to list all the variables even if they follow the default option.*

For long lists of variables, a useful statement is the IMPLICIT statement, which imparts declarations in very concise form. Suppose one wants to declare that all variables starting with alphabets A–H, O–Z are double precision real, and then the declaration

IMPLICIT DOUBLE PRECISION (A − H, O − Z)

serves the purpose. A complex variable Z whose real and imaginary parts are X and Y can be paired by the function CMPLX as follows:

$$Z = \text{CMPLX}(X, Y)$$

1.4.3 FORTRAN Arrays (Subscripted Variables) and DIMENSION

Arrays such as vectors and matrices are named in Fortran in the manner of variables. The *dimension* of an array (such as the number of elements of a vector or number of rows and columns of a matrix) is specified by the DIMENSION statement. Suppose there is a vector V with 10 integer components and a matrix A with 12 rows and 15 columns, the array initialisation syntax is

INTEGER DIMENSION(15) :: V

REAL DIMENSION(12, 15) :: A

In short, one can also use *type declarations* by omitting the keyword DIMENSION.

$$\text{INTEGER :: V(15)}$$
$$\text{REAL :: A(12, 15)}$$

Other variables and arrays can be included in the list of such type declarations.

Double precision arrays can be initiated by using say REAL(8) and COMPLEX(8) statements as explained earlier.

The *elements* of an array, that are written as *subscripts* in matrix algebra are written as the array name with the subscripts in parentheses. Thus, the Ith element of V is written as V(I), and similarly, the element on the Ith row and Jth column of matrix A is written as A(I,J).

1.4.4 Arithmetic Operations and the Assignment Statement

The arithmetic operations of Fortran are

+ for addition
− for subtraction
  for multiplication
/ for division
∗ for exponentiation

The notation for exponentiation stands out by the presence of two consecutive characters. The left and right parentheses (and) can be used for grouping of terms as in arithmetic. Use of *brackets* [ ] and *braces* { } is *not* permitted. Instead of these, nesting of parentheses is permitted to form arithmetic expressions. Such expressions look like *algebraic expressions* involving constants and variables (with or without subscripts), but in reality, they are arithmetic in nature. Here, the compiler does not manipulate algebraic symbols but instead computes the value of an expression for the current value of the variables.

The operators are given precedence pecking order, so as to reduce excessive use of (nested) parenthesis. The hierarchy is

()   innermost highest, outermost lowest
∗∗
∗ and /
+ and −

For operations with the same hierarchy, execution takes from left to right

Examples 1.

A/B∗C   is equivalent to   (A/B)∗C

A*B+C/D−E*F is equivalent to ((A*B)+(C/D))−E**F
(A*B+C/(D**E+F)−A)/B+C is equivalent to
((A*B)+(C/((D**E)+F))−A)/B+C

In the operator **, if the exponent is an integer, the expression is computed as a product. For example, $X**3$ is computed as $X*X*X$. On the other hand, if the exponent is real, the expression is computed by taking log and antilog, that is to say, $X**3$. is computed as $\exp(3 \ln X)$, consuming much more execution time. In other mixed operations, such as between integer and real, the former is converted to real. For instance, $X+3$ and $X*3$ are converted to $X+3.0$ and $X*3.0$, respectively, before execution.

The above features are for *real arithmetic*. *Complex arithmetic* has similar features, the result being always complex. *Integer arithmetic* always yields an integer. This fact has interesting consequence that division yields the integer part of the quotient. For instance, 1/2, 3/2, 11/3 and 25/5 are, respectively, computed as 0, 1, 3 and 5.

An **Assignment Statement** in Fortran has the syntax

Variable name=Arithmetic expression

Here, the value of the expression is computed and stored in the location of the variable in memory. If the location did not exist before, it is created with this statement. Clearly, in Fortran, = does not stand for equality but means an assignment.

Examples 2. Suppose that the values of A, B, C, D, V(J) and I, MU (integers) are given, then one can write

$$X=A/B*C$$
$$Y(I)=A*V(J)+ D−MU$$
$$X=X+Y(I)$$

X and Y(I) are computed in the first two statements. In the peculiar looking third statement, the two values are added and assigned to X.

If the initial values of variables are given as input, they can be generated by using assignments. For subscripted variables, the subscripts can be generated by using special character.

Examples 3. REAL:: A, B(2,3)
$$A = 5.0$$
$$B(1,:) = (/1, 7, −2/)$$
$$B(2,:) = (/3, 4, 6/)$$

means that the value of A is 5 and the elements of the two rows of B are 1, 7, −2 and 3, 4, 6, respectively.

An arithmetic expression may contain mathematical *intrinsic functions* of Fortran and also other defined functions. These topics are discussed later in Sect. 1.4.10.

Operators on array names are also possible. Let A and B be two arrays of equal dimensions and let X be a scalar, then

A+B   computes   sum of A and B
A−B   computes   difference of A and B
X+A   computes   array with X added to elements of A
X*A   computes   product of X and A
A**X   computes   array with elements of A raised to power X
MATMUL(A, B)   computes   product of matrices A and B
DOT_ PRODUCT(A, B) computes dot product of vectors A and B

1.4.5   Relational Operators and Logical Expressions

Relational operations like =, <, etc., are important in mathematical formulations. In Fortran, the relational operators are:

== or .EQ. meaning equal to
/ = or .NE. meaning not equal to
< or .LT. meaning less than
<= or .LE. meaning less than or equal to
> or .GT. meaning greater than
>= or .GE. meaning greater than or equal to

The operators have the same level of hierarchy. In addition to the above, logical NOT, AND and OR are, respectively, written as .NOT., .AND. and .OR. with *decreasing* order of hierarchy.

Logical expressions can be formed by using these operators.

Example 1. A/=B
$\qquad$ (A<B).AND.(I==J).OR.(X>=Y)

Logical expressions are either true or false. Such expressions may be assigned a **logical variable** that can assume only two **logical constants**, .TRUE. and .FALSE. The syntax of type declaration of logical variables is

$\qquad$ LOGICAL   logical variables list (separated by commas)

Example 2. LOGICAL Z
$\qquad$ statements
$\qquad$ Z=(A<B).AND.(I==J).OR.(X>=Y)
$\qquad$ statements

So far we have only considered sequential execution of statements in a program. The logic of algorithms often demands to alter such flow. We proceed to describe the *flow control statements*, the use of which accomplishes the tasks.

1.4.6 PAUSE and STOP Statements

The syntaxes of these one-word statements are

PAUSE

and

STOP

The former suspends execution, which can be resumed by pressing any key. The latter statement stops execution altogether. It is usually used in conjunction with the IF statement described in Sect. 1.4.8. It may also be used temporarily for program debugging. The statement PAUSE is considered obsolete in the latest version of Fortran.

1.4.7 The GOTO Statement

The syntax of the statement is

GOTO   label

where 'label' of a statement is usually a natural number followed by the statement. By the GOTO statement, the execution passes to the statement bearing the label stated in the GOTO statement. The statement bearing the stated label may precede or succeed the GOTO statement. *It should be sparingly used for clear understanding of the flow of the program.*

1.4.8 The IF Statement and the IF Construct

This statement is frequently indispensable in programming. Its construct comes in a variety of ways. The simplest is

IF(logical-expression) true statement

Here the true statement is executed if the logical-expression is true; otherwise, it is skipped. If there are more than one true statements then the construct is

IF(logical-expression) THEN

true group of statements

END IF

Entry into the true group of statements by a GOTO statement elsewhere is not permitted. Nesting of IF constructs is permissible, such as

IF(logical-expression__ 1) THEN

true group A

IF(logical-expression__ 2) THEN

true group B

END IF

true group C

END IF

If a false group of statements is also to be executed, then one can use the IF-ELSE construct:

IF(logical-expression THEN)

true group A

ELSE

false group B

END IF

Example 1. Compute

$$y = |x| \quad = x, \quad x \geq 0$$
$$= -x, \quad x < 0$$

for given x. Here, we can write

IF(X>=0.0) THEN

Y=X

ELSE

Y=−X

END IF

In case of multiple choices, the ELSE-IF construct

ELSE-IF(logical expression) THEN

can be employed.

Example 2 For given x compute y from the ramp function

$$y = 0, \quad x \leq 0$$
$$= x, \quad 0 < x < 1$$
$$= 1, \quad x \geq 1$$

For the above, we can write

      IF(X<=0.0)THEN

        Y=0.0

      ELSE IF(X<1.0) THEN

        Y=X

      ELSE

        Y=1.0

      END IF

1.4.9 The DO Statement

This statement is indispensable when repeated execution (looping) of statements is required. Its construct is

      DO loop-control

        Statements

      END DO

The loop-control may be a logical expression, so that looping is performed as long as the logical expression is true. The second form of the widely used loop-control is

      Index = initial value, limit, increment

where the index is an integer variable name and the initial value, limit and increment are of similar type. During looping, the index is first set equal to initial value. Next, the latter is changed to initial value + increment. If the number does not exceed the limit, the statements are again executed. In the contrary case, the looping terminates. The default value of increment is 1, that is to say, it need not be written if it is 1. If the initial value exceeds the limit, the increment must be a negative integer.

Entry into a DO loop is not permitted from outside by a GOTO statement, whereas exit and termination are permitted by using the statement EXIT. It is possible to skip a cycle of the loop by using the statement CYCLE. Obviously, EXIT and CYCLE must be conditioned by IF statements.

Nesting of DO loops is permissible, in which case, execution takes place from *inner most* to *outer most* loop.

1.4.10 Functions in FORTRAN

Functions can be intrinsic to the compiler, or user-defined subprograms separate from the main program.

(*a*) **Intrinsic Functions**. A long list of standard functions is available. A short list is given below (X, Y are real and Z is complex):

Function	FORTRAN name	Function	FORTRAN name
$\lvert X \rvert$	ABS(X)	$\sin^{-1}X$	ASIN(X)
$\sqrt{X}$	sqrt(X)	$\cos^{-1}X$	ACOS(X)
e^X	EXP(X)	$\tan^{-1}X$	ATAN(X)
		$(-\pi/2 \le X \le \pi/2)$	
$\ln X$	ALOG(X)	$\tan^{-1}(X/Y)$	ATAN(X,Y)
		$(-\pi \le X/Y \le \pi)$	
$\log_{10}X$	ALOG10(X)	$\sinh X$	SINH(X)
$\sin X$	SIN(X)	$\cosh X$	COSH(X)
$\cos X$	COS(X)	$\sin Z$	CSIN(Z)
$\tan X$	TAN(X)	$\cos Z$	CCOS(Z)
Z^*	CONJG(Z)	X mod Y	MOD(X,Y)
		(Integer X, Y)	
$Re(Z)$	REAL(Z)	$\text{Max}(X_1, X_2, \cdots)$	MAX($X_1, X_2, \cdots$)
$Im(Z)$	AIMAG(Z)	$\text{Min}(X_1, X_2, \cdots)$	MIN($X_1, X_2, \cdots$)
$\lvert Z \rvert$	CABS(Z)	$\sqrt{Z}$	CSQRT(Z)
e^Z	CEXP(Z)	$\ln Z$	CLOG(Z)

If the arguments are double precision, the function name or *procedure* starts with D, e.g. DABS(X) stands for $\lvert X \rvert$, where X is double precision.

(*b*) **Function Subprogram**. A *subprogram* is a separate program from the main and any other subprogram. Any function of one or more variables can be computed from a function subprogram. Its structure is

> FUNCTION function name($x_1, x_2, \cdots, x_n$)
> type declaration of f
> type declaration of $x_1, x_2, \cdots, x_n$
> statements
> function name=expression
> RETURN
> END FUNCTION function name

In the above $x_1, x_2, \cdots, x_n$ is a **dummy** argument list. Whenever the value of the function in the main program or any other subprogram for a prescribed set of argument list is referenced, execution passes to this subprogram, with $x_1, x_2, \cdots, x_n$ acquiring these prescribed values. Return to invoking function together with the computed value of function name takes place by the RETURN statement. There is only one output from the function subprogram, viz. the function name. If the output of the function subprogram is of integer type, then that can be specified by writing **INTEGER** before FUNCTION followed by the function name. Similarly for COMPLEX type output of a function.

(*c*) **Subroutine Subprogram**. When values of several dependent variables are required for a given set of independent variables, then the subroutine is a powerful tool. Its structure is

> SUBROUTINE subroutine-name(x_1, x_2 $\cdots$, x_m, y_1, y_2, $\cdots$, y_n)
> type declaration of x_1, x_2, $\cdots$, x_m, y_1, y_2, $\cdots$, y_n
> statements
> RETURN
> END SUBROUTINE subroutine-name

The input x_1, x_2, $\cdots$, x_m and the output y_1, y_2, $\cdots$, y_n are *dummy* variables. Execution of subroutine takes place when the calling statement

$$\text{CALL subroutine-name}(a_1, a_2, \cdots , a_m, b_1, b_2, \cdots , b_n)$$

is encountered in a calling program. A calling program may be the main or any other subprogram. With the CALL statement known values of a_1, a_2, $\cdots$, a_m are, respectively, linked to x_1, x_2, $\cdots$, x_m. With execution of the subroutine, the computed values of y_1, y_2, $\cdots$, y_n are, respectively, linked to b_1, b_2, $\cdots$, b_n as output. In special cases, values of the output variables can also be returned through the input variables, once the latter are completely used in the preceding statements of the subroutine.

Any argument in a list can also be an array name (without subscript). A subroutine may not call itself. Function names can also be passed to a subroutine, provided that they are declared EXTERNAL in the calling program. The syntax of the statement is

> EXTERNAL function-name list (separated by commas)

Sometimes it is necessary to share certain variable/array names between a calling program and some of the subroutines. Then such a task is easily accomplished by the use of the COMMON statement:

> COMMON variable/array-name list (separated by commas)

placed before executable statements of the program units. In some cases, parts of the name list of variables/arrays of a calling program are shared with different subroutines selectively, and then, the COMMON statement of the calling program can be split into blocks with specified names:

> COMMON /block-name-1/ name list of variables/arrays (separated by commas)
>
> COMMON /block-name-2/ name list of variables/arrays (separated by commas)

etc. Then, only the needed COMMON blocks needed for a particular shared subroutine is required to be stated in that subroutine. It may also be noted that the names in COMMON statement in two program units of a program may be different, as common data is shared between the two name lists.

In very special cases a call of the subroutine itself can be made by appending the command RECURSIVE before the SUBROUTINE subroutine-name statement. Only one recursion is allowed with an output; otherwise, it becomes infinite [32]. A recursive procedure can always be converted into an iterative procure and is preferable as recursion consumes more system resources and execution time. However, as some algorithms are intrinsically recursive, it is sometimes convenient to express such an algorithm as a recursive procedure.

Example 1 Write a subroutine for $n!$.

Solution.   RECURSIVE SUBROUTINE Factorial(n,value)
                    INTEGER :: n   ! (Input)
                    INTEGER ;; value   ! (Output)
                    IF(n==1) THEN
                    value=1   ! Exit Point
                    ELSE
                    CALL Factorial((n-1),value)
                    value=n*value
                    END IF
                    END SUBROUTINE Factorial

If the above subroutine is called by the main program for a given value of n , say $n = 5$, using the statement
                    CALL Factorial(5,value)
the output returned by the subroutine is value = 120.
Remark. Recursion of function subprograms is also possible [32], but is not used in the book.

1.4.11   Input and Output (I/O) Statements

Input and output are indispensable parts of algorithms and programs. Input should be user-friendly and output should be self-explanatory.

(*a*) **Input.** In simple cases, assignment statements can be used, as was stated in Sect. 1.4.4. The longer list of variables and arrays can be read by the statement

                    READ*, variables list (separated by commas)

Example 1 Read variables X, Y, array A with 10 components and complex variable Z. For this purpose, we write
                    READ*, X, Y, A, Z

By this statement, execution is halted for data entry from *keyboard* (represented by * in the statement). At this stage, data is entered for X and then Y, followed by those of the ten components of A, and the real and imaginary parts of Z.

(*b*) **Output.** The output of variables and arrays on the *monitor* is obtained by the statement
                    PRINT*, variables list (separated by commas)

The output is *format-free*, as stored in the computer. The *, in both READ and PRINT, is optional, in latest versions of Fortran.

It is sometimes necessary to incorporate messages. The messages could be included in the variables list of the output statement, by writing them within quote marks ' and '. Such character strings are called *character constants*.

The output can be obtained to required number of digits and in tabular form by writing a statement of the form

> PRINT format-label, variables list (separated by commas)
>
> format-label FORMAT (format-specification list)

The format specifications are as follows:

In	Integer to n places
Fm.n	Real in fractional form, to n decimal places contained in a total of m places
Em.n	Real in exponent form, with n digit mantissa contained in a total of m places
nX	Skip n places

Example 2 Obtain suitable output for integer I, real X, Y and complex Z.
One can write

> PRINT 100, I, X, Y, Z
> 100 FORMAT (5X, I3, 3X, F7.4, E9.5, 5X, 2F10.6)

Since Z is composed of two real constants, the two E formats F10.6 and F10.6 have been clubbed together as 2F10.6. Such clubbing to several number of times is permitted for I, F and E formats.

Output can also be obtained on other memory devices like the hard disks. For this purpose, the device must be given a **unit number** as a label. Then a data file for the output must be opened by the statement

> OPEN(unit number, file='name of data file')

output on the device can then be obtained by the statements

> WRITE(unit number,*) variable list

or   WRITE(unit number, format label) variables list

for format-free or formatted output. When outputting is over the file may be closed by the statement

> CLOSE(unit number)

We conclude this section, by noting that many other features of Fortran have been left untouched. Even then, it must be clear by now, that the language is like a super mind for handling complex computational algorithms by means of simple looking statements. In Sec. 1.4.13 examples are given to illustrate the usage of the statements. In these examples, mostly lower case alphabets are used for naming variables, as they are permitted in Fortran.

1.4.12   Other Statements

In this section, a few other statements are given, which Fortran programmers often use. These statements are however not used in the remaining chapters as the focus is on the development of special purpose short subroutines, for important different mathematical tasks.

1.4.12.1   INTENT Statement

The (dummy) arguments of a subroutine subprogram act either as input or output or sometimes both as input and output of the program unit. The intention with which they are used can be programmed in a subroutine, by using the **INTENT** statement in the type declarations of the subroutine. The syntax of type declarations of such a subroutine is therefore of the form:

$$\text{type\_declaration, INTENT(IN)} :: x_1, x_2, \cdots, x_m$$
$$\text{type\_declaration, INTENT(OUT)} :: y_1, y_2 \cdots, y_n$$
$$\text{type\_declaration, INTENT(INOUT)} :: z_1, z_2, \cdots, z_l$$

where IN stands for Input, Out for Output and INOUT for Input–Output both. The INTENT statement is *optional*, as was done in the description of the subroutine subprogram in Sect. 1.4.10. The Intent can also be alternatively clarified by Comment statements following the argument lists. In this book, the latter style is adopted for easy understanding of the subroutines developed in the subsequent chapters.

1.4.12.2   CASE Statement

The **CASE** statement allows multiway branching. In its general form, the syntax of the CASE statement is:

SELECT CASE (expression)

CASE (low_1 : high_1)
statements_1
CASE (low_2 : high_2)

statements_2

. . . .

END SELECT

The low and high in the above construct are the exact lower and upper bounds of the selected expression. If the evaluated expression lies between the two bounds low and high (both included), the statement block following it is executed; otherwise, the execution passes to the next CASE. When all the CASEs are checked and the statement blocks executed, the END SELECT statement is executed and the execution passes to the next statement following it. Exiting from a group of statements by a GOTO statement after the END SELECT statement is permitted. The lower bound can be blank, implying that the evaluated value of the expression must be lesser than the given higher bound. On the other hand, if the higher bound is left blank, then the evaluated value of the expression must be greater than the given lower bound. Thus, **CASE (:high)** and **CASE (low:)**, respectively, require that the value of the expression is $\leq$ the value of high, and in the second case, the value of the expression is $\geq$ the value of low.

1.4.12.3 DO WHILE Statement

The syntax of **DO WHILE** is:

DO WHILE (logical-expression)

statements

END DO

The logical expression is tested; if it is true all the executable statements following the DO WHILE up to the statement END DO are executed. The control is then passed to the top of the loop and a fresh cycle begins. The condition is tested at the beginning of each cycle. The execution continues till the condition is false and in that case, the control is passed to the statement following the END DO.

1.4.12.4 MODULE

A **module** is a sepaprate *global* program unit that is not executed directly, but contains data specifications and procedures that may be utilised by other program units via the **USE** statement. Hence, it can be written even before the main program, so that it can be used by the main program as well as by other program units that follow the main program. A simple module is of the following form:

MODULE module-name

type declarations

parameter lists

CONTAINS

FUNCTION FN1(argument list)

. . . .

FN1=· · · · · ·

END FUNCTION FN1

⋮
⋮

SUBROUTINE SUB1((argument list)

. . . .

END SUBROUTINE SUB1

⋮
⋮

END MODULE module-name

The program unit using a module, generates the input data for the module and the output generated by the module can be used by that program unit by the statement:

USE module name

at the beginning of the user program unit before the occurrence of executable statements of that unit. The output from a module can also be selectively generated by the statement:

USE module name, ONLY: a subprogram of the module

A module can use another module name. In this way, a user can make a library of subprograms, encapsulated in a module. Moreover, a subprogram in a module may be written in some other programming language like C and Python. The data interoperability in the module is maintained by the respective statements:

USE iso_c_binding   and   USE forpy_mod

where the modules *iso_c_binding* and *forpy_mod* can be incorporated in Fortran. For details see [32] and the internet.

Example 1 Construct a module of constants π, e, and the complex constant i and another module that contains a function subprogram $F(x) = x + x^2 + \pi$, and a subroutine SWAP that swaps two variables a and b. Use the two modules to compute $F(2)$ and swap two numbers 5 and 13. Also print the constants π, e and i.

Solution. The program reads as follows:

```
MODULE Myconstants
REAL, PARAMETER :: pi=3.141593, e=2.718282
COMPLEX,  PARAMETER :: ci=(0.0,1.0)
```

END MODULE Myconstants
!************************************

MODULE Mylibrary

USE Myconstants

CONTAINS

FUNCTION F(x)

F=x+x**2+pi

END FUNCTION F

SUBROUTINE SWAP(a,b)

temp=a; a=b; b=temp   ! algorithm for swapping

END SUBROUTINE SWAP
!**
!
PROGRAM MAIN

USE Mylibrary

PRINT*, pi, e
PRINT*, ci

x=2.0
phi=F(x)
PRINT*, phi

p=5.0; q=13.0
CALL SWAP(p,q)
PRINT*, p, q

END

On running the program, the output comes out as

```
3.141593                2.718282
    (0.000000E+00, 1.000000)
9.141593
13.000000               5.000000
```

as can be verified easily.   □

1.4.13　Fortran Programming Examples

Example 1. Write a program to compute
the side $a = \sqrt{b^2 + c^2 - 2bc \cos A}$
perimeter $p = a + b + c$
area $\Delta = \frac{1}{2}bc \sin A$
of a triangle ABC, given $b = 3\,cm$, $c = 5\,cm$ and angle A $= 50^\circ$.

Solution. In Fortran, the argument of trigonometric functions must be in radians. One can write the program as

```
PROGRAM TRIANGLE
REAL :: a, b, c, ANGLE_A, p, area
b=3.0; c=5.0
ANGLE_A=50*3.141593/180.0        !(Input)
a=sqrt(b**2+c**2-2*b*c*&         !& is for continuation in the next line
cos(ANGLE_A))
p=a+b+c
area=0.5*b*c*sin(ANGLE_A)
PRINT*, a, p, area          !(output)
END PROGRAM
```

Note that the expression for p contains A, so A must be computed before p. The unit of centimetre has no role in the program, but from dimensional consideration the computed a and p must have the same unit of centimetres. □

　　From now on, we shall drop writing the PROGRAM statement—it will automatically be named MAIN.

Example 2. Solve the quadratic equation $ax^2 + bx + c = 0$, for given real a, b, c as input.
Solution. If $b^2 \geq 4ac$, the real roots are

$$x_1 = \frac{1}{2}[-b - \text{sgn}(b)\sqrt{b^2 - 4ac}], \quad x_2 = c/(ax_1)$$

(see Sect. 1.2.2.2) and if $b^2 < 4ac$, the complex roots are

$$z_1 = \frac{1}{2}[-b + \sqrt{b^2 - 4ac}], \quad z_2 = c/az_1$$

A Fortran program is

```
REAL :: a, b, c, disc, a₂, x₁, x₂, sign_ b
COMPLEX:: z₁, z₂
READ*, a, b, c    ! (Input)
disc=b**2-4.0*a*c    ! Compute the discriminant
a₂=2.0*a
if(b>0.0) sign_b=1.0
if(b==0.0) sign_b=0.0
if(b<0.0) sign_b=−1.0
if(disc>=0.0) then
    x₁=(−b−sign_ b*sqrt(disc))/a₂; x₂=c/(a*x₁)
    PRINT*, 'Real Solution ', x₁, x₂    ! (Output)
else
    z₁=(−b+csqrt(cmplx(disc,0.0.)))/a₂; z₂=c/(a*z₁)
    PRINT*, 'Complex Solution ', z₁, z₂,    ! (Alternate Output)
end if
END
```

□

Example 3. (Computing a table of values of a function). Evaluate

$$y = 2.35x^3 - 3.478x^2 + 1.531x - 6 \quad \text{for} \quad x = -5(.1)5$$

Solution. We employ Horner's scheme

$$y = -6 + x(1.531 - x(3.478 - 2.35x))$$

A Fortran program is

```
real :: x, y, delt_x
x=−5.0
delta_x=0.1    ! Increment
DO(x<=5.0001)    ! Avoiding skipping of the value 5.0
                 ! that may be affected by roundoff
    y = -6. + x*(1.531 − x*(3.478 − 2.45*x))
    PRINT*, x, y
    pause
    x = x + delta_x
END DO
END
```

A PAUSE statement is inserted to view data, which are large in number.   □

Example 4. (Summing a power series). Compute sin x from the series

$$\sin x = x - \frac{x^3}{3!} + \frac{x^5}{5!} - \frac{x^7}{7!} + \cdots \infty$$

for $x = \pi$, using the termination criterion that a term added to the series is less than 10^{-10}.

Solution. The ith term is $t_i = (-1)^{i-1} \dfrac{x^{2i-1}}{(2i-1)!}$ and so $t_{i+1} = (-1)^i \dfrac{x^{2i+1}}{(2i+1)!}$

Hence $t_{i+1}/t_i = -x^2/[2i(2i+1)]$. Since the sine series is an alternating series with terms monotonically decreasing in modulus, the remainder R_n after n terms satisfies the condition

$$|R_n| \leq \text{modulus of the next } (n+1)\text{th term}$$

This justifies the convergence criterion. A Fortran program for the procedure is

```
INTEGER :: i
REAL :: x, t, s
x=3.141593     ! (Input)
i=1
t=x    ! First term
s=x     ! s is sum of the series
1 t=t*x**2/(2*i*(2*i+1)); s=s+t     ! Add the term to sum
if(t<1.0E-10) goto 2
i=i+1
goto 1
2 PRINT*, s
END
```

Observe that looping is achieved by GOTO, so is exiting from the loop.   □

Example 5. Generate the sequence $x_1, x_2, \cdots$ by iterating $x_{n+1} = (p+1)x_n - q$ for $p = 3$, $q = 1$, in (a) ascending order starting with $x_1 = q/p$ and (b) descending order with starting value $x_{30} = 10$. Obtain output up to x_{15}.

```
REA L :: x(35), x_do(35)
p=3.0; q=1.0
x(1)=q/p
DO n = 1,15
    x(n+1)=(p+1)*x(n) – q
END DO
x_do(31)=10.0
DO n=30,1,−1
    x_do=1.0/(p+1)*(x_do*(n+1)+q)
END DO
DO n=1,15
PRINT*, x(n), x_do
END DO
END
```

Upon execution of the program, the output will be obtained as tabulated in Sect. 1.2.2.2.   □

Example 6 (Computing series and product). Compute

$$S = \sum_{i=1}^{n} a_i \quad \text{and} \quad P = \prod_{i=1}^{n} a_i, \quad n \le 100 \text{ (say)}$$

Solution. We consider the elements a_i as components of a vector. Thus, we can write

```
REAL :: a(100)
READ*, a
s=0.; p=1.
do i=1,n
    S=S+a(i); P=P*a(i)
end do
PRINT*, S, P
END
```

In the above program, we have adopted some abbreviations. The real constants 0.0 and 1.0 have been written simply as 0. and 1. The variables S and P are accepted as real by default, and similarly, n is accepted as an integer variable. The DO and END DO have been written in lower case.

The sequence $\{a_i\}$ may be arranged in ascending order of magnitude before summation to diminish the errors due to roundoff (see Sect. 1.2.2.1). This can be done by adopting the technique of Example 8 given below.   □

Example 7. (Finding the largest and the smallest element of a sequence). Given $a_1, a_2, \cdots, a_n, n \le 100$ (say), find the largest and the smallest element.

Solution. We think of $a_1, a_2 \cdots, a_n$ in n storage locations. Create locations named BIG and SMALL for the largest and the smallest element, respectively. Set BIG = a_1 and compare it with a_2. If it is smaller than a_2, set BIG = a_2. Continuing the process, BIG will ultimately contain the largest element. For the smallest element, set SMALL = a_1. If it is smaller than a_2, set SMALL = a_2 and repeat the process, to finally obtain the smallest number.

```
real :: a(100)
BIG=a(1); SMALL=a(1)
do i=2,n
    if(BIG<a(i)) BIG=a(i)
    if(SMALL>a(i)) SMALL=a(i)
end do
PRINT*, BIG, SMALL
end
```

□

Example 8. (Ascending arrangement of sequence). Sort $a_1, a_2, \cdots, a_n, n \leq 100$ (say) in ascending order.

Solution. Compare two elements a_i and a_j $(i < j)$. If $a_i > a_j$, swap the two elements:

```
real :: a(100)
read*, n, a
do i=1,n−1
do j=i+1,n
if(a(i)>a(j)) then
    temp = a(j)      ! temp is a temporary location used for interchange
    a(j) = a(i)
    a(i) = temp
end if
end do
end do
print*, a
end
```

All the statements in the program are written in lower case for convenience of writing. *This is entirely permissible*. The complexity of the algorithm is $O(n^2)$, as the swapping operation is done $n \times n$ times.   □

Example 9. Given matrices A and B, compute the product

$$C := AB = \sum_{k=1}^{n} a_{ik}b_{kj} \quad (n \leq 10)$$

Solution. The summation generation technique is employed here:

```
real :: A(100,100), B(100,100), C(100,100)
read*, n, A, B        ! Input A and B are to be read columnwise
do i=1,n; do j=1,n
C(i,j)=0.
do k=1,n
    C(i,j)=C(i,j)+A(i,k)*B(k,j)
end do; end do; end do
print*, ((C(i,j),j=1,n),i=1,n)        ! Output is printed rowwise
end
```

The print statement is called an **IMPLIED DO**. It is equivalent to

```
   do i=1,n; do j=1,n
      print*, C(i,j)
   end do; end do
```

C can of course be obtained from intrinsic function matmul: C=MATMUL(A,B) □

Example 10. Compute the slope of the chord joining the points $x = 1/2$ and $x = 2$ of the curve

$$y = \sqrt{x}, \quad 0 \le x \le 1$$
$$= e^{1-x}, \quad x > 1$$

Solution. Here y is separately programmed as a function subprogram.

```
PROGRAM MAIN
slope=(y(0.5)−y(2.0))/(0.5−2.0)
print*, slope    ! (Output)
END PROGRAM
!**************************
FUNCTION y(x)
IF(x>=0..and.x<=1.) THEN
     y=sqrt(x)
ELSE-IF(x>1.) THEN
     y=exp(1.−x)
ELSE
     print*, 'Function not defined'
END IF
RETURN
END FUNCTION y
```

□

Example 11. Given a set of n experimental data $x_1, x_2, \cdots, x_n$, $n = 100$ (say), write a program that calls a subroutine named STAT to compute

$$\text{mean } \bar{x} = \frac{1}{n} \sum_{i=1}^{n} x_i \text{ and}$$

$$\text{standard deviation } s = \sqrt{\frac{\sum_{i=1}^{n}(x_i - \bar{x})^2}{n - 1}}$$

Solution. Using the technique for summation, we obtain

```
PROGRAM MAIN
REAL:: x(100)
n=100     ! (Input)
READ*, x    ! (Input)
CALL STAT(n,x,x_ bar,s_ d)
PRINT*, x_ bar, s_ d     ! (Output)
END PROGRAM
!*******************************
SUBROUTINE STAT(n,x,x_ bar,s_ d)
REAL:: x(n)
sum=0.
DO i=1,n
    sum=sum+x(i)
END DO
x_ bar=sum/n
sum=0.
DO i=1,n
    sum=sum+(x(i)−x_ bar)**2
END DO
s_ d=sqrt(sum/(n−1))
RETURN
END SUBROUTINE STAT
```

□

Example 12. (HeapSort Algorithm). Sort a sequence $a_1, a_2, \cdots, a_n$, $n \le 100$ (say) in ascending order by arranging the sequence into a tree graph of the form depicted below, writing a suitable subroutine for the procedure employed.

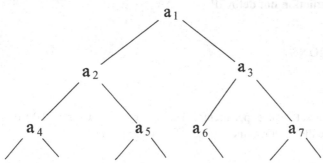

Solution. The tree graph shown above consists of a sequence of *binary trees*. A binary tree in the graph consists of a sequence of a parent element a_i, having a pair of *child*, if n is odd as is evident in the figure. If, on the other hand, n is even, then a_n is a single child of its parent a_i (verify). It follows from the figure that (i) the two children of parent a_i are the left child a_{2i} and the right child a_{2i+1}, and (ii) the parent element of a child a_j is $a_{[j]}$, where the symbol $[\cdot]$ represents the integer part of the argument. A tree graph in general is said to form a *Heap* if, $a_i \geq a_{2i}$, a_{2i+1}, and all the elements of the subtrees emanating from the two child nodes.

Thus, the largest element in the heap is stored at the nodal *root* a_i. The heapifying of the nodes of a binary subtree at indices i, $2i$, and $2i + 1$ is performed by a swapping procedure called *siftdown*, which turns the element at node i dominant with respect to the other two elements of the subtree. The process is iterated in the reverse order, beginning the bottom most index n moving up to the index 1. This iteration creates the heap of the full array of n elements, with the first element at index 1 as the largest element. In order to create sorting in ascending order, the first element at index 1 is pulled down to the bottom most position at n, that is to say, the current a_n is replaced by the current a_1 and the *siftdown* process applied again from bottom upwards up to the second element with index 2. The procedure is applied repeatedly to successively obtain a_{n-1}, a_{n-2}, $\cdots$, a_2 and the residual a_1 in the descending order.

```
PROGRAM MAIN
 REAL :: a(100)
 READ*, n, a
 PRINT*, a
 END PROGRAM

SUBROUTINE HEAPSORT(a,n)
REAL :: a(n)
INTEGER :: b
DO i=n/2, 1,-1                ! Builds the Heap of a(n)
CALL SIFTDOWN(a,n,i,n)
END DO
DO b=n,2,-1     ! Builds the decending order a(n),a(n-1),··· a(2),a(1)
                             ! of the sequence a
   temp=a(1); a(1)=a(b); a(b)=temp
CALL SIFTDOWN(a,n,1,b)
END DO
END SUBROUTINE HEAPSORT

SUBROUTINE SIFTDOWN(a,n,i,b)
INTEGER :: b
REAL :: a(n)
k=i                          ! k is root index of a binary subtree
```

```
    IF(2*k<b) THEN       ! i is a root
      lc=2*k              ! lc is a left child
      IF(lc+1<b) THEN
        IF(a(lc)<a(lc+1)) lc=lc+1
      END IF
      IF(a(k)<a(lc)) THEN
        temp=a(lc); a(lc)=a(k); a(k)=temp; k=lc
      ELSE
        RETURN
      END IF
    END IF
  END SUBROUTINE SIFTDOWN
```

The complexity of the HeapSort algorithm is $n \log_2 n$. For, if m is the number of binary subtrees in the heap, $2^m = n$ so that $m = 2^n$ or, $m = \log_2 n$, and the number of siftdown operations is of the order of n. Compared to the simple algorithm of Example 8 having complexity $O(n^2)$, the $O(n \log_2 n)$ complexity is much smaller when n is large. The algorithm, for this reason, has important application in the transmission of huge sized priority queues of data.   □

Exercises

1. The current in an AC circuit containing resistance, capacitance and inductance is given by

$$I = \frac{E}{\sqrt{R^2 + [2\pi/L - 1/(2\pi f C)]^2}}$$

where I = current (amps), E = voltage (volts), R = resistance (ohms), L = inductance (henrys), C = capacitance (farads) and f = frequency (hertz). Write a Fortran program to compute I for given values of E, R, L, C and f.

2. The earth pressure P on a retaining structure is given by the formula

$$P = \frac{wh^2}{2} \frac{1 - \sin\phi}{1 + \sin\phi}$$

where w = weight per unit volume of filled up earth = 513 kg/m^3
$\quad h$ = height of earth fill = 3 m
$\quad \phi$ = angle of repose of filled material = 30°.
Write a Fortran program to compute P.

3. The displacement x and velocity v of the piston of an internal combustion engine is given by

$$x = r(1 + n - \cos\theta - \sqrt{n^2 - \sin^2})$$

$$v = \omega r \left(\sin\theta + \frac{\sin 2\theta}{2\sqrt{n^2 - \sin^2\theta}} \right)$$

where n = rod length/crank radius = 4, ω = angular velocity = 2000 rpm and θ is the crank angle. Compute x and v for θ varying from 0° to 360° at 10° interval. [Use DO statement: DO i=0,360,10; theta=i*3.141593/180].

4. The allowable stress for a column of slenderness ratio R is given by

$$\begin{aligned} S &= 17000 - 0.485 R^2 \quad \text{if} \quad R \le 120 \\ &= \frac{18000}{1 + R^2/18000} \quad \text{if} \quad R > 120 \end{aligned}$$

Compute S for R increasing from 20 to 200 in increments of 5. [Use IF-ELSE and GOTO statement for looping].

5. Write a program to obtain the solution of the quadratic equation $ax^2 + bx + c = 0$, where all the quantities are complex.

6. Write programs to compute

(i) $S_1 = 1 + 2 + 3 + \cdots + n$
(ii) $S_2 = 1^2 + 2^2 + 3^2 + \cdots + n^2$
(iii) $S_3 = 1^3 + 2^3 + 3^3 + \cdots + n^3$
(iv) $n! = 1 \cdot 2 \cdot 3 \cdots n$

for $n = 100$ (say). What happens during execution of (iv) ? [Use technique of Example 6. Overflow].

7. Convert the algorithm for computing a polynomial

$$P(x) = a_0 x^n + a_1 x^{n-1} + \cdots + a_{n-1} x + a_n$$

by Horner's scheme into a Fortran program. [See Example 3].

8. Compute the following infinite series by writing Fortran program, assuming termination criterion as in Example 3 of Sect. 1.4.12:

(a) $\cos x = 1 - \dfrac{x^2}{2!} + \dfrac{x^4}{4!} - \dfrac{x^6}{6!} + \cdots \infty$
for $x = \pi/2$.

(b) $e^x = 1 + x + \dfrac{x^2}{2!} + \dfrac{x^3}{3!} + \cdots \infty$

restricting to the case $0 \le x < 1$ select $x = 1/2$. How will you treat the case $x \ge 1$?

(c) Let $x = 2^m z$, $\xi = (1 - z)/(1 + z)$, $(-1 \le z \le 1)$ then

$$\ln x = m \ln 2 - 2\left(\xi + \frac{\xi^3}{3} + \frac{\xi^5}{5} + \cdots\right)$$

Compute $\ln x$ for $x = 3$, using $\ln 2 = 0.6931472$.

(d) $\sin^{-1} x = x + \dfrac{x^3}{2 \cdot 3} + \dfrac{1 \cdot 3 \cdot x^5}{2 \cdot 4 \cdot 5} + \dfrac{1 \cdot 3 \cdot 5 \cdot x^7}{2 \cdot 4 \cdot 6 \cdot 7} + \cdots \infty$

$$= \sum_{k=0}^{\infty} \frac{(2k)! \; x^{2k+1}}{2^{2k}(k!)^2(2k+1)} \text{ for } x = 1/2 \text{ Hence compute the value of } \pi.$$

[(b) Set x=[x]+y].

9. Write separate programs to compute the vector and matrix norms: (a) $\|\mathbf{x}\|_2 =$

$$\sqrt{\sum_{i=1}^{n} |x_i|^2} \quad \text{and} \quad \|\mathbf{x}\|_\infty = \max |x_i| \ (1 \le i \le n), \text{ of a vector } \mathbf{x} = [x_1, x_2, \cdots, x_n].$$

(b) $\|A\|_2 = \sqrt{\displaystyle\sum_{i=1}^{n}\sum_{j=1}^{n} |a_{ij}|^2}$ and

$$\|A\|_\infty = \max \sum_{j=1}^{n} |a_{ij}| \ (1 \le i \le n),$$

of a matrix $A = [a_{ij}]$.

10. Given a sequence of integers $a_1, a_2, \cdots, a_n, n \le 100$ (say). Write a program to find the elements that are divisible by the integer 3.
[Use integer arithmetic. $a_i/3 \times 3$ then equals a_i only when a_i is divisible by 3].

11. Write a program to sort prime numbers between 1 and 1000.

[Prime numbers are odd integers. An odd integer not divisible by previous prime numbers is a prime number].

12. A one-dimensional array named y contains $n = 50$ experimental data. Perform *smoothing* of the data by replacing each data, except the first and the nth, by

$$y_i = \frac{y_{i-1} + y_i + y_{i+1}}{3}, \quad i = 2, 3 \cdots, 49$$

Write a Fortran program to obtain smoothed data.

13. Given two sets of experimental data $x_1, x_2, \cdots, x_n$ and $y_1, y_2, \cdots, y_n$ of variates x and y. By writing a Fortran program, compute the correlation coefficient ρ between the two variates defined by

$$\rho = \frac{n \sum_{i=1}^{n} x_i y_i - \sum_{i=1}^{n} x_i \cdot \sum_{i=1}^{n} y_i}{\sqrt{n \sum_{i=1}^{n} x_i^2 - \left(\sum_{i=1}^{n} x_i\right)^2} \sqrt{n \sum_{i=1}^{n} y_i^2 - \left(\sum_{i=1}^{n} y_i\right)^2}}$$

14. Cardano's solution of the cubic equation: Let the equation be

$$x^3 + a_1 x^2 + a_2 x + a_3 = 0$$

with real coefficients a_1, a_2, a_3. Let

$$Q = \frac{1}{9}(a_1^2 - 3a_2) \quad and \quad R = \frac{1}{54}(2a_1^3 - 9a_1 a_2 + 27 a_3)$$

If $Q^3 - R^2 \geq 0$, then the cubic equation has real roots

$$x_k = -2\sqrt{Q}\cos\left[\frac{\theta + 2\pi(k-1)}{3}\right] - \frac{a_1}{3}, \quad k = 1, 2, 3$$

where $\theta = \cos^{-1}(R/\sqrt{Q^3})$. If, on the other hand, $Q^3 - R^2 < 0$, then the equation has one real root and a pair of complex roots

$$x_1 = s + \frac{Q}{s} - \frac{a_1}{3}, \quad z_2 = \omega s + \omega^2 \frac{Q}{s} - \frac{a_1}{3}, \quad z_3 = \omega^2 s + \omega \frac{Q}{s} - \frac{a_1}{3}$$

where $s = -\,\mathrm{sgn}(R)(|R| + \sqrt{R^2 - Q^3})$ and $\omega = \frac{1}{2}(-1 + \sqrt{3}i)$ (cube root of unity).
Write a Fortran program to compute the roots.

15. In a *continued fraction*

$$f = b_0 + \cfrac{a_1}{b_1 + \cfrac{a_2}{b_2 + \cfrac{a_3}{b_3 + \cdots}}} = b_0 + \frac{a_1}{b_1+}\,\frac{a_2}{b_2+}\,\frac{a_3}{b_3+}\cdots \quad \text{(say)}$$

the nth convergent f_n $(:= p_n/q_n)$ by stopping at the nth quotient a_n/b_n, satisfy the Euler–Wallis recurrence relations

$$p_0 = b_0, \quad q_0 = 1, \quad p_1 = b_1 p_0 + a_1, \quad q_1 = b_1$$

$$\left.\begin{array}{l} p_n = b_n p_{n-1} + a_n p_{n-2} \\ q_n = b_n q_{n-1} + a_n q_{n-2} \end{array}\right\} \quad n \geq 2$$

Write a Fortran program to compute 20 convergents. To prevent overflow in computing p_n, q_n, rescale the present values and the preceding values p_{n-1}, q_{n-1} by q_n and then iterate.

Apply the program to verify the formula

$$\pi = \frac{4}{1+} \frac{1^2}{3+} \frac{2^2}{5+} \frac{3^2}{7+} \cdots \infty$$

Verify that the successive convergents always *bracket* the exact value.

16. Compute $\tan x$ $(0 \le x \le \pi/4)$, using Lambert's formula $\tan x = x/y$, where

$$y = 1 + \frac{-x^2}{3+} \frac{-x^2}{5+} \frac{-x^2}{7+} \cdots \frac{-x^2}{2n+1} \cdots \infty$$

for $x = 40^o$.
[In order to compute y accurate to 10^{-10} it is sufficient to take $n = 7$ and terminate the continued fraction].

17. For the integral

$$I_n = \int_0^1 \frac{x^n}{x+5} dx$$

prove the reduction formula $I_n = \dfrac{1}{n} - 5 I_{n-1}$ and obtain particular values $I_0 = \ln(6/5)$ and $I_\infty = 0$. Write a Fortran program to compute I_n for ascending values of n up to 15. Are you satisfied with the computed values? If not try evaluation in descending order taking $I_{30} \approx 0$.

Chapter 2
Equations

Solving an equation is one of the prime topics of elementary algebra, where it is a powerful tool for problem solving, whenever algebraic formulation is possible. Equations appear in advanced sciences as well in a variety of ways, mostly in complicated forms. Developing efficient computational schemes for solution in a general manner is therefore of paramount importance. It is emphasised that through these schemes only approximate numerical solutions are sought, that may never be exact. In a sense it is useless to think of exact solutions, because equations arising in practice always contain experimentally determined parameters that are always contaminated by errors of measurement. Even if it is assumed that an equation is exactly given, machine computation of the solution following any method, invites roundoff errors.

In this opening chapter, we consider a single unknown x to be determined as a solution from an equation of the general form

$$f(x) = 0 \tag{2.1}$$

where x is real or complex defined over some interval or region as the case may be. The interval or region may be finite or infinite. The function f is supposed to be *continuous* over the interval or the region. In certain cases we shall assume that the derivatives f' and f'' are also continuous.

Every finite value ξ of x (real or complex) for which $f(\xi) = 0$ is called a *root* of Eq. (2.1), or a *zero* of the function $f(x)$. A root of an equation may or may not exist ($1/x = 0$ and $e^x = 0$ for example, do not possess any root). When a root exists, it must either be real or complex. When several roots exist, their number may be finite or infinite. For instance if $f(x)$ is a *polynomial* of degree n, then the corresponding *algebraic equation* has exactly n roots (Fundamental Theorem of Algebra). Thus algebraic equations are very special. Equations containing other types of functions are called **transcendental equations**. Such equations may have finite or infinite number of roots.

© Springer Nature Singapore Pte Ltd. 2019
S. K. Bose, *Numerical Methods of Mathematics Implemented in Fortran*, Forum for Interdisciplinary Mathematics, https://doi.org/10.1007/978-981-13-7114-1_2

In approximate numerical solution of an equation one may be interested to compute all the roots of the equation. A study of this chapter shows that the roots in practice are mostly obtained one by one. Moreover, the search for real roots and their computation is easier compared to those for complex roots. Consequently, we mainly focus on real roots, treating complex roots only by working methods that work.

2.1   Real Roots

For real roots, the obvious procedure is to first isolate the roots in small intervals, followed by refinement of their values. The first task is relatively simple and is described below.

2.1.1   Isolation of a Real Root

Isolation of a root means that there is a neighbourhood which does not contain any other root. A straight forward method is to plot the graph of the function $f(x)$ for values of x we are interested in. The graphing could be done on a computer. Inspite of digitisation and small errors, one can read off small intervals containing the points where the graph of the function cuts, or in some special case, merely touches the x-axis. Such intervals isolate the roots of the equation.

Isolation by *trial* is also possible. It is based on the following theorem of Analysis:

Theorem 2.1 *If a continuous function $f(x)$ assumes values of opposite sign at the end points of an interval $[a, b]$, i.e.*

$$f(a) f(b) < 0 \tag{2.2}$$

then the interval will contain at least one real root ξ of $f(x) = 0$, i.e. $f(\xi) = 0$, $\xi \in (a, b)$. Moreover if $f(x)$ is strictly monotonic in $[a, b]$, ξ is unique.

The theorem is illustrated in Fig. 2.1a, b.

In the trial method, one finds the sign of $f(x)$ at a series of intermediate points $x = c_1, c_2, c_3, \cdots$, taking in to account the peculiarities of $f(x)$. If it turns out that $f(c_k) f(c_{k+1}) < 0$, then there is a root in (c_k, c_{k+1}). Evidently, in this method, there is a chance of missing a root like ξ_2 of Fig. 2.1a. The difficulty can be avoided by checking monotonicity of $f(x)$ in (c_k, c_{k+1}). If $f(x)$ lacks it, the interval $[c_k, c_{k+1}]$ must be broken up in to smaller segments.

A root like ξ_2 is a multiple root of even order. For example, if $f(x) = (x - 1)^2$, then the parabola touches the x-axis at $x = 1$, yielding a double root $x = 1, 1$ of the equation $(x - 1)^2 = 0$. In a similar manner $(x - 1)^3 = 0$ has a triple root

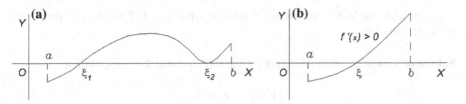

Fig. 2.1 a Root isolation. **b** Unique root

$x = 1,\ 1,\ 1$, but in this case the cubic parabola intersects the x-axis at $x = 1$ without being strictly monotonic.

Example 1. Isolate the real roots of the Wallis's equation

$$f(x) := x^3 - 2x - 5 = 0$$

(In the history of numerical methods, this particular equation was taken up for bench marking the efficiency of a newly discovered method of root computation).

Solution. The cubic equation has three roots. Computing $f(x)$ at suitably selected points in $(-\infty, \infty)$, one finds

$$f(-\infty) < 0$$
$$\vdots$$
$$f(-1) = -4 < 0$$
$$f(0) = -5 < 0$$
$$f(1) = -6 < 0$$
$$f(2) = -1 < 0$$
$$f(3) = 16 > 0$$
$$\vdots$$
$$f(\infty) > 0$$

Thus there must be a real root in $[2, 3]$ since there is a change of sign in $f(x)$. The remaining two roots appear to be complex as there is no possibility of change of sign in any other interval.   □

John Wallis (1616–1703), English mathematician. He introduced the familiar symbol of infinity. His ingenious works singularly exerted important influence on Newton.

Example 2. Equations of the form

$$f(x) := e^{ax} - ax - b = 0$$

occur in heat conduction problems. Supposing that $a = 1/2$ and $b = 9$, isolate its positive real roots.

Solution. This is an example of transcendental equation. By computation

$$f(0) = -8 < 0$$
$$f(1) = -7.85 < 0$$
$$f(2) = -7.28 < 0$$
$$f(3) = -6.02 < 0$$
$$f(4) = -3.61 < 0$$
$$f(5) = 0.68 > 0$$
$$f(6) = 8.08 > 0$$
$$f(7) = 20.62 > 0$$

Apparently there is only one real root in [4, 5].   □

2.1.2 Refinement: Computation of a Real Root

We suppose that a real root ξ of Eq. (2.1) has been isolated in a neighbourhood $[a, b]$. The approximate computation of ξ can be accomplished in a number of ways described in the following. It may be noted that all the methods are *iterative* in nature, in which ξ is approached by successive approximations.

2.1.2.1 Bolzano Bisection Method

In this method, the interval $[a, b]$ is divided in to two halves by the point of bisection $x_m = (a + b)/2$. If $f(x_m) = 0$, then $\xi = x_m$ is the root. If not so, then choose that half $[a, x_m]$ or $[x_m, b]$, at the end points of which $f(x)$ has opposite signs. Denote the chosen half as $[a_1, b_1]$ and bisect it again applying the same test to generate a smaller neighbourhood $[a_2, b_2]$. In this way an infinite sequence of nested intervals $[a_1, b_1], [a_2, b_2], \cdots, [a_n, b_n], \cdots$ is generated such that

$$f(a_n) f(b_n) < 0 \tag{2.3a}$$

where

$$b_n - a_n = \frac{1}{2^n}(b - a) \tag{2.3b}$$

The sequence of points $a_1, a_2, a_3, \cdots$ is monotonic increasing and bounded above. So it has a limiting point ξ. Similarly the sequence $b_1, , b_2, b_3, \cdots$ is monotonic decreasing but bounded below, converging to the same limit ξ in view of the fact that $[a_n - b_n] \to 0$ as $n \to \infty$. Moreover, from Eq. (2.3a) passing to the limit $n \to \infty$, we must have $[f(\xi)]^2 \le 0$ whence $f(\xi) = 0$. Thus ξ is the required root of the equation.

The root ξ is always contained in the nested intervals. This aspect is often described as *bracketing* of the root.

Following the method, approximate computation of ξ can be performed by iteratively generating the intervals till $b_n - a_n < \epsilon$, where ϵ is a given *tolerance*. Then $\xi \approx a_n$ or b_n. The algorithm for the method would read like the following:

Algorithm 1. Bolzano Bisection method

Input: Real $f(x)$, a, b, ϵ

Output: Real a, b

Computation: 1. If $f(a) f(b) > 0$, Stop. No root.
2. Else set $x_m \leftarrow .5 \times (a + b)$
3. If $f(x_m) = 0$, Stop. x_m is the root.
4. If $f(a) f(x_m) < 0$, set $b \leftarrow x_m$
    else set $a \leftarrow x_m$
5. If $|b - a| < \epsilon$, Stop. a or b is the root.
else go to step 2

The algorithm usually requires large number of iterations, because no specific information about the shape of the function $f(x)$ is used. This lacuna is removed in the remaining methods.

2.1.2.2 Regula Falsi Method

When the end points of the graph of $f(x)$ in the interval $[a, b]$ containing a root ξ are joined by the chord (or secant), we obtain the latter's intersection with the x-axis as an approximation of ξ. This is the idea behind regula falsi or false position.

The equation of the chord joining $[a, f(a)]$, $[b, f(b)]$ is

$$\frac{x - a}{b - a} = \frac{y - f(a)}{f(b) - f(a)}$$

If it cuts the x-axis ($y = 0$) at $x = x'$, we get the first iterate

$$x' = a - \frac{(b - a) f(a)}{f(b) - f(a)} = \frac{a f(b) - b f(a)}{f(b) - f(a)}$$

Noting that one of $f(a)$ or $f(b)$ is negative, x' is the *weighted average* of a and b with weights $|f(b)|$ and $|f(a)|$. If $f(x') \neq 0$, we obtain a smaller interval $[a_1, b_1]$ containing ξ in the following manner. Without loss of generality we assume $f(x)$ to be concave; otherwise if $f(x)$ is convex, the equation can be written as $-f(x) = 0$, in which $-f(x)$ is concave. Since $f(a) f(b) < 0$, two cases arise; (*i*) $f(a) < 0$, $f(b) > 0$ and (*ii*) $f(a) > 0$, $f(b) < 0$. These are shown in Fig. 2.2a, b below:

Evidently for case (*i*) $a_1 = x'$, $b_1 = b$ and for case (*ii*) $a_1 = a$, $b_1 = x'$. If the process is continued we get a sequence of intervals $[a_1, b_1]$, $[a_2, b_2]$, $\cdots [a_n, b_n]$, $\cdots$,

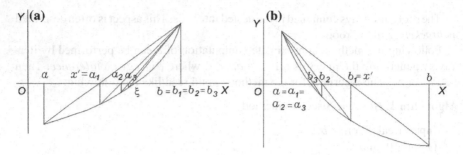

Fig. 2.2 Regula falsi iteration

where for case (i) $a_1 < a_2 < a_3 < \cdots < a_n < \cdots$, and $b_1 = b_2 = b_3 = \cdots = b_n = \cdots = b$, while for case (ii) $a_1 = a_2 = a_3 = \cdots = a_n = \cdots = a$, and $b_1 > b_2 > b_3 > \cdots > b_n > \cdots$. The sequences satisfy

$$a_{n+1} = \frac{a_n f(b) - b f(a_n)}{f(b) - f(a_n)}, \quad b_{n+1} = \frac{a f(b_n) - b_n f(a)}{f(b_n) - f(a)} \tag{2.4}$$

The two sequences are monotonic and bounded. Hence if $a_n \to \xi$ as $n \to \infty$, the first of the above two equations yields

$$\xi = \frac{\xi f(b) - b f(\xi)}{f(b) - f(\xi)}$$

Thus $f(\xi) = 0$ (since $\xi < b$), so that ξ is the required root. Similar conclusion can be drawn for the sequence $\{b_n\}$.

In this method bracketing of ξ is observed, but the length of the nth subinterval $b_n - a_n$ does not diminish to zero. This lacuna can be removed, and at the same time convergence accelerated by a computational trick, in which the secants are now replaced by chords of *decreasing slopes*, until x' falls to the opposite side of the root. This is achieved by reducing the ordinate at the fixed end of the interval by half of its value. This is illustrated in Fig. 2.3.

Fig. 2.3 Modified regula
falsi iteration

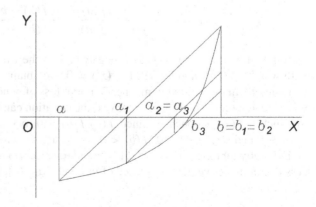

In the figure, the first secant is drawn as usual to obtain the iterate a_1 where $f(a) f(a_1) > 0$, while $b_1 = b$. So the next chord is drawn through the points $[a_1, f(a_1)]$ and $[b_1, f(b_1)/2]$, to yield the iterate a_2 and $b_2 = b$. Repeating the process, we find in the figure that, $a_3 = a_2$ and $b_3 < b_2$. Generalising the idea, given a tolerance ϵ of the root, we obtain the algorithm

Algorithm 2. Modified Regula Falsi method.

> Input: Real $f(x)$, a, b, ϵ
> Output: Real a, b
> Computation: 1. If $f(a) f(b) > 0$, Stop. No root.
> 2. Else set $F \leftarrow f(a)$, $G \leftarrow f(b)$, $x_0' \leftarrow a$
> 3. Set $x' \leftarrow (aG - bF)/(G - F)$
> 4. If $f(x') = 0$, Stop. x' is the root.
> 5. If $f(a) f(x') < 0$ set $b \leftarrow x'$, $G \leftarrow f(x')$;
>   if also $f(x_0') f(x') > 0$ set $F \leftarrow F/2$
>   else set $a \leftarrow x'$, $F \leftarrow f(x')$;
>   if also $f(x_0') f(x') > 0$ set $G \leftarrow G/2$
> 6. If $|b - a| < \epsilon$ Stop. a or b is the root.
> 7. Else set $x_0' \leftarrow x'$ and go to step 3.

2.1.2.3 Secant Method

This method, as the name suggests, is also based on drawing secants, but with a variation. We write $a = x_0$ and $b = x_1$. If a and b are close to the root ξ, x_0 and x_1 become approximations to ξ. The next approximation is obtained by drawing the secant as

$$x_2 = \frac{x_0 f(x_1) - x_1 f(x_0)}{f(x_1) - f(x_0)}$$

Using x_1 and x_2 a third iterate x_3 can be constructed in a similar manner. By continuing the process, we obtain a sequence of points $x_0, x_1, x_2, \cdots, x_n, \cdots$ satisfying

$$x_{n+1} = \frac{x_{n-1} f(x_n) - x_n f(x_{n-1})}{f(x_n) - f(x_{n-1})} = x_n - f(x_n) \frac{x_n - x_{n-1}}{f(x_n) - f(x_{n-1})} \tag{2.5}$$

The root may not be bracketed in general during iteration. So the sequence may diverge in some cases. If it converges to ξ, then the above equation in the limit becomes $\xi = \xi - f(\xi)/f'(\xi)$. Thus if $f'(\xi) \neq 0$, the limit ξ is the root of the equation.

To enforce bracketing and ensure convergence, we combine the method with that of bisection, adopting the following strategy. Changing notations which will be useful for programming, suppose that the current pair of iterates is a and b. We store the value a and $f(a)$, whenever $f(a)$ and $f(b)$ have opposite sign. Say the values are c and $f(c)$. As iteration proceeds with a and b, if $f(a)$ and $f(b)$ have the same sign then the point of bisection of c and b is chosen as the next iterate, otherwise it is obtained by the secant method. The following Fortran subroutine is developed on this basis.

SUBROUTINE SECANT_BISECTION(f,a,b,xtol,root)
! computes a root of f(x)=0 lying in [a,b]
! by combined secant and bisection methods.
! f(x) must be given by a separate function subprogram.
! xtol = tolerance allowed in the computation of root.   (Input)
! root = the desired root with tolerance xtol.   (Output)
aa=a; bb=b; c=aa; fa=f(aa); fb=f(bb); fab=fa*fb
IF(fab>0.0) then
**   print*, 'No Root'; RETURN**
END IF
10 IF(fab<0) d=(bb−aa)*(fb/(fa−fb))   !Shift from b in secant method
IF(d<a−bb .OR. d>b−bb) d=0.5*(c−bb)   !Shift in bisection method
IF(ABS(d)<xtol) THEN
**   root=bb+d; RETURN**
END IF
c=aa; fc=fa; aa=bb; fa=fb; bb=bb+d; fb=f(bb); fab=fa*fb
GOTO 10
END SUBROUTINE SECANT_BISECTION
    The question of convergence of pure secant based method is examined in Sect. 2.4.

2.1.2.4   Newton's Method

We now make the additional assumption that the derivative $f'(x)$ exists in $[a, b]$. If $f(x)$ is considered monotonic in $[a, b]$ $f'(x) \neq 0$. Suppose x_n is an approximation of the root ξ represented by point N on the x-axis. Then referring to Fig. 2.4, the point of intersection T of the tangent at P $[x_n, f(x_n)]$ may be a closer approximation x_{n+1} to ξ. Thus,

$$x_{n+1} = x_n - TN = x_n - \frac{PN}{\tan \angle PTN} = x_n - \frac{f(x_n)}{f'(x_n)} \qquad (2.6)$$

Fig. 2.4 Newton iteration

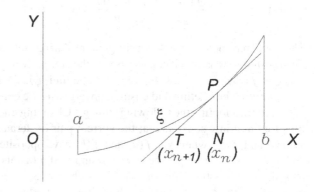

Starting from an approximation x_0 in $[a, b]$, one can generate the sequence $x_1, x_2, \cdots, x_n, \cdots$ by, iteration, provided that $f(x_n) \neq 0$. If the sequence converges to ξ, then it easily follows that $f(\xi) = 0$, i.e. ξ is a root of Eq. (2.1).

One can derive formula (2.6) without recourse to geometry. Let $\xi = x_n + h$, then $f(x_n + h) = 0$. Assuming atleast the continuity of $f''(x)$, we have by Taylor's theorem

$$0 = f(x_n + h) = f(x_n) + h\, f'(x_n) + O(h^2)$$

Thus to $O(h)$, $h = -f(x_n)/f'(x_n)$ yielding the formula. The success of the formula heavily depends on the behaviour of $f'(x_n)$, which in any case should be away from 0.

The method can also be viewed as a limiting case of the secant method. For, by Lagrange's Mean Value Theorem

$$f(x_n) - f(x_{n-1}) = (x_n - x_{n-1})f'(\eta)$$

where $x_{n-1} < \eta < x_n$. If x_{n-1} is close to x_n, $f'(\eta) \approx f'(x_n)$ and so

$$\frac{x_n - x_{n-1}}{f(x_n) - f(x_{n-1})} \approx \frac{1}{f'(x_n)}$$

Plugging the approximation in Eq. (2.5), one obtains Eq. (2.6).

Following Newton's method, a subroutine can easily be written down. One such is given below:

```
SUBROUTINE NEWTON(f,f1,x,maxiter,xtol,root)
! computes a root of f(x)=0 by Newton's method.
! f and derivative f1 must be declared external in the calling program.
! x = initial approximation of root      (Input).
! maxiter = maximum number of iterations allowed      (Input).
! xtol = tolerance in the computation of the root      (Input).
! root = the desired root with tolerance xtol      (Output).
DO i=1,maxiter
xnew=x−f(x)/f1(x)
IF(ABS|xnew−x|<xtol) THEN
   root=xnew; EXIT
ELSE
x=xnew; IF(i==maxiter) PRINT*, 'No Solution'
END IF
END DO
RETURN
END SUBROUTINE NEWTON
```

If for a particular f, a sufficiently large n does not deliver the root, there is convergence problem. Conditions for convergence are discussed in Sect. 2.4.

Historically, Newton (**Sir Isaac Newton (1643–1727)** foremost English Mathematician and physicist) developed a method for computing a root of a cubic equation, based on iterative linearisation process. He published his method as a means to solve Kepler's equation $x = 2e \sin x + t/T$ for planetary motion (x = "eccentric anomaly", e = eccentricity of orbit, t = time and T = time period of orbit). In about 1690, Joseph Raphson formulated Newton's ideas for the case of a polynomial equation in a form closer to Eq. (2.6). Thus the method is often referred to as **Newton–Raphson method**. A more accurate version of the Kepler problem is considered in Sect. 2.2, Exercises 2.3.

Remark **Modified Newton's Method.** If the derivative $f'(x)$ varies but slightly in $[a, b]$, then in Eq. (2.6), one can put $f'(x_n) \approx f'(x_0)$. This leads to iterations

$$x_{n+1} = x_n - \frac{f(x_n)}{f'(x_0)} \tag{2.6a}$$

for the root ξ of Eq. (2.1). The advantage accrued in Eq. (2.6a) is that the derivative of $f(x)$ need be computed only once in the whole range of iterations. It can be proved that if $f'(x)$ and $f''(x)$ have constant sign in $[a, b]$, the iterations converge to ξ.

Example 1. The equation $f(x) := e^{x/2} - x/2 - 9 = 0$ was considered earlier in Sect. 2.1.1. It has only one real root in the interval $[4, 5]$. Writing Fortran programs using Algorithms 1 and 2, the Secant and NEWTON, compute the root by the four methods, comparing the number of iterations required in the four methods.

Solution. Writing a program, the table of values given on the next page is obtained for the root. With $xtol = 0.1E - 6$, the root is indicated to be 4.873679, but it takes as many as 22 iterations in the case of bisection method. The modified regula falsi method takes 10 iterations in an oscillatory fashion and still does not completely settle to the final value. In comparison, the Secant and the Newton methods converge in nine and five iterations respectively. The subroutine SECANT_BISECTION yields the result with $xtol = 0.1E - 5$ only. $\Box$

n	Bisection method	Modified regula falsi	Secant method	Newton method
0	4.500000	4.000000	4.000000	4.500000
1	4.750000	4.841038	4.682076	4.915249
2	4.875000	4.893651	4.904085	4.874149
3	4.812500	4.873500	4.935609	4.873680
4	4.843750	4.873678	4.843255	4.873679
5	4.859375	4.873681	4.873688	4.873679
6	4.867188	4.873679	4.873697	
7	4.873697	4.873679	4.873670	
8	4.873047	4.873680	4.873679	
9	4.874023	4.873679	4.873679	
10	4.873535	4.873679		
11	4.873779			
12	4.873657			
13	4.873718			
$\vdots$	$\vdots$			
21	4.873679			
22	4.873679			

Example 2. Using the subroutine named secant_bisection, solve the equation

$$\sqrt{x} + \sqrt{x-2} + \sqrt{6-x} - 4 = 0$$

for its real roots.

Solution. Apparently, the function is defined for real x on the interval [2, 6] only, such that $f(2) = \sqrt{2} - 2 < 0$ and $f(6) = \sqrt{6} - 2 > 0$. Hence one can take $a = 2$, $b = 6$. If one uses the secant method, the iterations soon breakdown because the iterates lie outside [2, 6]. The subroutine secant_bisection avoids this difficulty by using bisection method for such cases. However, for starting values $a = 2, b = 6$, the compiler faces difficulty in computing $\sqrt{0} = 0^{0.5} = e^{0.5 \ln 0}$ and execution is aborted. One can get around this difficulty also by taking slightly different values for a and b, say $a = 2.01, b = 5.99$. This procedure yields the root as

$$\xi = 2.31251 \qquad \qquad \Box$$

Example 3. The critical buckling load of a fixed-pin ended bar is given by the equation $\tan x = x$, where $x = l\sqrt{\frac{P}{EI}}, l = $ length of the bar, $P = $ compressive load, E = Young's modulus, and I = moment of inertia of a cross section. Find the smallest positive root by Newton's method.

Solution. The real roots of the equation $\tan x = x$ are given by the points of intersection of the graphs of $y = \tan x$ and the line $y = x$. Now the graph of $\tan x$ consists of parallel monotonic curves from $-\infty$ to $+\infty$ having asymptotes $x = \pm(2n+1)\pi/2$, $(n = 0, 1, 2, \cdots)$. Except for $x = 0$, the line cuts these curves close to the asymptotes. Thus we obtain the approximation $x_0 \approx 3\pi/2 = 4.71239$ for the smallest non zero positive root. This fact can be verified by writing the equation in the form

$$f(x) := \sin x - x \cos x = 0$$

yielding $f(4) = 1.86 > 0$, $f(5) = -2.38 < 0$.
   For using Newton's method we also calculate

$$f'(x) = \cos x - \cos x + x \sin x = x \sin x$$

Using the two functions $f(x)$ and $f'(x)$ in subroutine NEWTON, one finds that the root is

$$\xi = 4.49341 \qquad \qquad \Box$$

Exercises

1. By computing the function at unit intervals between -10 and $+10$, isolate the real roots of the following equations:

(i) $x^3 - 3x - 5 = 0$ $[(2, 3)]$

(ii) $x^3 - 0.39x^2 - 10.5x + 11 = 0$ $[(-4, -3),\ (1, 2),\ (2, 3)]$

(iii) $x^4 - x - 10 = 0$ $[(-2, -1),\ (1, 2)]$

(iv) $2^x - 2x^2 - 1 = 0$ $[(0, 0),\ (0, 1),\ (6, 7)]$

(v) $2 \ln x - \dfrac{x}{2} + 1 = 0$ $[(0, 1),\ (11, 12)]$

(vi) $x - 0.2 \sin x - 0.5 = 0$ $[(0, 1)]$

2. Write a Fortran program for the modified regula falsi method for computing the real roots of the following equations after their isolation as in Exercise 1 above:

(i) $x^3 - 4x^2 + x + 6 = 0$ $[-1.0]$

(ii) $x = \tan 2(x - 1)$ $[0.21460,\ \text{requires large number of iterations}]$

(iii) $x^2 + \ln x = 2$ $[1.31410]$

3. Using the subroutine secant_bisection, determine the real roots of the equations given in Exercise 1.

$[(i)$ 2.27902 (ii) -3.50336, 1.14063, 2.75273 (iii) -1.69747, 1.85559 (iv) 0, 0.39928, 6.35235 (v) 0.72751, 11.90927 (vi) 0.61547$]$.

4. Find by subroutine named NEWTON, the real roots of:

(i) $x^3 - 1.5x^2 + 5.74x + 7 = 0$ $[-0.88984]$

(ii) $\cos x = 3x - 1$ $[0.60710]$

(iii) $x^2 - 10 \ln x = 0$ $[1.13836, 3.56556]$

5. Isolate the positive roots of

$$\tan x + \tanh x = 0$$

in unit intervals. Can you capture the first three roots following Newton's method by starting from the middle point of the sub-intervals?

[(1,2), (2,3), (4,5). No. The second root 2.36502 can be captured. The other two roots by bisection method are 1.57080, 4.71239].

6. (Chebyshev Method). As in the analytical derivation of the Newton iterations, if the second-order term is also retained in the Taylor expansion of $f(x)$, then show that the iteration formula becomes

$$x_{n+1} = x_n - \frac{f(x_n)}{f'(x_n)} - \frac{1}{2} \frac{[f(x_n)]^2 f''(x_n)}{[f'(x_n)]^3}$$

Apply the formula to compute the roots of

(i) $x^3 - x - 1 = 0$, lying in (1,2).   [1.32472]

(ii) $x \sin x + \cos x = 0$, lying in (2,3).   [2.79839]

7. Using Newton's method, prove the following iterative schemes:

(i) $x_{n+1} = \frac{1}{2}\left(x_n + \frac{a}{x_n}\right)$ for computing $\sqrt{a}$ (Hero's algorithm)

(ii) $x_{n+1} = \frac{1}{3}\left(2x_n + \frac{a}{x_n^2}\right)$ for computing $\sqrt[3]{a}$

(iii) $x_{n+1} = x_n(2 - ax_n)$ for computing a^{-1}.

Hero (or Heron) of Alexandria (A.D. 62) under which a number of works are known. They were mostly written in Greek.

8. (Generalised Newton's Method). If ξ is a root of $f(x) = 0$ with multiplicity p, then show that Newton's method can be generalised to

$$x_{n+1} = x_n - p \frac{f(x_n)}{f'(x_n)} \quad (n = 1, 2, 3, \cdots)$$

with given initial value x_0.

[In this case we can write $f(x) = (x - \xi)^p \phi(x)$, where $\phi(x)$ *varies slowly with* x in the neighbourhood of ξ, i.e. $\phi'(x)$ is small in the said neighbourhood. By logarithmic differentiation $f'(x_n)/f(x_n) = p/(x_n - \xi) + \phi'(x_n)/\phi(x_n) \approx p/(x_n - \xi)$. If $\xi := x_n + h, h \approx -pf(x_n)/f'(x_n)$].

9. (Fourier Method) Show that if the Newton iterations are modified as

$$x_{n+1} = x_n - \frac{f(x_n)}{f'(x_n)}, \quad z_{n+1} = z_n - \frac{f(z_n)}{f'(x_n)}$$

where $x_0 = a$ and $z_0 = b$, then x_n, $z_n \to \xi$ the root such that

$$\lim_{n \to \infty} \frac{x_{n+1} - z_{n+1}}{(x_n - z_n)^2} = \frac{f''(\eta)}{2 f'(\eta)}$$

implying quadratic convergence.

10. The velocity of propagation c of surface Rayleigh waves along the flat surface of a semi-infinite isotropic elastic solid is given by $c = \sqrt{x}\, \beta$, where x is a root of the cubic equation

$$x^3 - 8x^2 - 8 \frac{1 + 2\nu}{1 - 2\nu} x + \frac{16}{1 - 2\nu} = 0$$

in which ν = Poisson's ratio and β = velocity of *shear wave* propagation in the solid. Solve the equation by Newton's method for $\nu = 1/3$.
[$x = 0.86962$].

11. A load R_L is supplied from an A.C. source at voltage V_m through a resistor R_S in series with a diode. The capacity is sufficiently large to eliminate any ripple in the resulting D.C. voltage V_L across the load. The following approximation for the voltage ratio $V_L / V_m = x$ can be established

$$x = 1 - \left(\frac{\pi}{\sqrt{2}} \frac{R_S}{R_L} x \right)^{2/3}$$

Solve the equation for $\dfrac{\pi}{\sqrt{2}} \dfrac{R_S}{R_L} = 1$, by Newton's method.
[$x = 0.43019$].

12. The voltage drop v across a diode in the so-called biasing problem of an electronic circuit requires the solution of the equation

$$50 I_s \left(e^{vq/kT} - 1 \right) + v = 2$$

where $I_s = 10^{-9}$ amp for Si diode, k = Boltzman constant, T = temp. in $°K$ and q = charge of an electron. Solve the equation for $q/kT = 40$ by Newton's method.
[Write the equation as $5 \left(e^{40v} - 1 \right)/10^8 + v - 2 = 0$. $v = 0.43153$].

13. The displacement d of a cam follower is given by the equation

$$d = \frac{1}{2}[1 + e^{-x/2\pi} \sin x], \quad 0 \le x \le 2\pi$$

where x is the angle of rotation of the cam. Compute x for a desired displacement of $d = 0.75$ cm by the Newton method.
[$x = 2.33114$].

14. The friction factor $c_f = x^2$ for turbulent flow in a smooth pipe is given by the equation

$$x[1.74 \ln(Rx) - 0.4] = 1$$

where R is Reynolds' number of flow. Compute c_f for $R = 10^3$, using the trial $0.1 < x < 0.2$.
$[c_f = 0.01562]$.

15. The chemical equation in the production of methanol from CO and H_2 is

$$\frac{x(3 - 2x)^2}{(1 - x)^3} = 249.2$$

where x is the *equilibrium extent* of the reaction. Find x.
$[x = 0.81712]$.

16. The critical load of buckling of a typical portal frame with columns fixed at the base is given by

$$(i) \quad x \sin x + (4 + x^2) \cos x = 4 \quad \text{(for symmetric buckling)}$$

and $\qquad (ii) \quad \tan x = -\dfrac{x}{2} \quad$ (for antisymmetric buckling)

where $x = l\sqrt{\frac{P}{EI}}$, l = length of each side of the frame, P = compressive load, E = Young's modulus and I = moment of inertia of a cross-section. Find the smallest positive x in the two cases.
$[(i) \ x = 5.01819, (ii) \ x = 1.57079]$.

17. The frequency of vibration of a free–free bar is given by the equation

$$\cos x \cosh x + 1 = 0$$

where frequency $\omega = \sqrt{\dfrac{EI \, x^2}{4\rho \ l^2}}$, ρ is the density and E, I, l are defined as in Example 3, Sect. 2.1.2.4. Find the smallest positive root by Newton's method.
$[x = 1.87510]$.

2.2   The General Iteration Method

In the Newton method, the iterative formula (2.6) expresses x_{n+1} as a function x_n. The function contains $f(x_n)$ and $f'(x_n)$. This suggests that if it is possible to rewrite the original equation $f(x) = 0$ in such a form that yields x_{n+1} as a function of x_n, then the iterates x_n may yield the root directly.

Towards this end, suppose it is possible to rewrite the equation in the form

$$x = \phi(x) \tag{2.7}$$

then a root ξ of Eq. (2.7) would satisfy

$$\xi = \phi(\xi)$$

This means that the function ϕ maps ξ on to itself, or, in other words, ξ is a **fixed point** of ϕ.

Let x_0 be an initial approximation of ξ, then using this value in Eq. (2.7), the next approximation to ξ could be $x_1 = \phi(x_0)$. Continuing in this manner one can obtain iterates $x_0, x_1, x_2, \cdots, x_n, \cdots$ where

$$x_{n+1} = \phi(x_n) \tag{2.8}$$

Evidently, if the sequence of iterates converge to ξ, then following Eq. (2.8) in the limit, $\xi = \phi(\xi)$ and thus ξ is a root of Eq. (2.7). The question of convergence of the sequence is discussed later on in Sect. 2.4, where it is proved that if $|\phi'(x)| \le K < 1$ in an interval $[a, b]$ containing the root ξ, then the iterates do converge to ξ.

The rewriting of a given equation in the form (2.7) is by no means unique. The appropriate representation would be that one for which convergence takes place.

Example 1. Consider Wallis' equation

$$f(x) := x^3 - 2x - 5 = 0$$

Rewrite the equation in various iterative forms and find which ones converge.

Solution. One can rewrite the given equation as

(i) $\phi_1(x) := \frac{1}{2}(x^3 - 5)$

(ii) $\phi_2(x) := (2x + 5)^{1/3}$

(iii) $\phi_3(x) := (2x + 5)/x^2$

(iv) $\phi_4(x) := x - \lambda(x^3 - 2x - 5), \quad \lambda \ne 0$

The root of the equation lies in $[2, 3]$, and one can start with the bisection point $x_0 = 2.5$.

For (i), $\phi_1'(x) = \frac{3}{2}x^2 > \frac{3}{2} \times 4 = 6 > 1$ for $x \in [2, 3]$ and so the scheme (i) is unsuitable for iteration.

For (ii), $\phi_2'(x) = 2/[3(2x + 5)^{2/3}] < 2/[3(2 \times 3 + 5)^{2/3}] < 1$ and consequently the scheme is convergent. Writing a Fortran program allowing a maximum of 20 iterations, it is found that the root is $\xi = 2.09455$.

For (iii), $|\phi_3'(x)| = \dfrac{2}{x^2} + \dfrac{10}{x^3} > 1$ in $[2, 3]$ and does not converge.

The last scheme (iv) is $\phi_4(x) = x - \lambda f(x)$. For the scheme $|\phi_4'(x)| = |1 - \lambda f'(x)| < 1$ if $-1 < 1 - \lambda f'(x) < 1$ or $0 < \lambda f'(x) < 2$. The condition is satisfied by

$$\lambda = \frac{2}{\underset{x \in [2, 3]}{\text{Max}(3x^2 - 2)}} = \frac{2}{25}$$

With this value of λ, $\phi_4(x) = x - \frac{2}{25}(x^2 - 2x - 5)$ is convergent and as before the fixed point ξ is obtained as 2.09455. $\qquad\qquad\qquad\qquad\qquad\qquad\qquad\square$

A general procedure for rewriting Eq. (2.1) in form (2.8) by procedure of the last type is considered in Exercise 1.

Example 2. The equation $e^x - 4x^2 = 0$ has two roots in $(0, 1)$ and $(4, 5)$. Compute the roots by iteration.

Solution. One way of writing the equation is naturally

$$x = \frac{1}{2}e^{x/2}$$

For the root in $[0, 1]$, one may start with $x_0 = 0.5$. The iteration quickly yields the root as $\xi = 0.714806$. For the other root however, the starting value $x_0 = 4.5$ yields a divergent sequence. These facts on convergence can be seen from $\phi(x) = \frac{1}{2}e^{x/2}$, with

$$\phi'(x) = \frac{1}{4}e^{x/2} < 1 \quad \text{for} \quad x \in [0, 1]$$
$$> 1 \quad \text{for} \quad x \in [4, 5]$$

For the second root, one may try the form

$$x = 2\ln(2x)$$

The form converges to $\xi = 4.30659$. $\qquad\qquad\qquad\qquad\qquad\qquad\qquad\qquad\square$

Example 3. The iteration method is sometimes a useful tool in the approximate construction of analytical solution of equations in special cases. The Chebyshev method of Exercise 6 of the preceding section is an example. As another example, it is known that the frequencies of lateral vibrations of a free–free or clamped–clamped bar is given by the equation

$$\cos x \cosh x = 1$$

where frequency $\omega = \sqrt{\dfrac{EI}{\rho}\dfrac{x^2}{2l^2}}$, $l :=$ length, $\rho :=$ mass/length, $E :=$ Young's modulus and $I :=$ moment of inertia of a cross-section. Construct approximate expression for its roots.

Solution. If the equation is rewritten as $\cosh x = \sec x$, the roots are the points of intersection of $y = \cosh x$ and $y = \sec x$. The former has the form of of a *catenary* or the form of a suspended light string and the latter is a series of $\bigcup$ and $\bigcap$ shaped curves between vertical asymptotes $x = (2n + 1)\dfrac{\pi}{2}$, $(n = 1, 2, \cdots)$. Clearly the two curves intersect near $x = \pi/2, 3\pi/2, 5\pi/2, \cdots, (2n + 1)\pi/2, \cdots$, and the nth positive root of the equation can be written as

$$\xi_n = (2n + 1)\frac{\pi}{2} - (-1)^n \epsilon_n, \quad (n = 1, 2, 3, \cdots)$$

where ϵ_n is a small quantity. Inserting the expression in the given equation,

$$\cosh \xi_n = \frac{1}{\cos \xi_n} = \frac{1}{(-1)^n \cdot (-1)^n \sin \epsilon_n} = \frac{1}{\sin \epsilon_n}$$

and therefore $\sinh \xi_n = \sqrt{\cosh^2 \xi_n - 1} = \cot \epsilon_n$. These expressions yield

$$e^{-\xi_n} = \cosh \xi_n - \sinh \xi_n = \frac{1 - \cos \epsilon_n}{\sin \epsilon_n} = \tan \frac{\epsilon_n}{2}$$

This means that

$$\tan \frac{\epsilon_n}{2} = e^{-(2n+1)\pi/2} \cdot e^{(-1)^n \epsilon_n} =: \delta_n e^{(-1)^n \epsilon_n}$$

where $\delta_n := e^{-(2n+1)\pi/2}$ is small. In fact, the largest delta is $\delta_1 = 0.0089833$. Expanding the two sides of the above equation, we get

$$\frac{\epsilon_n}{2} + \frac{1}{3}\left(\frac{\epsilon_n}{2}\right)^3 - \cdots = \delta_n\left[1 + (-1)^n \epsilon_n + \frac{\epsilon_n^2}{2} + \cdots\right]$$

or

$$\epsilon_n = 2\delta_n\left[1 + (-1)^n \epsilon_n + \frac{\epsilon_n^2}{2} + \cdots\right] - \frac{\epsilon_n^3}{12} + \cdots$$

The approximation of the above iterative equation is $\epsilon_n = 2\delta_n$. The second approximation is

$$\epsilon_n = 2\delta_n[1 + (-1)^n 2\delta_n + 2\delta_n^2 + \cdots] - \frac{2}{3}\delta_n^3 + \cdots = 2\delta_n + (-1)^n 4\delta_n^2 + \frac{10}{3}\delta_n^3 + \cdots$$

The third approximation is

$$\epsilon_n = 2\delta_n[1 + (-1)^n\{2\delta_n + (-1)^n 4\delta_n^2 + \frac{1}{3}\delta_n^3 + \cdots\}$$

$$+ \frac{1}{2}\{2\delta_n + (-1)^n 4\delta_n^2 + \frac{10}{3}\delta_n^3 + \cdots\}^2 + \cdots]$$

$$- \frac{1}{12}\{2\delta_n + (-1)^n 4\delta_n^2 + \frac{10}{3}\delta_n^3\}^3 + \cdots$$

$$= 2\delta_n + (-1)^n 4\delta_n^2 + (8 + 4 - \frac{2}{3})\delta_n^3 + \cdots$$

$$\approx 2\delta_n + (-1)^n 4\delta_n^2 + \frac{34}{3}\delta_n^3$$

Thus, to the third order in δ_n, we get

$$\xi_n \approx (2n+1)\frac{\pi}{2} - 2(-1)^n\delta_n[1 + (-1)^n 2\delta_n + \frac{17}{3}\delta_n^2]$$

The solution yields $\xi_1 = 4.73004$, $\xi_2 = 7.85321$, $\xi_3 = 10.99561$ etc. $\qquad\square$

Exercises

1. An equation $f(x) = 0$ has a root $\xi \in [a, b]$ and $f'(x)$ retains sign in the interval. Prove that the equivalent form

$$x = x - \lambda f(x)$$

converges for $\lambda = 1/M$, where $M = \max\limits_{x \in [a, b]} \{f'(x)\}$, if $f'(x) > 0$.

What should be the choice for $f'(x) < 0$?

$[\phi(x) = x - \lambda f(x), |\phi'(x)| = |1 - \lambda f'(x)| < 1$, where $-1 < 1 - \lambda f'(x) < 1$ or $0 < \lambda f'(x) < 2$. For $f'(x) > 0, \lambda < \frac{2}{f'(x)} < \frac{2}{M}$. So select $\lambda = 1/M$. When $f'(x) < 0$, select $\lambda = -1/M$].

2. Solve by iteration method the following equations:

(i) $x^3 + x + 1 = 0$ $\hfill [-0.68233]$

(ii) $5x^3 - 20x + 3 = 0$ $\hfill [0.15136]$

(iii) $3x - \cos x - 1 = 0$ $\hfill [0.60710]$

(iv) $\sqrt{x+1} = \dfrac{1}{x}$ $\hfill [0.75488]$

(v) $x + \ln x = \frac{1}{2}$ $\hfill [\text{Write as } x = e^{(.5-x)}, 0.76625]$

(vi) $x^3 + 3x^2 - 3 = 0$ $\hfill [\text{Write as } x = x - \frac{1}{6}(x^3 + 3x^2 - 3), -2.53209]$

(vii) $e^{-x} = \sin x$ $\hfill [\text{Write as } x = x + e^{-x} - \sin x, 0.58853]$

3. The angular distance of a planet from the nearer end of the major axis of the elliptic orbit is θ. If t is the time of sweeping this angle then it is known to be given by the equation

$$\theta = nt + 2e \sin \theta - \frac{3}{4}e^2 \sin 2\theta$$

correct to $O(e^2)$, where n is the mean angular velocity of the planet. Show that correct to this order in e

$$\theta \approx nt + 2e \sin nt + \frac{5}{4} e^2 \sin 2nt.$$

2.3   Rate of Convergence

If x_n is the nth iterate of a root ξ of an equation, then the (absolute) error e_n in it is defined as $e_n := \xi - x_n$. For convergence e_n must tend to zero as $n \to \infty$. Assuming convergence, a convenient measure of the *rate* of convergence is a number $p \geq 1$, such that for the next iterate

$$e_{n+1} = O(e_n^p)$$

where O is the Landau symbol (see Sect. 1.2.1, Chap. 1). This means that the successive errors must satisfy

$$\left| \frac{e_{n+1}}{e_n^p} \right| = C \quad \text{for} \quad n \to \infty$$

where $C \neq 0$ is a constant called the *asymptotic error constant*. Evidently the higher the value of p is, the faster the iterates converge to ξ. If $p = 1$, the convergence is said to be *linear*, and if $p > 1$, it is called *superlinear*. We proceed to determine the rates of convergence of the different methods developed so far.

1°. Bolzano Bisection Method.

If x_{n-1} and x_n are two successive iterates (so that $f(x_{n-1}) f(x_n) < 0$), then the next iterate is

$$x_{n+1} = \frac{1}{2}(x_n + x_{n-1})$$

Therefore,

$$\xi - e_{n+1} = \frac{1}{2}(\xi - e_n + \xi - e_{n-1})$$

Hence

$$2e_{n+1} - e_n - e_{n-1} = 0$$

which is a *linear difference equation* of the second order. To solve it, let $e_n = \rho^n$, then inserting this expression of e_n in the above equation we get

$$\rho^n (2\rho - 1 - \rho^{-1}) = 0$$

or,

$$2\rho^2 - \rho - 1 = 0$$

whose solution is $\rho = [1 \pm \sqrt{1+8}]/4 = 1, -1/2$. Hence the general solution of the difference equation must be

$$e_n = C_1 \, 1^n + C_2 \left(-\frac{1}{2}\right)^n$$

where C_1, C_2 are arbitrary constants. As the method is convergent, we must have $C_1 = 0$ and therefore $e_n = (-1)^n C_2/2^n$. Hence

$$\left|\frac{e_{n+1}}{e_n}\right| = \frac{1}{2} \quad \text{for all } n$$

So the method is linear.

2°. General Iteration Method.

In this method the successive iterates satisfy $x_{n+1} = \phi(x_n)$ (see Eq. (2.8)) and so

$$\xi - e_{n+1} = \phi(\xi - e_n) = \phi(\xi) - e_n \, \phi'(\xi - \theta \, e_n) \quad 0 < \theta < 1$$

by the Mean Value Theorem. Since $\xi = \phi(\xi)$ and $e_n \to 0$ as $n \to \infty$ in case of convergence, the above reduces to

$$\left|\frac{e_{n+1}}{e_n}\right| = |\phi'(\xi - \theta \, e_n)| = |\phi'(\xi)| \quad \text{for } n \to \infty$$

assuming continuity of the derivative ϕ'. Hence if $\phi'(\xi) \neq 0$, this method is also linear in convergence.

If $\phi'(\xi) = 0$, then proceeding as above by Taylor's theorem of order two, one gets

$$\left|\frac{e_{n+1}}{e_n^2}\right| = \frac{1}{2}|\phi''(\xi)| \quad \text{for } n \to \infty$$

and convergence becomes superlinear, which is atleast quadratic.

3°. Newton's Method.

This method can be considered as a particular case of general iteration with

$$\phi(x) = x - \frac{f(x)}{f'(x)}$$

according to Eq. (2.6). By differentiation, one finds that

$$\phi'(x) = \frac{f(x) \, f''(x)}{[f'(x)]^2}, \quad \phi''(x) = \frac{f''(x)}{f'(x)} + f(x)\frac{d}{dx}\left[\frac{f''(x)}{\{f'(x)\}^2}\right]$$

Hence $\phi'(\xi) = 0$ (since $f(\xi) = 0$) and $\phi''(\xi) = f''(\xi)/f'(\xi) \neq 0$, in general. Thus, according to the discussion of general iteration given above, the convergence is *quadratic*.

One can arrive at the same conclusion proceeding ab initio by applying second-order Taylor's theorem to the error equation for Eq. (2.6). The details are left as an exercise to the reader.

4°. Secant Method.

The secant iterations are given by Eq. (2.8). Introducing the error e_n, it yields

$$e_{n+1} = e_n - \frac{(e_n - e_{n-1})f(\xi - e_n)}{f(\xi - e_n) - f(\xi - e_{n-1})} = \frac{e_{n-1}f(\xi - e_n) - e_n f(\xi - e_{n-1})}{f(\xi - e_n) - f(\xi - e_{n-1})}$$

By Taylor's theorem of order two, one can write noting that $f(\xi) = 0$,

$$f(\xi - e_n) = -e_n f'(\xi) + \frac{e_n^2}{2} f''(\xi - \theta_n e_n), \quad 0 < \theta_n < 1$$

Thus the above expression becomes

$$\frac{e_{n+1}}{e_n e_{n-1}} = \frac{1}{2} \frac{e_n f''(\xi - \theta_n e_n) - e_{n-1} f''(\xi - \theta_{n-1} e_{n-1})}{-(e_n - e_{n-1})f'(\xi) + \frac{1}{2}[e_n^2 f''(\xi - \theta_n e_n) - e_{n-1}^2 f''(\xi - \theta_{n-1} e_{n-1})]}$$

Therefore, letting $n \to \infty$, or taking the limit $e_{n-1} \to 0$ and $e_n \to 0$ in succession, one gets

$$\frac{e_{n+1}}{e_n e_{n-1}} \to -\frac{1}{2}\frac{f''(\xi)}{f'(\xi)} \quad \text{as} \quad n \to \infty$$

To find the rate of convergence p, let

$$\left|\frac{e_{n+1}}{e_n e_{n-1}}\right| =: c_n \to c_\infty := \frac{1}{2}\left|\frac{f''(\xi)}{f'(\xi)}\right| \quad \text{as} \quad n \to \infty$$

Therefore,

$$\left|\frac{e_{n+1}}{e_n^p}\right| = c_n |e_n|^{1-p}|e_{n-1}| = c_n \left|\frac{e_n}{e_{n-1}^p}\right|^\alpha$$

provided that $1 - p = \alpha$ and $1 = -\alpha p$. Eliminating α, one obtains the equation $p^2 - p - 1 = 0$, whose positive root is $p = \frac{1}{2}(1 + \sqrt{5}) \approx 1.618$.

With $\alpha = -1/p$, the above equation looks like a fixed point iteration $y_{n+1} = c_n y_n^{-1/p}$, whose fixed point η satisfies $\eta = c_\infty \eta^{-1/p}$ or, $\eta^{(1+1/p)} = c_\infty$ or, $\eta^p = c_\infty$, i.e. $\eta = c_\infty^{1/p}$. Therefore,

$$\left|\frac{e_{n+1}}{e_n^p}\right| \to c_\infty^{1/p} = \left[\frac{1}{2}\left|\frac{f''(\xi)}{f'(\xi)}\right|\right]^{1/p} \quad \text{as} \quad n \to \infty$$

and so the rate of convergence is $p \approx 1.618$. This rate is somewhat lower than that of the Newton method, but is superior to that of the bisection method.

Exercises

1. Show that if the modified Newton iteration $x_{n+1} = x_n - f(x_n)/f'(x_0)$ is convergent, then it is linear.

2. Show that if the generalised Newton's method for a root of multiplicity p given by $x_{n+1} = x_n - pf(x_n)/f'(x_n)$ is convergent, then it is quadratic.

3. Show that if the Chebyshev iteration a given in Exercise 6 of Sect. 2.1.2.4 is convergent, then it is cubic.

2.3.1 Acceleration of Convergence of Linearly Convergent Sequences: Aitken's Δ^2-process

Suppose we have linearly converging sequence $x_0, x_1, x_2, \cdots, x_n, \cdots$ as in the iteration method, then $e_{n+1} = K e_n$ for $n \to \infty$. This means that for large n

$$\xi - x_{n+1} = K(\xi - x_n)$$

replacing n by $n - 1$, we similarly have

$$\xi - x_n = K(\xi - x_{n-1})$$

Eliminating K, we get

$$(\xi - x_{n+1})(\xi - x_{n-1}) = (\xi - x_n)^2$$

which simplifies in to

$$\xi = \frac{x_{n+1}x_{n-1} - x_n^2}{x_{n+1} - 2x_n + x_{n-1}} = x_{n+1} - \frac{(x_{n+1} - x_n)^2}{x_{n+1} - 2x_n + x_{n-1}}$$

Thus given $x_0, x_1, x_2, \cdots$ one can construct a new sequence $\hat{x}_1, \hat{x}_2, \hat{x}_3 \cdots$ defined by

$$\hat{x}_n = x_{n+1} - \frac{(x_{n+1} - x_n)^2}{x_{n+1} - 2x_n + x_{n-1}} \tag{2.9}$$

The new sequence converges to ξ faster than the original sequence. In fact

$$\frac{\hat{x}_n - \xi}{x_n - \xi} = \frac{x_{n+1}(x_{n-1} - \xi) - (x_n - \xi)^2 - \xi(x_{n-1} - \xi)}{x_n - \xi}$$

$$= \frac{1}{K}(x_{n+1}\xi) - (x_n - \xi) \to 0 \quad \text{as} \quad n \to \infty$$

Hence the assertion.
If we define *finite differences* (see also Chap. 4),

$$\Delta x_n := x_{n+1} - x_n, \quad \Delta^2 x_n := \Delta(\Delta x_n)$$

then $\Delta^2 x_{n-1} = \Delta(x_n - x_{n-1}) = \Delta x_n - \Delta x_{n-1} = x_{n+1} - x_n - (x_n - x_{n-1}) = x_{n+1}$
$- 2x_n + x_{n-1}$ and Eq. (2.10) can be recast as

$$\hat{x}_n = x_{n+1} - \frac{(\Delta x_n)^2}{\Delta^2 x_{n-1}}$$

which explains the term "Δ^2–process".

Alexander Craig Aitken (1895–1967), New Zealand mathematician. Professor at University of Edinburgh. Displayed unusual genius not only in his published mathematical works, but by his extra ordinary skill in mental calculation.

Example 1. Solve $\quad x = \frac{1}{2}(3 + \sin x)$.

Solution. By fixed point iteration we obtain $\xi = 1.96219$ i 9 iterations, if we start with $x0 = 1.5$ in a straight forward way. For using Eq. (2.10) we write the Fortran program

```
x0=1.5
do n=1,10
x1=.5*(3+sin(x0)); x2=.5*(3+sin(x1))
x_hat=x2-(x2-x0)**2/(x2-2*x1+x0)
print*, n, x_hat
x0=x1
end do
end
```

Execution of the program yields the same result in 8 iterations.   □

Exercises

1. Apply Aitken's Δ^2–process to the iterative equations
(i) $x=\sqrt{2+x}$ [2.00000]

(ii) $x=\frac{1}{2}(3+\cos x)$ [1.52359]

(iii) $x=\frac{1}{2}(7+\ln x)$ [4.21990]

What is the reduction in the number of iterations in each case?

2.4 Convergence Theorems

During the development of Newton and secant methods, it was noted that convergence to the root ξ is not always ensured. The same is true of the general iteration method. In this section we give sufficient conditions which ensure convergence of these methods.

Theorem 2.2 *Let $f \in C_2[a, b]$, such that*

(i) $f(a) f(b) < 0$

(ii) $f'(x) \neq 0, \quad a < x < b$

(iii) $f''(x)$ is either > 0, or < 0, $\quad a < x < b$

and (iv) $\left| \dfrac{f(a)}{f'(a)} \right| < b - a; \quad \left| \dfrac{f(b)}{f'(b)} \right| < b - a$

then the iterations $x_0, x_1, x_2, \cdots, x_n, \cdots$ for
    (a) the Newton's method, Eq. (2.6) converges to root ξ
and (b) the Secant method, Eq. (2.5) converges to root ξ.

Proof Geometrically condition (i) implies existence of ξ in (a, b), while condition (ii) ensures uniqueness of ξ. Condition (iii) implies that f is either convex or concave and condition (iv) implies that the *tangents at the end points $x = a$ and $x = b$ cut the x-axis at points within (a, b).* For the sake of definiteness we assume the configurations of Figs. 2.4 and 2.2a for which (i) $f(a) < 0$, $f(b) > 0$ (ii) $f'(x) > 0$, $a < x < b$ (iii) $f''(x) > 0$, $a < x < b$ and (iv) $-f(a)/f'(a) < b - a$ and $f(b)/f'(b) < b - a$.

(a) **Newton's method.** Let $x_0 > \xi$ as in Fig. 2.4. The next iterate following Eq. (2.6) is

$$x_1 = x_0 - \frac{f(x_0)}{f'(x_0)} < x_0, \quad \text{since} \quad f(x_0) > 0, \ f'(x_0) > 0$$

Consider the function

$$\phi(x) = x - \frac{f(x)}{f'(x)}$$

then

$$\phi'(x) = \frac{f(x) f''(x)}{[f'(x)]^2} > 0, \quad \xi < x < b$$

Thus $\phi(x)$ is monotonic in $\xi < x < b$ and so $\phi(\xi) < \phi(x_0)$. Therefore,

$$\xi - \frac{f(\xi)}{f'(\xi)} < x_0 - \frac{f(x_0)}{f'(x_0)} = x_1$$

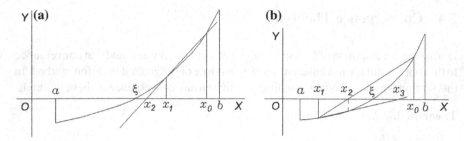

Fig. 2.5 Convergence pattern in secant method

Since $f(\xi) = 0$, we get $\xi < x_1 < x_0 < b$. By similar argument, $\xi < x_2 < x_1 < x_0 < b$ and generally

$$\xi < \cdots < x_n < \cdots < x_2 < x_1 < x_0 < b$$

Hence the sequence $x_0, x_1, x_2, \cdots, x_n, \cdots$ is monotonic decreasing and bounded with limiting point ξ.

Next let $x_0 < \xi$. In this case $\phi'(x) < 0$, $a < x < \xi$. Thus $\phi(x)$ is monotonic decreasing in $a < x < \xi$ and

$$\phi(a) > \phi(x_0) > \phi(\xi)$$

or

$$a - \frac{f(a)}{f'(a)} > x_0 - \frac{f(x_0)}{f'(x_0)} > \xi - \frac{f(\xi)}{f'(\xi)}$$

or

$$b > a - \frac{f(a)}{f'(a)} > x_1 > \xi$$

Thus, $\xi < x_1 < b$. The succeeding iterates $x_2, x_3, \cdots$ will follow the same pattern as that of the previous case, completing the proof. □

(b) Secant Method.

The convergence pattern can essentially be divided in to two types as shown in Fig. 2.5

As depicted in Fig. 2.5a, let $\xi < x_1 < x_0 < b$. The next iterate is

$$x_2 = \frac{x_0 f(x_1) - x_1 f(x_0)}{f(x_1) - f(x_0)} = x_1 - (x_1 - x_0) \times \frac{f(x_1)}{f(x_1) - f(x_0)}$$

$$= x_1 - \frac{f(x_1)}{f'(\eta)} < x_1, \quad x_1 < \eta < x_0$$

since $f(x_1) > 0$ and $f'(\eta) > 0$. We also prove that $x_2 > \xi$ or, since $f(x_1) < f(x_0)$

$$x_0 f(x_1) - x_1 f(x_0) < \xi[f(x_1) - f(x_0)]$$

or

$$\frac{f(x_1)}{x_1 - \xi} < \frac{f(x_0)}{x_0 - \xi}$$

For proving the above inequality consider $\phi(x) = f(x)/(x - \xi)$. By differentiation

$$\phi'(x) = \frac{(x - \xi) f'(x) - f(x)}{(x - \xi)^2} = \frac{1}{2} f''(\xi) > 0$$

by Taylor's theorem of second order. Thus $\phi(x)$ is monotonic increasing and for $x_1 < x_0$, $\phi(x_1) < \phi(x_0)$. proving the above mentioned inequality. Thus $\xi < x_2 < x_1 < x_0 < b$. Proceeding in this manner the sequence $x_0, x_1, x_2, \cdots, x_n, \cdots$ is monotonic decreasing satisfying

$$\xi < \cdots < x_n < \cdots < x_2 < x_1 < x_0 < b$$

with limiting point ξ.

In the second case, Fig. 2.5b, let $a < x_1 < \xi < x_0 < b$. As before the next iterate satisfies

$$x_2 = x_1 - (x_1 - x_0) \times \frac{f(x_1)}{f(x_1) - f(x_0)} = x_1 - \frac{f(x_1)}{f'(\eta)} > x_1, \quad x_1 < \eta < x_0$$

since $f(x_1) < 0$ and $f'(\eta) > 0$. We also have $x_2 < \xi$, since with $f(x_1) < 0, f(x_0) < 0, (x_0 - \xi) f(x_1) > (x_1 - \xi) f(x_0)$, or since $x_1 < \xi < x_0, f(x_1)/(x_1 - \xi) < f(x_0)/(x_0 - \xi)$ as proved before. Thus $a < x_1 < x_2 < \xi < x_0 < b$. In much the same manner one can prove that the next iterate x_3 satisfies $\xi < x_3 < x_0$. The details are left as an exercise for the reader. Thus the secant sequence for this case turns out to be

$$a < x_1 < x_2 < x_4 < \cdots < \xi < \cdots < x_5 < x_3 < x_0 < b$$

with limiting point ξ.   □

Remark. If $[a, b]$ is small enough condition (iv) will automatically be satisfied. In other words if x_0 (and x_1 for the secant method) are sufficiently close to ξ, conditions (i)–(iii) withstanding, the Newton and secant methods will always be convergent.

Theorem 2.3 (Fixed-Point Theorem) *Let ϕ be contractive mapping on $[a, b]$, that is for $x, z \in [a, b]$*

$$|\phi(x) - \phi(z)| \leq K|x - z|, \quad K < 1$$

then the iterates $x_{n+1} = \phi(x_n), n \geq 0$ (Eq. (2.8)) converge monotonically to a unique fixed point ξ.

Proof From the given condition on $\phi(x)$

$$|x_{n+1} - x_n| = |\phi(x_n) - \phi(x_{n-1})| \le K|x_n - x_{n-1}|$$

$$\le K^2|x_{n-1} - x_{n-2}| \le \cdots \le K^n|x_1 - x_0|$$

this implies that for $m > n$

$$|x_m - x_n| \le |x_m - x_{m-1}| + |x_{m-1} - x_{m-2}| + \cdots + |x_{m+1}|$$

$$\le (K^{m-1} + K^{m-2} + \cdots + K^n)|x_1 - x_0|$$

$$= K^n \frac{1 - K^{m-n}}{1 - K}|x_1 - x_0| \le \frac{K^n}{1 - K}|x_1 - x_0| < \epsilon, \quad n > n_0$$

since for $K < 1$, $K^n \to 0$ as $n \to \infty$. Thus $x_0, x_1, \cdots, x_n, \cdots$ is a Cauchy sequence with limiting point ξ. This point ξ is a fixed point since

$$|\xi - \phi(\xi)| \le |\xi - x_n| + |x_n - \phi(\xi)| = |\xi - x_n| + |\phi(x_{n-1}) - \phi(\xi)|$$

$$\le |\xi - x_n| + K|x_{n-1} - \xi| < \epsilon, \quad \text{for} \quad n > n_0$$

since $x_n \to \xi$ as $n \to \infty$. Hence $\xi = \phi(\xi)$. Moreover ξ is unique. For, if there is another fixed point η, $\eta = \phi(\eta)$, then

$$|\xi - \eta| = |\phi(\xi) - \phi(\eta)| \le K|\xi - \eta|$$

Since $K < 1$, this means that $\eta = \xi$.

The convergence is monotonic since

$$|\xi - x_n| = |\phi(\xi) - \phi(x_{n-1})| \le K|\xi - x_{n-1}| < |\xi - x_{n-1}|$$

$\square$

1°. **Error Bounds**.

In the development, letting $m \to \infty$, we obtain the bound

$$|\xi - x_n| \le \frac{K^n}{1 - K}|x_1 - x_0|$$

This bound after only one iteration indicates the number of iterations n that would be required for achieving a prescribed accuracy.

Alternatively, one obtains

$$|\xi - x_n| = |\phi(\xi) - \phi(x_{n-1})| \leq K|\xi - x_{n-1}| \leq K[|\xi - x_n| + |x_n - x_{n-1}|]$$

Hence

$$|\xi - x_n| \leq \frac{K}{1-K}|x_n - x_{n-1}|$$

This estimate can be used to carry out the iterations for prescribed accuracy ϵ. For $|\xi - x_n| < \epsilon$, one would require

$$|x_n - x_{n-1}| < \frac{1-K}{K}\epsilon$$

Theorem 2.4 *If $\phi \in C_1[a, b]$ such that*

$$|\phi'(x)| \leq \max_{x \in [a, b]} |\phi'(x)| \leq K < 1$$

then the fixed point theorem holds. Further, if (i) $0 < \phi'(x) < 1$, $x \in [a, b]$ the sequence $x_0, x_1, x_2, \cdots$ is monotonic, and if (ii) $-1 < \phi'(x) < 0$, $x \in [a, b]$ the sequence $x_0, x_1, x_2 \cdots$ oscillates about ξ.

Proof By Mean Value Theorem

$$|\phi(x) - \phi(z)| = |x - z| \cdot |\phi'(\zeta)| \leq K|x - z|, \quad x, z \in [a, b]$$

and so ϕ is a contractive mapping. This proves the first part of the theorem.
   For proving (i) consider two iterates x_n and x_{n+1}, then

$$x_{n+1} - \xi = \phi(x_n) - \phi(\xi) = (x_n - \xi)\phi'(\zeta)$$

If $x_n > \xi$, it follows that $x_{n+1} > \xi$. Since the convergence is monotonic it follows that

$$\xi < \cdots < x_{n+1} < x_n < \cdots < x_1 < x_0$$

Similarly if $x_n < \xi$, then $x_{n+1} < \xi$ and so

$$x_0 < x_1 < \cdots < x_n < x_{n+1} < \cdots < \xi$$

   For case (ii) since $\phi'(\zeta) < 0$, if $x_n > \xi$ then $x_{n+1} < \xi$ and if $x_n < \xi$ then $x_{n+1} > \xi$. Thus, either

$$x_1 < x_3 < \cdots < x_{n+1} < \cdots < \xi < \cdots < x_n < \cdots < x_2 < x_0$$

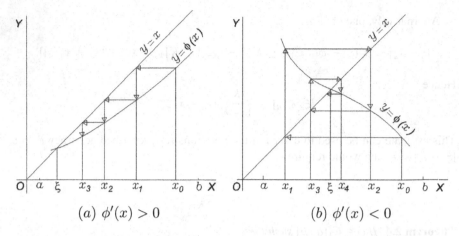

(a) $\phi'(x) > 0$ (b) $\phi'(x) < 0$

Fig. 2.6 Convergence pattern in general iteration

or

$$x_0 < x_2 < \cdots < x_n < \cdots < \xi < \cdots < x_{n+1} < \cdots < x_3 < x_1$$

The two cases of convergence are graphically depicted in Fig. 2.6.   □

Theorem 2.5 *If* $\phi \in C_1 [a, b]$ *such that*

$$\min_{x \in [a, b]} |\phi'(x)| > 1$$

then every iteration starting with $x_0 \neq \xi$ *diverges.*

Proof Proceeding as in the preceding theorem, by Mean Value Theorem

$$|\phi(x) - \phi(z)| = |x - z||\phi'(\zeta)| > |x - z|, \quad x, z \in [a, b]$$

Hence with ξ as a root of the equation $x = \phi(x)$,

$$|x_{n+1} - \xi| = |\phi(x_n) - \phi(\xi)| > |x_n - \xi|$$
$$> |x_{n-1} - \xi| > \cdots > |x_1 - \xi| > |x_0 - \xi| > 0$$

This means that $x_0, x_1, x_2, \cdots$ recede from ξ and the sequence is therefore divergent.   □

2.5 Complex Roots

A major difficulty in computing complex roots of a general equation of the form $f(x) = 0$ lies in the fact that isolation of a root is not possible by algebraic or analytic means. Even in the face of this difficulty, suppose an approximation x_0 (complex) to a complex root is some how obtained. Since in *complex analysis*, Taylor's theorem is valid, we conclude that Newton's and Chebyshev's methods are applicable for refinement of the complex root. Similarly, if two complex approximations x_0 and x_1 are available then the secant method in some way could be used, because the method depends on linear approximation of $f(x)$ in the neighbourhood of the root and such linear approximation is a valid process in the complex domain. In these methods, however, difficulty arises when the approximation to a complex root, viz. x_0 and x_1 are real-valued and so is the function f. The Newton, Chebyshev and secant iterations in such cases will never deliver a complex root.

Müller's method [D.E. Müller (1956)] described below was devised as an extension of the secant method for faster convergence. In this method $f(x)$ is approximated by a *quadratic* expression in the neighbourhood of a root, real or complex. In the process, a quadratic equation is solved, with possibility of real or complex roots. Thus even if the initial approximations are real there is no inherent difficulty in converging to a complex root. By generality of the method, it is evident that extraction of real roots is not precluded.

2.5.1 Müller's Method

Let ξ be a root real or complex and let x_{-1}, x_0, x_1 be three initial approximation to it. In the iterative steps, we successfully construct approximations $x_2, x_3, \cdots, x_{n-2}, x_{n-1}, x_n \cdots$, to the root ξ. Consider the small circular region in the Argand diagram enclosing the successive points x_{n-2}, x_{n-1}, x_n. Then $f(x)$ can be approximated by a quadratic passing through the points $[x_{n-2}, f(x_{n-2})]$, $[x_{n-1}, f(x_{n-1})]$, $[x_n, f(x_n)]$, that is

$$f(x) \approx a(x - x_n)^2 + b(x - x_n) + c$$

Since the quadratic is assumed to pass through the aforesaid points, one has

$$c = f(x_n)$$
$$a(x_{n-1} - x_n)^2 + b(x_{n-1} - x_n) + c = f(x_{n-1})$$
$$a(x_{n-2} - x_n)^2 + b(x_{n-2} - x_n) + c = f(x_{n-2})$$

Hence, eliminating c from the second equation

$$b = (x_n - x_{n-1})\, a + \frac{f(x_n) - f(x_{n-1})}{x_n - x_{n-1}}$$

where by elimination of b and c from the third equation

$$a = \frac{1}{x_n - x_{n-2}}\left[\frac{f(x_n) - f(x_{n-1})}{x_n - x_{n-1}} - \frac{f(x_{n-1}) - f(x_{n-2})}{x_{n-1} - x_{n-2}}\right]$$

The next iterate x_{n+1} satisfies $f(x_{n+1}) \approx 0$ and so solving the approximate quadratic equation

$$a\,(x_{n+1} - x_n)^2 + b\,(x_{n+1} - x_n) + c \approx f(x_{n+1}) \approx 0$$

one obtains

$$x_{n+1} \approx x_n + \frac{-b \pm \sqrt{b^2 - 4ac}}{2a} = x_n - \frac{2c}{b \pm \sqrt{b^2 - 4ac}} \qquad (2.10)$$

Between the two signs in the denominator, the one yielding larger value in magnitude should be selected, in order to minimise roundoff error (see Sect. 1.2.2.2, Chap. 1). Based on the method, one can tentatively write:

Algorithm 3. Müller's Method:

<u>Input</u>: Real (or complex) $f(x)$, x_{-1}, x_0, x_1, ϵ_1, ϵ_2, n_{max}
<u>Output</u>: Complex x_{n+1}
<u>Computation</u>: 1. Set $n \leftarrow 1$
2. $h_{n-1} \leftarrow x_{n-1} - x_{n-2}$; $h_n \leftarrow x_n - x_{n-1}$;
   $f[x_{n-1}, x_{n-2}] \leftarrow (f(x_{n-1}) - f(x_{n-2}))/h_{n-1}$;
   $f[x_n, x_{n-1}] \leftarrow (f(x_n) - f(x_{n-1})/h_n$;
   $a \leftarrow (f[x_n, x_{n-1}] - f[x_{n-1}, x_{n-2}])/(h_{n-1} + h_n)$;
   $b \leftarrow a \cdot h_n + f[x_n, x_{n-1}])$; $c \leftarrow f(x_n)$
3. $h_{n+1} \leftarrow -2c/(b \pm \sqrt{b^2 - 4ac})$,   selecting   sign   to   gi
   maximum magnitude to the denominator
4. Set $x_{n+1} \leftarrow x_n + h_{n+1}$
5.   If $|x_{n+1} - x_n| < \epsilon_1|x_n|$ else   if $|f(x_{n+1})| < \epsilon_2$,   Stc
   x_{n+1} is a root
6. If $n + 1 < n - n_{max}$ set $n \leftarrow n + 1$ and go to step 2.

The algorithm uses two criteria for convergence to a root, with tolerances ϵ_1 and ϵ_2. It leads to the following subroutines together with the main program.

```
PROGRAM MAIN
COMPLEX :: roots(3)
EXTERNAL fn
roots(1)=0.; roots(2)=0.; roots(3)=0.
CALL MULLER(fn,roots,3,200,.1e-5,.1e-6)
PRINT*, (roots(i), i=1,3)
END PRORAM MAIN
```

```
!*****************************************

SUBROUTINE fn(x,fx)
COMPLEX :: x, fx
fx=x**3-2*x-5     ! Solving Wallis' equation
RETURN
END SUBROUTINE fn
!***********************************************************

SUBROUTINE MULLER(fn,roots,n,maxiter,epsilon1,epsilon2)
! fn=subroutine name, of the form fn(x,fx). Returns complex
! fx for complex x.    (Input)
! n=number of roots sought.    (Input)
! roots(1),···,roots(n)=first guess of the n roots. If you know
! nothing, 0 is as good as any other guess.    (Input)
! maxiter=maximum number of function evaluations allowed
! per root. (Input)
! epsilon1=maximum of relative h allowed.    (Input)
! epsilon2=maximumof |f(x)| allowed if epsilon1 criterion is not
! met.    (Input)
! roots(1),···,roots(n)=computed n roots.    (Output)
!***************************************************************

COMPLEX :: roots(n),x,fx,cprim,cprev,c,a,b,hprev,h,dd,ddp,sqr,den,den1
EXTERNAL fn
DO i=1,n
kount=0; x=roots(i)
CALL DEFLATE(fn,x+0.5,i,kount,fx,cprim,roots)
CALL DEFLATE(fn,x-0.5,i,kount,fx,cprev,roots)
CALL DEFLATE(fn,x,i,kount,fx,c,roots)
hprev=-1.0; h=0.5; ddp=(cprev-cprim)/hprev
10 dd=(c-cprev)/h
a=(dd-ddp)/(hprev+h); b=a*h+dd
sqr=CSQRT(b**2-4.0*a*c); den=b+sqr, den1=b-sqr
IF(CABS(den1)>CABS(den)) den=den1
IF(CABS(den)<=0.) den=1.0
h=-2*c/den; x=x+h; cprev=c; hprev=h
IF(kount>maxiter) THEN

   PRINT*, 'maximum iterations reached'

   RETURN

END IF
20 CALL DEFLATE(fn,x,i,kount,fx,c,roots)
IF(CABS(h)<epsilon1*CABS(x)) GOTO 30
IF(MAX(CABS(fx),CABS(c))<epsilon2) GOTO 30
```

```
IF(CABS(c)>10*CABS(cprev)) THEN
   h=0.5*h; x=x−h; GOTO 20
   ELSE
   GOTO 10
END IF
30 roots(i)=x
END DO
RETURN
END SUBROUTINE MULLER
!*********************************************

SUBROUTINE DEFLATE(fn,x,i,kount,fx,fxdfl,roots)
COMPLEX :: x,fx,fxdfl,den,roots(i)
EXTERNAL fn
kount=kount+1
CALL fn(x,fx)
fxdfl=fx; IF(i==1) RETURN
DO j=2,i
den=x−roots(j−1)
IF(CABS(den)==0.0) THEN
   roots(i)=x*1.001
   RETURN
ELSE
   fxdfl=fxdfl/den
END IF
END DO
RETURN
END SUBROUTINE DEFLATE
```

The subroutine MULLER incorporates some computational tricks to avoid divergence or total failure and the roots are sought in increasing order of magnitude to minimise roundoff error growth, as the roots are computed one by one.

The subroutine uses another subroutine named DEFLATE. *Deflation* is a procedure to find more than one root. If for example, a root ξ_1 has already been computed, the subroutine computes the deflated value of the function

$$f_1(x) = \frac{f(x)}{x - \xi_1}$$

Proceeding in this manner if roots $\xi_1, \xi_2, \cdots, \xi_r$ have been computed, it computes

$$f_r(x) = \frac{f_{r-1}(x)}{x - \xi_r}$$

Deflation causes roundoff errors in the deflated equation, and may affect accuracy of the subsequent roots.

Exercises

Solve the following equations by Müller's method:

1. $x^3 - 2x - 5 = 0$ (Wallis' equation)

$[(2.09455, -.78044E - 6), (-1.04728 \pm 1.13594i)]$.

2. Equation for Rayleigh wave velocity for Poisson's ration $\nu = 1/3$ (Exercise 8, Sect. 2.1.2)

$$x^3 - 8x^2 + 20x - 12 = 0$$

$[0.86960, 3.56520 \pm 1.04343i]$

3. $x^4 - 7x^3 + 19x^2 - 13x - 10 = 0$.
$[-0.43906, 2 + i0.77876E - 12, (2.71953 \pm 1.99806i)]$.

4. $x^3 - 3x^2 + 3x - 1 = 0$
$[(0.99960 - i0.46404E - 1), (1.01480 + i0.10261E - 1), (1.00110 + i0.45276E - 2)$, epsilon1=.1$E - 6$, epsilon2=.1$E - 3$, maxiter=800].

5. Three roots of $\cos x = 2$. $[\pm 1.31696i, 6.28319 - 1.31696i]$.

6. Three nonzero roots of $\cos x \cosh x = 1$. $[4.73004, 7.85321, 10.99561]$.

We now discuss in brief the *rate of convergence* of the method. The Müller iterations are given by Eq. (2.10). Introducing the error $e_n := \xi - x_n$ in the nth iterate x_n of the root ξ, one can prove after some labour that

$$a = \frac{1}{2}f''(\xi) - \frac{1}{6}(e_n + e_{n-1} + e_{n-2})f'''(\xi) + \cdots$$

$$b = f'(\xi) - e_n f''(\xi) + \frac{1}{6}(2e_n^2 + e_n e_{n-1} + e_n e_{n-2} - e_{n-1}e_{n-2})f'''(\xi) + \cdots$$

$$c = f(x_n) = -e_n f'(\xi) + \frac{e_n^2}{2}f''(\xi) - \frac{e_n^3}{6}f'''(\xi) + \cdots$$

These expressions yield

$$b^2 - 4ac = f'^2 - \frac{1}{3}(e_n e_{n-1} + e_n e_{n-2} + e_{n-1}e_{n-2})f'(\xi)f'''(\xi) + \cdots$$

So Eq. (2.10) ultimately yields

$$\frac{e_{n+1}}{e_n e_{n-1} e_{n-2}} \rightarrow -\frac{1}{6}\frac{f'''(\xi)}{f'(\xi)} \quad \text{as} \quad n \rightarrow \infty$$

In order to find the rate of convergence, let $|e_{n+1}/e_n^p| \rightarrow C$ as $n \rightarrow \infty$. This means that $|e_n/e_{n-1}^p| \rightarrow C$ and $|e_{n-1}/e_{n-2}^p| \rightarrow C$ as $n \rightarrow \infty$. Consequently

$$\left|e_{n+1}\right| \to C\left|e_n\right|^p, \quad \left|e_{n-1}\right| \to \left(\frac{\left|e_n\right|}{C}\right)^{1/p}, \quad \left|e_{n-2}\right| \to \left(\frac{\left|e_n\right|^{1/p}}{C^{1+1/p}}\right)^{1/p}$$

and therefore

$$\left|\frac{e_{n+1}}{e_n e_{n-1} e_{n-2}}\right| \to C\left|e_n\right|^{p-1} \cdot C^{1/p}\left|e_n\right|^{-1/p} \cdot C^{1/p+1/p^2}\left|e_n\right|^{-1/p^2}$$

$$= \frac{1}{6}\left|\frac{f'''(\xi)}{f'(\xi)}\right| \quad \text{as} \quad n \to \infty$$

Thus one must have, $p - 1 - 1/p - 1/p^2 = 0$ or,

$$\phi(p) := p^3 - p^2 - p - 1 = 0$$

Since $\phi(1) = -2 < 0$ and $\phi(2) = 1 > 0$ the real root of this equation lies in $[1,2]$. Its real solution by Müller's method is $p = 1.84$. The constant C must satisfy

$$C^{1+2/p+1/p^2} = \frac{1}{6}\left|\frac{f'''(\xi)}{f'(\xi)}\right|$$

or,

$$C^{2.38} = \frac{1}{6}\left|\frac{f'''(\xi)}{f'(\xi)}\right|.$$

To conclude, the rate of convergence of the method is $p = 1.84$, which is slightly less than 2, viz. the rate for Newton's method; but this rate is superior to that of the secant method.

2.6 Algebraic Equations

A function f is said to be *algebraic* if in defining the value of $f(x)$ at x, only the basic arithmetic operations $+$, $-$, $\times$, $\div$ and rational exponentiation are allowed.

Examples: $f_1(x) := x^3 + 15x^2 - 12x + \sqrt{7}$

$$f_2(x) := \frac{\sqrt[3]{3}}{x+7} + \frac{4x+3}{3x^2+5}$$

$$f_3(x) := \frac{3x^2 - 4x + 3\sqrt{x-1}}{7x-4}$$

are all algebraic functions.

If an algebraic function f contains only integral powers of x then it is called *rational function*. If root extraction of x is also allowed, then it is called an *irrational function*. In the above examples, according to these criteria, f_1 and f_2 are rational, whereas f_3 is irrational. If x does not appear as a divisor as in the case of f_1, then the function is called *rational–integral* or *entire*. It is evident that entire functions are such that $f(x)$ is a *polynomial*.

Given an algebraic equation $f(x) = 0$, it can always be converted by algebraic methods to an equation of the form

$$P(x) := a_0 x^n + a_1 x^{n-1} + a_2 x^{n-2} + \cdots + a_{n-1} x + a_n = 0 \qquad (2.11)$$

This form is therefore considered as the standard form of algebraic equations. Since the left hand side of Eq. (2.11) is a polynomial, an algebraic equation is also called a *polynomial equation*.

Polynomial equations arise frequently in physical applications. The solution of such equations can be attempted by any of the foregoing methods. Müller's method is specially suitable as it is capable of delivering complex roots. The general polynomial Eq. (2.11) fascinated mathematicians for centuries, whose findings constitute the subject of *Theory of Equations*. It deals with results on the number of real and complex roots, domain of their location and questions on exact and approximate evaluation. We state some of them for our purpose.

(1) Fundamental Theorem of Algebra (Gauss). *An algebraic equation of degree n has exactly n roots, real or complex, provided that each root is counted according to its multiplicity.*

Thus if $\xi_1, \xi_2, \cdots, \xi_n$ are the n roots, one can write

$$P(x) = a_0 (x - \xi_1)(x - \xi_2) \cdots (x - \xi_n)$$

The fundamental theorem of algebra was proved by Euler for polynomials with real coefficients. Gauss in 1799 completed the proof for complex coefficients in his famous doctoral dissertation. Since then several proofs have been provided by multitude of authors.

Algebraic equations with real coefficients are very important in practice. Such equations are sometimes called *real polynomial equations*. The rest of the theorems of this section deal with such equations.

(2) *If an algebraic equation with real coefficients has a complex root $\xi = \alpha + i\beta$ of multiplicity m, then its conjugate $\xi^* = \alpha - i\beta$ is also a root of the same multiplicity m.* As a consequence:

(3) *An algebraic equation with real coefficients and odd degree has at least one real root.* In particular a cubic or a quintic equation with real coefficients has at least one real root.

(4) *An even degree algebraic equation with real coefficients such that $a_0 > 0$ and $a_n < 0$ has at least one positive and at least one negative root.* For, in this case $f(0) = a_n < 0$ and $f(\infty), f(-\infty) > 0$. Thus there are real roots in $(-\infty, 0)$ and $(0, \infty)$.

(5) DeCartes' Rule of Signs (Upper bounds on the number of positive and negative roots). *The number of positive roots of an algebraic equation with real coefficients (a root of multiplicity m being counted as m roots), is equal to the number of variations in sign from + to − or − to + in the sequence of coefficients*

$$a_0, a_1, a_2, \cdots, a_n$$

(where the coefficients equal to 0 are ignored) or less than that number by an even integer.

Similar information on number of negative roots can be obtained by considering the equation $P(-x) = 0$. *However, if the equation is complete with none of the coefficients equal to zero, then the number of negative roots (counting multiplicities) is equal to the number of continuations of sign in the above sequence of coefficients or less than that number by an even integer.*

Example 1. Consider $P(x) := x^5 - 18x^4 + 13x^3 + 6x^2 - x + 2 = 0$.

This equation has five roots, of which at least one root is real. The equation is complete and the sequence of signs of the coefficients is

$$+ - + + - +$$

The number of change of signs is four. Therefore, the number of positive roots is four, two or none. The number of continuation of sign is one and so the number of negative roots is exactly one.   □

Example 2. Consider $P(x) := x^6 - 3x^4 + 2x^3 + x^2 - 1 = 0$.

This equation has six roots. Since $a_0 > 0$ and $a_6 < 0$, it has at least one positive root and at least one negative root. The sequence of signs of the coefficients is

$$+ - + + -$$

The number of change of signs is three and consequently there are either three or one number of positive roots. For finding the maximum number of negative roots consider the equation

$$P(-x) = x^6 - 3x^4 - 2x^3 + x^2 - 1 = 0$$

The sequence of signs is

$$+ - - + -$$

having three changes of sign. Hence there can be either three or one negative root of $P(x) = 0$.   □

(6) Sturm's Theorem. *Let $P(x)$, $P_1(x)$, $P_2(x)$, $\cdots$, $P_m(x)$ be the Sturm sequence defined by $P_1(x) := P'(x)$, $-P_2(x) := the remainder when $P(x)$ is divided by $P_1(x)$,*

$-P_3(x) :=$ *the remainder when* $P_1(x)$ *is divided by* $P_2(x)$ *and so on until the last remainder is* $-P_m(x) :=$ *constant, then if Eq.* (2.11) *has distinct real roots, their number in an interval* $[a, b]$ *is* $V(a) - V(b)$, *where* $V(a)$ *and* $V(b)$ *are the numbers of variation in sign in the Sturm sequence for* $x = a$ *and* $x = b$ *respectively.*

Sturm's sequence can be used for isolation of real roots. The number of positive roots can be found by selecting $a = 0$, $b = +\infty$. Similarly the number of negative roots can be obtained by choosing $a = -\infty$, $b = 0$. By diminishing the length of interval $[a, b]$ so as to contain only one real root, isolation of such roots is achieved. The construction of Sturm sequence is tedious; but some simplification is possible by multiplying or dividing any of the functions by arbitrary positive numbers.

Example 3. Consider $P(x) := 5x^3 - 20x + 3$, $P_1(x) = 15x^2 - 20$. Therefore,

$$\frac{P(x)}{P_1(x)} = \frac{5x^3 - 20x + 3}{15x^2 - 20} \Rightarrow \frac{15x^3 - 60x + 9}{15x^2 - 20} = \frac{x(15x^2 - 20) - 40x + 9}{15x^2 - 9}$$

$$= x - \frac{40x - 9}{15x^2 - 9}$$

Therefore, $P_2(x) = 40x - 9$. Next,

$$\frac{P_1(x)}{P_2(x)} = \frac{15x^2 - 20}{40x - 9} \Rightarrow \frac{120x^2 - 160}{120x - 27} = \frac{x(120x - 27) + 27x - 160}{120x - 27}$$

$$\Rightarrow x + \frac{x - 160/27}{x - 9/40} = x + 1 + \frac{9/40 - 160/27}{x - 9/40}$$

Therefore,

$$P_3(x) = -\left(\frac{9}{40} - \frac{160}{27}\right) = \frac{6400 - 243}{40 \times 27} > 0$$

We now compile the table of the signs of the Sturm functions, noting change of $V(x)$.

x	$P(x)$	$P_1(x)$	$P_2(x)$	$P_3(x)$	$V(x)$
$-\infty$	$-$	$+$	$-$	$+$	3
-3	$-$	$+$	$-$	$+$	3
-2	$+$	$+$	$-$	$+$	2
-1	$+$	$-$	$+$	$+$	2
0	$+$	$-$	$-$	$+$	2
1	$-$	$-$	$+$	$+$	1
2	$+$	$+$	$+$	$+$	0

We can see from the table that the roots lie in $[-3, -2]$, $[0, 1]$ and $[1, 2]$. $\quad\square$

2.6.1 Root Bounding Theorems

Theorem 2.6 (Annulus Theorem) *In Eq. (2.11) with real or complex coefficients, let* $A = max\{|a_1|, |a_2|, \cdots, |a_n|\}$, $B = max\{|a_0|, |a_1|, \cdots, |a_{n-1}|\}$, *then all of the roots* ξ_k, $(k = 1, 2, \cdots, n)$ *of the equation lie in the annulus* $r < |\xi_k| < R$, *where*

$$r = \frac{1}{1 + B/|a_n|}; \quad R = 1 + \frac{A}{|a_0|}$$

Proof Setting $|x| > 1$, we have Eq. (2.11)

$$|P(x)| \geq |a_0 x^n| - (|a_1 x^{n-1}| + |a_2 x^{n-2}| + \cdots + |a_n|)$$

$$\geq |a_0||x|^n - A(|x|^{n-1} + |x|^{n-2} + \cdots + 1)$$

$$= |a_0||x|^n - A\frac{|x|^n - 1}{|x| - 1} > \left(|a_0| - \frac{A}{|x| - 1}\right)|x|^n$$

Thus if $|a_0| \geq A/[|x| - 1]$, or if

$$|x| \geq 1 + \frac{A}{|a_0|} = R$$

then $P(x) > 0$. Hence the roots ξ_k for which $P(\xi_k) = 0$, must satisfy the reversed inequality $|\xi_k| < R$.

For finding a lower bound on the modulus, set $x = 1/y$ in Eq. (2.11) to obtain

$$a_n y^n + a_{n-1} y^{n-1} + \cdots + a_0 = 0$$

The roots η_k of this equation satisfy

$$|\eta_k| = \frac{1}{|\xi_k|} < 1 + \frac{B}{|a_n|} = \frac{1}{r}$$

This means that $|\xi_k| > r$.   □

This theorem can fruitfully be applied to isolate the (real and) complex roots of Eq. (2.12) by computing $P(x)$ for complex values of x in small incremental steps of $Re(x)$ and $Im(x)$, such that $r < |x| < R$.

The next theorem supplies closer bounds for real roots of real polynomials and can be applied for isolation of real roots.

Theorem 2.7 (Lagrange) *Suppose $a_0 > 0$ and a_k ($k \geq 1$) is the first of the negative coefficients of the polynomial $P(x)$ and L is the largest absolute value of the negative coefficients of $P(x)$, then all positive roots ξ^+ satisfy $0 < \xi^+ < R$, where*

$$R = 1 + \sqrt[k]{\frac{L}{a_0}}$$

Proof Set $x > 1$. If in $P(x)$ each of the nonnegative coefficients $a_1, a_2, \cdots, a_{k-1}$ is replaced by zero and each of the remaining coefficients $a_k, a_{k+1}, \cdots, a_n$ is replaced by negative number $-L$, then the value of $P(x)$ will only diminish. Hence one obtains

$$P(x) \geq a_0 x^n - L(x^{n-k} + x^{n-k-1} + \cdots + 1) = a_0 x^n - L\frac{x^{n-k+1} - 1}{x - 1}$$

$$> a_0 x^n - \frac{L}{x - 1} x^{n-k+1} = \frac{x^{n-k+1}}{x - 1}\left[a_0 x^{k-1}(x - 1) - L\right]$$

$$> \frac{x^{n-k+1}}{x - 1}\left[a_0(x - 1)^k - L\right]$$

Consequently for $x \geq 1 + \sqrt[k]{L/a_0} = R$, $P(x) > 0$. Thus all the positive roots ξ^+ will satisfy $\xi^+ < R$.

Remark If there is no such coefficient as a_k, all the coefficients of the equation are nonnegative. Such equation has no positive roots. □

The lower bound of the positive roots can be obtained by considering the equation $P(1/x) = 0$. If r is the upper bound of this equation obtained as above, then $1/r$ is the lower bound of $P(x) = 0$.

The bounds on the negative roots of Eq. (2.11) can also be obtained. If R_1 and r_1 are the upper bounds of $P(-x) = 0$ and $P(-1/x) = 0$ respectively then $-R_1$ and $-1/r_1$ are the lower and upper bounds of the negative roots of $P(x) = 0$.

Example 1. Consider $P(x) := 8x^4 - 8x^2 - 32x + 1 = 0$.

Solution. Here $a_0 = 8$, $a_1 = 0$, $a_2 = 8$, $a_3 = -32$, $a_4 = 1$. The equation has four roots.

For the annulus theorem, $A = max\{0, 8, 32, 1\} = 32$ and $B = max\{8, 0, 8, 32\} = 32$. Hence

$$R = 1 + \frac{32}{8} = 5, \quad r = \frac{1}{1 + 32/1} = \frac{1}{33} = 0.0303$$

Thus all the four roots lie in the annulus $0.0303 < |x| < 5$. If the equation has real roots, then they must lie in $[0.0303, 5]$ and $[-5, -0.0303]$.

Next we apply Lagrange's theorem. For this purpose, $k = 2$ (the number of the first negative coefficient) and $L = 32$. Therefore,

$$R = 1 + \sqrt{32/8} = 3$$

Also, for the polynomial equation

$$x^4 P(1/x) = x^4 - 32x^3 - 8x^2 + 8 = 0$$

$k = 1$ and $L = 32$. Thus

$$r = 1 + 32 = 33$$

Consequently, the positive roots if any must lie $[1/33, \ 3]$ or $[-.0303, \ 3]$. For negative roots, consider

$$P(-x) = 8x^4 - 8x^2 + 32x + 1 = 0$$

so that $k = 2$, $L = 8$, $a_0 = 8$ with $R = 1 + \sqrt{8/8} = 2$. Finally we have

$$x^4 P(-1/x) = x^4 + 32x^3 - 8x^2 + 8 = 0$$

for which $k = 2$, $L = 8$, $a_0 = 1$. Therefore, $r = 1 + \sqrt{8/1} = 1 + 2\sqrt{2} = 3.832$. Consequently, the negative roots if any must lie in $[-2, \ -1/3.832]$ or $[-2, \ -0.261]$. Note that Lagrange's method yields improved bounds on real roots as compared to the annulus theorem.   □

We conclude that there are several other results on bounds of roots, but these results are more difficult to apply. Researchers are interested in this topic even to the present times.

2.6.2 Application of Newton's Method

We assume that approximation of a desired root of a given algebraic equation is obtained by methods of the general theory and the preceding subsection. We use the approximation and refine it by Newton's iteration, Eq. (2.6) which requires computation $P(x)$ and $P'(x)$. Both these quantities can be computed economically by Horner's scheme (Exercise 2, Sect. 1.3, Chap. 1) by writing

$$P(x) = a_n + x(a_{n-1} + x(a_{n-2} + \cdots + x(a_1 + a_0 x)))$$

which can be iteratively written as

$$b_0 := a_0$$
$$b_1 := b_0 x + a_1$$
$$b_2 := b_1 x + a_2$$
$$\cdots \quad \cdots$$
$$P(x) = b_n := b_{n-1} x + a_n$$

Interestingly the quantities b_0, b_1, b_2, $\cdots$, b_n are not constants but help in computing $P'(x)$. For,

$$P'(x) = na_0 x^{n-1} + (n-1)a_1 x^{n-2} + (n-2)a_2 x^{n-3} + \cdots + a_{n-1}$$

$$= a_0 x^{n-1} + (a_0 x + a_1)x^{n-2} + ((a_0 x + a_1)x + a_2)x^{n-3} + \cdots$$

$$+(((a_0 x + a_1)x + a_2)x + \cdots + a_{n-2})x + a_{n-1}$$

$$=: b_0 x^{n-1} + b_1 x^{n-2} + b_3 x^{n-3} + \cdots + b_{n-1}$$

whose iterative scheme is as before

$$c_0 := b_0$$
$$c_1 := c_0 x + b_1$$
$$c_2 := c_1 x + b_2$$
$$\cdots \quad \cdots$$
$$P'(x) = c_{n-1} := c_{n-2} x + b_{n-1}$$

With the above tricks of computing $P(x)$ and $P'(x)$ and taking $a_0 = 1$ without loss of generality, subroutine NEWTON following Eq. (2.6) can be modified as:

SUBROUTINE NEWTON_ HORNER(x,n,a,xtol,maxiter,root)

! computes a real or complex root of a polynomial equation of degree n.

! x=initial approximation of root, real or complex.   (Input)

! n=degree of polynomial.   (Input)

! a(1),a(2),$\cdots$,a(n)=coefficients of the polynomial.   (Input)

! xtol=tolerance in the computation of the root.   (Input)

! root=the desired root with tolerance xtol.   (Output)

!**

COMPLEX :: a(n),b(n),c(n),x,xnew,root
DO k=1,maxiter

```
b(1)=x+a(1); c(1)=x+b(1)
DO i=2,n−1
b(i)=b(i−1)*x+a(i); c(i)=c(i−1)*x+b(i)
END DO
b(n)=b(n−1)*x+a(n)
xnew=x−b(n)/c(n−1)
IF(CABS(xnew−x)<xtol) THEN
root=xnew; EXIT
END IF
x=xnew
IF(k==maxiter) PRINT*, 'no solution'
END DO
RETURN
END SUBROUTINE NEWTON_HORNER
```

The famous Newton method, Eq. (2.6) was in fact discovered for polynomial equations by Newton and Raphson. The general result (2.6) is due to Lagrange.

Example 1. Wallis' equation $x^3 - 2x - 5 = 0$.

The equation has a real root in [2, 3] and a pair of complex conjugate roots. Taking initial approximation as $x = 2.5 + i \cdot 0$ and maxiter $= 20$, the subroutine yields root $= 2.09455 + i \cdot 0$. For the complex root above the real axis, the initial value $x = -1 + i$ yields root $= -1.04728 + 1.13594i$. The third must be $-1.04728 - 1.13594i$.   □

Example 2. The equation

$$x^4 - 7x^3 + 19x^2 - 13x - 10 = 0$$

has complex roots (see Example 4, Sect. 2.6). Find all the roots.

The equation has real roots in $[-1, 0]$ and exactly at 2. Using initial approximation of $-0.5 + 0 \cdot i$, the subroutine yields the root as -0.43906. Since the sum of the roots is 7,

$$-0.43906 + 2 + 2u = 7$$

where $u \pm iv$, are the complex roots. This yields $u = 2.72$. With a trial of $2.72 + 2i$, the complex root in the upper half plane is found to be $2.71953 + 1.99805i$. Thus the roots are

$$-0.43906, \ 2, \ 2.71953 \pm 1.99805i$$

□

Exercises

1. Apply the subroutine NEWTON_HORNER to solve

(i) $x^3 - 6x^2 + 21x - 26 = 0$.

(ii) $x^4 + 4x^3 - 3x^2 - 11x - 9 = 0$.

(iii) $x^5 - 3x^3 + x^2 - 1 = 0$.

$[(i)$ 2, $2 \pm 3i$, (ii) -4.08766, 1.98154, $-0.94694 \pm 0.46301i$, (iii) -1.83929, -0.61803, 1.61803, $0.41964 \pm 0.60629i]$.

2.6.3 Bairstow's Method

In order to extract a real or complex root of Eq. (2.12), Bairstow's method given in the year 1920, attempts to extract a quadratic factor of the form $x^2 + px + q$ from the polynomial $P(x)$. In general if $P(x)$ is divided by $x^2 + px + q$ we obtain a quotient of degree $n - 2$ of the form

$$b_0 x^{n-2} + b_1 x^{n-3} + \cdots + b_{n-3}x + b_{n-2}$$

and a linear remainder of the form $Rx + S$. Our problem then is to find p and q such that

$$R(p, q) = 0; \quad S(p, q) = 0$$

Starting with a guess of p and q we find corrections Δp and Δq, so that

$$R(p + \Delta p, q + \Delta q) = 0; \quad S(p + \Delta p, q + \Delta q) = 0$$

Expanding in Taylor's series and truncating after the first-order terms, as in Newton's method, we get

$$R(p, q) + \frac{\partial R}{\partial p}\Delta p + \frac{\partial R}{\partial q}\Delta q = 0, \quad S(p, q) + \frac{\partial S}{\partial p}\Delta p + \frac{\partial S}{\partial q}\Delta q = 0$$

The pair of equations yield expressions for Δp and Δq. Supposing that Δp and Δq have been computed, the procedure is repeated for the corrected values $p + \Delta p$ and $q + \Delta q$.

In order to compute the coefficients $b_0, b_1, b_2, \cdots, b_{n-2}, R$ and S, we use the identity

$$a_0 x^n + a_1 x^{n-1} + \cdots + a_n \equiv (x^2 + px + q)(b_0 x^{n-2} + b_1 x^{n-3} + \cdots + b_{n-2}) + Rx + S$$

Comparing like powers of x on the two sides, we obtain

$$a_0 = b_0$$
$$a_1 = b_1 + pb_0$$
$$a_2 = b_2 + pb_1 + qb_0$$
$$\cdots \quad \cdots\cdots$$
$$a_k = b_k + pb_{k-1} + qb_{k-2}$$
$$\cdots \quad \cdots\cdots$$
$$a_{n-2} = b_{n-2} + pb_{n-3} + qb_{n-4}$$
$$a_{n-1} = R + pb_{n-2} + qb_{n-3}$$
$$a_n = S + qb_{n-2}$$

The quantities $b_0, b_1 \cdots, b_n$, R and S can be obtained recursively from these equations. Defining

$$b_{-2} := b_{-1} := 0; \quad b_{n-1} =: R; \quad b_n + pb_{n-1} =: S$$

the recursive solution is then

$$b_k = a_k - pb_{k-1} - qb_{k-2}, \quad (k = 0, 1, 2, \cdots, n) \tag{2.12}$$

If we insert the newly defined expressions for R and S in the equations for Δp and Δq, we obtain

$$\frac{\partial b_{n-1}}{\partial p} \Delta p + \frac{\partial b_{n-1}}{\partial q} \Delta q + b_{n-1} = 0$$

$$\left(\frac{\partial b_n}{\partial p} + p \frac{\partial b_{n-1}}{\partial p} + b_{n-1} \right) \Delta p + \left(\frac{\partial b_n}{\partial q} + p \frac{\partial b_{n-1}}{\partial q} \right) \Delta q + b_n + pb_{n-1} = 0$$

The second equation reduces owing to the first, leading to the pair

$$\frac{\partial b_{n-1}}{\partial p} \Delta p + \frac{\partial b_{n-1}}{\partial q} \Delta q + b_{n-1} = 0$$

and

$$\left(\frac{\partial b_n}{\partial p} + b_{n-1} \right) \Delta p + \frac{\partial b_n}{\partial q} \Delta q + b_n = 0$$

The partial derivatives of b_k are given by

$$\frac{\partial b_k}{\partial p} = -b_{k-1} - p \frac{\partial b_{k-1}}{\partial p} - q \frac{\partial b_{k-2}}{\partial p}; \quad \frac{\partial b_0}{\partial p} = \frac{\partial b_{-1}}{\partial p} = \frac{\partial b_{-2}}{\partial p} = 0$$

and

$$\frac{\partial b_k}{\partial q} = -p \frac{\partial b_{k-1}}{\partial q} - b_{k-2} - q \frac{\partial b_{k-2}}{\partial q}; \quad \frac{\partial b_0}{\partial q} = \frac{\partial b_{-1}}{\partial q} = \frac{\partial b_{-2}}{\partial q} = 0$$

Eliminating b_{k-1} from the above two equations

$$\frac{\partial b_k}{\partial p} + p \frac{\partial b_{k-1}}{\partial p} + q \frac{\partial b_{k-2}}{\partial p} = \frac{\partial b_{k+1}}{\partial q} + p \frac{\partial b_k}{\partial q} + q \frac{\partial b_{k-1}}{\partial q}$$

This relation holds for arbitrary p and q and so equating the coefficients one must have

$$\frac{\partial b_k}{\partial q} = \frac{\partial b_{k-1}}{\partial p} =: -c_{k-2}, \quad (k = 0, 1, 2, \cdots, n)$$

With the introduction of quantities c_k, the two equations for the partial derivatives of b_k pass in to the forms

$$c_{k-1} = b_{k-1} - p c_{k-2} - q c_{k-3}; \quad c_{k-2} = b_{k-2} - p c_{k-3} - q c_{k-4}$$

or equivalently

$$c_k = b_k - p c_{k-1} - q c_{k-2}, \quad (k = 0, 1, 2, \cdots, n) \tag{2.12a}$$

with $c_{-2} = c_{-1} = 0$. Equation (2.12a) has a form similar to Eq. (2.12a) and c_k is computed from b_k in exactly the same way as b_k from a_k. With the partial derivatives of b_k determined in this manner, the pair of equations for Δp and Δq become

$$c_{n-2} \Delta p + c_{n-3} \Delta q = b_{n-1}$$

$$(c_{n-1} - b_{n-1}) \Delta p + c_{n-2} \Delta q = b_n$$

Solving the two equations we finally obtain the corrections

$$\Delta p = \frac{b_{n-1} c_{n-2} - b_n c_{n-3}}{c_{n-2}^2 - c_{n-3}(c_{n-1} - b_{n-1})}, \quad \Delta q = \frac{b_n c_{n-2} - b_{n-1}(c_{n-1} - b_{n-1})}{c_{n-2}^2 - c_{n-3}(c_{n-1} - b_{n-1})} \tag{2.12b}$$

It is easy to program the computation of b_k and c_k following Eqs. (2.12) and (2.12a). If the starting values of p and q are not too bad the iterations will converge quadratically as in the Newton's method. The termination of iterations can be based on the smallness of $|R|$ and $|S|$, or that of $\sqrt{b_{n-1}^2 + b_n^2}$. If nothing is known about the starting values of p and q, a search operation can be performed over range of values and select the pair that makes $\sqrt{b_{n-1}^2 + b_n^2}$ the smallest.

On successfully separating the quadratic factor, the *deflated* polynomial $b_0 x^{n-2} + b_1 x^{n-3} + \cdots + b_{n-2} = 0$ can similarly be treated for finding all the roots of the original equation $P(x) = 0$.

Leonard Bairstow (1880–1963), English aeronautical engineer engaged in aeroplane research held a chair at Imperial College, London. The method appeared in the appendix of his book *Applied Aerodynamics*.

The following subroutine implements Bairstow's method. For convenience it is assumed that $a_0 = 1 = b_0$ in the equation. It is also assumed that n is even for applicability of the method; otherwise if n is odd, one can multiply the given equation by x and discard the root $x = 0$ after application of the program.

```
SUBROUTINE BAIRSTOW(a,b,c,z1,z2,n)
REAL :: a(n),b(n),c(n)
COMPLEX :: z1,z2
M=100                       ! Search for p and q begins.
small=0.8388607E38     ! Largest Number that can be stored in the computer.
p=-M                        ! Search for p=-M-1 (1) M+1.
DO i=1,2*M+1
q=-M                        ! Search for q=-M-1 (1) M+1.
DO j=1,2*M+1
DO k=1,n
IF(k==1) b(k)=a(k)-p
IF(k==2) b(k)=a(k)-p*b(k-1)-q
IF(k>2) b(k)=a(k)-p*b(k-1)-q*b(k-2)
END DO
sb=sqrt(b(n-1)**2+b(n)**2)         ! Serach for the smallest value of sb.
IF(sb<small) THEN
small=sb; ii=i; jj=j
END IF
q=q+1
END DO
p=p+1
END DO
p=-M-1+ii; q=-M-1+jj
DO i=1,20                              ! 20 Iteration for p, q.
DO k=1,n
IF(k==1) b(k)= a(k)-p
IF(k==2) b(k)=a(k)-p*b(k-1)-q
IF(k>2) b(k)=a(k)-p*b(k-1)-q*b(k-2)
END DO
DO k=1,n
IF(k==1) c(k)=b(k)-p
IF(k==2) c(k)=b(k)-p*c(k-1)-q
IF(k>2) c(k)=b(k)-p*c(k-1)-q*c(k-2)
END DO
delta=c(n-2)**2-c(n-3)*(c(n-1)-b(n-1))
deltap=(b(n-1)*c(n-2)-b(n)*c(n-3))/delta
deltaq=(b(n)*c(n-2)-b(n-1)*(c(n-1)-b(n-1)))/delta
```

```
IF(SQRT(deltap**2+deltaq**2)<0.1E-6) EXIT
p=p+deltap; q=q+deltaq
END DO
disc=p**2-4.0*q
z1=0.5*(-p+CSQRT(CMPLX(disc,0.0)))
z2=0.5*(-p-CSQRT(CMPLX(disc,0.0)))
RETURN
END SUBROUTINE BAIRSTOW
```

Example 1. Find the quadratic factors of

$$x^4 + 5x^3 + 4x^2 - 5x - 9 = 0$$

by Bairstow's method and find the roots.

Solution. In this equation $n = 4$ and $a_1 = 5$, $a_2 = 4$, $a_3 = -5$, $a_4 = -9$. The program for a pair of roots is thus

```
PROGRAM MAIN
REAL :: a(1)=5; a(2)=4; a(3)=-5; a(4)=-9
COMPLEX :: z1, z2
n=4
CALL BAIRSTOW(a,b,c,z1,z2,n)
PRINT*, z1, z2, (b(i), i=1,n-2)
END
```

Appending the subroutine and running the program, the two roots are found to be $z1 = 1.15353$ and $z2 = -3.74765$ (both real). The coefficients of the deflated quadratic are found to be b(1)=2.40588 and b(2)=2.08188. The remaining two roots are thus roots of the quadratic $x^2 + 2.40588x + 2.08188 = 0$ whose roots are $-1.20294 \pm 0.79676\,i$ (complex conjugate). $\qquad\square$

Exercises

1. Solve the following equations using Bairstow's method:
(i) $x^4 - 3x^3 - 4x^2 - 2x + 8 = 0$.

$[b_1 = 2,\ b_2 = 2,\ z_1 = 4,\ z_2 = 1,\ z_3,\ z_4 = -1 \pm i]$.

(ii) $x^4 - 7x^3 + 19x^2 - 13x - 10 = 0$.

$[b_1 = -5.43906,\ b_2 = 11.38805,\ z_1 = 2,\ z_2 = -0.43906,$
$z_3,\ z_4 = 2.719528 \pm 1.99805i]$.

(iii) $x^4 - x^3 + 6x^2 + 5x + 10 = 0$.

$[b_1 = 1.14457, \ b_2 = 1.42193, \ z_1, \ z_2 = 1.07229 \pm 2.42547i$,
$z_3, z_4 = -0.57229 \pm 1.04614i]$.

(iv) $x^3 - 2x - 5 = 0$.

[Multiply the equation by x and take $a_4 = 0$ in the program and discard the root $x = 0$. $b_1 = 2.09455$, $b_2 = 2.38715$, $z_1 = 2.09455$, $(z_2 = 0)$,
$z_3, z_4 = -1.04728 \pm 1.13594i]$.

2.6.4 Method of Eigenvalues

As in the preceding, we assume $a_0 = 1$ in Eq. (2.11) and consider the $n \times n$ *Frobenius matrix*

$$A = \begin{bmatrix} -a_1 & -a_2 & -a_3 & \cdots & -a_{n-1} & -a_n \\ 1 & 0 & 0 & \cdots & 0 & 0 \\ 0 & 1 & 0 & \cdots & 0 & 0 \\ \cdots & \cdots & \cdots & \cdots & \cdots & \cdots \\ 0 & 0 & 0 & \cdots & 1 & 0 \end{bmatrix}$$

The eigenvalues of A are given by $det[A - \lambda I] = 0$, i.e.

$$\begin{vmatrix} -a_1 - \lambda & -a_2 & -a_3 & \cdots & -a_{n-1} & -a_n \\ 1 & -\lambda & 0 & \cdots & 0 & 0 \\ 0 & 1 & -\lambda & \cdots & 0 & 0 \\ \cdots & \cdots & \cdots & \cdots & \cdots & \cdots \\ 0 & 0 & 0 & \cdots & 1 & -\lambda \end{vmatrix} = 0$$

Expanding the determinant in terms of the elements of the first row, we have

$$-(a_1 + \lambda)(-\lambda)^{n-1} + a_2(-\lambda)^{n-2} - a_3(-\lambda)^{n-3} + \cdots + (-1)^n a_n = 0$$

or, $\lambda^n + a_1\lambda^{n-1} + a_2\lambda^{n-2} + a_3\lambda^{n-3} + \cdots + a_n = 0$

This shows that the eigenvalues λ are the roots of Eq. (2.11), and so the roots of Eq. (2.11) can be found by finding the eigenvalues of the matrix A! This procedure may seem strange, but there exists an efficient computing technique, called the QR method by which the eigenvalues of any given matrix can be efficiently found (see Chap. 9).

2.6.5 Ill-Conditioned Equations

We now discuss in brief, the problem of sensitivity of roots with respect to coefficients $a_0, a_1, \cdots, a_n$ of a polynomial equation and our focus is on such equations. An equation is *ill-conditioned* if small changes in the coefficients result in large changes in the roots. A celebrated example is the polynomial equation with roots ξ_i, $i = 1, 2, \cdots, 20$:

$$(x - 1)(x - 2) \cdots (x - 20) = x^{20} - 210x^{19} + 20615x^{18} - \cdots + 20! = 0$$

If the coefficient $a_1 = -210$ is replaced by $a_1 = -(210 + 2^{-23}) = -210.00000$ $0119 \cdots$ there is a change in the 10th significant figure caused by converting the input from decimal to binary system leaving the other coefficients unchanged. The roots of this slightly altered equation change drastically since they become complex. For example it has been found that

$$\xi_{14}, \xi_{15} = 13.992 \pm 2.519i, \quad \xi_{16}, \xi_{17} = 16.731 \pm 2.813i$$

rounded to three decimal places. The first seven roots ξ_i ($i = 1, 2, \cdots, 7$) however, are not affected within 4-significant-figure accuracy.

In regard to polynomials it can be proved that

(i) The existence of several roots having ratio close to unity always implies ill-conditioning. of those roots, and

(ii) Multiple roots are ill-conditioned.

A suggested remedy of sorts is to always use double or higher precision arithmetic in computing the roots.

2.7 Choice of Methods

In applications, the function f of an equation $f(x) = 0$ is often a complicated function or a polynomial of high degree. The number of real and complex roots of such an equation may be unspecified. For computing possible real roots, in such a scenario, isolation of each root can be performed by drawing a graph of $f(x)$. Alternatively, one may compute $f(x)$ at specified intervals, say of unit length and note changes in its sign. A change of sign indicates existence of a root in the interval. Accurate determination of root so detected is not advisable by Bolzano Bisection or Modified Regula Falsi methods because of slowness of convergence. On the other hand, secant and Newton's method possess higher *rate* of convergence, but they do not always fulfill the primary requirement of convergence. In this respect, the combined secant–bisection method can be adopted as it fulfills both the conditions of convergence and an accelerated rate of the secant method.

Newton's method converges faster than the secant method, but has less computational efficiency in the sense that computation of $f'(x)$ is also required. It has also the disadvantage that the starting approximation x_0 must be close to the root for convergence and may not be readily available. One can, however, estimate x_0, using say, the Bolzano Bisection method with a few iterations and then sharpen the root by using Newton's method.

The general fixed-point iteration method converges linearly and stands no real competition with the secant and the Newton methods. Even Aitken's Δ^2–process leads to convergence that is at best quadratic at the cost of increased complexity of the algorithm. So the method is not regarded better than the secant method.

Polynomial equations belong to a special class and possess several elegant results on root isolation. These results can be used to confirm the domains of roots computed by Müller's method. These approximations can be used in the specialised Newton routine for polynomials to yield roots with high accuracy.

Bairstow's method is an elegant alternative that is computationally efficient. It determines real and complex roots in pairs. By repeated application, it delivers all the roots simultaneously. It is simple when the roots are real and distinct, but complications can arise in the case of repeated real and complex roots.

In practice it is desirable to compute the roots of an equation by different methods in order to ensure correctness of the computed roots.

Chapter 3
System of Equations

The problem of computing n unknowns from a system of n equations, where $n > 1$ is fundamentally important in numerical computations. Often n is large—of the order of 10, 100 or even 1000. Thus, an algorithm for root finding must be capable of treating cases of large n and must possess low complexity, in order to debar excessive computational time.

When each equation of the system is a *linear* function of the unknowns, the equations form a *linear system*. Such systems occur directly in many applied problems; as for example, in electrical networks and stresses in structural members of construction. The numerical treatment of problems of analysis also require the solution of such systems of equations. Approximation of functions, solution of ordinary and partial differential equations by finite differences or finite elements and solution of linear integral equations, depend on such systems for numeric computing. The subject matter of Chaps. 4, 6 and 7 constitute some of these topics. The study of linear systems of equations is therefore eminently important in numerical computations. We begin this topic, ending with a treatment of *non-linear systems* in Sect. 3.4.

3.1 Linear System of Equations

A *linear system of equations* in n unknowns $x_1, x_2, \cdots, x_n$ real or complex, is of the form

$$
\begin{aligned}
a_{11}x_1 + a_{12}x_2 + \cdots + a_{1n}x_n &= b_1 \\
a_{21}x_1 + a_{22}x_2 + \cdots + a_{2n}x_n &= b_2 \\
\cdots\cdots\cdots\cdots\cdots\cdots\cdots \\
a_{n1}x_1 + a_{n2}x_2 + \cdots + a_{nn}x_n &= b_n
\end{aligned}
\tag{3.1}
$$

© Springer Nature Singapore Pte Ltd. 2019
S. K. Bose, *Numerical Methods of Mathematics Implemented in Fortran*, Forum for Interdisciplinary Mathematics,
https://doi.org/10.1007/978-981-13-7114-1_3

In the above set of equations, the *coefficients* a_{ij} $(i, j = 1, 2, \cdots, n)$ and the right-hand side b_i $(i = 1, 2, \cdots, n)$ are supposed to be known real or complex constants. Using matrix notation, Eq. (3.1) can be written compactly as

$$Ax = b \tag{3.1a}$$

where

$$A = \begin{bmatrix} a_{11} & a_{12} & \cdots & a_{1n} \\ a_{21} & a_{22} & \cdots & a_{2n} \\ \cdots & \cdots & \cdots & \cdots \\ a_{n1} & a_{n2} & \cdots & a_{nn} \end{bmatrix}$$

$$\mathbf{x} = [x_1, x_2, \cdots, x_n]^T, \quad \mathbf{b} = [b_1, b_2, \cdots, n_n]^T$$

It is known from linear algebra that the solution of Eq. (3.1a) exists if $det[A] \neq 0$. If the condition is satisfied, the unique solution is given by *Cramer's rule* in the form of ratios of n, $n \times n$ determinants divided by $det[A]$. The complexity of computing an $n \times n$ determinant is $O(n!)$ and is prohibitively large. Hence, the rule cannot be applied for large n.

Computationally, efficient methods can be divided into two categories: those based on **direct systematic elimination** of the unknowns and those on **relaxation**. The latter approach described in Sect. 3.3 is essentially *iterative*. We first treat the direct methods.

3.1.1 Tridiagonal Matrices: Thomas Method

Before taking up the solution of Eq. (3.1) by direct elimination, we illustrate the method for *tridiagoanl system of equations*. For such system

$$a_{ij} = 0, \quad \text{for} \quad |i - j| > 1$$

and has the form

$$d_1 x_1 + a_1 x_2 \qquad\qquad\qquad\qquad\qquad = b_1 \tag{3.2}$$
$$c_2 x_1 + d_2 x_2 + a_2 x_3 \qquad\qquad\qquad = b_2 \tag{3.3}$$
$$c_3 x_2 + d_3 x_3 + a_3 x_4 \qquad\qquad = b_3 \tag{3.4}$$
$$\cdots\cdots\cdots \qquad\qquad \cdots\cdots\cdots\cdots \qquad \cdots$$

$$c_{n-1} x_{n-2} + d_{n-1} x_{n-1} + a_{n-1} x_n = b_{n-1} \tag{3.$n-1$}$$

$$c_n x_{n-1} + d_n x_n = b_n \tag{3.n}$$

The elements $d_1, d_2, \cdots, d_n$ constitute the *diagonal* of the matrix A. Similarly, $a_1, a_2, \cdots, a_{n-1}$ and $c_2, c_3, \cdots, c_n$ are respectively called the *superdiagonal* and *subdiagonal* elements of A. In the direct elimination method, the n equations are linearly combined to reduce the subdiagonal elements to zero, rendering the solution to a trivial matter. Thus in the first step, we eliminate x_1 from Eq. (3.3) by the operation Eq. (3.3)$- c_2 \times$ Eq. (3.2)$/d_1$. Thus Eq. (3.3) becomes

$$d_2' x_2 + a_2 x_3 = b_2' \qquad\qquad (3.3)'$$

where

$$d_2' := d_2 - c_2 \times \frac{a_1}{d_1}, \quad b_2' := b_2 - c_2 \times \frac{b_1}{d_1}$$

In the second step, we eliminate x_2 from Eq. (3.4) by the operation Eq. (3.4) $- c_3 \times$ Eq. (3.3)$'/d_2'$. The equation thus becomes

$$d_3' x_3 + a_3 x_4 = b_3' \qquad\qquad (3.4)'$$

where

$$d_3' := d_3 - c_3 \times \frac{a_2}{d_2'}, \quad b_3' := b_3 - c_3 \times \frac{b_2'}{d_2'}$$

Continuing in this way, we get in the kth step, the equation

$$d_{k+1}' x_{k+1} + a_{k+1} x_{k+2} = b_{k+1}' \qquad\qquad (3.k+1)'$$

where

$$d_{k+1}' := d_k - c_{k+1} \times \frac{a_k}{d_k'}, \quad b_{k+1}' := b_{k+1} - c_{k+1} \times \frac{b_k'}{d_k'}$$

When the final $(n-1)$th step is performed, the system reduces to

$$
\begin{aligned}
d_1' x_1 + a_1 x_2 &= b_1' \\
d_2' x_2 + a_2 x_3 &= b_2' \\
\cdots\cdots\cdots \quad\quad \cdots\cdots\cdots \quad & \cdots \\
d_{n-1}' x_{n-1} + a_{n-1} x_n &= b_{n-1}' \\
d_n' x_n &= b_n'
\end{aligned}
$$

where $d_1' := d_1$ and $b_1' := b_1$. We note that in the process, the superdiagonal elements remain unchanged. The transformed equations can now be solved in the reverse order $x_n, x_{n-1}, \cdots, x_1$ by a process called *back substitution*. Thus we get

$$x_n = \frac{b'_n}{d'_n}$$

$$x_{n-1} = \frac{b'_{n-1} - a_{n-1}x_n}{d'_{n-1}} \qquad (3.5)$$

etc. In general we have

$$x_i = \frac{b'_i - a_i x_{i+1}}{d'_i}, \quad i = n - 1, \, n - 2, \cdots, \, 1 \qquad (3.6)$$

which solves the problem. The method is called the Thomas method. Based on Eqs. $(3.k + 1)'$, (3.6) and (3.5), a Fortran subroutine can be written as in the following:

```
subroutine THOMAS(n,d,a,c,b)
! n=number of unknowns in a tridiagonal system.      (Input)
! d=diagonal vector.     (Input)
! a=superdiagonal vector with a(n)=0.0     (Input)
! c=subdiagonal vector with a(1)=0.0     (Input)
! b=right hand side vector.     (Input)
! The solution is returned in b.     (Output)
REAL :: d(n), a(n), c(n), b(n)
! Elimination begins
DO k=1,n−1
IF(d(k)==0.0) THEN
PRINT*, 'No Solution'; RETURN
END IF
ratio=c(k+1)/d(k)
d(k+1)=d(k+1)−ratio*a(k)
b(k+1)=b(k+1)−ratio*b(k)
END DO
IF(d(n)==0.0) THEN
PRINT*, 'No Solution'; RETURN
END IF
! Back substitution begins
b(n)=b(n)/d(n)
DO i=n−1,1,−1
b(i)=(b(i)−a(i)*b(i+1))/d(i)
END DO
RETURN
END SUBROUTINE THOMAS
```

From the program, it is clear that the number of arithmetic operations in the elimination process is $8(n - 1)$, while in back substitution, it is $1 + 4(n - 1)$. The total of the two is $12n - 11$ or $O(12n)$. This is far less than $O(n!)$ in Cramer's rule.

The elimination procedure can be viewed as decomposition of matrix A into lower and upper triangular matrices. A *lower triangular mtrix* is one in which all the superdiagonal elements are zero. Similarly an *upper triangular matrix* has all of its subdiagonal elements equal to zero. The matrix of the coefficients of the final transformed equations of the tridiagonal system is upper triangular. Hence if one looks for the factorisation

$$
A = \begin{bmatrix}
d_1 & a_1 & 0 & \cdots & 0 & 0 \\
c_2 & d_2 & a_2 & \cdots & 0 & 0 \\
\multicolumn{6}{c}{\cdots\cdots\cdots\cdots} \\
0 & 0 & 0 & \cdots & c_n & d_n
\end{bmatrix}
$$

$$
= \begin{bmatrix}
1 & 0 & 0 & \cdots & 0 & 0 \\
l_2 & 1 & 0 & \cdots & 0 & 0 \\
\multicolumn{6}{c}{\cdots\cdots\cdots\cdots} \\
0 & 0 & 0 & \cdots & l_n & 1
\end{bmatrix}
\times
\begin{bmatrix}
d_1' & a_1 & 0 & \cdots & 0 \\
0 & d_2' & a_2 & \cdots & 0 \\
\multicolumn{5}{c}{\cdots\cdots\cdots\cdots} \\
0 & 0 & 0 & \cdots & d_n'
\end{bmatrix}
:= LU
$$

then by equating the elements on the two sides, it follows that

$$
d_1' = d_1, \quad d_{k+1}' = d_{k+1} - l_{k+1}a_k, \quad l_{k+1} = c_{k+1}/d_k', \quad (k = 1, 2, \cdots, n-1)
$$
(3.7a)

Now the system $A\mathbf{x} = \mathbf{b}$ becomes $LU\mathbf{x} = \mathbf{b}$, which is equivalent to $U\mathbf{x} = \mathbf{b}'$ with $L\mathbf{b}' = \mathbf{b}$. Comparing the elements of the last relation one has

$$
b_1' = b_1, \quad b_{k+1}' = b_{k+1} - l_{k+1}b_k', \quad (k = 1, 2, \cdots, n-1) \tag{3.7b}
$$

The system $U\mathbf{x} = \mathbf{b}'$ is identical to the system obtained by the Thomas method, with expressions for d_i' and b_i' $(i = 1, 2, \cdots, n)$, given as before.

Llewellyn Hileth Thomas (1903–1992), American physicist. At Columbia University, he was also a staff member of a scientific computing laboratory established by IBM. He is best remembered for his model of the atom, concurrently developed by Enrico Fermi known as Thomas–Fermi Model.

Exercises

1. Show that by THOMAS elimination the linear system

$$
\begin{aligned}
x_1 +x_2 && = \tfrac{3}{2} \\
-x_1 +2x_2 +3x_3 && = 1 \\
4x_2 -x_3 +2x_4 &= 1 \\
2x_3 -x_4 &= 1
\end{aligned}
\qquad \text{reduces to} \qquad
\begin{aligned}
x_1 +x_2 && = \tfrac{3}{2} \\
3x_2 +3x_3 && = \tfrac{5}{2} \\
-5x_3 +2x_4 &= -\tfrac{7}{3} \\
-\tfrac{1}{5}x_4 &= \tfrac{1}{15}
\end{aligned}
$$

Hence solve the system.
$[x_1 = 1, \; x_2 = 1/2, \; x_3 = 1/3, \; x_4 = -1/3].$

2. Solve the system

$$
\begin{aligned}
2x_1 \ -x_2 \qquad\qquad\qquad &= 1 \\
-x_{i-1} \ +2x_i \ -x_{i+1} &= 0 \ (i = 2, 3, \cdots, n-1) \\
-x_{n-1} \ +2x_n &= 0
\end{aligned}
$$

by subroutine THOMAS, when $n = 10$ and compare the answer with the exact solution $x_i = (n + 1 - i)/(n + 1)$, $(i = 1, 2, \cdots, n)$.
[0.90909, 0.81818, 0.72728, 0.63637, 0.54546, 0.45455, 0.36364, 0.27273, 0.18182, 0.09091]

3. A uniform beam rests on $n + 1$ equidistant supports $A_0, A_1, \cdots, A_n$. The bending moments $M_1, M_2, \cdots, M_{n-1}$ at $A_1, A_2, \cdots, A_{n-1}$ by Clapeyron's equation of three moments satisfy the equations

$$
\begin{aligned}
4M_1 \ +M_2 \qquad\qquad\qquad &= -1 \\
M_{i-1} \ +4M_i \ +M_{i+1} &= -1 \ (i = 2, 3, \cdots, n-2) \\
M_{n-2} \ +4M_{n-1} &= -1
\end{aligned}
$$

Solve the system by subroutine THOMAS when $n = 8$.
[$M_1 = M_7 = 0.21134$, $M_2 = M_6 = -0.15464$, $M_3 = M_5 = 0.17010$, $M_4 = -0.16495$].

3.1.2   Gauss Elimination for General Linear System

As in the case of the tridiagonal system Eq. $(3.1 - n)$, the unknowns $x_1, x_2, \cdots, x_{n-1}$ are successively eliminated from the lower triangle of the general system of equations (1). Designating the n equations of the system as (1.1), (1.2), $\cdots$, (1.n), we first eliminate x_1 from Eqs. (1.2), (1.3), $\cdots$, (1.n) with the help of Eq. (1.1). So, we perform the operations Eq. (1.2)$-a_{21} \times$ Eq. (1.1)$/a_{11}$, Eq. (1.3)$-a_{31} \times$Eq. (1.1)$/a_{11}$ etc., Eq. (1.n) $- a_{n1} \times$Eq. (1.1)$/a_{11}$. The divider a_{11} is called the **pivot** and Eq. (1.1) the **pivotal equation**. The system is thus transformed in to

$$
\begin{aligned}
a_{11}^{(1)} x_1 \ +a_{12}^{(1)} x_2 \ +a_{13}^{(1)} x_3 \ +\cdots \ +a_{1n}^{(1)} x_n &= b_1^{(1)} \\
a_{22}^{(1)} x_2 \ +a_{23}^{(1)} x_3 \ +\cdots \ a_{2n}^{(1)} x_n &= b_2^{(1)} \\
\cdots \qquad \cdots \qquad\quad \cdots \qquad\quad \cdots & \\
a_{n2}^{(1)} x_1 \ +a_{n3}^{(1)} x_3 \ +\cdots \ +a_{nn}^{(1)} x_n &= b_n^{(1)}
\end{aligned}
$$

where $a_{ij}^{(1)} := a_{ij}$, $(j = 1, 2, \cdots, n)$, $b_1^{(1)} := b_1$ and

$$
a_{ij}^{(1)} = a_{ij} - \frac{a_{i1}}{a_{11}} \times a_{1j}, \quad b_i^{(1)} := b_i - \frac{a_{i1}}{a_{11}} \times b_1, \quad (i, j = 2, 3, \cdots, n)
$$

In the second step, x_2 is similarly eliminated from the last $n - 2$ equations using $a_{22}^{(1)}$ as the pivot. We thus obtain the system

$$a_{11}^{(2)} x_1 + a_{12}^{(2)} x_2 + a_{13}^{(2)} x_3 + \cdots \quad a_{1n}^{(2)} x_n = b_1^{(2)}$$
$$a_{22}^{(2)} x_2 + a_{23}^{(2)} x_3 + \cdots + a_{2n}^{(2)} x_n = b_2^{(2)}$$
$$a_{33}^{(2)} x_3 + \cdots + a_{3n}^{(2)} x_n = b_3^{(2)}$$
$$\cdots \quad \cdots \quad \cdots \quad \cdots$$
$$a_{n3}^{(2)} x_3 + \cdots \quad a_{nn}^{(2)} x_n = b_n^{(2)}$$

where $a_{1j}^{(2)} := a_{1j}^{(1)}$, $(j = 1, 2, \cdots, n)$, $a_{2j}^{(2)} := a_{2j}^{(1)}$, $(j = 2, 3, \cdots, n)$ and

$$a_{ij} := a_{ij}^{(1)} - \frac{a_{i2}^{(1)}}{a_{22}^{(1)}} \times a_{2j}^{(1)}, \quad b_i^{(2)} := b_i^{(1)} - \frac{a_{i2}^{(1)}}{a_{22}^{(1)}} \times b_2^{(1)}, \quad (i, j = 3, 4, \cdots, n)$$

Proceeding in this manner, in the general kth step, x_k is eliminated from the last $n - k$ equations to obtain a reduced system in which the coefficients of the first k equations remain unchanged, while the remaining coefficients are changed to

$$a_{ij}^k := a_{ij}^{(k-1)} - \frac{a_{ik}^{(k-1)}}{a_{kk}^{(k-1)}} \times a_{kj}^{(k-1)}, \quad b_i^{(k)} := b_i^{(k-1)} - \frac{a_{ik}^{(k-1)}}{a_{kk}^{(k-1)}} \times b_k^{(k-1)} \qquad (3.8)$$

where $(i, j = k + 1, \cdots, n)$. In the final $(n - 1)$th step, the system reduces to the triangular form

$$a_{11}^{(n-1)} x_1 + a_{12}^{(n-1)} x_2 + \cdots \qquad \cdots \qquad + a_{1n}^{(n-1)} x_n = b_1^{(n-1)}$$
$$a_{22}^{(n-1)} x_2 + \cdots \qquad \cdots \qquad + a_{2n}^{(n-1)} x_n = b_2^{(n-1)}$$
$$\cdots \qquad \cdots \qquad \cdots$$
$$a_{n-1,n-1}^{(n-1)} x_{n-1} + a_{n-1,n}^{(n-1)} x_n = b_{n-1}^{(n-1)}$$
$$a_{nn}^{(n-1)} x_n = b_n^{(n-1)}$$

in which the only change in coefficients from the preceding step are

$$a_{nn}^{(n-1)} := a_{nn}^{(n-2)} - \frac{a_{n,n-1}^{(n-2)}}{a_{n-1,n-1}^{(n-2)}} \times a_{n-1,n}^{(n-2)}, \quad b_n^{(n-1)} := b_n^{(n-2)} - \frac{a_{n,n-1}^{(n-2)}}{a_{n-1,n-1}^{(n-2)}} \times b_{n-1}^{(n-2)}$$

The triangular system yields the required solution by *back substitution*. Proceeding from the last equation upwards, one obtains

$$x_n = b_n^{(n-1)} / a_{nn}^{(n-1)}, \quad x_{n-1} = (b_{n-1}^{(n-1)} - a_{n-1,n}^{(n-1)} x_n) / a_{n-1,n-1}^{(n-1)} \qquad (3.9a)$$

and in general

$$x_i = \left(b_i^{(n-1)} - \sum_{j=i+1}^{n} a_{ij}^{(n-1)} x_j \right) / a_{ii}^{(n-1)}, \quad (i = n - 1, n - 2, \cdots, 1) \qquad (3.9b)$$

Carl Friedrich Gauss (1777–1855), German mathematician. He is famous for both the breadth and depth of his works in every area of mathematics and is regarded as the most influential mathematician of the nineteenth century. He also made important contributions to geodesy, astronomy and physics.

The procedure would work nicely if none of the *pivots* a_{11}, $a_{22}^{(1)}$, $a_{33}^{(2)} \cdots$, $a_{nn}^{(n-1)}$ is zero. The difficulty can be circumvented in the following way. In the beginning of the kth step ($k = 1, 2, \cdots, n - 1$), the residual system is

$$
\begin{aligned}
a_{kk}^{(k-1)} x_k \quad &+a_{k,k+1}^{(k-1)} x_{k+1} \quad + \cdots +a_{kn}^{(k-1)} x_n = b_k^{(k-1)} \\
a_{k+1,k}^{(k-1)} x_k \quad &+a_{k+1,k+1}^{(k-1)} x_{k+1} \quad + \cdots +a_{k+1,n} x_n = b_{k+1}^{(k-1)} \\
&\cdots \\
a_{nk}^{(k-1)} x_k \quad &+a_{n,k+1}^{(k-1)} \quad + \cdots \quad a_{nn}^{(k-1)} x_n = b_n^{k-1)}
\end{aligned}
$$

Instead of straight away selecting $a_{kk}^{(k-1)}$ as the pivot for elimination, one can select the *largest in magnitude* among the coefficients of x_k, viz. $a_{kk}^{(k-1)}$, $a_{k+1,k}^{(k-1)}, \cdots, a_{nk}^{(k-1)}$ as the pivot. In that case the entire *pivotal equation* has to be moved to the first position of the residual system, in exchange of the first equation of the system. This procedure is called **partial pivoting**. Theoretically the procedure appears infallible if the system has a solution.

Example 1. Solve the system of equations

$$
\begin{aligned}
2x_1 -7x_2 +4x_3 &= 9 \\
x_1 +9x_2 -6x_3 &= 1 \\
-3x_1 +8x_2 +5x_3 &= 6
\end{aligned}
$$

by Gaussian elimination, using partial pivoting.

Solution. The elimination process can be conveniently carried out by considering the *augmented matrix*

$$
\begin{bmatrix}
2 & -7 & 4 & 9 \\
1 & 9 & -6 & 1 \\
\boxed{-3} & 8 & 5 & 6
\end{bmatrix}
$$

Let R_1, R_2, R_3 denote row 1, row 2 and row 3, respectively. In the first column (the elements of the coefficient of x_1), the maximum element in magnitude is -3. So we interchange R_3 with R_1 and obtain

$$
\begin{bmatrix}
\boxed{-3} & 8 & 5 & 6 \\
1 & 9 & -6 & 1 \\
2 & -7 & 4 & 9
\end{bmatrix}
$$

(Interchange or **permutation** of rows is permissible in the sense that the position of the corresponding equations is interchanged). For elimination, *normalising the pivot* (which is equivalent to dividing the equation by -3), we obtain

$$\begin{bmatrix} 1 & -8/3 & -5/3 & -2 \\ 1 & 9 & -6 & 1 \\ 2 & -7 & 4 & 9 \end{bmatrix}$$

The row operations $R_2 - R_1$ and $R_2 - 2 \times R_1$, reduce the above to

$$\begin{bmatrix} 1 & -8/3 & -5/3 & -2 \\ 0 & \boxed{35/3} & -13/3 & 3 \\ 0 & -5/3 & 22/3 & 13 \end{bmatrix}$$

In the second column leaving out R_1, the maximum element in magnitude is 35/3, which lies on the second row and can be taken as the second pivot. Normalising it we get

$$\begin{bmatrix} 1 & -8/3 & -5/3 & -2 \\ 0 & 1 & -13/35 & 9/35 \\ 0 & -5/3 & 22/3 & 13 \end{bmatrix}$$

Performing the row operation $R_3 + \frac{5}{3} \times R_2$, we have

$$\begin{bmatrix} 1 & -8/3 & -5/3 & -2 \\ 0 & 1 & -13/35 & 9/35 \\ 0 & 0 & \boxed{141/21} & 94/7 \end{bmatrix}$$

In the third column, leaving out R_1 and R_2, the third pivot is automatically 141/21. Normalising it, the system is equivalent to

$$x_1 - \frac{8}{3}x_2 - \frac{5}{3}x_3 = -2$$

$$x_2 - \frac{13}{35}x_3 = \frac{9}{35}$$

$$x_3 = 2$$

By back substitution, we obtain

$$x_3 = 2$$

$$x_2 = \frac{9}{35} + \frac{13}{35} \times 2 = 1$$

$$x_3 = -2 + \frac{8}{3} \times 1 - \frac{5}{3} \times 2 = 4$$

Hence the solution is $x_1 = 4$, $x_2 = 1$, $x_3 = 2$. □

A computational complication can be visualised in which a candidate pivotal equation is such that the entire equation is multiplied while the pivot is actually small. A small pivot sneaking in this way may lead to loss of accuracy in elimination and back substitution. Examples have been constructed in the literature to show that the resultant solution is totally inaccurate. A remedy is to scale the coefficients of each equation of the residual system before a comparison is made for selecting the pivot. The scale of the coefficients of an equation may be defined as

$$S_i := \max_{1 \le j \le n} |a_{ij}| \quad i = 1, 2, \cdots, n$$

In *scaled partial pivoting*, the coefficients a_{ik} $(i = k, k + 1, \cdots, n)$ of x_k are measured relative to the scale S_i, i.e. a_{ik} is divided by S_i. The largest among them, in magnitude, is now selected as the kth pivot. Evidently, a small element cannot now sneak in as a pivot. The accuracy gained this way, however, is not known in general terms.

Adopting the above-mentioned procedure, a subroutine named GAUSS is given below. We note that in programming the procedure, the superscripts appearing in the theoretical discussion are superfluous on account of using assignment statements in succeeding steps.

```
SUBROUTINE GAUSS(n,a,b)
! n=number of unknowns and equations.    (Input)
! a=n×n matrix of the coefficients.       (Input)
! b=right hand side vector.               (Input)
! The solution is returned in vector b.   (Output)
! s(i)=size of the row i of matrix a.
REAL :: a(n,n), b(n), s(n)
INTEGER :: p, q
! Initialise s
DO i=1,n
s(i)=0.0
DO j=1,n
absaij=ABS(a(i,j)); IF(absaij>s(i)) s(i)=absaij
END DO; END DO
! Elimination begins
DO k=1,n−1
   ! Determine pivot row p
   absaks=abs(a(k,k))/s(k); p=k
   DO i=k+1,n
   absais=ABS(a(i,k))/s(i)
   IF(absais<=absaks) EXIT
   absaks=absais; p=i
   END DO
```

```
IF(absaks==0.0) THEN
PRINT*, 'No Solution'; RETURN
END IF
IF(p/=k) THEN        ! Interchange row k and row p.
DO q=k,n
temp=a(k,q); a(k,q)=a(p,q); a(p,q)=temp
END DO
temp=b(k); b(k)=b(p); b(p)=temp
END IF
! Eliminate x(k)
DO i=k+1,n
ratio=a(i,k)/a(k,k); b(i)=b(i)−ratio*b(k)
DO j=k+1,n
a(i,j)=a(i,j)−ratio*a(k,j)
END DO; END DO
END DO
IF(a(n,n)==0.0) THEN
PRINT*, 'No Solution'; RETURN
END IF
Back Substitution begins
b(n)=b(n)/a(n,n)
DO i=n−1,1,−1
sum=0.0
DO j=i+1,n
sum=sum+a(i,j)*b(j)
END DO
b(i)=(b(i)−sum)/a(i,i)
END DO
RETURN
END SUBROUTINE GAUSS
```

The number of floating point operations in the elimination part is

$$\sum_{k=1}^{n-1}\left[1+\sum_{i=k+1}^{n}\left(1+3+\sum_{j=k+1}^{n}2\right)\right]=\frac{1}{3}(2n+3)(n^2-1)$$

In back substitution it is

$$1+\sum_{i=1}^{n-1}\left(2+\sum_{j=i+1}^{n}2\right)=n^2+n-1$$

Hence in total, the complexity of the Gauss method is $O(\frac{2}{3}n^3)$ emanating solely from the elimination process. The order is sharply lower than that of Cramer's rule.

Exercises

1. Show by Gauss elimination with partial pivoting that the linear system:

(i) $2x_1 + x_2 + x_3 = 10$ reduces to $x_1 + \frac{2}{3}x_2 + x_3 = 6$

$\quad 3x_1 + 2x_2 + 3x_3 = 18$ $\qquad\qquad\qquad x_2 + \frac{12}{5}x_3 = 2$

$\quad x_1 + 4x_2 + 9x_3 = 16$ $\qquad\qquad\qquad x_3 = 5$

(ii) $3x_1 + 2x_2 + 4x_3 = 7$ reduces to $x_1 + \frac{2}{3}x_2 + \frac{4}{3}x_3 = \frac{7}{3}$

$\quad 2x_1 + x_2 + x_3 = 4$ $\qquad\qquad\qquad x_2 + \frac{11}{7}x_3 = -\frac{1}{7}$

$\quad x_1 + 3x_2 + 5x_3 = 2$ $\qquad\qquad\qquad x_3 = \frac{5}{8}$

$\quad 2x_1 + 2x_2 + x_3 + 2x_4 = 7$ reduces to $x_1 - \frac{1}{3}x_2 - \frac{2}{3}x_3 - \frac{1}{3}x_4 = 1$

(iii) $\begin{aligned} x_1 - 2x_2 \quad\quad - x_4 &= 2 \\ 3x_1 - x_2 - 2x_3 - x_4 &= 3 \\ x_1 \quad\quad\quad -2x_4 &= 0 \end{aligned}$ $\qquad \begin{aligned} x_2 + \frac{7}{8}x_3 + x_4 &= \frac{15}{8} \\ x_3 + \frac{8}{17}x_4 &= \frac{33}{17} \\ x_4 &= \frac{40}{37} \end{aligned}$

Hence solve the three sets of equations.

[(i) $x_1 = 7$, $x_2 = -9$, $x_3 = 5$ (ii) $x_1 = 9/4$, $x_2 = -9/8$, $x_3 = 5/8$

(iii) $x_1 = 80/37$, $x_2 = -17/37$, $x_3 = 53/37$, $x_4 = 40/37$].

2. Solve the following linear systems by using subroutine GAUSS:

(i) $\begin{aligned} x_1 + x_2 + 2x_3 &= 4 \\ 3x_1 + x_2 - 3x_3 &= -4 \\ 2x_1 - 3x_2 - 5x_3 &= -5 \end{aligned}$ $\qquad [x_1 = 1,\ x_2 = -1,\ x_3 = 2].$

(ii) $\begin{aligned} 5x_1 - x_2 + x_3 &= 10 \\ x_1 + 2x_2 - 3x_3 &= 6 \\ x_1 + x_2 + 5x_3 &= -1 \end{aligned}$ $\qquad [x_1 = 2.29167,\ x_2 = 0.66667,\ x_3 = -0.79167].$

(iii) $\begin{aligned} 4.11x_1 - 9.68x_2 + 2.01x_3 &= 4.94 \\ 1.87x_1 - 4.62x_2 + 5.50x_3 &= 3.10 \\ 1.10x_1 - 0.96x_2 + 2.71x_3 &= 4.02 \end{aligned}$ $\qquad \begin{aligned} [x_1 &= 4.19851, \\ x_2 &= 1.32383, \\ x_3 &= 0.24816]. \end{aligned}$

(iv) $\begin{bmatrix} 1 & 2 & 3 & 4 \\ 7 & 10 & 5 & 2 \\ 13 & 6 & 2 & -3 \\ 11 & 14 & 8 & -1 \end{bmatrix} \begin{bmatrix} x_1 \\ x_2 \\ x_3 \\ x_4 \end{bmatrix} = \begin{bmatrix} 10 \\ 40 \\ 34 \\ 64 \end{bmatrix}$ $\qquad [x_1 = 1,\ x_2 = 2,\ x_3 = 3,\ x_4 = -1].$

(v) $\begin{bmatrix} 1 & 2 & -12 & 8 \\ 5 & 4 & 7 & -2 \\ -3 & 7 & 9 & 5 \\ 6 & -12 & -8 & 3 \end{bmatrix} \begin{bmatrix} x_1 \\ x_2 \\ x_3 \\ x_4 \end{bmatrix} = \begin{bmatrix} 27 \\ 4 \\ 11 \\ 39 \end{bmatrix}$ $\qquad \begin{aligned} [x_1 &= 2.63778,\ x_2 = -1.28527, \\ x_3 &= 0.67135,\ x_4 = 4.37362]. \end{aligned}$

3. An electrical resistor network leads to the following equations following Kirchoff's Laws:

$$R_1 I_1 + R_2(I_1 - I_3) + R_3(I_1 - I_2) = 0$$
$$R_4 I_2 + R_3(I_2 - I_1) + R_5(I_2 - I_3) = v_1$$
$$R_6 I_3 + R_5(I_3 - I_2) + R_2(I_3 - I_1) = v_2$$

If the resistances $R_1, \cdots, R_6$ are given by $R_1 = R_2 = R_3 = 2$, $R_4 = R_5 = R_6 = 3$ and the voltages by $v_1 = v_2 = 5$; find the currents I_1, I_2, I_3.
[$I_1 = 0.90909$, $I_2 = I_3 = 1.36364$].

4. The tensions T_1, $T_2, \cdots$, T_5 in a seven member *symmetric* truss are given by the equations

$$\begin{bmatrix} \cos\theta & 1 & 0 & 0 & 0 \\ 0 & \sin\theta & 0 & 0 & 1 \\ 0 & 0 & 2\sin\theta & 0 & 0 \\ 0 & -\cos\theta & \cos\theta & 1 & 0 \\ 0 & \sin\theta & \sin\theta & 0 & 0 \end{bmatrix} \begin{bmatrix} T_1 \\ T_2 \\ T_3 \\ T_4 \\ T_5 \end{bmatrix} = \begin{bmatrix} 0 \\ 0 \\ 1 \\ 0 \\ 0 \end{bmatrix}$$

Determine $T_1, \cdots, T_5$ for $\theta = 53^o$.
[$T_1 = 1.04030$, $T_2 = -0.62607 = -T_3$, $T_4 = -0.75355$, $T_5 = 0.50000$].

5. Modify subroutine GAUSS for handling complex matrices. Apply it to find the voltage v at nodes of an electrical Chebyshev filter governed by the equation $A\mathbf{v} = \mathbf{E}_R$, where A is 5×5 tridiagonal matrix $[a_{ij}]$ with elements
$a_{11} = 1/R_1 + 1/(iL_1\omega)$, $a_{12} = a_{21} = -1/(iL_1\omega)$, $a_{22} = 1/(iL_1\omega) + iC_1\omega$,
$a_{23} = a_{32} = -iC_1\omega$, $a_{33} = 1/(iL_2\omega) + i\omega(C_1 + C_2 + C_3)$, $a_{34} = a_{43} = -iC_3\omega$,
$a_{44} = 1/(iL_3\omega) + iC_3\omega$, $a_{45} = a_{54} = -1/(iL_3\omega)$, $a_{55} = 1/(iL_3\omega) + 1/R_5$ (the remaining elements being zero), and

$\mathbf{v} = [v_1, \cdots, v_5]^T$, $\mathbf{E}_R = [E/R_1, 0, 0, 0, 0]^T$, $\omega = 2\pi f$, $f = $ frequency.

Solve the system for $f = 500$, given

$$R_1 = R_5 = 50, \ L_1 = L_3 = 0.127 \times 10^{-6}, \ L_2 = 0.829 \times 10^{-8},$$

$$C_1 = 0.22 \times 10^{-11}, \ C_2 = 0.349 \times 10^{-10}, \ C_3 = 0.228 \times 10^{-11}, \ E = 1$$

[$v_1 = 0$, $v_2 = 1 - 7.980 \times 10^{-6}i$, $v_3 = -1.800 \times 10^{-3} + 1.436 \times 10^{-18}i$,
$v_4 = 7.400 \times 10^{-25} - 9.274 \times 10^{-20}i$, $v_5 = -7.400 \times 10^{-25} - 9.274 \times 10^{-20}i$].

6. (Jordan Elimination). In this method [due to **Wilhelm Jordan (1842–1899)**], after the lower triangular elements are annihilated, as in Gauss elimination, the upper triangular elements are also annihilated, leaving a simple diagonal matrix. The complexity is about 50% higher than in Gauss method. Solve by the method, the system

$$\begin{aligned} 2x_1 + 4x_2 - 6x_3 &= -8 \\ x_1 + 3x_2 + x_3 &= 10 \\ 2x_1 - 4x_2 - 2x_3 &= -12 \end{aligned} \qquad [x_1 = 1, \ x_2 = 2, \ x_3 = 3].$$

3.1.3 LU Decomposition Method

In some applications, it is sometimes necessary to solve the linear system (3.1) for the same matrix A but with different right-hand side vectors. For instance, the computation of the inverse A^{-1} requires such a procedure (see Sect. 3.1.4). In the Gaussian method, the elimination process finally converts A into the *upper triangular* matrix

$$
U =: \begin{bmatrix}
a_{11}^{(n-1)} & a_{12}^{(n-1)} & \cdots & a_{1n}^{(n-1)} \\
 & a_{22}^{(n-1)} & \cdots & a_{2n}^{(n-1)} \\
 & & \cdots & \cdots \\
 & & & a_{nn}^{(n-1)}
\end{bmatrix} \tag{3.10}
$$

The triangularisation procedure may be viewed as premultiplication of A by a *lower triangular* matrix. For the first step of Gaussian elimination, it can be written in matrix notations as

$$
\begin{bmatrix}
1 & 0 & 0 & \cdots & 0 \\
-\dfrac{a_{21}}{a_{11}} & 1 & 0 & \cdots & 0 \\
-\dfrac{a_{31}}{a_{11}} & 0 & 1 & \cdots & 0 \\
\cdots & & & & \\
-\dfrac{a_{n1}}{a_{11}} & 0 & 0 & \cdots & 1
\end{bmatrix}
\times
\begin{bmatrix}
a_{11} & a_{12} & a_{13} & \cdots & a_{1n} \\
a_{21} & a_{22} & a_{23} & \cdots & a_{2n} \\
a_{31} & a_{32} & a_{33} & \cdots & a3n \\
\cdots & & & & \\
a_{n1} & a_{n2} & a_{n3} & \cdots & a_{nn}
\end{bmatrix}
$$

$$
=
\begin{bmatrix}
a_{11}^{(1)} & a_{12}^{(1)} & a_{13}^{(1)} & \cdots & a_{1n}^{(1)} \\
0 & a_{22}^{(1)} & a_{23}^{(1)} & \cdots & a_{2n}^{(1)} \\
0 & a_{32}^{(1)} & a_{33}^{(1)} & \cdots & a_{3n}^{(1)} \\
\cdots & & & & \\
0 & a_{n2}^{(1)} & a_{n3}^{(1)} & \cdots & a_{nn}^{(1)}
\end{bmatrix}
$$

where $a_{ij}^{(1)}$ on the right-hand side are identical to those defined in the preceding Sect. 3.1.2. By inversion, the above equation yields

$$
A =
\begin{bmatrix}
1 & 0 & \cdots & 0 \\
a_{21}/a_{11} & 1 & \cdots & 0 \\
a_{31}/a_{11} & 0 & \cdots & 0 \\
\cdots & & & \\
a_{n1}/a_{11} & 0 & \cdots & 1
\end{bmatrix}
\times
\begin{bmatrix}
a_{11}^{(1)} & a_{12}^{(1)} & \cdots & a_{1n}^{(1)} \\
0 & a_{22}^{(1)} & \cdots & a_{2n}^{(1)} \\
0 & a_{32}^{(1)} & \cdots & a_{3n}^{(1)} \\
\cdots & & & \\
0 & a_{n2}^{(1)} & \cdots & a_{nn}^{(1)}
\end{bmatrix}
$$

Similarly, it may be verified that the second step of elimination yields the factorisation

$$
A =
\begin{bmatrix}
1 & 0 & \cdots & 0 \\
a_{21}/a_{11} & 1 & \cdots & 0 \\
a_{31}/a_{11} & a_{32}^{(1)}/a_{22}^{(1)} & \cdots & 0 \\
\cdots & & & \\
a_{n1}/a_{11} & a_{n2}^{(1)}/a_{22}^{(1)} & \cdots & 1
\end{bmatrix}
\times
\begin{bmatrix}
a_{11}^{(2)} & a_{12}^{(2)} & a_{13}^{(2)} & \cdots & a_{1n}^{(2)} \\
0 & a_{22}^{(2)} & a_{23}^{(2)} & \cdots & a_{2n}^{(2)} \\
0 & 0 & a_{33}^{(2)} & \cdots & a_{3n}^{(2)} \\
\cdots & & & & \\
0 & 0 & a_{n3}^{(2)} & \cdots & a_{nn}^{(2)}
\end{bmatrix}
$$

Proceeding in this way, we get in the final $(n-1)$th step

$$
A = LU \tag{3.11}
$$

where

$$L = \begin{bmatrix} 1 & & & & \\ a_{21}/a_{11} & 1 & & & \\ a_{31}/a_{11} & a_{32}^{(1)}/a_{22}^{(1)} & 1 & & \\ \cdots & \cdots & \cdots & \cdots & \cdots \\ a_{n1}/a_{11} & a_{n2}^{(1)}/a_{22}^{(1)} & \cdots & a_{n,n-1}^{(n-2)}/a_{n-1,n-1}^{(n-2)} & 1 \end{bmatrix} \tag{3.12}$$

where U is defined as in Eq. (3.10). Obviously L is lower triangular with unit diagonal. Comparing with Eq. (3.8), it is interesting to note that the elements of L are simply the multipliers of the pivotal equations in the Gaussian method. The elements of U are as before, generated by the elimination process.

For solving a linear system like Eq. (3.1) for a given right- hand side vector $\mathbf{b}$, we have $LU\mathbf{x} = \mathbf{b}$ which splits into two systems

$$L\mathbf{y} = \mathbf{b} \quad \text{where} \quad U\mathbf{x} = \mathbf{y} \tag{3.13}$$

The solution of the first system by *forward substitution* is

$$y_1 = b_1, \quad y_2 = b_2 - \frac{a_{21}}{a_{11}} \times y_1 \tag{3.14a}$$

and in general

$$y_i = b_i - \sum_{j=1}^{i-1} \frac{a_{ij}^{(j-1)}}{a_{jj}^{(j-1)}} \times y_j \tag{3.14b}$$

Having computed $\mathbf{y}$, $\mathbf{x}$ can be computed from the second of the Eq. (3.13) by back substitution, taking $\mathbf{y}$ as the right- hand side vector.

The method is essentially due to T. Banachiewicz.

Tadeusz Banachiewwicz (1882–1994), Polish astronomer and mathematician. He is known for his modified method of determining parabolic orbits and inventing a theory of 'Cracovians' (a special kind of matrix algebra). After working for several years in Germany and Russia, he became the director of Cracow Observatory in Poland.

The entire procedure is divided into two subroutines named BANACHIEWICZ and SOLVE. In the first subroutine, A is decomposed into L and U, in which the elements of U are generated by Gaussian elimination and the pivotal multipliers form the elements of L. While programming, the elements of U are stored in the upper triangle of A and those of L in the lower triangle, with the tacit understanding that the diagonal of L is unity. This type of allocation is possible because in Gaussian elimination, the elements of the lower triangle of A (except the diagonal) are vacated. In the subroutine BANACHIEWICZ, we also introduce scaled partial pivoting. In general, there may be a permutation (interchange) of the rows of A by the process. A track of the number of the pivotal row is kept by storing it in a vector named ipivot.

This vector helps in keeping track of the particular element of **b** in the forward substitution, Eq. (3.14a, 3.14b).

The second subroutine named SOLVE follows Eq. (3.14a, 3.14b) and (3.9a, 3.9b) for forward and back substitutions in order to obtain the solution **x**. It uses subroutine BANACHIEWICZ and the vector ipivot.

The subroutine BANACHIEWICZ also computes $det(A)$ as an output. Here, $det(A) = det(LU) = det(L) \times det(U) = 1 \times a_{11}^{(n-1)} \times a_{22}^{(n-1)} \times \cdots \times a_{nn}^{(n-1)}$. Since partial pivoting permutes the rows, a variable iflag is introduced, whose value is $+1$ or -1 according to even or odd permutations undergone. With these observations the subroutines follow, drawing heavily from the subroutine GAUSS.

```
SUBROUTINE BANACHIEWICZ(n,a,ipivot,det)
! n=Order of the matrix.     (Input)
! a=square matrix of order n, to be decomposed as LU.     (Input)
! L and U are returned through matrix a,
! with the unit diagonal suppressed.     (Output)
! ipivot(k)= The row number used in the k–th step
! of factorisation.     (Output)
! det=determinant of matrix a.     (Output)
! ************************************************
REAL :: a(n,n), s(n)
INTEGER :: ipivot(n), p, q
! Initialise iflag, ipivot and scale s
DO i=1,n
ipivot(i)=i; s(i)=0.0
DO j=1,n
absaij=ABS(a(i,j))
IF(absaij>s(i)) s(i)=absaij
END DO; END DO
iflag=1
! Factorisation Begins
DO k=1,n-1
   ! Determine pivot row p
   absaks=ABS(a(k,k))/s(k); p=k
   DO i=k+1,n
   absais=ABS(a(i,k))/s(i)
   IF(absais<absaks) EXIT
   absaks=absais; p=i
   END DO
   IF(absaks==0.0) THEN
   PRINT*, 'No Solution'; RETURN
   END IF
   IF(p>k) THEN     ! Interchange row k with row p.
   DO q=1,n
   temp=a(k,q); a(k,q)=a(p,q); a(p,q)=temp
```

```
    END DO
    itemp=ipivot(k); ipivot(k)=ipivot(p); ipivot(p)=itemp
    temp=s(k); s(k)=s(p); s(p)=temp; iflag=−iflag
    END IF
    DO i=k+1,n
    a(i,k)=a(i,k)/a(k,k); ratio=a(i,k)
    DO j=k+1,n
    a(i,j)=a(i,j)−ratio*a(k,j)
    END DO; END DO
END DO
IF(a(n,n)==0.0) PRINT*, 'No Solution'
det=iflag
DO i=1,n
det=det*a(i,i)
END DO
RETURN
END SUBROUTINE BANACHIEWICZ

SUBROUTINE SOLVE(n,a,b,x,ipivot)
! Solves the linear system of equations Ax=b.
! a=square matrix of order n, as factored by
! subroutine BANACHIEWICZ.    (Input)
! b=right hand side vector.    (Input)
! x=solution vector.    (Output)
! ipivot(k)=Row number which was used in the kth step of
! decomposition in BANACHIEWICZ.    (Input)
! ***********************************************
REAL :: a(n,n), b(n), x(n), s(n)
INTEGER :: ipivot(n)
! Solution of Ly=b, y overwrites x
ip=ipivot(1); x(1)=b(ip)
DO i=2,n
sum=0.0
DO j=1,i−1
sum=sum+a(i,j)*x(j)
END DO
ip=ipivot(i); x(i)=b(ip)−sum
END DO
! Solution of Ux=y
x(n)=x(n)/a(n,n)
DO i=n−1,1,−1
sum=0.0
DO j=i+1,n
sum=sum+a(i,j)*x(j)
END DO
```

```
x(i)=(x(i)-sum)/a(i,i)
END DO
RETURN
END SUBROUTINE SOLVE
```

From a general stand point, L and U can be considered as

$$L =: \begin{bmatrix} l_{11} & 0 & \cdots & 0 \\ l_{21} & l_{22} & \cdots & 0 \\ \cdots & \cdots & \cdots & \cdots \\ l_{n1} & l_{n2} & \cdots & l_{nn} \end{bmatrix} \quad \text{and} \quad U =: \begin{bmatrix} u_{11} & u_{12} & \cdots & u_{1n} \\ 0 & u_{22} & \cdots & u_{2n} \\ \cdots & \cdots & \cdots & \cdots \\ 0 & 0 & \cdots & u_{nn} \end{bmatrix}$$

so that $l_{ij} = 0$, $(i < j)$ and $u_{ij} = 0$, $(i > j)$. The number of non-zero elements of L and U is thus $2 \times (1 + 2 + \cdots + n) = n(n + 1)$. They can be determined in terms of the n^2 elements of A if n other conditions are available. There are two simple choices due to Banachiewicz and Crout.

1° Banaciewicz's Method. Assume $l_{11} = l_{22} = \cdots = l_{nn} = 1$, then

$$a_{ij} = \sum_{k=1}^{n} l_{ik} u_{kj} = \sum_{k=1}^{i} l_{ik} u_{kj} = u_{ij} + \sum_{k=1}^{i-1} l_{ik} u_{kj}$$

$$= \sum_{k=1}^{j} l_{ik} u_{kj} = l_{1j} u_{jj} + \sum_{k=1}^{j-1} l_{ik} u_{kj}$$

Thus u_{ij}, l_{ij} are given recursively as

$$u_{1j} = a_{1j}, \quad u_{ij} = a_{ij} - \sum_{k=1}^{i-1} l_{ik} u_{kj}, \quad (1 < i \le j)$$

$$l_{i1} = \frac{a_{i1}}{u_{11}}, \quad l_{ij} = \frac{1}{u_{jj}} \left(a_{ij} - \sum_{k=1}^{j-1} l_{ik} u_{kj} \right), \quad (i > j > 1) \tag{3.15a}$$

The subroutine BANACHIEWICZ essentially determines these elements recursively.

2° Crout's Method. Assume $u_{11} = u_{22} = \cdots = u_{nn} = 1$. As in the preceding method, it can be proved that the elements of L and U are given by

$$l_{i1} = a_{i1}, \quad l_{ij} = a_{ij} - \sum_{k=1}^{j-1} l_{ik} u_{kj}, \quad (i \ge j > 1)$$

$$u_{1j} = \frac{a_{1j}}{l_{11}}, \quad u_{ij} = \frac{1}{l_{ii}} \left(a_{ij} - \sum_{k=1}^{i-1} l_{ik} u_{kj} \right), \quad (1 < i < j) \tag{3.15b}$$

The above two methods are essentially the same. Suppose by Crout's method $A = LU$, then one can write $L = L_1 D$, where L_1 is lower triangular with unit diagonal and D is diagonal matrix with elements $l_{11}, l_{22}, \cdots, l_{nn}$ (and zero nondiagonal elements). Thus $A = L_1 DU = L_1(DU)$ and is of Banachiewicz's form.

Banachiewicz's method is sometimes known as asymmetric Cholesky's method. It is also sometimes called Dolittle method.

Example 1. Decompose the matrix

$$A := \begin{bmatrix} 4 & 1 & 2 \\ 2 & -3 & 8 \\ -1 & 11 & 4 \end{bmatrix}$$

in Banachiewicz's form. Hence solve the system $A\mathbf{x} = [4, -12, 17]^T$.

Solution. In this method we have ab initio

$$\begin{bmatrix} 1 & 0 & 0 \\ l_{21} & 1 & 0 \\ l_{21} & l_{22} & 1 \end{bmatrix} \begin{bmatrix} u_{11} & u_{12} & u_{13} \\ 0 & u_{22} & u_{23} \\ 0 & 0 & u_{33} \end{bmatrix} = \begin{bmatrix} 4 & 1 & 2 \\ 2 & -3 & 8 \\ -1 & 11 & 4 \end{bmatrix}$$

Equating the elements, for $i = 1$,

$$u_{11} = 4, \quad u_{12} = 1, \quad u_{13} = 2$$

For $i = 2$, $l_{21}u_{11} = 2$, $l_{21}u_{12} + u_{22} = -3$, $l_{21}u_{13} + u_{23} = 8$
Therefore

$$l_{21} = \frac{2}{4} = \frac{1}{2}, \quad u_{22} = -3 - \frac{1}{2} \times 1 = -\frac{7}{2}, \quad u_{23} = 8 - \frac{1}{2} \times 2 = 7$$

For $i = 3$, $l_{31}u_{11} = -1$, $l_{31}u_{12} + l_{32}u_{22} = 11$, $l_{31}u_{13} + l_{32}u_{23} + u_{33} = 4$
and so

$$l_{31} = -\frac{1}{4}, \quad l_{32} = \frac{11 - (-1/4) \times 1}{-7/2} = -\frac{45}{14}, \quad u_{33} = 4 - \left(-\frac{1}{4}\right) \times 2 - \left(-\frac{45}{14}\right) \times 7 = 27$$

Thus

$$L = \begin{bmatrix} 1 & 0 & 0 \\ 1/2 & 1 & 0 \\ -1/4 & -45/14 & 1 \end{bmatrix}, \quad U = \begin{bmatrix} 4 & 1 & 2 \\ 0 & -7/2 & 7 \\ 0 & 0 & 27 \end{bmatrix}$$

Next, in order to solve the given system, we set $U\mathbf{x} = \mathbf{y}$, so that $L\mathbf{y} = [4, -12, 17]^T$
or,

$$\begin{array}{rcl} y_1 & = & 4 \\ \frac{1}{2}y_1 + y_2 & = & -12 \\ -\frac{1}{4}y_1 - \frac{45}{14}y_2 + y_3 & = & 17 \end{array}$$

whose solution is $y_1 = 4$, $y_2 = -12 - \frac{1}{2} \times 4 = -14$, $y_3 = 17 + \frac{1}{4} \times 4 + \frac{45}{14} \times (-14) = -27$. Therefore $\mathbf{x}$ is given by

$$
\begin{aligned}
4x_1 + x_2 + 2x_3 &= 4 \\
-\tfrac{7}{2}x_2 + 7x_3 &= -14 \\
27x_3 &= -27
\end{aligned}
$$

Thus $x_3 = -1$, $x_2 = -\frac{2}{7}[-14 - 7 \times (-1)] = 2$, $x_1 = \frac{1}{4}[4 - 2 - 2 \times (-1)] = 1$.

$\square$

Exercises

1. Prove that the matrix $\begin{bmatrix} 0 & 1 \\ 1 & 0 \end{bmatrix}$ cannot be LU decomposed, but decomposition is possible after a permutation.

2. Prove that the sum and the product of two lower/upper triangular matrices is lower/upper triangular and the inverse of a lower/upper triangular matrix is lower/upper triangular.

3. Decompose the following matrices in Banachiewicz's form:

$(i)\ A_1 := \begin{bmatrix} 3 & 2 & 1 \\ 2 & 3 & 2 \\ 1 & 2 & 3 \end{bmatrix}$ $(ii)\ A_2 := \begin{bmatrix} 8 & -3 & 2 \\ 4 & 11 & -1 \\ 6 & 3 & 12 \end{bmatrix}$

Hence, compute $det(A_1)$, $det(A_2)$ and solve the linear systems $A_1\mathbf{x} = [8,\ 10,\ 8]^T$ and $A_2\mathbf{x} = [20,\ 33,\ 36]^T$.

$[(i)\ L_1 := \begin{bmatrix} 1 & 0 & 0 \\ 2/3 & 1 & 0 \\ 1/3 & 4/5 & 1 \end{bmatrix}$ $U_1 := \begin{bmatrix} 3 & 2 & 1 \\ 0 & 5/3 & 4/3 \\ 0 & 0 & 8/5 \end{bmatrix}$ $det(A_1) = 8$

$(ii)\ L_2 := \begin{bmatrix} 1 & 0 & 0 \\ 1/2 & 1 & 0 \\ 3/4 & 21/50 & 1 \end{bmatrix}$ $U_2 := \begin{bmatrix} 8 & -3 & 2 \\ 0 & 25/2 & -2 \\ 0 & 0 & 567/50 \end{bmatrix}$ $det(A_2) = 1134$

$(i)\ x_1 = 1$, $x_2 = 2$, $x_3 = 1$ $(ii)\ x_1 = 1$, $x_2 = 3$, $x_3 = 5]$.

4. Decompose the following matrices in Crout's form:

$(i)\ A_1 := \begin{bmatrix} 2 & -6 & 8 \\ 5 & 4 & -3 \\ 3 & 1 & 2 \end{bmatrix}$ $(ii)\ A_2 := \begin{bmatrix} 3 & -1 & 2 \\ 1 & 2 & 3 \\ 2 & -2 & -1 \end{bmatrix}$

and solve the system $A_1\mathbf{x} = [24,\ 2,\ 16]^T$, $A_2\mathbf{x} = [12,\ 11,\ 2]^T$.

$[(i)L_1 := \begin{bmatrix} 2 & 0 & 0 \\ 5 & 19 & 0 \\ 3 & 10 & 40/19 \end{bmatrix}$ $U_1 := \begin{bmatrix} 1 & -3 & 4 \\ 0 & 1 & -23/19 \\ 0 & 0 & 1 \end{bmatrix}$

$(ii)L_2 := \begin{bmatrix} 3 & 0 & 0 \\ 1 & 7/3 & 0 \\ 2 & -4/3 & -1 \end{bmatrix}$ $U_2 := \begin{bmatrix} 1 & -1/3 & 2/3 \\ 0 & 1 & 1 \\ 0 & 0 & 1 \end{bmatrix}$

$(i)\ x_1 = 3$, $x_2 = 2$, $x_3 = 1$ $(ii)\ x_1 = 3$, $x_2 = 1$, $x_3 = 2]$.

5. Use subroutines BANACHIEWICZ and SOLVE to compute the roots of the equations:

$$(i) \quad \begin{bmatrix} 5 & 2 & 1 \\ -1 & 4 & 2 \\ 2 & -3 & 10 \end{bmatrix} \begin{bmatrix} x_1 \\ x_2 \\ x_3 \end{bmatrix} = \begin{bmatrix} -12 \\ 20 \\ 3 \end{bmatrix}$$

$$(ii) \quad \begin{bmatrix} 1 & 2 & 3 & 4 \\ 13 & 6 & 2 & -3 \\ 7 & 10 & 5 & 2 \\ 11 & 14 & 8 & -1 \end{bmatrix} \begin{bmatrix} x_1 \\ x_2 \\ x_3 \\ x_4 \end{bmatrix} = \begin{bmatrix} 10 \\ 34 \\ 40 \\ 64 \end{bmatrix}$$

$$(iii) \quad \begin{bmatrix} 3 & -2 & -5 & 1 \\ 2 & -3 & 1 & 5 \\ 1 & 2 & 0 & -4 \\ 1 & -1 & -3 & 9 \end{bmatrix} \begin{bmatrix} x_1 \\ x_2 \\ x_3 \\ x_4 \end{bmatrix} = \begin{bmatrix} -5 \\ 7 \\ -1 \\ -4 \end{bmatrix}$$

Also compute the determinants of the matrices on the left-hand side of the equations.

[(i) $x_1 = -4$, $x_2 = 3$, $x_3 = 2$; $det = 253$
(ii) $x_1 = 1$, $x_2 = 2$, $x_3 = 3$, $x_4 = -1$; $det = -896$
(iii) $x_1 = 1$, $x_2 = -1$, $x_3 = 2$, $x_4 = 0$; $det = -364$].

3.1.4 Matrix Inversion

The inverse of a given $n \times n$ square matrix A is defined to be the matrix X of the same dimension, such that

$$AX = XA = I \tag{3.16}$$

where I is the $n \times n$ unit matrix consisting of unit diagonal elements and zero off-diagonal elements. It is known that X exists only when $det(A) \neq 0$. It is customary to denote X as A^{-1}.

Computation of A^{-1} is sometimes required in direct physical application. For instance if A is a *resistance* or *impedance matrix* of an electrical network, then A^{-1} is the *conductance* or *admittance matrix* of the same network. Similarly if A is the *stiffness matrix* of a linked structure, then A^{-1} is the *compliance matrix* of the structure.

In principle, one may use A^{-1} to compute the solution of the linear system (1), noting that $\mathbf{x} = A^{-1}\mathbf{b}$. However, the procedure is advisable only when the solution is required for several values of the right-hand side vector $\mathbf{b}$.

To compute A^{-1} from Eq. (3.16), we regard X and I to be *partitioned* along the columns, that is to say, we write

$$AX_j = I_j, \quad (j = 1, 2, \cdots, n) \tag{3.16a}$$

For solving the above system, we use LU-decomposition method, factorising A into LU and the repeatedly solve Eq. (3.16a) for $j = 1, 2, \cdots, n$. In the following subroutine for A^{-1}, the elements of X_j are first stored in a one-dimensional array named **x** and later returned in the location for A.

SUBROUTINE MATINV(n,A)
! This subroutine must be accompanied by subroutines
! BANACHIEWICZ and SOLVE.
! n=Order of the matrix to be inverted.   (Input)
! A=Square matrix to be inverted.   (Input)
! The inverse is returned in A.   (Output)
REAL :: A(n,n), b(n), x(n∗n)
INTEGER :: ipivot(n)
CALLl BANACHIEWICZ(n,A,ipivot,det)
DO i=1,n
b(i)=0.0
END DO
jinv=1
DO j=1,n
b(j)=1.0
CALL SOLVE(n,A,b,x(jinv),ipivot)
b(j)=0.0
jinv=jinv+n
END DO
DO i=1,n
DO j=1,n
A(i,j)=x(n*(j−1)+i)
END DO; END DO
RETURN
END SUBROUTINE MATINV

Exercises

1. Invert the following matrices using subroutine MATINV:

(i) $\begin{bmatrix} 1 & 2 & 6 \\ 2 & 5 & 15 \\ 6 & 15 & 46 \end{bmatrix}$ (ii) $\begin{bmatrix} 8 & -1 & -3 \\ -5 & 1 & 2 \\ 10 & -1 & -4 \end{bmatrix}$

(iii) $\begin{bmatrix} 3 & 7 & 8 & 15 \\ 2 & 5 & 6 & 11 \\ 2 & 6 & 10 & 19 \\ 4 & 11 & 19 & 38 \end{bmatrix}$ (iv) $\begin{bmatrix} 0.32 & 0.52 & -0.42 & 0.23 \\ 0.44 & -0.25 & 0.36 & -0.51 \\ -1.06 & 0.74 & -0.83 & 0.48 \\ 0.96 & 0.82 & 0.55 & 0.38 \end{bmatrix}$

$[(i)$ $\begin{bmatrix} 5 & -2 & 0 \\ -2 & 10 & -3 \\ 0 & -3 & 1 \end{bmatrix}$ (ii) $\begin{bmatrix} 2 & 1 & -1 \\ 0 & 2 & 1 \\ 5 & 2 & -3 \end{bmatrix}$

$$(iii) \begin{bmatrix} 25 & -41 & 16 & -6 \\ 16 & 27 & -11 & 4 \\ 16 & -27 & 13 & -5 \\ -6 & 10 & -5 & 2 \end{bmatrix} \quad (iv) \begin{bmatrix} 1.20 & -0.32 & -0.82 & -0.12 \\ -0.17 & 1.61 & 1.25 & 0.69 \\ -1.75 & 0.16 & 0.32 & 0.87 \\ -0.12 & -2.91 & -1.09 & 0.17 \end{bmatrix}].$$

3.1.5 Cholesky's Method for Symmetric Matrices

Suppose we have a linear system $A\mathbf{x} = \mathbf{b}$, in which the matrix $A = [a_{ij}]$ is a **symmetric matrix**; then $a_{ij} = a_{ji}$, $(i, j = 1, 2, \cdots, n)$, $A^T = [a_{ji}] = A$ and A can be decomposed as

$$A = LL^T \tag{3.17}$$

where L is a *lower triangular matrix* of the type

$$L = \begin{bmatrix} l_{11} & 0 & \cdots & 0 \\ l_{21} & l_{22} & \cdots & 0 \\ \cdots & \cdots & \cdots & \cdots \\ l_{n1} & l_{n2} & \cdots & l_{nn} \end{bmatrix} \quad \text{and} \quad L^T = \begin{bmatrix} l_{11} & l_{12} & \cdots & l_{1n} \\ 0 & l_{22} & \cdots & l_{2n} \\ 0 & 0 & \cdots & l_{nn} \end{bmatrix} \tag{3.18}$$

where $l_{ij} = l_{ji}$ and $l_{ij} = 0$ for $i < j$ in L. Proceeding as in Banachiewicz's method, we have from Eqs. (3.17), (3.18)

$$a_{ij} = \sum_{k=1}^{n} l_{ik}l_{jk} = \sum_{k=1}^{j} l_{ik}l_{jk} = l_{ij}l_{jj} + \sum_{k=1}^{j-1} l_{ik}l_{jk}$$

where the property $l_{jk} = 0$ for $k > j$ is used. In particular for $i = j$,

$$a_{jj} = \sum_{k=1}^{j} l_{jk}^2 = l_{jj}^2 + \sum_{k=1}^{j-1} l_{jk}^2$$

Hence we find in succession

$$l_{11} = \sqrt{a_{11}}, \quad l_{i1} = a_{i1}/l_{11}, \quad (2 \le i \le n)$$

$$l_{jj} = \sqrt{a_{jj} - \sum_{k=1}^{j-1} l_{jk}^2} \quad (2 \le i \le n)$$

$$l_{ij} = \frac{1}{l_{jj}} \left[a_{ij} - \sum_{k=1}^{j-1} l_{ik}l_{jk} \right] \quad (i > j) \tag{3.19}$$

It may be noted that some of the elements starting with a diagonal element may turn out to be complex. Thus in such cases, L and L^T become complex matrices.

The system of equations has a unique solution if $l_{ii} \neq 0$ ($i = 1, 2, \cdots, n$). This is so because

$$det\,A = det\,L \cdot det\,(L^T) = (l_{11}l_{22} \cdots l_{nn})^2 \neq 0$$

The condition is always satisfied in the important case of *positive definite matrices*. Assume $l_{ii} \neq 0$ ($1 \leq i \leq n$), the given linear system is equivalent to

$$L\mathbf{y} = \mathbf{b} \quad \text{where} \quad L^T\mathbf{x} = \mathbf{y}$$

The solution of the two systems by forward and back substitutions yield

$$y_1 = \frac{b_1}{l_{11}}, \quad y_i = \frac{1}{l_{ii}}\left[b_i - \sum_{j=1}^{i-1} l_{ij}y_j\right] \quad (i = 2, 3, \cdots, n) \tag{3.20a}$$

and

$$x_n = \frac{y_n}{l_{nn}}, \quad x_i = \frac{1}{l_{ii}}\left[y_i - \sum_{j=i+1}^{n} l_{ji}x_j\right] \quad (i = n-1, \cdots, 2, 1) \tag{3.20b}$$

completing the solution of the equations.

The complexity of the algorithm can be shown to be $\frac{1}{3}(n^3 + 3n^2 - n) = O(\frac{1}{3}n^3)$ together with n square root evaluations. This is one-half of that of Gauss elimination.

Example 1. Factorise the symmetric matrix

$$A := \begin{bmatrix} 4 & 1 & 1 \\ 1 & 5 & 2 \\ 1 & 2 & 3 \end{bmatrix}$$

by Cholesky's method.

Solution. We write

$$\begin{bmatrix} l_{11} & 0 & 0 \\ l_{21} & l_{22} & 0 \\ l_{31} & l_{32} & l_{33} \end{bmatrix} \begin{bmatrix} l_{11} & l_{21} & l_{31} \\ 0 & l_{22} & l_{32} \\ 0 & 0 & l_{33} \end{bmatrix} = \begin{bmatrix} 4 & 1 & 1 \\ 1 & 5 & 2 \\ 1 & 2 & 3 \end{bmatrix}$$

Equating elements of each row, we have for $i = 1$

$$l_{11}^2 = 4 \text{ or, } l_{11} = 2, \quad l_{11}l_{21} = 1 \text{ or, } l_{21} = \frac{1}{2}, \quad l_{11}l_{31} = 1 \text{ or, } l_{31} = \frac{1}{2}$$

For $i = 2$,

$$l_{21}^2 + l_{22}^2 = 5 \text{ or, } l_{22} = \sqrt{5 - 1/4} = \frac{\sqrt{19}}{2}$$

$$l_{21}l_{31} + l_{22}l_{32} = 2 \text{ or, } l_{32} = \frac{2}{\sqrt{19}}\left(2 - \frac{1}{2} \times \frac{1}{2}\right) = \frac{7}{2\sqrt{19}}$$

For $i = 3$,

$$l_{31}^2 + l_{32}^2 + l_{33}^2 = 3 \text{ or, } l_{33} = \sqrt{3 - \frac{1}{4} - \frac{49}{4 \times 19}} = \sqrt{\frac{40}{19}} = 2\sqrt{\frac{10}{19}}$$

Hence

$$L = \begin{bmatrix} 2 & 0 & 0 \\ 1/2 & \sqrt{19}/2 & 0 \\ 1/2 & 7/(2\sqrt{19}) & 2\sqrt{10/19} \end{bmatrix}$$

□

André-Louis Cholesky (1875–1918), French army major. He was involved in geodesy and surveying from 1906 to 1909 during the International Occupation of Crete, and later in North Africa. He developed the method named after him to compute the solutions of least squares data fitting problems (see the Chap. 8 on approximation of functions).

The treatment of symmetric positive definite matrices has important application in *finite element methods*. Cholesky's method is a useful tool in such applications. The following subroutine implements the algorithm (3.19) and (3.20a, 3.20b):

```
SUBROUTINE CHOLESKY(n,a,el,b)
! n=Order of matrix.          (Input)
! a=symmetric square matrix of order n,
! to be decomposed by Cholesky method.     (Input)
! n=Order of matrix to be decomposed.          (Input)
! el=Cholesky factored lower triangular matrix.        (Output)
! b= Right hand side vector.       (input)
! The solution of Ax=b is returned in vector b.        (Output)
! *****************************************************
REAL :: a(n,n), el(n,n), b(n), y(n)
el(1,1)=sqrt(a(1,1))
DO i=2,n
el(i,1)=a(i,1)/el(1,1)
END DO
DO i=1,n
DO j=1,i
sum1=0.0; sum2=0.0
DO k=1,j-1
IF(i==j) THEN
sum1=sum1+el(j,k)**2
el(j,j)=SQRT(a(i,j)-sum1)
ELSE IF(i>j) THEN
```

```
sum2=sum2+el(i,k)*el(j,k)
el(i,j)=(a(i,j)-sum2)/el(j,j)
ELSE
el(i,j)=0.
END IF
END DO
END DO
END DO
! ****************************************************
y(1)=b(1)/el(1,1)              ! Forward Substitution for solving Ly=b.
DO i=2,n
sum=0.0
DO j=1,i-1
sum=sum+el(i,j)*y(j)
END DO
y(i)=(b(i)-sum)/el(i,i)
END DO
b(n)=y(n)/el(n,n)              ! Backward Substitution for solving L^T x=y.
DO i=n-1,1,-1
sum=0.0
DO j=i+1,n
sum=sum+el(j,i)*b(j)
END DO
b(i)=(y(i)-sum)/el(i,i)
END DO
RETURN
END SUBROUTINE CHOLESKY
```

In case the matrix L is complex, the subroutine can be trivially converted for that case, by using complex arithmetic and square root extraction.

Exercises

1. Factorise by Cholesky's method

$$(i) A_1 := \begin{bmatrix} 1 & 2 & 3 \\ 2 & 8 & 22 \\ 3 & 22 & 82 \end{bmatrix} \quad (ii) A_2 := \begin{bmatrix} 4 & -6 & 2 \\ -6 & 10 & -7 \\ 2 & -7 & 21 \end{bmatrix} \quad (iii) A_3 := \begin{bmatrix} 1 & 2 & 3 & 4 \\ 2 & 1 & 2 & 3 \\ 3 & 2 & 1 & 2 \\ 4 & 3 & 2 & 1 \end{bmatrix}$$

Hence solve the equations: (a) $A_1 x = [6, 32, 107]^T$, (b) $A_2 x = [8, -9, -12]^T$, (c) $A_3 x = [5, 1, 1, -5]^T$.

$$[(i) L_1 = \begin{bmatrix} 1 & 0 & 0 \\ 2 & 2 & 0 \\ 3 & 8 & 3 \end{bmatrix} \quad (ii) L_2 = \begin{bmatrix} 2 & 0 & 0 \\ -3 & 1 & 0 \\ 1 & -4 & 2 \end{bmatrix} \quad (iii) L_3 = \begin{bmatrix} 1 & 0 & 0 & 0 \\ 2 & \sqrt{3}i & 0 & 0 \\ 3 & 4i/\sqrt{3} & i\sqrt{8/3} & 0 \\ 4 & 5i/\sqrt{3} & 8i/\sqrt{6} & 2 \end{bmatrix}$$

3.2 Error: Matrix Norms and Condition Number

The solution of a linear system by any of the preceding methods must be considered as approximate solution due to round-off errors. If $\tilde{\mathbf{x}}$ is a computed solution of the system $A\mathbf{x} = \mathbf{b}$, then the absolute error $\mathbf{e}$ is defined by

$$\mathbf{e} := \mathbf{x} - \tilde{\mathbf{x}}$$

This error is unknown, since $\mathbf{x}$ is unknown and one may be tempted to compute the *residual*

$$\mathbf{r} := A(\mathbf{x} - \tilde{\mathbf{x}}) = \mathbf{b} - A\tilde{\mathbf{x}}$$

and judge the accuracy of $\tilde{\mathbf{x}}$ on the smallness of $\mathbf{r}$? But this need not always be true since $\mathbf{e} = A^{-1}\mathbf{r}$ implies that the smallness of $\mathbf{e}$ depends on the matrix A as well. For further development, we require the definition of **norm**. The norm of an algebraic entity x is defined as a number $\|x\|$ that satisfies the axioms

(i) $\|x\| > 0$, for $x \neq 0$ and $\|0\| = 0$
(ii) $\|cx\| = |c| \cdot \|x\|$, where c is an arbitrary number; and
(iii) $\|x + y\| \leq \|x\| + \|y\|$ for any two elements x, y (triangle inequality).

This definition is now adopted for vectors and matrices.

1°. **Vector Norm:** The norm of an n-vector $\mathbf{x}$ can be defined in various ways. The most commonly used norms are:
(a) Euclidean norm (*2-norm*):

$$\|\mathbf{x}\|_2 = \sqrt{\sum_{i=1}^{n} |x_i|^2} \tag{3.21a}$$

(b) Laplace's norm (*1-norm*):

$$\|\mathbf{x}\|_1 = \sum_{i=1}^{n} |x_i| \tag{3.21b}$$

(c) Chebyshev–Laplace norm (*maximum norm*):

$$\|\mathbf{x}\|_\infty = \max_{1 \leq i \leq n} |x_i| \tag{3.21c}$$

All the three norms can be shown to satisfy axioms (i)–(iii). We prefer to treat the maximum norm (c), as it is easiest to handle. Evidently, it satisfies axioms (i) and (ii). For verifying axiom (iii) consider two n-vectors $\mathbf{x}$ and $\mathbf{y}$. then there is a number k such that

$$\max_{1 \leq i \leq n} |x_i + y_i| = |x_k + y_k|$$

Consequenly

$$\max_{1 \le i \le n} |x_i + y_i| \le |x_k| + |y_k| \le \max_{1 \le i \le n} |x_i| + \max_{1 \le i \le n} |y_i|$$

This means that $\|\mathbf{x} + \mathbf{y}\|_\infty \le \|\mathbf{x}\|_\infty + \|\mathbf{y}\|_\infty$.

2°. **Matrix Norm.** The norm of an $n \times n$ matrix A is defined as

$$\|A\| := \max_{\mathbf{x} \neq 0} \frac{\|A\mathbf{x}\|}{\|\mathbf{x}\|}$$

where $\|\mathbf{x}\|$ and $\|A\mathbf{x}\|$ are vector norms already defined. In particular if $A = I$, $\|A\mathbf{x}\| = \|I\mathbf{x}\| = \|\mathbf{x}\|$ and so $\|I\| = 1$.

Depending on the choice of the vector norm of $\|\mathbf{x}\|$ and $\|A\mathbf{x}\|$. We have definitions of the norms $\|A\|_2$, $\|A\|_1$ and $\|A\|_\infty$. The matrix norms satisfy axioms (i)– (iii) and as before, we prefer to give th details of proof regarding $\|A\|_\infty$ only. It is evident that axioms (i) and (ii) are satisfied, while for (iii)

$$\|A + B\|_\infty = \max_{\mathbf{x} \neq 0} \frac{\|(A + B)\mathbf{x}\|_\infty}{\|\mathbf{x}\|_\infty} \le \max_{\mathbf{x} \neq 0} \frac{\|A\mathbf{x}\|_\infty + \|B\mathbf{x}\|_\infty}{\|\mathbf{x}\|_\infty}$$
$$\le \|A\|_\infty + \|B\|_\infty$$

A matrix norm also satisfies the estimates

(iv) $\|A\mathbf{x}\| \le \|A\|\|\mathbf{x}\|$, in which equality holds only for a particular vector $\mathbf{x}$. And
(v) $\|AB\| \le \|A\|\|B\|$, for any two matrices A and B.

The proof of (iv) follows immediately from the definition of $\|A\|$. Consequently

$$\|AB\| = \max_{\mathbf{x} \neq 0} \frac{\|AB\mathbf{x}\|}{\|\mathbf{x}\|} \le \max_{\mathbf{x} \neq 0} \frac{\|A\|\|B\mathbf{x}\|}{\|\mathbf{x}\|} = \|A\| \cdot \|B\|$$

An expression for the maximum norm $\|A\|_\infty$ in terms of the elements only is given by the following theorem:

Theorem 3.1 *The maximum norm $\|A\|_\infty$ of an $n \times n$ matrix A satisfies*

$$\|A\|_\infty = \max_{1 \le i \le n} \sum_{j=1}^{n} |a_{ij}| \tag{3.22}$$

Proof The definition of $\|A\|$ implies estimate (iv), in which the equality holds for *some particular* $\mathbf{x}^0$. Now from the definition (c) of vector norm

$$\|A\mathbf{x}\|_\infty = \max_{1 \le i \le n} \left| \sum_{j=1}^{n} a_{ij} x_j \right| \le \max_{1 \le i \le n} \sum_{j=1}^{n} |a_{ij}||x_j|$$

$$\leq \max_{1\leq i\leq n}\left[\max_{1\leq j\leq n}|x_j|\sum_{j=1}^{n}|a_{ij}|\right] = \|\mathbf{x}\|_\infty \cdot \max_{1\leq i\leq n}\sum_{j=1}^{n}|a_{ij}| \qquad (3.23)$$

Next, we seek $\mathbf{x}^0$. Let k be an integer between 1 and n such that the maximum in Eq. (3.23) is reached for the value k of i. If we consider

$$x_j^0 = \begin{cases} 1 & \text{if } a_{kj} \geq 0 \\ -1 & \text{if } a_{kj} < 0 \end{cases}$$

then clearly $\|x^0\| = 1$ and $a_{kj}x_j^0 = |a_{kj}|$, $(j = 1, 2, \cdots, n)$. Thus

$$\|A\mathbf{x}^0\| = \max_{1\leq i\leq n}\left|\sum_{j=1}^{n}a_{ij}x_j^0\right| \geq \left|\sum_{j=1}^{n}a_{kj}x_j^0\right| = \sum_{j=1}^{n}|a_{kj}|$$

$$= \|\mathbf{x}^0\| \cdot \max_{1\leq i\leq n}\sum_{j=1}^{n}|a_{ij}| \qquad (3.24)$$

Eqs. (3.23) and (3.24) establish the vaidity of Eq. (3.22).   □

Similarly $\|A\|_1$ and $\|A\|_2$ are given by

Theorem 3.2

$$\|A\|_1 = \max_{1\leq j\leq n}\sum_{i=1}^{n}|a_{ij}|, \quad \|A\|_2 = \sqrt{\rho(A^T A)} \qquad (3.25)$$

where the spectral radius ρ of a matrix B is defined by

$$\rho(B) := \max_{1\leq i\leq n}|\lambda_i|$$

$\lambda_1, \lambda_2, \cdots, \lambda_n$ *being the eigenvalues of B.*

3°. Condition Number of a Matrix. This quantity is a measure that expresses the relation between the error norm $\|\mathbf{e}\|$ and the residual norm $\|\mathbf{r}\|$. Since $\mathbf{e} = A^{-1}\mathbf{r}$, we must have

$$\|\mathbf{e}\| \leq \|A^{-1}\| \cdot \|\mathbf{r}\|$$

Thus dividing by $\|\mathbf{x}\|$, where $\|\mathbf{b}\| = \|A\mathbf{x}\| \leq \|A\|\|\mathbf{x}\|$, the *relative error* is

$$\frac{\|\mathbf{e}\|}{\|\mathbf{x}\|} \leq \frac{\|A^{-1}\| \cdot \|\mathbf{r}\|}{\|\mathbf{x}\|} \leq \|A\| \cdot \|A^{-1}\|\frac{\|\mathbf{r}\|}{\|\mathbf{b}\|} \qquad (3.26)$$

The quantity

$$cond(A) := \|A\| \cdot \|A^{-1}\| \qquad (3.27)$$

is called the *condition number* of the matrix A. Eqs. (3.26) and (3.27) state that *the relative error can be as large as* cond(A) *times the relative residual* $\|\mathbf{r}\|/\|\mathbf{b}\|$. Thus if cond($A$) is not large, a small relative residual will result in a small relative error in the solution. However, a large cond(A) may imply large error in solution, inspite of a small residual. In the former case, the matrix A is called **well conditioned** and in the latter case, it is called **ill conditioned**. More precisely, the two terms must be used in conjunction with precision of computation.

The number cond(A) depends on the norm selected and can vary considerably with change of norm. However, whatever be the norm

$$cond(A) = \|A\| \cdot \|A^{-1}\| \geq \|AA^{-1}\| = \|I\| = 1$$

Example 1. Find the condition number of the matrix

$$A := \begin{bmatrix} 4 & 2 & 1 \\ -1 & 3 & 2 \\ 2 & 3 & 5 \end{bmatrix}$$

Solution. The exact inverse of A can be shown to be

$$A^{-1} = \frac{1}{45} \begin{bmatrix} 9 & -7 & 1 \\ 9 & 18 & -9 \\ -9 & -8 & 14 \end{bmatrix}$$

The maximum norms of the two matrices are

$$\|A\|_\infty = \max\{4 + 2 + 1, \ 1 + 3 + 2, \ 2 + 3 + 5\} = 10$$

$$\|A^{-1}\|_\infty = \max \frac{1}{45}\{9 + 7 + 1, \ 9 + 18 + 9, \ 9 + 8 + 14\} = \frac{36}{45}$$

Hence

$$cond(A) = \|A\|_\infty \cdot \|A^{-1}\|_\infty = 10 \times \frac{36}{45} = 8$$

which is not a large number and A is well conditioned. □

Example 2. Consider the linear system with $\mathbf{b} = [-1, -1, \cdots, -1, 1]^T$ and

$$A = \begin{bmatrix} 1 & -1 & -1 & \cdots & -1 & -1 \\ 0 & 1 & -1 & \cdots & -1 & -1 \\ \cdots & \cdots & \cdots & \cdots & \cdots & \cdots \\ 0 & 0 & 0 & \cdots & 1 & -1 \\ 0 & 0 & 0 & \cdots & 0 & 1 \end{bmatrix}$$

So that $det(A) = 1 \neq 0$. Determine the condition of the system.

Solution. Written in detail the system of equations is

$$
\begin{aligned}
x_1 -x_2 -x_3 \cdots -x_{n-1} -x_n &= -1 \\
x_2 -x_3 \cdots -x_{n-1} -x_n &= -1 \\
\cdots \cdots \quad\quad \cdots \quad\quad \cdots \\
x_{n-1} -x_n &= -1 \\
x_n &= 1
\end{aligned}
$$

The exact solution of the system by back substitution is evidently $\mathbf{x} = [0, 0, \cdots, 0, 1]^T$. Suppose that a unique absolute error ϵ is committed in the computation of x_n from the nth equation, so that residual $\mathbf{r} = [0, 0, \cdots, 0, \epsilon]^T$. If $\mathbf{e}$ is the error in the computation of $\mathbf{x}$, the computed value is $\tilde{\mathbf{x}} = \mathbf{x} - \mathbf{e}$, where $\mathbf{e}$ satisfies the system of equations $A\mathbf{e} = \mathbf{r}$, or

$$
\begin{aligned}
e_1 -e_2 -e_3 \cdots -e_n &= 0 \\
e_2 -e_3 \cdots -e_n &= 0 \\
e_{n-1} -e_n &= 0 \\
e_n &= \epsilon
\end{aligned}
$$

Hence we obtain

$$
\begin{aligned}
e_n &= \epsilon \\
e_{n-1} &= e_n = \epsilon \\
e_{n-2} &= e_{n-1} + e_n = 2\epsilon \\
e_{n-3} &= e_{n-2} + e_{n-1} + e_n = 2^2\epsilon \\
\cdots & \cdots \cdots \cdots \cdots \cdots \cdots \\
e_1 &= e_{n-(n-1)} = 2^{n-2}\epsilon
\end{aligned}
$$

We have therefore $\|\mathbf{e}\|_\infty = 2^{n-2}|\epsilon|$. Also $\|\mathbf{r}\|_\infty = |\epsilon|$, $\|\mathbf{b}\|_\infty = 1$ and $\|\mathbf{x}\|_\infty = 1$. Hence from Eq. (3.26)

$$
cond(A) \geq \frac{2^{n-2}|\epsilon|/1}{|\epsilon|/1} = 2^{n-2}
$$

If n is large, say $n = 102$, $cond(A) \geq 2^{100} > 10^{30}$, which is a huge number. Thus the relative error in $\mathbf{x}$ exceeds 10^{30} times the relative residual. The system of equations is therefore ill conditioned.

We conclude with the observation that since $\|A\|_\infty = n$, $\|A^{-1}\| \geq 2^{n-2}/n$. $\quad\square$

Exercises

1. Find the condition number and condition of the following matrices:

(i) $\begin{bmatrix} 1.01 & 0.99 \\ 0.99 & 1.01 \end{bmatrix}$ $\quad$ (ii) $\begin{bmatrix} 1 & 4 & 7 \\ 2 & 5 & 8 \\ 3 & 6 & 9 \end{bmatrix}$

[(i) Cond. No.=100, Ill conditioned, (ii) Cond. No.=30.23, Well conditioned].

2. (T.S. Wilson) Show that for the linear system

$$
\begin{aligned}
10x_1 +7x_2 +8x_3 +7x_4 &= 32 \\
7x_1 +5x_2 +6x_3 +5x_4 &= 23 \\
8x_1 +6x_2 +10x_3 +9x_4 &= 33 \\
7x_1 +5x_2 +9x_3 +10x_4 &= 31
\end{aligned}
$$

$x_1 = 6$, $x_2 = -7.2$, $x_3 = 2.9$, $x_4 = -0.1$ and $x_1 = 1.5$, $x_2 = 0.18$, $x_3 = 1.19$, $x_4 = 0.89$ are 'approximate' solutions, while the exact solution is $x_1 = x_2 = x_3 = x_4 = 1$. What is your inference? Computing A^{-1} by subroutine MATINV, determine $cond(A)$.
[Ill conditioned. $cond(A) = 4488$].

3.3   Relaxation Methods

In these methods, the solution of an unknown x_i $(i = 1, 2, \cdots, n)$ of the linear system (3.1) is expressed as a function of the remaining unknowns. Thus, if $a_{ii} \neq 0$, one can write

$$
x_i = -\frac{1}{a_{ii}}\sum_{\substack{j=1 \\ (j \neq i)}}^{n} a_{ij}x_j + \frac{b_i}{a_{ii}} \qquad (i = 1, 2, \cdots, n) \tag{3.28}
$$

In matrix notations, Eq. (3.28) can be writen as

$$
\mathbf{x} = A'\mathbf{x} + \mathbf{b}' \tag{3.28a}
$$

where $A' = [a'_{ij}]$, $\mathbf{b}' = [b'_i]^T$ such that

$$
a'_{ij} = \begin{cases} 0, & i = j \\ -a_{ij}/a_{ii}, & i \neq j \end{cases} \qquad b'_i = b_i/a_{ii} \qquad (i, j = 1, 2, \cdots, n)
$$

Such representation for a linear system is evidently not unique. For example, the equations may be permuted (shifted in position). More generally one may premultiply it by an arbitrary $n \times n$ matrix C, to obtain a new system

$$
CA\mathbf{x} = C\mathbf{b}
$$

The system can then be rewritten in the form

$$
\mathbf{x} = \mathbf{x} - C(A\mathbf{x} - \mathbf{b}) = (I - CA)\mathbf{x} + C\mathbf{b} =: A'\mathbf{x} + \mathbf{b}' \tag{3.28b}
$$

where I is the $n \times n$ unit matrix. Eq. (3.28a) is a particular case of Eq. (3.28b), if one takes $C_{ii} = 1/a_{ii}$ $(a_{ii} \neq 0)$, $C_{ij} = 0$ $(i \neq j)$.

It appears that if the elements $|a'_{ij}| << 1$ $(i \neq j)$, the x_j's on the right-hand side of Eq. (3.28a) may be replaced by some initial approximation in one *sweep*, without much error. In other words, the equation is 'relaxed' in its exactness. Relaxation yields a new set of approximation of the unknowns, which may be used for further *cycle* of sweeping of the variables. Repeated cycles of sweeps may result in settling of the values of x_i, yielding the solution. The term *relaxation* is due to Southwell, but the idea dates back to Gauss. The condition $|a'_{ij}| << 1$ or $|a_{ij}| << |a_{ii}|$ $(i \neq j)$ is directly satisfied in many applications and thus relaxation methods have proved to be very useful in practice.

3.3.1 Jacobi's Method

We consider the linear system to be written as Eqs. (3.28) or (3.28a), in which we may *suppose that $|a'_{ij}| << 1$ or, $|a_{ij}| << |a_{ii}|$* $(i, j = 1, 2, \cdots, n)$. Let $\mathbf{x}^0$ be a starting zeorth approximation of $\mathbf{x}$. Altogether neglecting $\mathbf{x}$ on the right-hand side of Eq. (3.28) one may take $\mathbf{x}^0 = \mathbf{b}'$. Starting with this value, the succeeding relaxations of Eq. (3.28a) yield

$$\mathbf{x}^{(k)} = A'\mathbf{x}^{(k-1)} + \mathbf{b}' \qquad (k = 1, 2, \cdots) \tag{3.29}$$

or, using Eq. (3.28)

$$x_i^{(k)} = \frac{1}{a_{ii}} \left[b_i - \sum_{\substack{j=1 \\ (j \neq i)}}^{n} a_{ij} x_j^{(k-1)} \right] \qquad (i = 1, 2, \cdots, n) \tag{3.29a}$$

If the sequence $\{\mathbf{x}^k\}$ converges to $\mathbf{x}$, then evidently $\mathbf{x}$ is the required solution. In this event, the iterations can be carried out until

$$\max_{1 \leq i \leq n} \left| x_i^{(k)} - x_i^{(k-1)} \right| < \epsilon$$

where ϵ is a prescribed small number.

Equation (3.29) can be viewed as an iteration procedure

$$\mathbf{x}^{(k)} = H_J \mathbf{x}^{(k-1)} + \mathbf{c}_J \tag{3.30}$$

in which decomposing matrix A into lower and upper triangular matrices (with zero diagonal) L and U together with diagonal matrix D, i.e. $A = L + D + U$

$$H_J := A' = -D^{-1}(L + U), \qquad \mathbf{c}_J := \mathbf{b}' = D^{-1}\mathbf{b}$$

Karl Gustav Jacob Jacobi (1804–51), German mathematician who taught at University of Königsberg from 1826 until his death. He is famed for his study of the functional determinant called the Jacobian (which was actually introduced by Cauchy in 1815). He founded the theory of elliptic functions based on four theta functions. In function theory, he proved that a doubly periodic single-valued function must have an imaginary ratio of the periods. He also carried out important research in partial differential equations of the first order and applied them to differential equations of dynamics.

3.3.2 Seidel's Method

In this method, for accelerated convergence, Eq. (3.28) is written as

$$x_i = \frac{1}{a_{ii}} \left[b_i - \sum_{j=1}^{i-1} a_{ij} x_j - \sum_{j=1+1}^{n} a_{ij} x_j \right] \tag{3.31}$$

The initial approximation by dropping x_j's may be taken as $x_i^{(0)} = b_i / a_{ii}$ ($i = 1, 2, \cdots, n$). In the kth approximation ($k = 1, 2, 3, \cdots$), we note that the values of $x_j^{(k)}$ ($j = 1, 2, \cdots, i-1$) are available in the current iteration while only $x_j^{(k-1)}$ ($j = i+1, i+2, \cdots, n$) are currently available. Hence one can consider relaxation of Eq. (3.31) in the form

$$x_i^{(k)} = \frac{1}{a_{ii}} \left[b_i - \sum_{j=1}^{i-1} a_{ij} x_j^{(k)} - \sum_{j=i+1}^{n} x_j^{(k-1)} \right] \tag{3.32}$$

where $k = 1, 2, 3, \cdots$. If the sequence $\{x_i^{(k)}\}$ converges to x_i ($i = 1, 2, \cdots, n$), then obviously these values constitute the desired solution. The convergence criterion can be adopted as in Jacobi's method.

The process (3.32) can also be viewed as an iteration. For Eq. (3.32) in matrix notations is

$$(D + L)\mathbf{x}^{(k)} = -U\mathbf{x}^{(k-1)} + \mathbf{b}$$

This equation leads to the form of Eq. (3.30), with

$$H_S := -(D + L)^{-1} U, \quad \mathbf{c}_S := (D + L)^{-1} \mathbf{b}$$

Philipp Ludwig von Seidel (1821–1896), German mathematician who studied under Dirichlet, Bessel, Jacobi and Neumann. He wrote on dioptics and mathematical analysis. His work on lenses identified mathematically five coefficients, describing the aberration of a lens called 'Seidel Sums'. He introduced the concept of non-uniform convergence and applied probability to astronomy.

Example 1. Solve by Seidel iteration

$$
\begin{aligned}
10x +2\,y + z &= 9 \\
2x +20y - 2z &= -44 \\
-2x + 3y +10z &= 22
\end{aligned}
$$

correct to 2 decimal places.

Solution. The system is diagonally dominant. Rewriting it for iteration

$$
\begin{aligned}
x &= 0.9 - 0.2y - 0.1z \\
y &= -2.2 - 0.1x + 0.1z \\
z &= 2.2 + 0.2x - 0.3y
\end{aligned}
$$

The *zeroth approximation* is $x^{(0)} = 0.9$, $y^{(0)} = -2.2$, $z^{(0)} = 2.2$.
First Approx:

$$
\begin{aligned}
x^{(1)} &= 0.9 - 0.2 \times (-2.2) - 0.1 \times 2.2 = 1.12 \\
y^{(1)} &= -2.2 - 0.1 \times (1.12) + 0.1 \times 2.2 = -2.09 \\
z^{(1)} &= 2.2 + 0.2 \times 1.12 - 0.3 \times (2.09) = 3.05
\end{aligned}
$$

We note that for calculating $y^{(1)}$, the current iterate $x^{(1)}$ is available, while only $z^{(0)}$ is available. For $z^{(1)}$ however both $x^{(1)}$ and $y^{(1)}$ are available. These values are used in the current iteration.
Second Approx: In a similar manner

$$
\begin{aligned}
x^{(2)} &= 0.9 - 0.2 \times (-2.09) - 0.1 \times (3.05) = 1.013 \\
y^{(2)} &= -2.2 - 0.1 \times (1.013) + 0.1 \times (3.05) = -1.996 \\
z^{(2)} &= 2.2 + 0.2 \times (1.013) - 0.3 \times (-1.996) = 3.001
\end{aligned}
$$

Third Approx: In this approximation

$$
\begin{aligned}
x^{(3)} &= 0.9 - 0.2 \times (1.013) - 0.1 \times (3.001) = 0.9991 \\
y^{(3)} &= -2.2 - 0.1 \times (0.9991) + 0.1 \times (3.001) = -1.9998 \\
z^{(3)} &= 2.2 + 0.2 \times (0.9991) - 0.3 \times (-1.9998) = 2.9998
\end{aligned}
$$

Comparing the last two iterates, correct to 2 decimal places

$$
x = 1.00, \quad y = -2.00, \quad z = 3.00
$$

that are incidentally of exact value. $\qquad \square$

Following the method, subroutine SEIDEL is given below, which solves the linear system $Ax = b$. In the subroutine, apart for A and $\mathbf{b}$, the tolerance *epsilon* and the maximum number of *cycles* must be supplied as inputs in the calling program.

```
SUBROUTINE SEIDEL(n,a,b,x,epsilon,cycles)
! Solves the system of equations Ax=b.
! n=Number of unknowns and equations.    (Input)
! a=Matrix of the n×n coefficients.    (Input)
! b=Right hand side vector.    (Input)
! x=Solution vector.    (Output)
! epsilon=Tolerance in two successive iterates.    (Input)
! cycles=Maximum number of iterations allowed    (Input)
REAL :: a(n,n), b(n), x(n)
INTEGER :: cycles
DO i=1,n
x(i)=b(i)/a(i,i)
END DO
DO k=1,cycles
error=0.0
DO i=1,n
sum=0.0
DO j=1,n
IF(a(i,j)/=0.0) sum=sum+a(i,j)*x(j)      ! Avoids unnecessary multiplication
END DO
temp=(b(i)−sum+a(i,i)*x(i))/a(i,i)
err=ABS(x(i)−temp)
IF(error<err) error=err
x(i)=temp
END DO
IF(error<epsilon) RETURN
IF(k==cycles) PRINT*, 'Maximum No. of cycles reached'
END DO
RETURN
END SUBROUTINE SEIDEL
```

In many applications the matrix A is sparse, that is to say, many off-diagonal elements are zero. In accumulating the variable named sum, multiplication by such zero elements is avoided, making convergence faster.

Exercises

1. Solve the following equations by Seidel iteration:
(*i*) A beam ABC clamped at A and C is simply supported at B. The bending moments M_1, M_2, M_3 at A, B, C satisfy the equations

$$2M_1 + M_2 \qquad\quad = 1$$
$$M_1 + 6M_2 + 2M_3 = 9$$
$$M_2 + 4M_3 = 8$$

Find M_1, M_2, M_3. $[M_1 = 0,\ M_2 = 1,\ M_3 = 3/2]$.

$$\begin{aligned} &-x_1 + 10x_2 - 2x_3 = 7\\ (ii)\ &5x_1 - x_2 - x_3 = 3 \qquad\qquad [x_1 = 1,\ x_2 = 1,\ x_3 = 1].\\ &-x_1 - x_2 + 10x_3 = 8\\ &2x_1 - 7x_2 - 10x_3 = -17\\ (iii)\ &5x_1 + x_2 + 3x_3 = 14 \qquad\quad [x_1 = 1,\ x_2 = -3,\ x_3 = 4].\\ &x_1 + 10x_2 + 9x_3 = 7 \end{aligned}$$

2. Use subroutine SEIDEL to compute the roots of:

$$(i)\ \begin{aligned} 83x_1 + 11x_2 - 4x_3 &= 95\\ 7x_1 + 52x_2 + 13x_3 &= 104\\ 3x_1 + 8x_2 + 29x_3 &= 71 \end{aligned} \qquad \begin{bmatrix} x_1 = 1.05793\\ x_2 = 1.36717\\ x_3 = 1.96169 \end{bmatrix}.$$

$$(ii)\ \begin{aligned} 6.1x_1 + 0.7x_2 - 0.05x_3 &= 6.97\\ -1.3x_1 - 2.05x_2 + 0.87x_3 &= 0.10\\ 2.5x_1 - 3.12x_2 - 5.03x_3 &= 2.04 \end{aligned} \qquad \begin{bmatrix} x_1 = 1.21394\\ x_2 = -0.58157\\ x_3 = 0.55852 \end{bmatrix}.$$

$$(iii)\ \begin{aligned} 8.7x_1 - 3.1x_2 + 1.8x_3 - 2.2x_4 &= -9.7\\ 2.1x_1 + 6.7x_2 - 2.2x_3 &= 13.1\\ 3.2x_1 - 1.8x_2 - 9.5x_3 - 1.9x_4 &= 6.9\\ 1.2x_1 + 2.8x_2 - 1.4x_3 - 9.9x_4 x_4 &= 25.1 \end{aligned} \qquad \begin{bmatrix} x_1 = -0.75288\\ x_2 = 1.90591\\ x_3 = -0.86885\\ x_4 = -1.96470 \end{bmatrix}.$$

3.3.3 Young's Over Relaxation Method

This method often abbreviated as SOR (Successive Over Relaxation) method aims at improving the convergence rate of the Seidel method. For the latter method, Eq. (3.32) can be rewritten as

$$x_i^{(k)} = x_i^{(k-1)} - \left[\left(\sum_{j=1}^{i-1} a_{ij} x_j^{(k)} + \sum_{j=i}^{n} a_{ij} x_j^{(k-1)} - b_i \right) \Big/ a_{ii} \right]$$

The quantity within the square brackets is responsible for improving the $(k-1)$th approximation to yield the kth approximation. Hence for acceleration in the convergence, one may multiply the quantity by a factor $\omega > 1$ to get

$$x_i^{(k)} = x_i^{(k-1)} - \omega \left[\left(\sum_{j=1}^{i-1} x_j^{(k)} - \sum_{j=i}^{n} a_{ij} x_j^{(k-1)} - b_i \right) \Big/ a_{ii} \right] \qquad (3.33)$$

A theory of the optimal choice of ω does exist, but is difficult to implement in practice. Experience shows that it should lie between 1.2 and 1.6. A theorem due to Kahan

(William Mortan Kahan (1933–)) states that a necessary condition for convergence is $0 < \omega < 2$.

As in the case of Jacobi and Seidel methods, Eq.(3.33) is equivalent to iteration of the form (3.30) where

$$H_Y := (D + \omega L)^{-1}[(1 - \omega)D - \omega U], \quad c_Y := \omega(D + \omega L)^{-1}\mathbf{b}$$

The convergence of the three methods is discussed in Sect. 3.3.4.

David M. Young (1923–2008) was a professor of computer sciences at the University of Texas at Austin, U.S.A. His Ph.D thesis in 1950 led to the development of the SOR method in the year (1954).

Exercise

1. Write a subroutine for Young's over relaxation method. Test it for the problems of Exercise 2 of Sect. 3.3.2.

3.3.4 Convergence Theorems

Theorem 3.3 *If for any matrix norm $\|H\| < 1$, in the case of Jacobi, Seidel and Young's SOR methods, then the respective methods converge to the solution $\mathbf{x}$, for any initial vector $\mathbf{x}^0$.*

Proof Let the absolute error vector in the kth cycle be $\epsilon^k = \mathbf{x} - \mathbf{x}^{(k)}$, then since $\mathbf{x}$ satisfies the equation $\mathbf{x} = H\mathbf{x} + \mathbf{c}$ and $\mathbf{x}^{(k)} = H\mathbf{x}^{(k-1)} + \mathbf{c}$, $\epsilon^{(k)} = H\epsilon^{(k-1)}$. Hence

$$\|\epsilon^{(k)}\| \le \|H\|\|\epsilon^{(k-1)}\| \le \|H\|^2\|\epsilon^{(k-2)}\| \le \cdots \le \|H\|^k\|\epsilon^{(0)}\|$$

where $\epsilon^{(0)}$ is the initial absolute error vector. Since $\|H\| < 1$, $\|H\|^k \to 0$ as $k \to \infty$ and $\|\epsilon^{(k)}\| \to 0$, proving convergence. $\qquad\square$

Example 1. For the iterative system $\mathbf{x} = H\mathbf{x} + \mathbf{c}$, if $\|H\| < 1$, then prove that the condition number of the system is $\le (1 + \|H\|)/(1 - \|H\|)$.

Solution. Here $A = I - H$. Hence $\|H\| = \|I - H\| \le \|I\| + \|H\| = 1 + \|H\|$. Since the system is $A\mathbf{x} = \mathbf{c}$, $\mathbf{x} = A^{-1}\mathbf{c}$ and so $\|\mathbf{x}\| \le \|A^{-1}\|\|\mathbf{c}\|$, where *there exists a $\mathbf{c}$ for which equality holds.* Now

$$\|\mathbf{x}\| = \|H\mathbf{x} + \mathbf{c}\| \le \|H\|\|\mathbf{x}\| + \|\mathbf{c}\|$$

or, $\|\mathbf{x}\| \le \|\mathbf{c}\|/(1 - \|H\|)$. If $\mathbf{c}$ is selected such that $\|\mathbf{x}\| = \|A^{-1}\|\|\mathbf{c}\|$, then it follows that $\|A^{-1}\| \le 1/(1 - \|H\|)$. Hence

$$cond(A) = \|A\| \cdot \|A^{-1}\| \leq \frac{1 + \|H\|}{1 - \|H\|} < \frac{2}{1 - \|H\|}$$

$\square$

This example shows that if $\|H\|$ is not very close to unity, matrix A for the iterative system is well conditioned.

Theorem 3.4 *If the matrix A is non-singular and strictly diagonally dominant satisfying the strong row-sum criterion*

$$\sum_{\substack{j=1 \\ j \neq i}}^{n} \left| \frac{a_{ij}}{a_{ii}} \right| < 1$$

then the Jacobi and the Seidel iterations converge for any starting vector, to the solution.

Proof We apply the preceding theorem. For the Jacobi method, we have $H_J = -D^{-1}(L + U)$. Hence (or otherwise from Eq. (3.29a))

$$\|H_J\|_\infty = \max_{1 \leq j \leq n} \sum_{\substack{j=1 \\ j \neq i}}^{n} \frac{|a_{ij}|}{|a_{ii}|} < 1$$

proving the convergence of the Jacobi method, by application of Theorem 3.3.

For the Seidel method, consider an arbitrary vector $\mathbf{y}$ and let $\mathbf{z} := H_S \mathbf{y}$, $H_S = -(D + L)^{-1} U$. We prove that the components z_i of the vector $\mathbf{z}$ satisfy

$$|z_i| \leq \sum_{\substack{i=1 \\ j \neq i}}^{n} \left| \frac{a_{ij}}{a_{ii}} \right| \|\mathbf{y}\|_\infty$$

First, we have using Eq. (3.32),

$$|z_1| \leq \sum_{j=2}^{n} \left| \frac{a_{1j}}{a_{11}} \right| |y_j| \leq \sum_{j=2}^{\infty} \left| \frac{a_{1j}}{a_{11}} \right| \|\mathbf{y}\|_\infty \leq \|\mathbf{y}\|_\infty$$

Next, by complete induction, using the same equation

$$|z_i| \leq \frac{1}{|a_{ii}|} \left[\sum_{j=1}^{i-1} |a_{ij}||z_j| + \sum_{j=i+1}^{n} |a_{ij}||y_j| \right]$$

$$\leq \frac{1}{|a_{ii}|} \left[\sum_{j=1}^{i-1} |a_{ij}| + \sum_{j=i+1}^{n} |a_{ij}| \right] \|\mathbf{y}\|_\infty \leq \sum_{j=1}^{n} \left| \frac{a_{ij}}{a_{ii}} \right| \|\mathbf{y}\|_\infty$$

This completes the assertion. This means that $\|z\|_\infty \le \|H_J\|\|y\|_\infty$, or

$$\|H_S\|_\infty = \max_{y\neq 0} \frac{\|H_S y\|_\infty}{\|y\|_\infty} = \max_{y\neq 0} \frac{\|z\|_\infty}{\|y\|_\infty} \le \|H_J\| < 1$$

proving convergence of the Seidel method, by application of Theorem 3.3.   □

Remark If the strictly diagonally dominant matrix A satisfies the column-sum criterion

$$\sum_{\substack{i=1 \\ i\neq j}}^{n} \left|\frac{a_{ij}}{a_{jj}}\right| < 1$$

then also the Jacobi and Seidel methods can be proved to converge. For this case one has to use the 1-norm.

We state some other theorems omitting proofs because of technical details.

Theorem 3.5 *If the matrix A is symmetric and positive definite, then Seidel relaxations converge for every initial value.*

Theorem 3.6 (Young) *For strictly diagonally dominant matrices, the SOR method converges for all $0 < \omega < 2$ and the Seidel method converges twice as fast as the Jacobi method.*

Exercises

1. Show by appropriate examples that the strong row-sum criterion and the strong column-sum criterion are not equivalent.

$[A = \begin{bmatrix} 1 & 0.7 & 0.2 \\ 0.7 & 2 & 0.5 \\ 0.5 & 0.5 & 2 \end{bmatrix}$ satisfies strong row-sum criterion but not the other. A^T exemplifies the reverse case].

2. Show using the subroutine SEIDEL that the system

$$\begin{array}{rcl}
2x_1 - x_2 & = 1 \\
-x_1 + 2x_2 - x_3 & = 0 \\
- x_2 + 2x_3 - x_4 & = 0 \\
\cdots \quad \cdots \quad \cdots \quad \cdots & \\
-x_{n-1} + 2x_n & = 1
\end{array}$$

converges for $n = 2, 3, \cdots, N$ to the solution $[1, 1, \cdots, 1]$ although it is not diagonally dominant.

3.4 Non-linear System of Equations

A *non-linear system of equations* in n unknowns $x_1, x_2, \cdots, x_n$, real or complex, is of the form

$$
\begin{aligned}
f_1(x_1, x_2, \cdots, x_n) &= 0 \\
f_2(x_1, x_2, \cdots, x_n) &= 0 \\
\cdots\cdots\cdots\cdots\cdots\cdots \\
f_n(x_1, x_2, \cdots, x_n) &= 0
\end{aligned}
\tag{3.34}
$$

where $f_1, f_2, \cdots, f_n$ are functions of n arguments. In vector notations, the system can be compactly written as

$$
\mathbf{f}(\mathbf{x}) = 0
\tag{3.34a}
$$

where $\mathbf{f} = [f_1(\mathbf{x}),\ f_2(\mathbf{x}), \cdots, \mathbf{f}_n(\mathbf{x})]^T$ with $\mathbf{x} = [x_1, x_2, \cdots, x_n]^T$.

In the general form, direct sequential elimination is not possible to obtain a single non-linear equation in one unknown. In this context, the extension of iterative methods of Chap. 2, have proved useful. In Sects. 3.4.1 and 3.4.2, we develop such extensions for Newton's method and general iteration. Of the two methods, the former is important for practical purpose, while the latter is of greater theoretical interest. At practical level, the development is restricted to the computation of real roots only.

The iterative methods require initial approximation of the roots. If $n = 2$ and one is seeking only real roots of $f_1(x_1, x_2) = 0$, $f_2(x_1, x_2) = 0$, then a simple procedure could be to draw the graphs of the two equations in the x_1, x_2 plane and note their points of intersection. The x_1, x_2 values of such points approximate the real roots of the equations. A more general method is to consider the function

$$
F(x_1, x_2, \cdots, x_n) := f_1^2 + f_2^2 + \cdots + f_n^2 \ge 0
$$

The minimum value of F is obviously 0, corresponding to the solution of the system (3.34). Starting from some initial point $(x_1^0, x_2^0, \cdots, x_N^0)$, we compute F varying only the first variable x_1 at suitable intervals and locate a minimum $x_1 = x_1^1$ (atleast one minimum exists as F is bounded below). Repeating the whole procedure, varying $x_2, x_3, \cdots, x_n$ singly in succession, we obtain a definite number of minima. The minimum of the minima yields approximation to a real root. Alternatively, one may consider F defined by

$$
F(x_1, x_2, \cdots, x_n) := |f_1| + |f_2| + \cdots + |f_n|
$$

and proceed as before adopting simple search without any use of derivatives. After a root has been isolated and approximated, we next treat methods of refinement of a root.

3.4.1 Newton's Method for n Equations

Suppose $\mathbf{x}^{(0)}$ is an initial approximation to a root ξ of Eq. (3.34a). To find the successive iterates $\mathbf{x}^{(1)}$, $\mathbf{x}^{(2)}$, $\cdots$, we first note that Taylor's theorem for n variables applied to component f_i can be written as

$$f_i(\mathbf{x} + \mathbf{h}) = f_i(\mathbf{x}) + [\nabla f_i(\mathbf{x})]^T \cdot \mathbf{h} + O(\|\mathbf{h}\|^2)$$

where f_i is assumed to be continuous with continuous partial derivatives up to the second order. Compiling such equations for $i = 1, 2, \cdots, n$, we obtain in vector form

$$\mathbf{f}(\mathbf{x} + \mathbf{h}) = \mathbf{f}(\mathbf{x}) + \mathbf{f}'(\mathbf{x})\mathbf{h} + O(\|\mathbf{h}\|^2)$$

where $\mathbf{f}'$ is the *Jacobian matrix* of $\mathbf{f}$ at $\mathbf{x}$, given by

$$\mathbf{f}'(\mathbf{x}) := \left[\frac{\partial f_i}{\partial x_j}\right], \qquad (i, j = 1, 2, \cdots, n)$$

Thus, for an iterate $\mathbf{x}^{(k)}$, if $\xi = \mathbf{x}^{(k)} + \mathbf{h}$, we obtain by ignoring $O(\|\mathbf{h}\|^2)$ terms

$$\mathbf{f}'(\mathbf{x}^{(k)})\,\mathbf{h} = -\mathbf{f}(\mathbf{x}^{(k)}) \tag{3.35}$$

Hence

$$\mathbf{x}^{(k+1)} = \mathbf{x}^{(k)} - [\mathbf{f}'(\mathbf{x}^{(k)})]^{-1}\mathbf{f}(\mathbf{x}^{(k)}) \tag{3.36}$$

Computationally, however, it is more convenient to solve the *linear system* for $\mathbf{h}$ given by Eq. (3.35), by a direct method like Gauss elimination, rather than invert the Jacobian matrix.

If the initial approximation, $\mathbf{x}^{(0)}$ is close enough to ξ, the method converges quadratically. However, two types of difficulty may crop up in practical implementation of the method. First, if the components of $\mathbf{f}$ have complicated form, the calculation of the Jacobian $\mathbf{f}'(\mathbf{x})$ may prove to be tedious. In such cases, numerical differentiation using finite differencing can be used with equal effect. Although the topic of numerical differentiation is treated in Chap. 5, the partial derivatives appearing in the Jacobian matrix for the present purpose can be written by Taylor expansion as

$$\frac{\partial f_i}{\partial x_j} = \frac{f_i(x_1, \cdots, x_j + h_j, \cdots, x_n) - f_i(x_1, \cdots, x_j - h_j, \cdots, x_n)}{2h_j} + O(h_j^4), \quad \text{as } h_j \to 0$$

More importantly from practical point of view, convergence may fail because of poor initial guess of $\mathbf{x}^{(0)}$. To address this point, we note that the Newton direction $\mathbf{h} = -[\mathbf{f}'(\mathbf{x})]^{-1}\mathbf{f}(\mathbf{x})$ is a direction of *descent* for the function

$$F(\mathbf{x}) := f_1^2(\mathbf{x}) + f_2^2(\mathbf{x}) + \cdots + f_n^2(\mathbf{x}) =: \|\mathbf{f}(\mathbf{x})\|_2^2$$

and therefore points towards the solution of the problem. By differentiation one can write

$$\nabla F(\mathbf{x}) = 2\mathbf{f}'(\mathbf{x})^T \mathbf{f}(\mathbf{x})$$

Hence $\left(\nabla F(\mathbf{x})\right)^T \mathbf{h} = \left[2\mathbf{f}^T(\mathbf{x})\mathbf{f}'(\mathbf{x})\right]\left[-\{\mathbf{f}'(\mathbf{x})\}^{-1}\mathbf{f}(\mathbf{x})\right] = -2\|\mathbf{f}(\mathbf{x})\|_2^2 < 0$ which means that $\mathbf{h}$ is obtuse to the increasing direction of $F(\mathbf{x})$. In other words, $\mathbf{h}$ is a direction of descent. Thus if lack of monotone convergence is indicated by increase in the residual of $\mathbf{f}$ during iterations, viz. $\|\mathbf{f}(\mathbf{x}^{(k+1)})\|_2 > \|\mathbf{f}(\mathbf{x}^{(k)})\|_2$, then convergence can be forced by reducing the step size by considering $\mathbf{x}^{(k)} + \mathbf{h}/2^m$, $m = 1, 2, \cdots$ and take $\mathbf{x}^{(k+1)}$ to be the first such vector for which the residue is less than $\|\mathbf{f}(\mathbf{x}^{(k)})\|_2$. The procedure is called *Damped Newton's Method*. The following subroutine accomplishes this procedure:

```
SUBROUTINE DAMPED_NEWTON(x,n,xtol,maxiter)
! Computes a root of f(1)(x(1), x(2),....., x(n))=0
!                     f(2)(x(1), x(2), ....., x(n))=0, etc.
!                     f(n)(x(1), x(2), ......, x(n))=0
! given by a subroutine fn(x,f,n).
! n = number of unknowns and equations.    (Input)
! x(1), x(2), ....., x(n)      (Initial approximation of a root).    (Input)
! The root is returned in x(1), x(2), :..., x(n).    (Output)
! maxiter = maximum number of iterations allowed.    (Input)
!*********************************************************************
REAL :: x(n), f(n), h(n), yacobian(n,n), xnew(n)
DO k=1,maxiter
CALL fn(x,f,n)
sn=0.0
DO i=1,n
sn=sn+f(i)**2
END DO
CALL JACOBIAN(x,yacobian,n)
CALL GAUSS(n,yacobian,f)
DO i=1,n
h(i)=-f(i)
END DO
DO m=0,maxiter
DOi=1,n
xnew=x(i)+h(i)/2**m
END DO
amax=0.0
DO i=1,n
absdiff=ABS(xnew(i)-x(i))
IF(amax<absdiff) amax=absdiff
END DO
```

```
IF(amax<xtol .AND. sn<0.1E-10) RETURN
CALL fn(xnew,f,n)
snu=0.0
DO i=1,n
snu=snu+f(i)**2
END DO
IF(snu<sn) EXIT
END DO
DO i=1,n
x(i)=xnew(i)
END DO
END DO
END SUBROUTINE DAMPED_NEWTON
```

As illustration of the procedure consider the following example:

Example 1. The equations

$$f_1(x, y) := x + 3\ln|x| - y^2 = 0$$

$$f_2(x, y) := 2x^2 - xy - 5x + 1 = 0$$

has several solutions as revealed by computing $f_1^2 + f_2^2$ for a range of values of x and y. Solve the system for initial approximations (*i*) $x = 5$, $y = 1$, (*ii*) $x = 5$, $y = 0$ and (*iii*) $x = 2$, $y = -2$.

Solution. The Jacobian of f_1, f_2 is

$$\begin{bmatrix} 1 + 3/x & -2y \\ 4x - y - 5 & -x \end{bmatrix}$$

A Fortran program implementing the damped Newton method for the problem is:

```
PROGRAM MAIN
REAL :: x(2)
x(1)=5; x(2)=1        !First initial value.
! x(1)=5; x(2)=0      !Second initial value.
! x(1)=2; x(2)=-2     !Third initial value.
n=2; xtol=0.1E-6; maxiter=20
CALL DAMPED_NEWTON(x,n,xtol,maxiter)
PRINT*, (x(i), i=1,2)
END
!*****************************************

SUBROUTINE fn(x,f,n)
REAL :: x(n), f(n)
```

```
f(1)=x(1)+3*ALOG(ABS(x(1)))−x(2)**2
f(2)=2*x(1)**2−x(1)*x(2)−5*x(1)+1
RETURN
END SUBROUTINE fn
!*****************************************

SUBROUTINE JACOBIAN(x,yacobian,n)
REAL :: x(n), yacobian(n,n), fplus(n), fminus(n)
! yacobian(1,1)=1+3/x(1); yacobian(1,2)=−2*x(2)
! yacobian(2,1)=4*x(1)−x(2)−5; yacobian(2,2)=−x(1)
hj=0.0001
DO i=1,n
DO j=1,n
xj=x(j); x(j)=xj+hj; CALL fn(x,fplus,n)
x(j)=xj-hj; CALL fn(x,fminus,n); x(j)=xj
yacobian(i,j)=(fplus(i)-fminus(i))/(2*hj)
END DO
END DO
END SUBROUTINE JACOBIAN
!*****************************************
```

! Append SUBROUTINE DAMPED_NEWTON
! Append SUBROUTINE GAUSS

In the last segment, subroutine GAUSS from Sect. 3.1.2 must be appended as stated at the end of the program. For the initial approximation (i) $x = 5$, $y = 1$ the output of the program gives

$$x = 3.75683, \quad y = 2.77985$$

without going through the damping procedure (that is to say with $m = 0$). For the case (ii) $x = 5$, $y = 0$ however, another solution

$$x = 1.37348, \quad y = -1.52497$$

is obtained after damping is resorted to in the iterations $(m \geq 1)$. The smallness of sn2 confirms the accuracy of the roots. The third case (iii) yields the same root as in case (ii). $\qquad \square$

Exercises

Solve the following equations:

1.
$$4x^2 + y^2 + 2xy - y - 2 = 0$$
$$2x^2 + 3xy + y^2 - 3 = 0$$
Assume $x^{(0)} = 0.1$, $y^{(0)} = 0.1$. $\qquad\qquad [x = 0.5, \ y = 1.0]$

2.
$$x^6 - 5x^2y^2 + 136 = 0$$
$$y^4 - 3x^4y + 80 = 0$$
Assume $x^{(0)} = 1$, $y^{(0)} = 2$. $[x = 2.08838, \ y = 3.16875]$

3.
$$y \cos(xy) + 1 = 0$$
$$\sin(xy) + x - y = 0$$
Assume (i) $x^{(0)} = y^{(0)} = 0$, (ii) $x^{(0)} = y^{(0)} = 1$, (iii) $x^{(0)} = 1$, $y^{(0)} = 2$.
$[(i)$ $x = -3.20904, \ y = -2.30780$ (ii) $x = -4.22486, \ y = -3.27269$
(iii) $x = 1/08619, \ y = 1.94369]$

3.4.2 Broyden's Method

This method is an extension of the *Secant Method* for numerically solving a single non-linear equation to the case of a system of n non-linear equations. The objective of the method is to avoid repeated evaluation of the Jacobian matrix $\mathbf{f}'(\mathbf{x})$ in the Newton method, that may be of very complicated form. In the secant iterations (Sect. 2.1.2.3, Chap. 2), if x_0, x_1, x_2 are three iterates of a real zero ξ of the equation $f(x) = 0$, then writing $f_0 = f(x_0)$, $f_1 = f(x_1)$ for convenience, the next iterate is given by the relation

$$x_2 = x_1 - f(x_1) \left(\frac{x_1 - x_2}{f(x_1) - f(x_0)} \right) = x_1 - (f_1')^{-1} f_1$$

where $f_1' = (f_1 - f_0)/(x_1 - x_0)$ is the secant approximation of the derivative $f'(x_1) \approx f_1'$. The expression for f_1' can be put in the iterative form

$$f_1' = f_0' + \frac{[f_1 - f_0 - f_0'(x_1 - x_0)] (x_1 - x_0)}{(x_1 - x_0)^2} \tag{3.37}$$

in terms of the approximate derivative at the preceding point f_0', and the function values f_0, f_1 at the points x_0 and x_1.

In multidimension n, the secant approximation of Eq. (3.36) for $k = 1$ becomes

$$\mathbf{x}^{(2)} = \mathbf{x}^{(1)} - [\mathbf{f}'^{(1)}]^{-1} \mathbf{f}^{(1)} \tag{3.38}$$

with

$$\mathbf{f}'^{(1)} (\mathbf{x}^{(1)} - \mathbf{x}^{(0)}) = \mathbf{f}^{(1)} - \mathbf{f}^{(0)} \tag{3.39}$$

Equation (3.39) means that the Jacobian matrix $\mathbf{f}'^{(1)}$ maps the vector $\mathbf{x}^{(1)} - \mathbf{x}^{(0)}$ on to $\mathbf{f}^{(1)} - \mathbf{f}^{(0)}$. Now, any vector such as $\mathbf{f}^{(1)} - \mathbf{f}^{(0)}$ can be represented as a linear combination of $\mathbf{x}^{(1)} - \mathbf{x}^{(0)}$ and a vector $\mathbf{v}$ orthogonal to it, which in matrix notations, satisfies the orthogonality condition

$$[\mathbf{x}^{(1)} - \mathbf{x}^{(0)}]^T \mathbf{v} = 0$$

Suppose that the vector $\mathbf{v}$ is such that

$$\mathbf{f}'^{(1)} \mathbf{v} = \mathbf{f}'^{(0)} \mathbf{v}$$

then it is implied that $\mathbf{v}$ remains unaffected by updation of the Jacobian from $\mathbf{f}'^{(0)}$ to $\mathbf{f}'^{(1)}$. By direct substitution it can be verified that the above two conditions and Eq. (3.39) are uniquely satisfied if

$$\mathbf{f}'^{(1)} = \mathbf{f}'^{(0)} + \frac{[\mathbf{f}^{(1)} - \mathbf{f}^{(0)} - \mathbf{f}'^{(0)} (\mathbf{x}^{(1)} - \mathbf{x}^{(0)})] [\mathbf{x}^{(1)} - \mathbf{x}^{(0)}]^T}{||\mathbf{x}^{(1)} - \mathbf{x}^{(0)}||_2^2} \tag{3.40}$$

Evidently, Eq. (3.40) reduces to Eq. (3.37) in the case of one dimension $n = 1$. In general, the iteration (3.38) is

$$\mathbf{f}'^{(k)} (\mathbf{x}^{(k+1)} - \mathbf{x}^{(k)}) = -\mathbf{f}^{(k)} \tag{3.41}$$

while (3.40) generalises to

$$\mathbf{f}'^{(k)} = \mathbf{f}'^{(k-1)} + \frac{[\mathbf{f}^{(k)} - \mathbf{f}^{(k-1)} - \mathbf{f}'^{(k-1)} (\mathbf{x}^{(k)} - \mathbf{x}^{(k-1)})] [\mathbf{x}^k - \mathbf{x}^{(k-1)}]^T}{||\mathbf{x}^{(k)} - \mathbf{x}^{(k-1)}||_2^2} \tag{3.42}$$

The scheme (3.41), (3.42) is implemented below in which the initial approximation of the root $\mathbf{x}^{(o)}$ is assumed to be given. The next approximation $\mathbf{x}^{(1)}$ is obtained by Newton's method using the Jacobian value by finite differencing. Starting with these two approximations, $\mathbf{x}^{(2)}$ is obtained from the formulae (3.41) and (3.42), and the iterations are carried on until convergence is indicated. As in Damped Newton method, the step size is reduced by powers of 2 in the step length, if lack of convergence is indicated from the condition $||\mathbf{f}^{(k+1)}||_2 > ||\mathbf{f}^{(k)}||_2$.

Remark 3.1 In the implementation of the Broyden Method, a procedure for obtaining the inverse $[\mathbf{f}'^{(k)}]^{-1}$ using a matrix inversion formula due to Sherman - Morrison is often suggested to obtain $\mathbf{x}^{(k+1)}$ from $\mathbf{x}^{(k)}$ in formula (3.41). This however makes the subroutine much longer.

Charles George Broyden (1933–2011), British mathematician, who specialised in optimisation problems and numerical linear algebra. While working on a problem of physics in the English Electric Company, he developed the method in 1965 that bears his name. Later on, he joined Aberystwyth University, Wales moving to Essex University, and subsequently to University of Bologna, Italy.

```
SUBROUTINE BROYDEN(x,n,xtol,maxiter,mmax)
! Computes a root of f(1)(x(1), x(2),....., x(n))=0
!                     f(2)(x(1), x(2), ....., x(n))=0, etc.
!                     f(n)(x(1), x(2), ......, x(n))=0
! given by a subroutine fn(x,f,n).
! n = number of unknowns and equations.    (Input)
! x(1), x(2), ....., x(n)        (Initial approximation of a root).    (Input)
! The root is returned in x(1), x(2), :..., x(n).    (Output)
! maxiter = maximum number of iterations allowed.    (Input)
! mmax = maximum number of step length halving employed.    (Input)
!*************************************************************
REAL :: x(n), x1(n), f(n), f1(n), x2(n), f2(n), yacobian(n), fp(n), fm(n)
hj=0.0001
DO i=1,n; DO j=1,n
xj=x(j); x(j)=xj+hj; CALL fn(x,fp,n)
x(j)=xj-hj; CALL fn(x,fm,n); x(j)=xj
yacobian(i,j)=(fp(i)-fm(i))/(2*hj)
END DO; END DO
DO k=1,maxiter
CALL fn(x,f,n)
   IF(k==1) THEN
   CALL GAUSS(n,yacobian,f)
   DO i=1,n
   x1(i)=x(i)-f(i)
   END DO
   CALL fn(x1,f1,n)
   END IF
DO i=1,n
fp(i)=x1(i)-x(i)
END DO
sn=0.0; fpn=0.0
DO i=1,n
sn=sn+f1(i)**2
fpn=fpn+fp(i)**2
END DO
DO i=1,n
fm(i)=0.0
DO j=1,n
fm(i)=fm(i)+yacobian(i,j)*fp(j)
END DO; END DO
DO i=1,n; DO j=1,n
yacobian(i,j)=yacobian(i,j)+(f1(i)-f(i)-fm(i))*fp(j)/fpn
END DO; END DO
CALL GAUSS(n,yacobian,f1)
DO i=1,n
```

```
fp(i)=-f1(i)
END DO
DO m=0,mmax
DO i=1,n
x2(i)=x1(i)+fp(i)/2**m
END DO
amax=0.0
DO i=1,n
absdiff=ABS(x2(i)-x1(i))
IF(amax<absdiff) amax=absdiff
END DO
IF(amax<0.1E-6 .OR. sn<0.1E-10) RETURN
CALL fn(x2,f2,n)
snu=0.0
DO i=1,n
snu=snu+(f2(i))**2
END DO
IF(snu<sn) EXIT
END DO
DO i=1,n
x(i)=x1(i); x1(i)=x2(i); f(i)=f1(i); f1(i)=f2(i)
END DO
END DO
END SUBROUTINE BROYDEN
```

Example 1. Solve the equations of Example 1 of the preceding section employing subroutine BROYDEN.

Solution. A Fortran program is written as follows:

```
PROGRAM MAIN
REAL :: x(2)
x(1)=5; x(2)=1        !First initial value.
! x(1)=5; x(2)=0      !Second initial value.
! x(1)=2; x(2)=-2     !Third initial value.
n=2; xtol=0.1E-6; maxiter=20; mmax=20
CALL DAMPED_NEWTON(x,n,xtol,maxiter)
PRINT*, (x(i), i=1,2)
END
!*****************************************
SUBROUTINE fn(x,f,n)
REAL :: x(n), f(n)
f(1)=x(1)+3*ALOG(ABS(x(1)))−x(2)**2
f(2)=2*x(1)**2−x(1)*x(2)−5*x(1)+1
RETURN
```

END SUBROUTINE fn
!***

! Append SUBROUTINE BROYDEN
! Append SUBROUTINE GAUSS

Running the program, the answers are obtained exactly as from the application of the Damped Newton's Method. However, it is to be noted that for the first case x(1) = 5, x(2) = 1, as no subdivision of the step length is required, the maximum value of mmax should not be set greater than 16.

3.4.3   General Iteration for n Equations

Here the ideas of Sect. 2.2, Chap. 2 for a single equation are extended to the case of n equations. We suppose that it is possible to rewrite Eq. (3.34a) in the form

$$\mathbf{x} = \phi(\mathbf{x}) \tag{3.43}$$

A root ξ (real or complex), therefore satisfies $\xi = \phi(\xi)$ and so is a *fixed point* of ϕ. Let $\mathbf{x}^{(0)}$ be an initial approximation of ξ. starting with this value, one can construct the iterates $\mathbf{x}^{(1)}$, $\mathbf{x}^{(2)}, \cdots$ satisfying

$$\mathbf{x}^{(k+1)} = \phi(\mathbf{x}^{(k)}) \tag{3.44}$$

Sufficient conditions for convergence of the iterates to ξ is given by

Theorem 3.7 (Banach Fixed-Point Theorem) *Suppose ϕ maps a closed set X into itself and is contractive, that is to say, for* $\mathbf{x}, \mathbf{z} \in X$

$$\|\phi(\mathbf{x}) - \phi(\mathbf{z})\| \le K\|\mathbf{x} - \mathbf{z}\|, \ K < 1$$

then the iterates (3.44) converge to ξ, i.e. $\lim_{k \to \infty} \|\mathbf{x}^{(k)} - \xi\| = 0$ with error bounds

(i) $\|\xi - \mathbf{x}^{(k)}\| \le \dfrac{K^n}{1 - K}\|\mathbf{x}^{(1)} - \mathbf{x}^{(0)}\|$

(ii) $\|\xi - \mathbf{x}^{(k)}\| \le \dfrac{K}{1 - K}\|\mathbf{x}^{(k)} - \mathbf{x}^{(k-1)}\|.$

The proof is an extension of the proof of theorem 3 of Chap. 2, with the replacement of absolute values by vector norms.

Stefan Banach (1892–1945), Polish mathematician. He is the founder of the subject of functional analysis. The concept of normed linear spaces is due to Banach.

When ϕ is a linear function, the iteration (3.38), becomes identical to Jacobi relaxations. To accelerate convergence of iterations of (3.37), one may employ the procedure of Seidel relaxation, using currently available variables for updation.

Example 1. Solve the system of non-linear equations

$$x - 0.1y^2 + 0.05z^2 = 0.7$$
$$y + 0.3x^2 - 0.1xz = 0.5$$
$$z + 0.4y^2 + 0.1xy = 1.2$$

Solution. We write the equations in iterative form

$$x = 0.7 + 0.1y^2 - 0.05z^2$$
$$y = 0.5 - 0.3x + 0.1xz$$
$$z = 1.2 - 0.4y^2 - 0.1xy$$

starting with $x^{(0)} = 0.7$, $y^{(0)} = 0.5$, $z^{(0)} = 1.2$, we obtain the Seidel iterations

$$x^{(1)} = 0.65300 \quad y^{(1)} = 0.45044 \quad z^{(1)} = 1.08943$$
$$x^{(2)} = 0.66095 \quad y^{(2)} = 0.44095 \quad z^{(2)} = 1.09308$$
$$x^{(3)} = 0.65970 \quad y^{(3)} = 0.44155 \quad z^{(3)} = 1.09289$$
$$x^{(4)} = 0.65978 \quad y^{(4)} = 0.44151 \quad z^{(4)} = 1.09290$$
$$x^{(5)} = 0.65977 \quad y^{(5)} = 0.44152 \quad z^{(5)} = 1.09290$$
$$x^{(6)} = 0.65977 \quad y^{(6)} = 0.44152 \quad z^{(6)} = 1.09290$$

The last two iterates coincide, supplying the answer. □

3.4.4 Unconstrained Optimization of a Function

Optimization of a function of several variables of the form

$$z = F(x_1, x_2, \cdots, x_n) \tag{3.45}$$

where $x_1, x_2, \cdots, x_n$ are the independent variables, appear in some types of application in different sciences. When the points $(x_1, x_2, \cdots, x_n)$ are bounded by certain domain, the problem is called *Constrained Optimization*; otherwise it is called *Unconstrained Optimization*. The subject of *Operations Research* exclusively deals with the first type of optimization and does not belong to scope of this book. The latter type of optimization is generaliztion of the topic of *Maxima and Minima* of a function of one variable $y = f(x)$, albeit with associated difficulties of higher dimensions.

It is known from the calculus of functions of several variables that the points of maxima and minima of z defined by Eq. (3.45) are given by the system of equations

$$\frac{\partial F}{\partial x_1} = 0, \quad \frac{\partial F}{\partial x_2} = 0, \quad \cdots, \quad \frac{\partial F}{\partial x_n} = 0 \tag{3.46}$$

When the analytical form of the derivatives appearing in the left- hand side of these equations are known, the solution of the system (3.46) may be sought by the methods of Sects. 3.4.1–3.4.4. It is known from the theory of maxima and minima that the extremal points determined by Eq. (3.46) may also contain 'Saddle Points" that are maximum along certain sections and minimum along other sections, depending on the nature of the *Hessian matrix* formed by the second-order partial derivatives. The characterization of extremal points is not attempted because of computational difficulty entailed in that method.

In this section, an extremal point on the other hand, is *directly searched*. To focus attention, a local minimum of z is searched starting from a given starting point $(x_1^{(0)}, x_2^{(0)}, \cdots, x_n^{(0)})$, 'descending' along a slope. If a local maximum of z is required, the minimum of $-z$ is similarly searched. Now, in vector calculus it is known that the increase in the function $F(\mathbf{x})$, where $\mathbf{x} = [x_1, x_2, \cdots, x_n]^T$ is greatest in the direction of the gradient of F, or ∇F, and so the best choice of descent is along the direction of *steepest descent* $-\nabla F$. Hence from a starting point $\mathbf{x}^{(0)}$, if a small step $s > 0$ close to 0 is taken to reach the point

$$\mathbf{x}^{(1)} = \mathbf{x}^{(0)} - s \, \nabla F(\mathbf{x}^{(0)}) \tag{3.47}$$

then the problem is to find s so that $F(\mathbf{x}^{(1)})$ is significantly less than $F(\mathbf{x}^{(0)})$. A long step in general, can obviously spoil the search. The selection of s is therefore carried out by finding the minimum of the function

$$h(s) := F(\mathbf{x}^{(0)} - s \, \nabla(\mathbf{x}^{(0)})) \tag{3.48}$$

The seach process can then be carried out step by step to approach the minimum, tracing the line of *steepest decent* on the surface $z = F(\mathbf{x})$. However, as no analytical criterion for the minimum of z is introduced, *strict check on the iterations should be kept before accepting the end results.*

The minimum of $h(s)$ for small values of s is searched by approximation of $h(s)$ by a parabola passing through three nearby points $[s_1, h(s_1)], [s_2, h(s_2)]$ and $[s_3, h(s_3)]$ as was adopted in Muller's method (Sect. 2.5.1, Chap. 2), such that the parabola is concave upwards, possessing a minimum. Thus, for $0 = s_1 \le s_2 < s_3$, let

$$h(s) \approx a \, (s - s_2)^2 + b \, (s - s_2) + c \tag{3.49}$$

Since the parabola passes through the aforesaid points

$$c = h(s_2) \tag{3.50}$$
$$a(s_1 - s_2)^2 + b(s_1 - s_2) + c = h(s_1)$$
$$a(s_3 - s_2)^2 + b(s_3 - s_2) + c = h(s_3)$$

where, since $h(s)$ is supposed to be a dcreasing function $h(s_1) \geq h(s_2) > h(s_3)$. Eliminating c and b in succession from the above set of equations, we obtain

$$b = (s_2 - s_1) a + \frac{h(s_2) - h(s_1)}{s_2 - s_1}$$

$$a = \frac{1}{s_3 - s_1} \left[\frac{h(s_3) - h(s_2)}{s_3 - s_2} - \frac{h(s_2) - h(s_1)}{s_2 - s_1} \right]$$

In particular, if $s_2 \to s_1 = 0$, then using Eq. (3.48)

$$b \to h'(0) = -(\nabla F(\mathbf{x}^{(0)}) \cdot \nabla F(\mathbf{x}^{(0)})) = -\|\nabla F(\mathbf{x}^{(0)})\|_2^2 \tag{3.51}$$

and

$$a \to \frac{1}{s_3 - s_1} \left[\frac{h(s_3) - h(s_2)}{s_3 - s_2} - b \right] \tag{3.52}$$

The function $h(s)$ approximated by Eq. (3.49) is then minimum for

$$s = s_2 - \frac{b}{2a} = -\frac{b}{2a} \tag{3.53}$$

provided that $0 < s < s_3$; otherwise it is minimum at $s = s_3$.

The above procedure is implemented in the subroutine named STEEPEST_ DESCENT to follow. The step length s in the procedure is first scaled with respect to $\|\nabla F(\mathbf{x}^{(0)})\|_2$ in Eq. (3.48), or in other words $\nabla F(\mathbf{x}^{(0)})$ is first normalised by its norm. Next, at each step, it is assumed that $s_1 = s_2 = 0$ for which Eqs. (3.50), (3.51) and (3.52) hold, while s_3 is taken as 1. If $h(s_3) > h(s_2)$, then s_3 is reduced by half till $h(s_3) < h(s_2)$. Following Eq. (3.53), the minimum of $h(s)$ is attained at $s = min\{-b/2a, s_3\}$. The output is taken from the subroutine itself in order to track the progress of the iterations. It should be noted that failure at any step is possible because of the absence of a minimum in the neighbourhood of the starting point. Second, as the method is *solely dependent on the gradient of the function F*, possible rapid changes in the quantity may mar the nature of the iterations. The subroutine includes a high degree approximation for the computation of the gradient from function values, in case exact analytical formulae are not available for the components of the gradient. In the contrary case, the exact expressions for the components must be used for preserving greater accuracy in the iterations. The approximation of the gradient is based on the following finite difference formula that can be proved by Taylor series expansions:

$$\frac{\partial F}{\partial x_i} = \frac{1}{2h} \left[-F(x_i + 2h) + 8\,F(x_i + h) - 8\,F(x_i - h) + F(x_i - 2h) \right] + O(h^6)$$

where the arguments other than x_i have been suppressed for brevity in writing. In the subroutine, the function values of F are obtained from a separate function subprogram named **F(x,n)**.

```
SUBROUTINE STEEPEST_DESCENT(x,n,z,maxiter)
! Searches for a local minmum of the function z=F(x(1),x(2), .....,x(n)).
! x(1), x(2), ....., x(n)          (Initial approximation of a point of local
                                    minimum).    (Input)
! n = number of independent variables.    (Input)
! The point of minimum F is returned in x(1), x(2), .......,x(n).    (Output)
! The minimum value of F is retuned in z.    (Output)
! maxiter = maximum number of iterations allowed.    (Input)
!**************************************************************
REAL :: x(n), gradf(n)
z = F(x,n)
h=0.0001
DO k=1,maxiter
DO i=1,n
! Approximation for ∂F/∂xᵢ follows. Use exact expression if available.
xi=x(i); x(i)=xi+h; fplus1=F(x,n); x(i)=xi+2*h; fplus2=F(x,n)
x(i)=xi-h; fminus1=F(x,n); x(i)=xi-2*h; fminus2=F(x,n); x(i)=xi;
gradf(i)=(-fplus2+8*fplus1-8*fminus1+fminus2)/(12*h)
END DO
anorm=0.0
DO i=1,n
anorm=anorm+gradf(i)**2
END DO
anorm=SQRT(anorm)
DO i=1,n
gradf(i)=gradf(i)/anorm
END DO
IF(anorm==0.0) RETURN
s1=0; s2=0; s3=1
    h1=F(x-s1*gradf,n)
    h2=h1
10 h3=F(x-s3*grdaf,n)
IF(h3>h2) THEN
s3=s3/2; GOTO 10
END IF
b=-anorm**2; a=((h3-h2)/(s3-s2)-b)/(s3-s1)
s=-0.5*b/a
s=MIN(s,s3)
h=a*(s-s2)**2+b*(s-s2)+h2
DO i=1,n
x(i)=x(i)-s*gradf(i)
```

END DO
znew=F(x,n); hmin=MIN(h1,h2,h3)
amin=MIN(ABS(znew-z),ABS(znew-h),ABS(hmin-znew))
IF(amin<0.1E-10) EXIT
z=znew
PRINT*, k, x, z
END DO
RETURN
END SUBROUTINE STEEPEST_DESCENT
!
! Append FUNCTION F(x,n)

It is again emphasised that in any particular application, tracking of the iterations is essential for arriving at a minimum with reasonable accuracy, as can be seen in the following examples and exercises. The accurate point of minimum can be searched by separately computing the values of the function F in an element whose centre is the approximate value.

Example 1. Find the root of the following pair of equations by steepest descent search:

$$f_1(x, y) = x + 3 \ln |x| - y^2 = 0$$

$$f_2(x, y) = 2x^2 - xy - 5x + 1 = 0$$

considered earlier in Sect. 3.4.1; taking the starting point as $x = 4$, $y = 3$.

Solution. Writing a simple main program callings the subroutine STEEPEST_ DESCENT, taking $z = [f_1(x, y)^2 + f_2(x, y)^2]$ in the function subprogram F(x,n), and the finite difference approximation for the components of gradient of F in the subroutine, the minimum of z is found to be at x = 3.75746, y = 2.78203, with $z_{min} = 0.00014$ which is small. Hence the approximation of the root.   □

Example 2. Find the point where the function

$$F(x_1, x_2) = x_1^3 + x_2^3 - 2x_1^2 + 3x_2^2 - 5$$

is minimum, starting the search from the point $(3, -1)$

Solution. Writing a simple main program calling the subroutine STEEPEST_ DESCENT the minimum is found at $x_1 = 1.32807$, $x_2 = -0.01463$ using the exact expression for the components of the gradient of F, and $x_1 = 1.15553$, $x_2 = 0.00012$ from the finite difference approximation of the gradient of F. The exact point is $(x_1 = 4/3, x_2 = 0)$. Notice the error in the approximate method.   □

Exercises

Find the points of minimum value of the following functions, using both the exact
and the approximate expressions for the gradient of the functions:

1. $z = x^2 - xy + y^2 - 2x + y$, starting at $(1.5, -0.5)$.
[At $(0.99722, 0.00278)$ (for exact expression of gradient), and at $(1.00239, -0.00241)$ (for approximate expression of gradient). Exact point $(1, 0)$].

2. $z = 5x^2 + y^2 + 4xy - 14x - 6y + 20$, starting at $(0.5, 0.5)$.
[At $(1.06325, 0.71524)$ (for exact expression of gradient), and at $(1.06304, 0.71569)$ (for approximate expression of gradient). Exact point $(1, 1)$].

3. $z = 2x^2 - 2xy + y^2$, starting at the point $(2, 3)$.
[At $(0.87864, 0.22395)$ (for exact expression of gradient, and at $(0.12083, 0.25480)$ (for approximate expression of gradient). Notice that the minimum of z is reached before the iterations end. Exact point $(0, 0)$].

4. $z = x^3 y^3 - 3x - 3y$, starting at $(2, 2)$.
[At $(0.99382, 0.99382)$ (for exact expression of gradient), and at $(0.99394, 0.99394)$ (for approximate expression of gradient). Exact point $(1, 1)$. The point is actually a saddle point and not a minimum point in the strict sense].

5. $u = x^3 - 2x^2 + y^2 + z^2 - 2xy + xz - yz + 3z$, starting at $(2, 1, 0)$.
[At $(2.01496, 1.11808, -1.96995)$ (for exact expression of gradient), and at $(2.05479, 1.13437, -1.88389)$ (for approximate expression for gradient in 14 iterations. Exact point $(2, 1, 0)$].

Chapter 4
Interpolation

Approximation of a function by a simpler function is a fundamental topic in numerical computing. Looking back, the solution of an equation $f(x) = 0$ by the regula falsi, secant and Newton methods depended on local approximation of $f(x)$ by linear function (Chap. 2). The powerful Müller's method utilises approximation of $f(x)$ by a local quadratic function. The same approach is adopted for solving a system of nonlinear equations by the general Newton and the Broyden methods. Function minimization by the steepest-decent method is also based on the same technique. In the next chapter, the problems of *numerical differentiation and integration* are systematically developed by this very approach.

Given a continuous function f of a real variable $x \in [a, b]$, one can construct a **table** of distinct discrete values (x_i, y_i), where $y_i = f(x_i)$, $x_i \in [a, b]$ and $(i = 0, 1, 2, \cdots, n)$. The **interpolation points** or **nodes** x_i may be arranged in a net: $a = x_0 < x_1, < \cdots < x_n$ inorder to focus on their distribution. *Mathematical Tables* of transcendental functions like $\sin(x)$, $\cos(x)$, e^x, $\ln(x)$, present the functions in such tabular form. Results of laboratory experiments may be similarly presented, with the difference that the analytical expression for f may not be known. The functional relation between two variables x and y is thus represented in a discrete form. The concept can easily be generalised to functions of more than one independent variable.

Let a (known or unknown) function $f \in C[a, b]$ be approximated by a simpler function $\phi \in C[a, b]$. If the function ϕ also passes through the given **data points** $(x_i, y_i = f(x_i))$ $(i = 0, 1, \cdots, n)$, i.e. $\phi(x_i) = y_i$ then the approximation is called **interpolation**. Evidently, for $x \in [a, b]$, $f(x) \approx \phi(x)$ in some sense, where $\phi(x)$ is easily computable. If ϕ is a simple polynomial of some degree, the interpolation is called **polynomial interpolation**. Similarly, if ϕ is of trigonometric nature containing sine and cosine functions, we have **trigonometric interpolation**.

In general, the theory of construction of function ϕ requires use of Haar spaces (**Alfred Haar (1885–1933)**). Suppose that $\phi_0, \phi_1, \cdots, \phi_n \in C[a, b]$ are *linearly independent functions* such that every nonzero element ϕ spanned by $(\phi_0, \phi_1, \cdots, \phi_n)$ has atmost n distinct zeros in $[a, b]$, then the *function space*

© Springer Nature Singapore Pte Ltd. 2019
S. K. Bose, *Numerical Methods of Mathematics Implemented in Fortran*, Forum for Interdisciplinary Mathematics,
https://doi.org/10.1007/978-981-13-7114-1_4

$U := span(\phi_0, \phi_1, \cdots, \phi_n)$ is called a *Haar space*. Given a set of data points (x_i, y_i) $(i = 0, 1, \cdots, n)$ at the **nodes** x_i we can select an element $\phi \in U$ such that

$$\phi = a_0 \phi_0 + a_1 \phi_1 + \cdots + a_n \phi_n$$

where $a_0, a_1, \cdots, a_n$ are some constants. ϕ is an interpolating function if it passes through the distinct data points (x_i, y_i) satisfying the linear system of equations

$$a_0 \phi_0(x_i) + a_1 \phi_i(x_i) + \cdots + a_n \phi_n(x_i) = y_i$$

for $i = 0, 1, \cdots, n$. The system possesses a unique solution for $a_0, a_1, \cdots, a_n$, since $\phi_0, \phi_1, \cdots, \phi_n$ are linearly independent. Using this method, we proceed to discuss polynomial interpolation.

4.1   Polynomial Interpolation

This type of interpolation is most important in numerical analysis. In this case, ϕ is a polynomial, which in this chapter is taken in the notationally convenient form

$$\phi(x) = a_0 + a_1 x + a_2 x^2 + \cdots + a_n x^n \tag{4.1}$$

(like a truncated power series). The following theorem proves the existence of such ϕ:

Theorem 4.1 *Given a set of distinct nodes $x_i \in [a, b]$ with associated values $y_i = f(x_i)$ $(i = 0, 1, \cdots, n)$, there exists a unique polynomial ϕ of degree at most $n \in U$, which takes on these values.*

Proof A polynomial of degree at most n is of the type (4.1). Since it must pass through the data points (x_i, y_i) $(i = 0, 1, \cdots, n)$,

$$a_0 + a_1 x_i + \cdots + a_n x_i^n = y_i$$

for $i = 0, 1, \cdots, n$. The determinant of this linear system for $a_0, a_1, \cdots, a_n$ is the Vandermonde determinant(**Alexandre Theóphile Vandermonde (1735–1796)**)

$$V := \begin{vmatrix} 1 & x_0 & \cdots & x_0^n \\ 1 & x_1 & \cdots & x_1^n \\ \cdots\cdots\cdots\cdots\cdots \\ 1 & x_n & \cdots & x_n^n \end{vmatrix} = \prod_{0 \le j < i \le n} (x_i - x_j)$$

Thus, $V \neq 0$, yielding a unique solution for $a_0, a_1, \cdots, a_n$. ϕ is therefore uniquely determined. Evidently, $\phi \in U$.   □

Remark If $a_n = 0$ the degree of $\phi(x)$ is less than n, otherwise it is exactly n.

Example 1. Let $f(x) = \sin x$, $-\pi/2 \le x \le +\pi/2$. Find $\phi(x)$ for three data points $(-\pi/2, -1)$, $(0, 0)$ and $(\pi/2, 1)$.

Solution. For three data points the degree of $\phi(x)$ must be 2, that is $\phi(x) = a_0 + a_1 x + a_2 x^2$. Since it must through the three given points,

$$a_0 - a_1\frac{\pi}{2} + a_2\frac{\pi^2}{4} = -1, \quad a_0 = 0, \quad a_0 + a_1\frac{\pi}{2} + a_2\frac{\pi^2}{4} = +1$$

Hence $a_0 = a_2 = 0$, $a_1 = 2/\pi$. Thus $\phi(x) = \frac{2}{\pi}x$, which is linear in x. □

Although the method of proof of Theorem 1 is constructive, practical implementation is done by simpler alternative ways.

4.1.1 Lagrange's Method

As before, let a function $f \in C[a, b]$ be given as a set of $(n+1)$ distinct data points $(x_i, y_i = f(x_i))$ $(i = 0, 1, \cdots, n)$. In Lagrange's method, the approximating interpolation polynomial ϕ of degree n is written as

$$\begin{aligned}
\phi(x) = {}& A_0(x - x_1)(x - x_2)\cdots(x - x_n)\\
& + A_1(x - x_0)(x - x_2)\cdots(x - x_n)\\
& + \cdots\cdots\cdots\cdots\cdots\cdots\cdots\cdots\\
& + A_n(x - x_0)(x - x_1)\cdots(x - x_{n-1})
\end{aligned}$$

The right-hand side of the above expression contains unknown coefficients A_0, $A_1 \cdots$, A_n instead of a_0, $a_1, \cdots$, a_n of Eq. (4.1). Since it must pass through the given data points, setting $x = x_0$ $\phi(x_0) = y_0$, we get $y_0 = A_0(x_0 - x_1)(x_0 - x_2)\cdots(x_0 - x_n)$, directly yielding the coefficient A_0 as

$$A_0 = \frac{y_0}{(x_0 - x_1)(x_0 - x_2)\cdots(x_0 - x_n)}$$

Similarly setting $x = x_1$, $\phi(x_1) = y_1$; $\cdots$; $x = x_n$, $\phi(x_n) = y_n$, we get

$$A_1 = \frac{y_1}{(x_1 - x_0)(x_1 - x_2)\cdots(x_1 - x_n)}$$

$$\cdots\cdots\cdots\cdots\cdots\cdots\cdots\cdots$$

$$A_n = \frac{y_n}{(x_n - x_0)(x_n - x_1)\cdots(x_n - x_{n-1})}$$

Thus we obtain

$$\phi(x) = \frac{(x - x_1)(x - x_2)\cdots(x - x_n)}{(x_0 - x_1)(x_0 - x_2)\cdots(x_0 - x_n)}y_0$$

$$+ \frac{(x - x_0)(x - x_2) \cdots (x - x_n)}{(x_1 - x_0)(x_1 - x_2) \cdots (x_1 - x_n)} y_1$$

$$+ \cdots\cdots\cdots\cdots\cdots\cdots\cdots\cdots\cdots$$

$$+ \frac{(x - x_0)(x - x_1) \cdots (x - x_{n-1})}{(x_n - x_0)(x_n - x_1) \cdots (x_n - x_{n-1})} y_n \tag{4.2}$$

Compactly, Eq. (4.2) can be written as

$$\phi(x) = \sum_{i=0}^{n} l_i(x) \, y_i \tag{4.2a}$$

where

$$l_i(x) = \prod_{\substack{j=0 \\ j \neq i}}^{n} \frac{x - x_j}{x_i - x_j}$$

Joseph Louis Lagrange (1736–1813), Italy born French mathematician who made fundamental contribution to all fields of analysis, number theory, analytical and celestial mechanics. Regarded as the greatest mathematician of the eighteenth century. He was invited to the court of the king of Germany, and later to the court of the French emperor, and appointed a Senator after the French Revolution. His theorems in Calculus are too well known and his development of Calculus of Variations led to the creation of *Analytical Mechanics*, that reduced mechanics to a study of analysis. He also laid the foundations of Group Theory, and proved a number of theorems in Number Theory. In celestial mechanics, he studied the three-body problem and found special case solutions that are known as *Lagrangian Points*. A lunar crater is named after Lagrange.

Apparently formula (4.2) is easily applicable, except for two disadvantages. First, it contains several subtractions in both the numerator and the denominator, which may cause loss of significant figures due to roundoff errors. Second, if the number of data points is increased to assess the accuracy of the result, then there is no way to make use of the existing polynomial, which has already been computed. The method to follow in the next subsection is better in this latter respect.

Example 1. The function $y = f(x)$ is tabulated as follows:

x	1	2	4	7
y	1	1.26	1.59	1.91

Find its value at $x = 5$ by Lagrange interpolation.

Solution. Applying Eq. (4.2)

$$f(5) \approx \phi(5) = \frac{(5-2)(5-4)(5-7)}{(1-2)(1-4)(1-7)} \times 1 + \frac{(5-1)(5-4)(5-7)}{(2-1)(2-4)(2-7)} \times 1.26$$
$$+ \frac{(5-1)(5-2)(5-7)}{(4-1)(4-2)(4-7)} \times 1.59 + \frac{(5-1)(5-2)(5-4)}{(7-1)(7-2)(7-4)} \times 1.91$$
$$= 1.70 \qquad\qquad \Box$$

Exercises

1. Find the interpolating polynomial for the function $y = \sin \pi x$, choosing the points $x_0 = 0$, $x_1 = 1/6$, and $x = 1/2$.

$$[\phi(x) = \frac{x}{2}(7 - 6x)].$$

2. Find the Lagrange interpolating polynomial passing through the data points $(-1, -6)$, $(1, 1)$, $(2, 3/2)$ and $(3, 2)$.

$$[\phi(x) = \frac{1}{4}(x^3 - 6x^2 + 13x - 4)].$$

3. Obtain approximate value of y by interpolation when $x = 1.6$, from the following table:

x	1.2	2.0	2.5	3.0
y	1.36	0.58	0.34	0.20

[0.89].

4. Determine by Lagrange's interpolation, the percentage number of patients of over 40 years age, using the following data:

Age over (x) years	30	35	45	55
% number (y) of patients	148	96	68	34

[74.7].

5. The following table gives the form factor Y and the number of teeth Z for the design of a gear:

Z	20	22	24	30	40	50
Y	0.320	0.330	0.335	0.358	0.390	0.408

Calculate the value of Y when $Z = 35$.

[0.386].

6 (Hermite Interpolation). Let f and its derivative f' be given as data points (x_i, y_i, y'_i), $(i = 0, 1, \cdots, n)$. Prove that the polynomial of degree $2n + 1$, satisfying the data is

$$\phi(x) = \sum_{i=0}^{n} \left[\{1 - 2l_i'(x_i)(x - x_i)\}l_i^2(x) \, y_i + (x - x_i) \, l_i^2(x) y_i' \right]$$

[As in Lagrange interpolation, let

$$\phi(x) = \sum_{i=0}^{n} [u_i(x) \, y_i + v_i(x) \, y_i']$$

For it to pass through the given data points, one gets the nodal conditions

$$u_i(x_j) = \begin{cases} 1 & \text{for } i = j \\ 0 & \text{for } i \neq j \end{cases}, \qquad u_i'(x_j) = 0$$

$$v_i(x_j) = 0, \qquad v_i'(x_j) = \begin{cases} 1 & \text{for } i = j \\ 0 & \text{for } i \neq j \end{cases}$$

$u_i(x)$, $v_i(x)$ must be of degree $2n + 1$ to pass through the given $2n + 2$ data. Hence put

$$u_i(x) = a_i(x) \, l_i^2(x), \quad v_i(x) = b_i(x) \, l_i^2(x)$$

where $a_i(x)$, $b_i(x)$ are linear in x. The nodal conditions give

$$a_i(x_i) = 1, \quad a_i'(x_i) = -2l_i'(x_i)$$

$$b_i(x_i) = 0, \quad b_i'(x_i) = 1$$

Hence $a_i(x) = 1 - 2l_i'(x_i) (x - x_i)$, $b_i(x) = x - x_i$].

Charles Hermite (1822–1901). French mathematician and scholar of Sanskrit and Persian. In 1873, he proved for the first time that e is a transcendental number. Using similar methods **Lindemann** in the year 1882 proved that π was also a transcendental number. He applied elliptic functions to solve the general quintic equation. Hermite is regarded as successor of Gauss and Cauchy in higher arithmetic and analysis.

4.1.2   Newton's Method: Divided Differences

Given a function $f \in C[a, b]$ at a set of data points $(x_i, f(x_i)$, $(i = 0, 1, \cdots, n)$, Newton (Sir Isaac Newton (1642–1727)), represented the interpolating polynomial ϕ of degree at most n, in terms of increasing number of the given nodal points in the form

$$\phi(x) = B_0 + B_1(x - x_0) + B_2(x - x_0)(x - x_1) + \cdots + B_n(x - x_0)(x - x_1) \cdots (x - x_{n-1})$$

$$(4.3)$$

Since $\phi(x)$ passes through the data points $(x_0, f(x_0))$, $(x_1, f(x_1))$, $\cdots$, $(x_n, f(x_n))$, we must have

$$f(x_0) = B_0$$
$$f(x_1) = B_0 + B_1(x_1 - x_0)$$
$$f(x_2) = B_0 + B_1(x_2 - x_0) + B_2(x_2 - x_0)(x_2 - x_1)$$
$$\cdots\cdots\cdots\cdots\cdots\cdots\cdots\cdots\cdots\cdots\cdots$$
$$f(x_n) = B_0 + B_1(x_n - x_0) + B_2(x_n - x_0)(x_n - x_1) + \cdots$$
$$+ B_n(x_n - x_0)\cdots(x_n - x_{n-1})$$

The step by step solution of the system can be written as

$$B_0 = f(x_0) =: f[x_0]$$
$$B_1 = \frac{f(x_1) - f(x_0)}{x_1 - x_0} =: f[x_0, x_1]$$
$$B_2 = \frac{f[x_1, x_2] - f[x_0, x_1]}{x_2 - x_0} =: f[x_0, x_1, x_2] \quad \text{etc.}$$

where the defined quantities on the right hand sides are respectively called **divided differences** of order 0, 1, 2, etc. With a little more effort, we can determine B_3 and guess that

$$B_k = f[x_0, x_1, \cdots, x_k] := \frac{f[x_1, x_2, \cdots, x_k] - f[x_0, x_1, \cdots, x_{k-1}]}{x_k - x_0} \quad (4.4)$$

for $k = 1, 2, \cdots, n$. To establish Eq. (4.4), let $\phi_k(x)$ denote the polynomial of at most degree k up to the coefficient B_k. Then one can write

$$\phi_k(x) = \frac{x_k - x}{x_k - x_0}\phi_{k-1}(x) + \frac{x - x_0}{x_k - x_0}\psi_{k-1}(x)$$

where $\psi_{k-1}(x)$ is a polynomial of degree at most $k - 1$. Since $\phi_k(x)$ and $(x_k - x)\phi_{k-1}(x)$ pass through the points $(x_1, f(x_1))$, $\cdots$, $(x_k, f(x_k))$, $\psi_{k-1}(x)$ must also pass through these points. By uniqueness of polynomials, equating the coefficients of x^k on the two sides of the equation, we obtain Eq. (4.4).

In Newton's formula, the order of the nodes $x_0, x_1, \cdots, x_n$ is immaterial and the polynomial is unique. A divided difference $f[x_0, x_1, \cdots, x_k]$, $(k \le n)$ is the coefficient of x^k in the polynomial and solely depends on the numbers $x_0, \cdots, x_k$. Hence the nodal points in any divided difference can be permuted without change of value. In other words, the *divided differences are symmetric functions of their arguments*.

For calculating the divided differences by hand, it is convenient to arrange them in a table of the following form:

$k = 0$	1	2	3
x_0 $f[x_0]$			
	$f[x_0, x_1]$		
x_1 $f[x_1]$		$f[x_0, x_1, x_2]$	
	$f[x_1, x_2]$		$f[x_0, x_1, x_2, x_3]$
x_2 $f[x_2]$		$f[x_1, x_2, x_3]$	$\vdots$
	$f[x_2, x_3]$	$\vdots$	
x_3 $f[x_3]$	$\vdots$		
$\vdots$	$\vdots$		

The entries in the table can be made columnwise (in the order $k = 0, 1, 2, \cdots$) by inspection; for instance $f[x_0] = f(x_0)$, $f[x_1, x_2] = (f[x_2] - f[x_1])/(x_2 - x_1)$, $f[x_1, x_2, x_3] = (f[x_2, x_3] - f[x_1, x_2])/(x_3 - x_1)$. The divided differences required in the interpolation formula (4.3) together with the coefficients B_k given by Eq. (4.4), occur along the forward slope of the table.

Example 1. Solve Example 1, Sect. 4.1.1 by Newton's divided difference method.

Solution. As above we first form the divided difference table

	0	1	2	3
1	1			
		$\dfrac{1.26 - 1}{2 - 1} = 0.26$		
2	1.26		$\dfrac{0.165 - 0.26}{4 - 1}$ $= -0.03167$	
		$\dfrac{1.59 - 1.26}{4 - 2} = 0.165$		$\dfrac{-0.01167 + 0.03167}{7 - 1}$ $= 0.00333$
4	1.59		$\dfrac{0.10667 - 0.165}{7 - 2}$ $= -0.01167$	
		$\dfrac{1.91 - 1.59}{7 - 4} = 0.10667$		
7	1.91			

Thus, formula (4.3) with $x = 5$ yields the interpolated value

$$\phi(5) = 1 + 0.26(5 - 1) - 0.03167(5 - 1)(5 - 2) + 0.00353(5 - 1)(5 - 2)(5 - 4)$$
$$= 1.69992 = 1.70 \text{ (to two decimal places)}$$

as obtained by Lagrange's method.   □

Example 2. Show that $f[x_0, x_1, \cdots, x_k] = \sum\limits_{i=0}^{k} \dfrac{f(x_i)}{\omega'_k(x_i)}$, where $\omega_k(x) = (x - x_0) \cdots (x - x_k)$.

Solution. Note that $\omega'_k(x_i) = (x_i - x_0) \cdots (x_i - x_{i-1})(x_i - x_{i+1}) \cdots (x_i - x_k) = \prod\limits_{\substack{j=0 \\ j \neq i}}^{k} (x_i - x_j)$.

For $k = 1$, $f[x_0, x_1] = \dfrac{f(x_0)}{x_0 - x_1} + \dfrac{f(x_1)}{x_1 - x_0}$.

We continue the proof by induction, assuming the formula to be true for $1 \leq r < k$. Then

$$f[x_0, \cdots, x_{r+1}] = \frac{f[x_1, \cdots, x_{r+1}] - f[x_0, \cdots, x_r]}{x_{r+1} - x_0}$$

$$= \frac{1}{x_{r+1} - x_0}\left[\sum_{i=1}^{r+1} \frac{f(x_i)}{\prod\limits_{\substack{j=1 \\ j \neq i}}^{r+1} (x_i - x_j)} - \sum_{i=0}^{r} \frac{f(x_i)}{\prod\limits_{\substack{j=0 \\ j \neq i}}^{r} (x_i - x_j)} \right]$$

$$= \frac{f(x_0)}{\prod\limits_{j=1}^{r+1}(x_0 - x_j)} + \sum_{i=1}^{r} \frac{f(x_i)}{\prod\limits_{\substack{j=0 \\ j \neq i}}^{r+1} (x_i - x_j)} + \frac{f(x_{r+1})}{\prod\limits_{j=0}^{r}(x_{r+1} - x_j)}$$

$$= \sum_{i=0}^{r+1} \frac{f(x_i)}{\prod\limits_{\substack{j=0 \\ j \neq i}}^{r+1} (x_i - x_j)} \qquad \square$$

Example 3. Show that $f[x_0, x_1, \cdots, x_k] = \dfrac{f^{(k)}(\xi)}{k!}$.

Solution. Let $\phi_k(x)$ be the interpolating polynomial of $f(x)$ through $x_0, x_1, \cdots, x_k$. Consider the *error* $R(x) = f(x) - \phi_k(x)$, then $R(x)$ vanishes at $k + 1$ distinct points $(x_0, x_1, \cdots, x_k)$ in $[a, b]$. By Rolle's theorem $R'(x)$ must vanish at least at k points in $[a, b]$, $R''(x)$ must similarly vanish at least at $k - 1$ points in $[a, b]$ and so on. Finally, $R^k(x)$ must vanish at least at one point ξ in $[a, b]$, i.e.

$$f^{(k)}(\xi) = \phi^{(k)}(\xi) = f[x_0, x_1, \cdots, x_k] k!$$

since $f[x_0, \cdots, x_k]$ is the coefficient of x^k in $\phi_k(x)$.   □

Example 4. Show that

$$f(x) = f[x_0] + (x - x_0)f[x_0, x_1] + (x - x_0)(x - x_1)f[x_0, x_1, x_2] + \cdots$$
$$+ (x - x_0) \cdots (x - x_{n-1})f[x_0, x_1, \cdots, x_{n-1}] + (x - x_0) \cdots (x - x_n)f[x, x_0, x_1, \cdots, x_n]$$

Solution. From the definition and Eq. (4.4)

$$f(x) = f[x_0] + (x - x_0)f[x, x_0]$$
$$f[x, x_0] = f[x_0, x_1] + (x - x_1)f[x, x_0, x_1]$$
$$f[x, x_0, x_1] = f[x_0, , x_1, x_2] + (x - x_2)f[x, x_0, x_1, x_2]$$
$$\cdots\cdots\cdots\cdots\cdots\cdots\cdots\cdots\cdots\cdots\cdots\cdots\cdots\cdots\cdots$$
$$f[x, x_0, \cdots, x_{n-1}] = f[x_0, x_1, \cdots, x_n] + (x - x_n)f[x, x_0, x_1, \cdots, x_n]$$

Multiplying these equations by 1, $(x - x_0)$, $(x - x_0)(x - x_1)$, $\cdots$, $(x - x_0)(x - x_1) \cdots (x - x_{n-1})$ and adding, we obtain the required identity.   □

Remark So far we have defined divided differences at distinct nodes only. The definition can be extended to coincident nodes in the following way. We regard $f[x_0, x_0]$ as the limit of $f[x_0, x_1]$ when x_1 tends to x_0. Thus

$$f[x_0, x_0] := \lim_{x_1 \to x_0} \frac{f(x_1) - f(x_0)}{x_1 - x_0} = f'(x_0)$$

by Lagrange's Mean Value theorem or by Example 3. This example also shows that

$$f[x_0, x_0, x_0] := \frac{f''(x_0)}{2!} \quad etc.$$

$$f[x_0, \cdots, x_0] \ (k \ \text{arguments each equal to} \ x_0) := \frac{f^{(k)}(x_0)}{k!}.$$

The divided difference table is helpful in writing a simple subroutine. First we note that only the first element of each column of the table is required for interpolation. Thus if we suppose that the points are $(x_1, f[x_1]), \cdots, (x_n, f[x_n])$ and the divided differences are computed in the reverse bottom-up order, then *a one dimensional array* suffices for the purpose. For instance, if there are only three points, then $f(3) \leftarrow (f(3) - f(2))/(x(3) - x(2))$, $f(2) \leftarrow (f(2) - f(1))/(x(2) - x(1))$ leaves $f(1)$ unaltered, yielding the first-order divided difference in $f(2)$ and $f(3)$. Similar is the feature of computation of higher order divided differences. Second, the computation of the polynomial (4.3) should be performed by Horner's scheme:

$$\phi(x) = f[x_1] + (x - x_1)(f[x_1, x_2] + (x - x_2)(f[x_1, x_2, x_3] + \cdots$$
$$+ (x - x_{n-1})(f[x_1, x_2, \cdots, x_n])))$$

Following these observations one can write the following subroutine:

SUBROUTINE DIVDIFF(n,x,f,xbar,phi)
! n=number of nodal points.   (Input)
! x(1),$\cdots$,x(n)=abscissa of nodal points.   (Input)
! f(1),$\cdots$,f(n)=function value at the nodal points.   (Input)
! Divided differences are returned in these elements.   (Output)
! xbar=abscissa of the given point for interpolation.   (Input)
! phi=interpolated value of the function.   (Output)
!
REAL :: x(n), f(n)
DO k=2,n
DO i=n,k,$-$1
f(i)=(f(i)$-$f(i$-$1))/(x(i)$-$x(i$-$k+1))
END DO; END DO
phi=f(n)
DO k=n$-$1,1,$-$1
phi=f(k)+(xbar$-$x(k))*phi
END DO
RETURN
END SUBROUTINE DIVDIFF

Exercises

1. Using the following data find $f(x)$ as divided difference polynomial in x and
hence find $f(3)$:

x	0	1	2	5
y	2	3	12	147

$[f(x) = x^3 + x^2 - x + 2, f(3) \approx 35]$.

2. Solve Exercise 1, Sect. 4.1.1 by Newton's divided difference method.

3. Compile a table of divided differences for the function given as a table

x	-3	0	1	2	3
y	-15	-7	5	25	47

Calculate y for $x = -1$.
$[-10.33]$.

4. For the function $y = f(x)$ given as a table

x	1.08	1.16	1.23	1.26	1.33	1.39
y	2.945	3.190	3.421	3.525	3.781	4.015

Calculate the value of y at $x = 1.30$ by divided difference method. Verify the result by using subroutine DIVDIFF.
[3.669].

5. For the function $f(x) = 1/x$, prove that

$$f[x_0, x_1, \cdots, x_n] = (-1)^n/(x_0 x_1 \cdots x_n)$$

[Use induction].

6. Prove that a polynomial $P_n(x)$ of degree n satisfies the relation

$$P_n[x, x_0, \cdots, x_n] = 0$$

[Use definition to show that $P_n[x, x_0]$, $P_n[x, x_0, x_1]$, $\cdots$, $P_n[x, x_0, \cdots, x_{n-1}]$ are polynomials of degree $n - 1$, $n - 2$, $\cdots$, 0 respectively. Now use Eq. (4.4)].

7. Modify subroutine DIVDIFF to compute $f(x)$ at m points xbar(1), xbar(2),$\cdots$, xbar(m).

8. In the cold rolling of an aluminium strip the dimensionless reduction R in thickness at a point and the resistance to plane homogeneous deformation S in tons/sq. in. are given in the following table:

R	0	8.0	15.5	22.0	32.0
S	3.00	7.60	8.84	9.50	10.32

Calculate S for $R = 10.0$, by divided difference method.
[8.08].

9. For a soil test a plate load test was conducted on a 30 cm square plate which gave the following result:

Load (ton)	0.5	1.0	2.0	3.0	4.0	5.0
Settlement (mm)	1.5	3.0	6.0	12.0	18.0	24.0

Calculate the settlement for a load between 2.5 ton and 4.5 ton. You may use sub-routine DIVDIFF.
[8.7, 20.6].

4.1.3 Neville's Scheme

In certain applications of interpolation, one is interested in interpolating a function f at a fixed point $\bar{x}$, and at no others. Neville's scheme aims at approximating $f(\bar{x})$ by a sequence of polynomials of increasing degree, without actually computing the coefficients of the polynomial as in Newton's method. This allows one to add interpolation points until $f(\bar{x})$ is approximated to some desired accuracy.

Let the interpolation points be $(x_i, f(x_i) = y_i)$, $(i = 0, 1, \cdots, n)$. Let $\phi_{ij}(x)$ denote the polynomial of degree $j - i$, which interpolates $f(x)$ at the points $x_i, x_{i+1}, \cdots, x_j, (i \leq j)$. Then evidently $\phi_{ii}(x) = f(x_i)$ and

$$\phi_{ij}(x) = \frac{1}{x_j - x_i} \begin{vmatrix} \phi_{i,j-1}(x) & x_i - x \\ \phi_{i+1,j}(x) & x_j - x \end{vmatrix} \quad \text{for} \quad i < j \qquad (4.5)$$

To verify the relation (4.5), we first note that the degree of the right-hand side is $(j - i - 1) + 1 = j - i$. Moreover by definition, $\phi_{i,j-1}(x_k) = \phi_{i+1,j}(x_k) = y_k$ for $k = i + 1, \cdots, j - 1$ and so $\phi_{ij}(x_k) = y_k$ by calculation of (4.5). At $x = x_i$, $\phi_{i,j-1}(x_i) = y_i$ and at $x = x_j$, $\phi_{i+1,j}(x_j) = y_j$ and so $\phi_{ij}(x_i) = y_i$ and $\phi_{ij}(x_j) = y_j$ from Eq. (4.5). Thus the right-hand side of (4.5) has all the characteristics of $\phi_{ij}(x)$ and so the identity holds because of uniqueness of polynomials.

Given $\bar{x}$, a triangular table can be formed whose entries are generated columnwise following Eq. (4.5) for increasing degree k of the interpolating polynomial

	$k = 0$	1	2	$\cdots$	n
x_0	$\phi_{00}(\bar{x}) = f(x_0)$				
x_1	$\phi_{11}(\bar{x}) = f(x_1)$	$\phi_{01}(\bar{x})$			
x_2	$\phi_{22}(\bar{x}) = f(x_2)$	$\phi_{12}(\bar{x})$	$\phi_{02}(\bar{x})$		
$\vdots$	$\vdots$	$\vdots$	$\vdots$	$\ddots$	
$\vdots$	$\vdots$	$\vdots$	$\vdots$		
x_n	$\phi_{nn}(\bar{x}) = f(x_n)$	$\phi_{n-1,n}(\bar{x})$	$\phi_{n-2,n}(\bar{x})$	$\cdots$	$\phi_{0,n}(\bar{x})$

In the above table, by Eq. (4.5), the entries are generated thus

$$\phi_{01}(\bar{x}) = \frac{1}{x_1 - x_0} \begin{vmatrix} \phi_{00}(\bar{x}) & x_0 - \bar{x} \\ \phi_{11}(\bar{x}) & x_1 - \bar{x} \end{vmatrix}, \quad \phi_{12}(\bar{x}) = \frac{1}{x_2 - x_1} \begin{vmatrix} \phi_{11}(\bar{x}) & x_1 - \bar{x} \\ \phi_{22}(\bar{x}) & x_2 - \bar{x} \end{vmatrix}$$

etc.

$$\phi_{02}(\bar{x}) = \frac{1}{x_2 - x_0} \begin{vmatrix} \phi_{01}(\bar{x}) & x_0 - \bar{x} \\ \phi_{12}(\bar{x}) & x_2 - \bar{x} \end{vmatrix}, \quad \phi_{13}(\bar{x}) = \frac{1}{x_3 - x_1} \begin{vmatrix} \phi_{12}(\bar{x}) & x_1 - \bar{x} \\ \phi_{23}(\bar{x}) & x_3 - \bar{x} \end{vmatrix}$$

The computation is continued till desired accuracy is reached in a column of interpolated values, or until the data points are exhausted.

Eric Harold Neville (1889–1961) was a professor of mathematics at the University of Reeding, U.K.

An earlier scheme due to Aitken proceeds in a slightly different manner. In this scheme first a similar linear interpolation is carried through points (x_0, x_1), (x_0, x_2), $\cdots$, (x_0, x_n). Then second degree interpolation is carried through points (x_0, x_1, x_2), (x_0, x_1, x_3), $\cdots$, (x_0, x_1, x_n). Similarly higher degree interpolants are constructed until stability of calculated values is reached in the same degree interpolants or until all the data points are exhausted. The scheme is now considered obsolete.

Example 1. Solve Example 1, Sect. 4.1.1 by Neville s scheme.

Solution. Here $\bar{x} = 5$. The table obtained by the scheme is

	0	1	2	3
1	1			
		2.04		
2	1.26		1.66	
		1.755		1.70
4	1.59		1.72	
		1.69666		
7	1.91			

The entries in the table are made as follows. The entries of column 0 are the y values. The entries of column 1 are:

$$\phi_{01}(5) = \frac{1}{2-1} \begin{vmatrix} 1 & 1-5 \\ 1.26 & 2-5 \end{vmatrix} = 2.04$$

$$\phi_{12}(5) = \frac{1}{4-2} \begin{vmatrix} 1.26 & 2-5 \\ 1.59 & 4-5 \end{vmatrix} = 1.755$$

$$\phi_{23}(5) = \frac{1}{7-4} \begin{vmatrix} 1.59 & 4-5 \\ 1.91 & 7-5 \end{vmatrix} = 1.69666$$

The entries of column 2 are

$$\phi_{02}(5) = \frac{1}{4-1} \begin{vmatrix} 2.04 & 1-5 \\ 1.755 & 4-5 \end{vmatrix} = 1.66$$

$$\phi_{13}(5) = \frac{1}{7-2} \begin{vmatrix} 1.755 & 2-5 \\ 1.69666 & 7-5 \end{vmatrix} = 1.72$$

The entry of column 3 is obtained as

$$\phi_{03}(5) = \frac{1}{7-1} \begin{vmatrix} 1.66 & 1-5 \\ 1.72 & 7-5 \end{vmatrix} = 1.70$$

Hence the interpolated value of y for $x = 5$ is 1.70, in agreement with the value obtained by Lagrange interpolation as well as by divided differences (Example 1, Sect. 4.1.2). □

Neville's scheme is implemented in the following subroutine, especially noting that the indices of the nodal point vary from 0 to n, as in the theory:

```
SUBROUTINE NEVILLE(n,x,f,xbar,fphi)
! n+1 = number of nodal points.
! x(0), x(1), ···, x(n) = abscissa of the nodal points.     (Input)
! f(0), f(1), ···,f(n) = function value at the nodal points.     (Input)
! xbar = abscissa of the given point of interpolation.     (Input)
! fphi = interpolated value of the function at xbar.     (Output)
! phi(n+1, n+1) = interpolated values according to Neville's scheme.
                    (Output from the subroutine, if desired)
!********************************************************************
REAL :: x(n+1), f(n+1), phi(n+1,n+1)
DO k=0,n
DO j=0,n
phi(j,k)=0.0
IF(j==k) phi(j,k)=f(j)
END DO
END DO
DO k=1,n
DO j=k,n
IF(j>=k) phi(j-k,j)=(phi(j-k,j-1)*(x(j)-xbar)-phi(j-k+1,j)*(x(j-k)-xbar))/&
                    &(x(j)-x(j-k))
END DO; END DO
fphi=phi(0,n)          ! Output interpolated value of the function using the full
                       ! Neville's Table
!********************************************************************
DO k=0,n
DO j=k,n
PRINT*, 'k=', k, phi(j-k,j)          ! Output Neville's Table if desired;
                                     ! otherwise to be blocked
END DO
END DO
!********************************************************************
DO k=1,n-1
DO j=k,n
IF(ABS(phi(j-k,j)-phi(j-k-1,j-1))<0.00001) fphi=phi(j-k,j)
END DO
END DO
```

RETURN
END SUBROUTINE NEVILLE

Exercises

1. Show that
(i) $\phi_{01}(x) = \dfrac{x - x_1}{x_0 - x_1} y_0 + \dfrac{x - x_0}{x_1 - x_0} y_1$

(ii) $\phi_{02}(x) = \dfrac{(x - x_1)(x - x_2)}{(x_0 - x_1)(x_0 - x_2)} y_0 + \dfrac{(x - x_0)(x - x_2)}{(x_1 - x_0)(x_1 - x_2)} y_1 + \dfrac{(x - x_0)(x - x_1)}{(x_2 - x_0)(x_2 - x_1)} y_2$

and verify that
(i) $\phi_{01}(x) = f[x_0] + (x - x_0) f[x_0, x_1]$

(ii) $\phi_{02}(x) = f[x_0] + (x - x_0) f[x_0, x_1] + (x - x_0)(x - x_1) f[x_0, x_1, x_2]$

2. Interpolate by Neville's scheme the value of y when $x = 2$, given the data

x	1	3	4	6
y	−3	9	30	132

[0].

3. Interpolate $y = \log_{10} x$ for $x = 305$ given that

x	300	302	304	307	310
y	2.4771	2.4800	2.4829	2.4871	2.4914

[2.4843].

4. Work out Exercise 8, Sect. 4.1.2 by Neville's scheme.

5. Work out Exercise 5, Sect. 4.1.1 by Neville's scheme.

4.1.4 Error in Polynomial Interpolation

When a function $f \in C[a, b]$ is approximated by a polynomial ϕ of degree at most n, the *remainder* R at any point $x \in [a, b]$ is defined by the relation

$$f(x) = \phi(x) + R(x)$$

The remainder $R(x)$ represents the error in approximating $f(x)$ by $\phi(x)$. It depends on many factors such as the properties of the function f, and the position of the nodal points $x_0, x_1, \cdots, x_n$. It also depends on the position of the point x. If a given point $\bar{x}$ is under consideration, then an estimate of the absolute error $|R(\bar{x})|$ is possible, based on the following theorem.

Theorem 4.2 *Let (i) the interpolation points $x_0, x_1, \cdots, x_n \in [a, b]$ be distinct and (ii) $f \in C^{(n+1)}[a, b]$; then for a given point $\bar{x} \in [a, b]$ there exists a point $\xi \in (a, b)$ such that*

$$R(\bar{x}) = \frac{f^{(n+1)}(\xi)}{(n+1)!} \omega_n(\bar{x}) \tag{4.6}$$

where $\omega_n(\bar{x}) = (\bar{x} - x_0)(\bar{x} - x_1) \cdots (\bar{x} - x_n)$.

Proof If $\bar{x}$ coincides with any one of the nodal points then, $\omega_n(\bar{x}) = 0 \Rightarrow R(\bar{x}) = 0$, which is true. If $\bar{x} \neq x_i$, $(i = 0, 1, \cdots, n)$, then consider the auxiliary function

$$\psi(x) := f(x) - \phi(x) - k\,\omega_n(x) \tag{4.7}$$

where k is a constant, so chosen that $\psi(\bar{x}) = 0$, that is

$$k = \frac{f(\bar{x}) - \phi(\bar{x})}{\omega_n(\bar{x})} = \frac{R(\bar{x})}{\omega_n(\bar{x})}$$

By this choice of k, ψ vanishes in $[a,\,b]$ at the $(n+2)$ points $x_0, x_1, \cdots, x_n$ and $\bar{x}$. Hence by Rolle's theorem of differential calculus, ψ' vanishes at least $(n+1)$ times in $[a,\,b]$, consequently ψ'' vanishes at least n times in $[a,\,b]$, and so on, up to the derivative $\psi^{(n+1)}$ vanishing at least at one point $\xi \in [a,\,b]$. Thus by differentiation of ψ defined by Eq. (4.7), $(n+1)$ times, one obtains by using $\phi^{(n+1)}(x) \equiv 0$

$$0 = \psi^{(n+1)}(\xi) = f^{(n+1)}(\xi) - 0 - k(n+1)!$$

Thus, $$k = \frac{f^{(n+1)}(\xi)}{(n+1)!} = \frac{R(\bar{x})}{\omega_n(\bar{x})}$$

which proves the theorem. □

The remainder formula (4.6) is of limited practical utility, since $f^{(n+1)}(x)$ is seldom known and the point ξ is never known. However if a bound on $|f^{(n+1)}(x)|$ is known, say $|f^{(n+1)}(x)| < M_{n+1}$, then one has the estimate

$$|R(\bar{x})| \leq \frac{M_{n+1}}{(n+1)!} |\omega_n(\bar{x})|$$

of the error at $\bar{x}$. Also whatever be $x \in [a,\,b]$, a uniform estimate over the entire interval $[a,\,b]$ is

$$\max_{x \in [a,\,b]} |R(x)| \leq \frac{M_{n+1}}{(n+1)!} \max_{x \in [a,\,b]} |\omega_n(x)| \tag{4.8}$$

Let h be the *maximal distance between neighbouring data points* and let $x_i < x_j$ be two neighbouring data points containing x, i.e. $x_i < x < x_j$. Then $|(x - x_i)(x - x_j)|$ is maximum at $x = (x_i + x_j)/2$ with a value of $(x_i - x_j)^2/4 \leq h^2/4$. Hence by straight forward estimation of the other factor yields

$$\max_{x \in [a,\,b]} |\omega_n(x)| \leq \frac{h^2}{4} \cdot nh \cdot (n-1)h \cdots h = \frac{n!}{4} h^{n+1}$$

Thus one obtains the estimate

$$\max_{x \in [a, b]} |R(x)| \le \frac{M_{n+1}}{4(n + 1)} h^{n+1}$$

Such estimates are generally crude in the sense that the actual error is much smaller than the estimate. This happens because of estimating $f^{(n+1)}(\xi)$ by M_{n+1}.

4.2   Equally Spaced Points: Finite Differences

In the preceding sections, we considered interpolation on an arbitrary set of interpolation points. In many cases, the information about the function is given on *equally spaced points*. Before the advent of computers, such mathematical tables of functions (and of physical quantities) were in widespread use. The topic therefore gained much importance in the early development of numerical methods and several interesting formulae were discovered. Here we shall consider the principal ones, derived from Newton's difference formula. In this special case, not only the forms of the interpolation polynomial becomes simpler, but the required computations diminish, requiring simple computations.

Let the equally spaced points be denoted by $x_i = x_0 + i h$, $(i = 0, \pm 1, \pm 2, \cdots)$ where h is the *step size*. Three types of notations Δ, ∇ and δ are introduced regarding the *finite differences* of the function values $f(x_i) = y_i$. Thus, the *first-order finite differences* are defined by

$$\begin{aligned} y_1 - y_0 =: &\ \Delta y_0 =: \nabla y_1 =: \delta y_{1/2} \\ y_2 - y_1 =: &\ \Delta y_1 =: \nabla y_2 =: \delta y_{3/2} \\ &\ \cdots\cdots\cdots\cdots\cdots \\ y_{i+1} - y_i =: &\ \Delta y_i =: \nabla y_{i+1} =: \delta y_{i+1/2} \\ &\ \cdots\cdots\cdots\cdots\cdots \end{aligned}$$

or alternately

$$\Delta y_i := y_{i+1} - y_i, \quad \nabla y_i := y_i - y_{i-1}, \quad \delta y_{i+1/2} := y_{i+1} - y_i \qquad (4.9)$$

In a similar manner, the *second-order finite differences* are defined by

$$\Delta^2 y_i := \Delta y_{i+1} - \Delta y_i, \quad \nabla^2 y_i := \nabla y_i - \nabla y_{i-1}, \quad \delta^2 y_i := \delta y_{i+1/2} - \delta y_{i-1/2} \tag{4.10}$$

Finite differences of order k are similarly defined as

$$\begin{aligned} \Delta^k y_i := &\ \Delta^{k-1} y_{i+1} - \Delta^{k-1} y_i, \quad \nabla^k y_i := \nabla^{k-1} y_i - \nabla^{k-1} y_{i-1} \\ \delta^k y_{i+1/2} := &\ \delta^{k-1} y_{i+1} - \delta^{k-1} y_i \quad \text{for} \quad k \ge 1 \quad \text{and odd} \\ \delta^k y_i := &\ \delta^{k-1} y_{i+1/2} - \delta^{k-1} y_{i-1/2} \quad \text{for} \quad k \ge 2 \quad \text{and even} \end{aligned} \tag{4.11}$$

The differences Δ, ∇ and δ are respectively called **forward, backward** and **central differences**. They can be arranged in suitable tabular forms, such as

Forward Difference Table

x	y	Δ	Δ^2	Δ^3	Δ^4	$\cdots$	Δ^n
x_0	y_0						
		Δy_0					
x_1	y_1		$\Delta^2 y_0$				
		Δy_1		$\Delta^3 y_0$			
x_2	y_2		$\Delta^2 y_1$		$\Delta^4 y_0$		
		Δy_2		$\Delta^3 y_1$	$\vdots$	$\ddots$	$\Delta^n y_0$
x_3	y_3		$\Delta^2 y_2$	$\vdots$	$\Delta^4 y_{n-4}$		
		Δy_3	$\vdots$	$\Delta^3 y_{n-3}$			
x_4	y_4	$\vdots$	$\Delta^2 y_{n-2}$				
$\vdots$	$\vdots$	Δy_{n-1}					
x_n	y_n						

Backward Difference Table

x	y	∇	∇^2	∇^3	∇^4	$\cdots$	∇^n
x_0	y_0						
$\vdots$	$\vdots$	∇y_1					
x_{n-4}	y_{n-4}	$\vdots$	$\nabla^2 y_2$				
		∇y_{n-3}	$\vdots$	$\nabla^3 y_3$			
x_{n-3}	y_{n-3}		$\nabla^2 y_{n-2}$	$\vdots$	$\nabla^4 y_4$		
		∇y_{n-2}		$\nabla^3 y_{n-1}$	$\vdots$	$\ddots$	$\nabla^n y_n$
x_{n-2}	y_{n-2}		$\nabla^2 y_{n-1}$		$\nabla^4 y_n$		
		∇y_{n-1}		$\nabla^3 y_n$			
x_{n-1}	y_{n-1}		$\nabla^2 y_n$				
		∇y_n					
x_n	y_n						

Central Difference Table

x	y	δ	δ^2	δ^3	$\delta^4 \cdots \delta^n$
$\vdots$	$\vdots$				
x_{-2}	y_{-2}	$\vdots$			
		$\delta y_{-3/2}$	$\vdots$		
x_{-1}	y_{-1}		$\delta^2 y_{-1}$	$\vdots$	
		$\delta y_{-1/2}$		$\delta^3 y_{-1/2}$	$\vdots$ $\ddots$
x_0	y_0		$\delta^2 y_0$		$\delta^4 y_0$
		$\delta y_{1/2}$		$\delta^3 y_{1/2}$	$\vdots$
x_1	y_1		$\delta^2 y_1$	$\vdots$	
		$\delta y_{3/2}$	$\vdots$		
x_2	y_2	$\vdots$			
$\cdot$	$\cdot$				

In a mathematical table, the data points are usually closely spaced roundedoff numbers. When differencing is performed, subtraction of nearly equal numbers exacerbates error, as was seen in Chap. 1. If there is (roundoff) error ϵ in only one data viz. y_i, then the growth of error takes place as shown below:

x	y	Δ	Δ^2	Δ^3	$\cdots\cdot$
$\cdot$	$\cdot$	$\cdot$	$\cdot$	$\cdot$	$\cdot$ $\cdot$
x_{i-3}	y_{i-3}				$\cdot$ $\cdot$
		Δy_{i-3}	$\cdot$	$\cdot$	$\cdot$ $\cdot$
x_{i-2}	y_{i-2}		$\Delta^2 y_{i-3}$		$\cdot$ $\cdot$
		Δy_{i-2}		$\Delta^3 y_{i-3} +$	$\cdot$ $\cdot$
x_{i-1}	y_{i-1}		$\Delta^2 y_{i-2} +$		$\cdot$ $\cdot$
		$\Delta y_{i-1} +$		$\Delta^3 y_{i-2} - 3$	$\cdot$ $\cdot$
x_i	$y_i + \epsilon$		$\Delta^2 y_{i-1} - 2$		$\cdot$ $\cdot$
		$\Delta y_i -$		$\Delta^3 y_{i-1} + 3$	$\cdot$ $\cdot$
x_{i+1}	y_{i+1}		$\Delta^2 y_i +$		$\cdot$ $\cdot$
		Δy_{i+1}		$\Delta^3 y_i -$	$\cdot$ $\cdot$
x_{i+2}	y_{i+2}		$\Delta^2 y_{i+1}$	$\cdot$	$\cdot$ $\cdot$
		Δy_{i+2}	$\cdot$		$\cdot$ $\cdot$
x_{i+3}	y_{i+3}			$\cdot$	$\cdot$ $\cdot$
$\cdot$	$\cdot$	$\cdot$	$\cdot$	$\cdot$	

In the worst possible case, if there is error of $+\epsilon$ and $-\epsilon$ alternately in the data, then differencing causes doubling of error for every new difference introduced (check this). Hence, as a practical rule *as soon as a single nonzero digit appears in a difference column, higher order differences are ignored* as they consist of accumulated errors. The ignored differences are essentially incorrect.

Example 1. The value of the elliptic integral

$$K(m) = \int_0^{\pi/2} \frac{d\theta}{\sqrt{1 - m \sin^2 \theta}}$$

is tabulated below in the solution. Calculate the differences that are correct.

Solution. Set $m = x$ and $y = K(m)$. The difference table is

x	y	Δ	Δ^2	Δ^3	Δ^4	Δ^5
0.20	1.65462					
		0.00508				
0.21	1.66470		0.00007			
		0.00515		0		
0.22	1.66985		0.00007		0.00001	
		0.00522		0.00001		−0.00002
0.23	1.67507		0.00008		−0.00001	
		0.00530		0		
0.24	1.68037		0.00008			
		0.00538				
0.25	1.68575					

In the above table, single nonzero digit appears in the second-order differences. Hence, only the first- and second-order differences Δ and Δ^2 are considered correct. The higher order differences are also calculated, but they are incorrect.   □

With the above brief description of finite differences, we are now in a position to derive some of the important interpolation formulae. At the outset, we remark that all such formulae are actually identical, except for the form of writing them.

4.2.1 Gregory–Newton Forward Difference Formula

Let, the interpolation data be $(x_i, y_i = f(x_i))$, $(i = 0, 1, 2, \cdots, n)$, such that $x_{i+1} - x_i = h$, $(i = 0, 1, \cdots, n - 1)$, where h is constant. As in the case of Newton's divided difference formula, the interpolating polynomial through these points can be written as

$$\phi(x) = B_0 + B_1(x - x_0) + B_2(x - x_0)(x - x_1) + \cdots + B_n(x - x_0)(x - x_1) \cdots (x - x_{n-1})$$

Since $\phi(x_i) = f(x_i) = y_i$, $(i = 0, 1, \cdots, n)$, we must have

$$y_0 = B_0 \quad \text{or,} \quad B_0 = y_0$$

$$y_1 = B_0 + B_1(x_1 - x_0) = y_0 + B_1 h \quad \text{or,} \quad B_1 = \frac{y_1 - y_0}{h} = \frac{\Delta y_0}{h}$$

$$y_2 = B_0 + B_1(x_2 - x_0) + B_2(x_2 - x_0)(x_2 - x_1)$$

$$= y_0 + \frac{y_1 - y_0}{h} \cdot 2h + B_2 \cdot 2h \cdot h$$

$$\text{or,} \quad B_2 = \frac{y_2 - 2y_1 + y_0}{2h^2} = \frac{y_2 - y_1 - (y_1 - y_0)}{2h^2} = \frac{\Delta y_1 - \Delta y_0}{2h^2} = \frac{\Delta^2 y_0}{2! \, h^2}$$

etc. Thus, one would expect

$$B_k = \frac{\Delta^k y_0}{k! \, h^k} \quad (k = 1, 2, \cdots, n) \tag{4.12}$$

To fully justify the above form, we note that as in the divided difference formula $B_k = f[x_0, x_1, \cdots, x_k]$ and so by Eq. (4.4)

$$B_{k+1} = f[x_0, x_1, \cdots, x_{k+1}] = \frac{f[x_1, x_2, \cdots, x_{k+1}] - f[x_0, x_1, \cdots, x_k]}{x_{k+1} - x_0}$$

$$= \frac{\dfrac{\Delta^k y_1}{k! \, h^k} - \dfrac{\Delta^k y_0}{k! \, h^k}}{(k+1)h} = \frac{\Delta^{k+1} y_0}{(k+1)! \, h^{k+1}}$$

Thus, by induction Eq. (4.12) must hold for the expression of B_k.

Setting $t := (x - x_0)/h$, the expression of $\phi(x)$ thus becomes

$$\phi(x) = y_0 + \frac{\Delta y_0}{h} \, th + \frac{\Delta^2 y_0}{2! \, h^2} \, th \, (th - h) + \cdots$$

$$+ \frac{\Delta^n y_0}{n! \, h^n} \, th \, (th - h)(th - 2h) \cdots [th - (n-1)h]$$

$$= y_0 + t \, \Delta y_0 + \frac{t(t-1)}{2!} \, \Delta^2 y_0 + \cdots + \frac{t(t-1) \cdots [t - (n-1)]}{n!} \, \Delta^n y_0 \tag{4.13}$$

The formula exclusively uses the descending forward differences $\Delta y_0, \Delta^2 y_0, \cdots, \Delta^n y_0$ (see the table for forward differences Δ). Thus it is *useful for interpolation of function f at a point $\bar{x}$ near the beginning of the table of values of the function*. The remainder (error) of the formula from Eq. (4.6) is

$$R(\bar{x}) = \frac{f^{(n+1)}(\xi)}{(n+1)!} \, h^{n+1} \bar{t}(\bar{t} - 1) \cdots (\bar{t} - n)$$

where $\bar{t} := (\bar{x} - x_0)/h$ and $x_0 < \xi < x_n$.

4.2.2 Gregory–Newton Backward Difference Formula

Let the interpolating polynomial $\phi(x)$ be written in the bottom-up manner of the abscissa, then

$$\phi(x) = B_0 + B_1(x - x_n) + B_2(x - x_n)(x - x_{n-1}) + \cdots + B_n(x - x_n)(x - x_{n-1}) \cdots (x - x_1)$$

Since it passes through the data points $(x_i, \ y_i = f(x_i))$, $(i = n, \ n - 1, \cdots, \ 0)$,

$$y_n = B_0 \quad \text{or,} \quad B_0 = y_n$$

$$y_{n-1} = B_0 + B_1(x_{n-1} - x_n) = y_n + B_1(-h), \ \text{or,} \ B_1 = \frac{y_n - y_{n-1}}{h} = \frac{\nabla y_n}{h}$$

$$y_{n-2} = B_0 + B_1(x_{n-2} - x_n) + B_2(x_{n-2} - x_n)(x_{n-2} - x_{n-1})$$

$$= y_n + \frac{y_n - y_{n-1}}{h} \cdot (-2h) + B_2 \cdot (-2h)(-h)$$

or $\ B_2 = \dfrac{y_n - 2y_{n-1} + y_{n-2}}{2\,h^2} = \dfrac{\nabla^2 y_n}{2!\,h^2}.$

Proceeding in this manner, one has

$$B_k = \frac{\nabla^k y_n}{k!\,h^k} \quad (k = 1, \ 2, \cdots, \ n)$$

Full justification of the above can be given by induction as in the case of the forward difference formula. Setting $t := (x - x_n)/h \ (< 0)$, the formula becomes

$$\phi(x) = y_n + \frac{\nabla y_n}{h}\,th + \frac{\nabla^2 y_n}{2!\,h^2}\,th(th + h) + \cdots$$

$$+ \frac{\nabla^n y_n}{n!\,h^n}\,th(th + h)(th + 2h) \cdots [th + (n - 1)h]$$

$$= y_n + t\,\nabla y_n + \frac{t(t + 1)}{2!}\,\nabla^2 y_n + \cdots + \frac{t(t + 1) \cdots [t + (n - 1)]}{n!}\,\nabla^n y_n$$

$$(4.14)$$

The formula now uses the ascending backward differences $\nabla y_n, \nabla^2 y_n, \cdots, \nabla^n y_n$ (see table of backward differences ∇). It is useful for interpolation at a point $\bar{x}$ near the end x_n of the table. The remainder (error) of the formula is

$$R(\bar{x}) = \frac{f^{(n+1)}(\xi)}{(n + 1)!}\,h^{n+1}\,\bar{t}(\bar{t} + 1) \cdots (\bar{t} + n)$$

where $\bar{t} := (\bar{x} - x_n)/h$ and $x_0 < \xi < x_n$.

James Gregory (1638–1675), Scottish mathematician. In 1675, he became the first professor of mathematics at the University of Edinburgh. He is best remembered for his series expansion of $\tan^{-1}x$, $\tan x$, $\sin^{-1}x$, $\sec^{-1}x$, etc., although the well-known series expansion of $\tan^{-1}x$ was discovered much earlier by Indian mathematician **Madhav (1350–1425)**.

4.2.3   Stirling's Central Difference Formula

If the point $\bar{x}$ at which interpolation is sought, is near the middle of a table, then it is natural to consider interpolation points lying on both sides of the point. This is justified because the error, Eq. (4.6) becomes least because of low value of $(\bar{x} - x_0) \cdots (\bar{x} - x_n)$. Suppose we designate the node closest to $\bar{x}$ by x_0 $(\bar{x} > x_0)$ and consider the interpolation points $x_{-k} < \cdots < x_{-1} < x_0 < x_1 < \cdots < x_k$. The interpolating polynomial is then of degree at most $2k$. Setting $t := (x - x_0)/h$ it can be written in the form

$$\phi(t) = B_0 + \frac{B_1}{1!}t + \frac{B_2}{2!}t^2 + \frac{B_3}{3!}t(t^2 - 1^2) + \frac{B_4}{4!}t^2(t^2 - 1^2) + \cdots$$

$$+ \frac{B_{2k-1}}{(2k-1)!}t(t^2 - 1^2)\cdots(t^2 - (k-1)^2) + \frac{B_{2k}}{(2k)!}t^2(t^2 - 1^2)\cdots(t^2 - (k-1)^2)$$

The function must pass through the points $(x_0, y_0), (x_{\pm 1}, y_{\pm 1}), \cdots, (x_{\pm k}, y_{\pm k})$. Hence looking at the *table of central differences*

$$y_0 = B_0 \quad \text{or,} \quad B_0 = y_0$$
$$y_1 = B_0 + B_1 + \frac{B_2}{2}, \quad y_{-1} = B_0 - B_1 + \frac{B_2}{2}$$

or, $B_1 = \dfrac{y_1 - y_{-1}}{2} = \dfrac{(y_1 - y_0) + (y_0 - y_{-1})}{2} = \dfrac{1}{2}(\delta y_{1/2} + \delta y_{-1/2}) =: \bar{\delta} y_0$

$$B_2 = y_1 - 2y_0 + y_{-1} = \delta^2 y_0$$

Moreover,

$$y_2 = B_0 + 2B_1 + 2B_2 + B_3 + \frac{B_4}{2}$$
$$y_{-2} = B_0 - 2B_1 + 2B_2 - B_3 + \frac{B_4}{2}$$

that yield

$$B_3 = \frac{1}{2}(y_2 - 2y_1 + 2y_{-1} - y_{-2})$$

$$= \frac{1}{2}[(y_2 - 3y_1 + 3y_0 - y_{-1}) + (y_1 - 3y_0 + 3y_{-1} - y_{-2})]$$

$$= \frac{1}{2}(\delta^3 y_{1/2} + \delta^3 y_{-1/2}) =: \bar{\delta}^3 y_0$$

$$B_4 = y_2 - 4y_1 + 6y_0 - 4y_{-1} + y_{-2} = \delta^4 y_0, \quad \text{etc.}$$

Hence proceeding in this manner we obtain

$$\phi(t) = y_0 + t\,\bar{\delta}y_0 + \frac{t^2}{2!}\delta^2 y_0 + \frac{t(t^2 - 1^2)}{2!}\bar{\delta}^3 y_0 + \frac{t^2(t^2 - 1^2)}{4!}\delta^4 y_0 + \cdots$$

$$+ \frac{t(t^2 - 1^2)\cdots(t^2 - (k-1)^2)}{(2k-1)!}\bar{\delta}^{2k-1}y_0 + \frac{t^2(t^2 - 1^2)\cdots(t^2 - (k-1)^2)}{(2k)!}\delta^{2k}y_0$$

$$(4.15)$$

In the above expression, the *mean differences* are defined as $\bar{\delta}^m := \frac{1}{2}(\delta^m_{1/2} + \delta^m y_{-1/2})$ for $m \geq 1$ and odd. The central differences occurring in Eq. (4.15) are seen to lie along the horizontal line through the data (x_0, y_0) in the central difference table. The remainder in the formula is given by

$$R(\bar{x}) = \frac{f^{(2k+1)}(\xi)}{(2k+1)!}h^{2k+1}(t^2 - 1^2)\cdots(t^2 - k^2), \quad x_{-k} < \xi < x_k$$

James Stirling (1692–1770), Scottish mathematician educated at Oxford. The Stirling numbers are named after him.

Remark If one wishes to use a computer for interpolation of equally spaced data, then subroutine DIVDIFF can be used for the purpose. But it is a good idea to form a finite difference table also, as a table in which elements of a column do not have fluctuation of sign, is indicative of accuracy of the data. Also, decreasing magnitude of the differences with increasing order, is indicative of accuracy of interpolation.

Example 2. In Example 1, calculate $K(0.215)$, $K(0.225)$ and $K(0.243)$.

Solution. For $x = 0.215$, we apply the forward difference formula. If we use $x_0 = 0.20$, $t = (0.215 - 0.200)/0.01 = 1.5$. Using the table of finite differences (up to Δ^2) given in Example 1, formula (4.13) yields

$$K(0.215) \approx 1.65962 + 1.5 \times 0.00508 + \frac{1.5 \times (1.5 - 1)}{2!} \times 0.00007$$

$$= 1.66727$$

For $x = 0.225$ we apply Stirling's formula since the data occurs in the middle portion of the table. Taking $x_0 = 0.22$, $t = (0.225 - 0.22)/0.01 = 0.5$, formula (4.15) gives

$$K(0.225)7 \approx 1.66985 + 0.5 \times \frac{1}{2}(0.00515 + 0.00522) + \frac{0.5^2}{2!} \times 0.00007$$
$$= 1.67245$$

For $x = 0.243$, we apply the backward difference formula (4.14), since the point is near the end of the table. Taking $x_n = 0.25$, $t = (0.243 - 0.25)/0.01 = -0.7$. Hence applying Eq. (4.14)

$$K(0.243) \approx 1.68575 + (-0.7) \times 0.00538 + \frac{(-0.7)(-0.7 + 1)}{2!} \times 0.00008$$
$$= 1.68198$$

In the case of $x = 0.215$, one may alternatively select $x_0 = 0.21$, $t = (0.215 - 0.21)/0.01 = 0.5$, Eq. (4.13) then yields

$$K(0.215) \approx 1.66470 + 0.5 \times 0.00515 + \frac{0.5 \times (0.5 - 1)}{2!} \times 0.00007 = 1.66727$$

as before.   □

Example 3. Experiment has yielded contraction of a spring x as a function of the load P carried by the spring as under:

x(mm)	5	10	15	20	25	30
P(kg)	49	105	172	253	352	474

Find the load that yields a contraction of the spring by 17 mm.

Solution. The finite difference table for the problem is

x	y	Δ	Δ^2	Δ^3	Δ^4	Δ^5
5	49					
		56				
10	105		11			
		67		3		
15	172		14		1	
		81		4		0
20	253		18		1	
		99		5		
25	352		23			
		122				
30	474					

We wish to employ the full difference table for highest accuracy; hence we can adopt any interpolation formula. Adopting the forward difference formula with $x_0 = 5$, $t = (17 - 5)/5 = 2.4$, we have

$$P(x = 17) \approx 49 + 2.4 \times 56 + \frac{2.4 \times (2.4 - 1)}{2!} \times 11 + \frac{2.4 \times (2.4 - 1) \times (2.4 - 2)}{3!} \times 3$$
$$+ \frac{2.4 \times (2.4 - 1) \times (2.4 - 2) \times (2.4 - 3)}{4!} \times 1 + 0$$
$$= 202.5\text{kg}. \qquad \square$$

Exercises

1. The following table contains one error in the value of y:

x	2.5	3.0	3.5	4.0	4.5	5.0	5.5
y	4.32	4.83	5.27	5.47	6.26	6.79	7.23

Detect the erroneous data by forming the finite difference table.

[Error in 5.47]

2. Given the table of Bessel function $J_0(x)$:

x^2	0	1/3	2/3	1
$J_0(x)$	1.0000	0.9184	0.8402	0.7652

Show that in $[-1, 1]$, $J_0(x)$ can be approximated by $1 - 0.397x^2 + 0.0153x^4$.

[Use forward difference formula with $t = (x^2 - 0)/(1/3) = 3x^2$, dropping the inaccurate Δ^3 term].

3. The population of a town in the census is as given in the data. Estimate the population in the year 1965 and 1998.

Year	1961	1971	1981	1991	2002
Population (in 1000's)	46	66	81	93	101

[54.853, 99.200].

4. The following table gives the amount of chemical dissolved in water:

Temp.°C	10	15	20	25	30	35
Solubility	19.97	21.51	22.47	23.52	24.52	25.39

What is the solubility at 22 °C?

[22.88].

5. Given an abridged table of common logarithms

x	300	310	320	330	340	350
$\log x$	2.47712	2.49136	2.50515	2.51851	2.53148	2.54407

Calculate log(315) and log(336).

[2.49831, 2.52634].

6. The probability integral or the error function $\text{erf}(x) = \dfrac{2}{\sqrt{\pi}} \displaystyle\int_0^x e^{-t^2} dt$ is given as a
table

x	1.0	1.2	1.4	1.6	1.8	2.0
$\text{erf}(x)$	0.84270	0.91031	0.95229	0.97635	0.98909	0.99532

Find the value of erf(1.433). Can you apply Stirling's formula. Give reason.

[0.95730, by Gregory–Newton forward difference formula. Yes, to use all finite differences].

7. The following table of $\tan x$ near 90° is given:

x	89°21′	89°23′	89°25′	89°27′	89°29′
$\tan x$	88.14	92.91	98.22	104.17	110.90

Calculate 89°26′.

[101.11, by Stirling's formula].

8. The function f is displayed in the following table. It is known that $f(x)$ behaves
as $1/x$ as $x \to 0$. We want to calculate $f(0.15)$. If direct interpolation is used, what
value is obtained? Alternately if the data is retabulated for $g(x) = x f(x)$, what is the
interpolated value of $f(0.15)$? Which answer would you consider more accurate?
Give reason.

x	0.1	0.2	0.3	0.4	0.5
$f(x)$	20.02502	10.05013	6.74211	5.10105	4.12706

[$f(0.15) = 13.73547$ and 13.37089 respectively by the two methods. Second answer, since the finite differences diminish by an order of magnitude with increasing order of the finite differences].

9. (Bessel's Interpolation Formula). Given a net of $2k + 1$ nodes $x_{-k} < \cdots < x_{-1} < x_0 < x_1 < \cdots < x_{k+1}$, prove *Bessel's interpolation formula*

$$\phi(t) = \bar{y}_{1/2} + \frac{\delta y_{1/2}}{1!}\left(t - \frac{1}{2}\right) + \frac{\bar{\delta}^2 y_{1/2}}{2!} t(t-1) + \frac{\delta^3 y_{1/2}}{3!} t(t-1)\left(t - \frac{1}{2}\right) + \cdots$$

$$+ \frac{\bar{\delta}^{2k} y_{1/2}}{(2k)!} t(t^2 - 1^2) \cdots (t^2 - (k-1)^2)(t-k)$$

$$+ \frac{\delta^{2k+1} y_{1/2}}{(2k+1)!} t(t^2 - 1^2) \cdots (t^2 - (k-1)^2)(t-k)\left(t - \frac{1}{2}\right)$$

where $t := (x - x_0)/h$, $\bar{y}_{1/2} := \frac{1}{2}(y_0 + y_1)$ and $\bar{\delta}^m y_{1/2} := \frac{1}{2}(\delta^m y_0 + \delta^m y_1)$ for $m \geq 2$ and even. Show that the remainder relative to t is

$$R(\bar{x}) = \frac{f^{(2k+2)}(\xi)}{(2k+2)!} h^{2k+2} \prod_{i=-k}^{k+1}(t - i), \quad x_{-k} < \xi < x_{k+1}$$

[Proceed as in the derivation of Stirling's formula assuming a form of $\phi(t)$ suggested by the result].

Friedrich Wilhelm Bessel (1784–1846), German astronomer, mathematician and geodesist. He was the first astronomer to determine the distance of a fixed star. He is best known for the function known after him, which he systematically investigated in the year 1824. He came across the function in the treatment of perturbation of a planetary orbit by another planet. The function was however encountered earlier in 1732 by Jakob Bernoulli, in the study of oscillations of a heavy chain and by Euler in 1744, in the study of a circular membrane.

4.3 Best Interpolation Nodes: Chebyshev Interpolation

Let there be given a function $f \in C^{(n+1)}[a, b]$. The function can be approximated by different interpolating polynomials ϕ, that depend on the interpolation nodes $a \leq x_0, x_1, \cdots, x_n \leq b$. There arises then a question as to how to choose these nodes so that the maximum interpolation error for f is minimal on the given interval. This problem is complicated due to lack of knowledge regarding the behaviour of f, except for its continuity. Noting that the estimate of the error $|R(x)|$ given in Eq. (4.8) also depends on $\omega_n(x) = (x - x_0) \cdots (x - x_n)$, one seeks the arrangement of $x_0, x_1, \cdots, x_n$ so that $\max\limits_{x \in [a, b]} |\omega_n(x)|$ is minimal. The answer to this restricted problem is offered by *Chebyshev polynomials* on the standard interval $[-1, 1]$.

Chebyshev Polynomials. These polynomials of degree $n (\geq 0)$ are denoted by $T_n(x)$, $-1 \leq x \leq 1$ as defined by the formula

$$T_n(x) = \cos(n \cos^{-1} x) \tag{4.16}$$

In particular, for $n = 0$ and 1 one has $T_0(x) = \cos(0) = 1$, $T_1(x) = \cos(\cos^{-1} x) = x$. Further, from the identity

$$\cos(n + 1)\theta = 2\cos\theta \cos n\theta - \cos(n - 1)\theta$$

setting $\theta = \cos^{-1} x$, one obtains the *recursive relations*

$$T_{n+1}(x) = 2x\, T_n(x) - T_{n-1}(x) \tag{4.17}$$

for $n = 1, 2, \cdots$. $T_n(x)$ is actually a polynomial of degree n. Using the expression for $T_0(x)$ and $T_1(x)$, one has

$$\begin{array}{ll} T_2(x) = 2x^2 - 1, & T_3(x) = 4x^3 - 3x \\ T_4(x) = 8x^4 - 8x^2 + 1, & T_5(x) = 16x^5 - 20x^3 + 5x, \quad \text{etc.} \end{array} \tag{4.18}$$

The letter T for denoting the Chebyshev polynomials is in recognition of the Russian spelling TCHEBYCHEFF of the mathematician's name. **Pafnutii Lvovitsch Chebyshev (1821–1894)** worked in St. Petersberg. His contribution to mathematics is universal. These include number theory, probability theory, the theory of orthogonal functions and theoretical mechanics. He is considered a pathbreaker in constructive function theory, which includes the theory of uniform approximation (see Chap. 8).

Properties of Chebyshev polynomials

1. $T_n(x)$ is an even or odd function of x, according as n is even or odd.
2. The leading coefficient of the polynomial $T_n(x)$ for $n \geq 1$ is equal to 2^{n-1}.

The validity of the two properties is evident from the preceding discussion.
3. The polynomial $T_n(x)$ has n real zeros given by

$$x_i = \cos\left[\frac{(2i + 1)\pi}{2n}\right], \quad (i = 0, 1, \cdots, n - 1)$$

Indeed $\quad T_n(x_i) = \cos(n \cos^{-1} x_i) = \cos\dfrac{(2i + 1)\pi}{2} = 0 \text{ for } i = 0, 1, \cdots, n - 1.$

4. $|T_n(x)| \le 1$ and $T_n(x_m) = (-1)^m$ where $x_m = \cos(m\pi/n)$ for $m = 0, 1, \cdots, n$.

The property follows from the definition (4.16).

Remark When it is imperative to indicate the order of a polynomial, we use the suffixed notation, e.g. T_n, P_n etc.

Theorem 4.3 (Chebyshev). *Let* $\bar{P}_n(x)$ *be any polynomial of degree n, with the leading coefficient equal to 1, and let* $\bar{T}_n(x) = 2^{1-n} T_n(x)$. *Then*

$$\begin{array}{cc} max & |\bar{P}_n(x)| \ge \\ x \in [-1, 1] & \end{array} \quad \begin{array}{c} max \\ x \in [-1, 1] \end{array} |\bar{T}_n(x)| = 2^{1-n} \qquad (4.19)$$

i.e. $\bar{T}_n(x)$ *has the least maximum modulus value among all polynomials of degree n with leading coefficient equal to unity.*

Proof Assume that $\bar{P}_n(x) = x^n + a_2 x^{n-1} + \cdots + a_n$ does not satisfy (4.19), i.e.

$$\begin{array}{cc} max & |\bar{P}_n(x)| < \\ x \in [-1, 1] & \end{array} \quad \begin{array}{c} max \\ x \in [-1, 1] \end{array} |\bar{T}_n(x)| = 2^{1-n} \qquad (4.20)$$

Now by Property 2, $\bar{T}_n(x)$ is also a polynomial of degree n with leading coefficient equal to 1. Hence the difference $Q_{n-1}(x) := \bar{T}_n(x) - \bar{P}_n(x)$ is a polynomial of degree not higher than $n - 1$, and by assumption (4.20) it does not vanish identically. At the $n + 1$ points $x_m = \cos(m\pi/n)$, $m = 0, 1, \cdots, n$ on the basis of Property 4 and Eq. (4.20), *this difference attains nonzero values with alternating signs.* This means that $Q_{n-1}(x)$ vanishes at n points, which is impossible.   □

Remark It is possible to prove that *if* $\bar{P}_n(x) = x^n + a_1 x^{n-1} + \cdots + a_n$, $(n \ge 1)$ *is*

such that $\quad \begin{array}{c} max \\ x \in [-1, 1] \end{array} |\bar{P}_n(x)| = 2^{1-n}$, *then* $\bar{P}_n(x) = \bar{T}_n(x) = 2^{1-n} T_n(x)$.

Theorem 4.3 enables us to select the nodal points x_i, $(i = 0, 1, \cdots, n)$ in the best possible way. If we choose them as the zeros of $T_{n+1}(x)$, then by Property 3,

$$x_i = \cos \frac{(2i + 1)\pi}{2n + 2} \quad (i = 0, 1, \cdots, n) \qquad (4.21)$$

on the interval $[-1, 1]$. With this choice

$$\omega_n(x) = 2^{-n} T_{n+1}(x)$$

as the sides have identical zeros and both are polynomials with leading coefficient equal to 1. The error estimate (4.8) of Sect. 4.1.4, thus becomes

$$\max_{x \in [-1, 1]} |R(x)| \le \frac{M_{n+1}}{(n+1)! 2^n} \max_{x \in [-1, 1]} |T_n(x)| = \frac{M_{n+1}}{(n+1)! 2^n} \tag{4.22}$$

where Property 4 is used.

By virtue of Theorem 4.3, it is impossible to improve the error estimate (4.22) on $[-1, 1]$ by other choice of interpolation nodes. By such alternative choice, the maximum of the error estimate will be worse, that is, *the interpolation nodes* (4.21) *are optimal* for the maximum error estimate (4.8) on the interval $[-1, 1]$.

Reverting to the nodes of interpolation on an arbitrary interval $[a, b]$, one can transform it in to the interval $[-1, 1]$ by the linear transformation

$$x = \frac{b+a}{2} + \frac{b-a}{2} t, \quad \text{or,} \quad t = \frac{2x - b - a}{b - a}$$

The zeros of the Chebyshev polynomial $T_{n+1}(t)$, thus yield the points

$$x_i = \frac{b+a}{2} + \frac{b-a}{2} \cos \frac{(2i+1)\pi}{2n+2}, \quad (i = 0, 1, \cdots, n) \tag{4.23}$$

which are the optimal interpolation nodes in $[a, b]$. For these points

$$\omega_n(x) = (x - x_0)(x - x_1) \cdots (x - x_n)$$

$$= \frac{(b-a)^{n+1}}{2^{n+1}} \left(t - \cos \frac{\pi}{2n+2} \right) \left(t - \cos \frac{3\pi}{2n+2} \right) \cdots \left(t - \cos \frac{(2n+1)\pi}{2n+2} \right)$$

$$= \frac{(b-a)^{n+1}}{2^{n+1}} \bar{T}_{n+1}(t) = \frac{(b-a)^{n+1}}{2^{2n+1}} T_{n+1}(t)$$

Hence for the interval $[a, b]$, the maximum error estimate (4.8) becomes

$$\max_{x \in [a, b]} |R(x)| \le \frac{M_{n+1}}{(n+1)!} \frac{(b-a)^{n+1}}{2^{2n+1}} \max_{t \in [-1, 1]} T_{n+1}(t)$$

$$= \frac{M_{n+1}}{(n+1)!} \frac{(b-a)^{n+1}}{2^{2n+1}} \tag{4.24}$$

Example 1. (Runge's Example) Let

$$f(x) = \frac{1}{1 + 25x^2}, \quad -1 \le x \le 1$$

Let the interpolation points x_i, $(i = 0, 1, \cdots, n)$, where $n = 4, 5, \cdots, 20$ be selected as either (a) Chebyshev points, Eq. (4.21) or (b) equally spaced points $x_i = -1 + 2i/n$. Write a Fortran program to compute the interpolated value $\phi(\bar{x}_j)$ at points $\bar{x}_j = -1 + j/10$, where $j = 0, 1, \cdots, 20$ to compute the maximum error $\max_{0 \le j \le 20} |f(\bar{x}_j) - \phi(\bar{x}_j)|$ in the two cases using the subroutine DIVDIFF given in Sect. 4.1.2.

Solution. The Fortran program for solving the problem is as follows:

```
PROGRAM RUNGE
REAL :: x(21), f(21)
DO n=4,20
errmax=0.
DO j=1,21
DO i=1,n+1
x(i)=cos(((2*(i-1)+1)*3.141593)/(2*n+2))   ! (a) Chebyshev Points
! x(i)=-1+2.*(i-1)/n    ! (b) Equally Spaced Points
f(i)=1/(1+25*x(i)**2)
END DO
xbar=-1+(j-1)/10
fxbar=1./(1+25*xbar**2)
CALL DIVDIFF(n+1,x,f,xbar,phi)
! PRINT*, fxbar, phi
error=ABS(fxbar-phi)
errmax=MAX(error,errmax)
END DO
PRINT*, n, errmax
END DO
END PROGRAM RUNGE
```

The subroutine DIVDIFF given in Sect. 4.1.2 must be appended to this program. The statement for equally spaced $x(i)$ must be activated for the case (b). The output of the program in the two cases is

n	errmax (a)	errmax (b)
4	$3.956862E - 01$	$4.381339E - 01$
5	$5.559114E - 01$	$4.326923E - 01$
6	$2.618566E - 01$	$6.075863E - 01$
7	$3.917403E - 01$	$2.473586E - 01$
8	$1.670828E - 01$	1.007122
9	$2.691785E - 01$	$2.770456E - 01$
10	$9.103656E - 02$	1.531661
11	$1.827583E - 02$	$4.053842E - 01$
12	$5.860662E - 02$	2.130535
13	$1.233977E - 01$	$5.374568E - 01$
14	$4.651782E - 02$	2.665879
15	$8.310699E - 02$	$6.290064E - 01$
16	$2.815479E - 02$	2.859874
17	$5.590761E - 02$	$5.967625E - 01$
18	$1.834667E - 02$	2.224079
19	$3.759038E - 02$	$3.059536E - 01$
20	$1.364753E - 02$	$2.601263E - 02$

It is clear from the table that in case (a) (Chebyshev points), the maximum error more or less decreases with increasing n, while in the case of (b) (equally spaced points) this is not the case.

The above example illustrates that interpolation with equally spaced points does not necessarily improve accuracy with increasing order of the polynomial.

Exercises

1. Draw the graphs of $T_2(x)$, $T_3(x)$, $T_4(x)$, $T_5(x)$ in $[-1, 1]$.

2. Prove the following elementary identities:

(a) $T_m[T_n(x)] = T_{mn}(x)$

(b) $2T_m(x) T_n(x) = T_{m+n}(x) + T_{|m-n|}(x)$

(c) $T_{2n}(x) = T_n(2x^2 - 1)$

3. Prove that $T_n(x)$ satisfies the second order differential equation

$$(1 - x^2)T_n''(x) - x T_n'(x) + n^2 T_n(x) = 0$$

4. Prove that $\int_{-1}^{1} T_n(x)dx = -\dfrac{1 + (-1)^n}{n^2 - 1}$.

[Use definition of $T_n(x)$ and put $x = \cos \theta$].

4.4 Piecewise–Polynomial Spline Interpolation

Suppose a sufficiently smooth function f is defined on an interval $[a,\ b]$ and the interval is subdivided by points $a \le x_0 < x_1 < \cdots < x_{n-1} < x_n \le b$. In spline theory the points $x_0,\ x_1,\ \cdots,\ x_n$ are especially called **knots**. A **spline** is a function which, together with its several derivatives, is continuous on $[x_0,\ x_n]$ and is such that on each separate subinterval $[x_i,\ x_{i+1}]$ it is some algebraic polynomial. The terminology was introduced by I. J. Schoenberg in 1946, although the idea was used earlier informally by several authors. Technically a 'spline' is a draughtsman's flexible implement which is used to draw 'smooth' curves through a series of points.

A simple example of spline approximation of f is piecewise linear interpolation over the data points $(x_i,\ f(x_i))$, $(i = 0,\ 1,\ \cdots,\ n)$. This technique is of frequent use in reading from *tables* of values. In this case the approximating curve to f is however not smooth at the knots.

The most commonly used smooth splines are of third degree or **cubic splines**. If S is the spline on $[x_0,\ x_n]$ in which S_i is cubic on $[x_i,\ x_{i+1}]$, then $S = \cup_{i=0}^{n-1} S_i$. Now, a cubic can be determined by interpolating over four points. Considering these four points as $x_i,\ x_i,\ x_{i+1},\ x_{i+1}$, one has by the divided difference formula

$$S_i = f(x_i) + (x - x_i)\, f[x_i,\ x_i] + (x - x_i)^2 f[x_i,\ x_i,\ x_{i+1}]$$

$$+ (x - x_i)^2 (x - x_{i+1})\, f[x_i,\ x_i,\ x_{i+1},\ x_{i+1}]$$

Now,

$$f[x_i,\ x_i] = f'(x_i)$$

$$f[x_i,\ x_i,\ x_{i+1}] = \frac{f[x_i,\ x_{i+1}] - f[x_i,\ x_i]}{\Delta x_i} = \frac{f[x_i,\ x_{i+1}] - f'(x_i)}{\Delta x_i}$$

$$f[x_i,\ x_i,\ x_{i+1},\ x_{i+1}] = \frac{f[x_i,\ x_{i+1},\ x_{i+1}] - f[x_i,\ x_i,\ x_{i+1}]}{\Delta x_i}$$

$$= \frac{f'(x_{i+1}) - 2\, f[x_i,\ x_{i+1}] + f'(x_i)}{(\Delta x_i)^2}$$

where $\Delta x_i := x_{i+1} - x_i$. Hence writing $x - x_{i+1} = (x - x_i) - \Delta x_i$, one obtains

$$S_i = a_i + b_i\, (x - x_i) + c_i\, (x - x_i)^2 + d_i\, (x - x_i)^3 \qquad (4.25)$$

where

$$a_i = f(x_i) = y_i$$

$$b_i = f'(x_i)\ \ \text{(slope at } x_i\text{)}$$

$$c_i = \frac{f[x_i,\ x_{i+1}] - b_i}{\Delta x_i} - d_i\, \Delta x_i \qquad (4.26)$$

$$d_i = \frac{b_{i+1} - 2\, f[x_i,\ x_{i+1}] + b_i}{(\Delta x_i)^2}$$

Thus to define S_i, one requires the data $(x_i, f(x_i) = y_i)$ as well as the slopes b_i at the knots. The spline is evidently continuous on $[x_0, x_n]$. The derivative is also continuous, since

$$S_i'(x_{i+1}) = b_i + 2c_i\Delta x_i + 3d_i\Delta x_i^2$$
$$= b_i + 2(f[x_i, x_{i+1}] - b_i) + (b_{i+1} - 2f[x_i, x_{i+1}] + b_i)$$
$$= b_{i+1} = f'(x_{i+1}) = S_{i+1}'(x_{i+1})$$

To determine the slope b_i we now assume that S is continuous up to the second derivatives. This means that $S_i''(x_i) = S_{i-1}''(x_i)$ for $i = 1, 2, \cdots, n-1$. Now from Eqs. (4.25), (4.26)

$$S_i''(x_i) = 2c_i = -\frac{4b_i}{\Delta x_i} - \frac{2b_{i+1}}{\Delta x_i} + \frac{6f[x_i, x_{i+1}]}{\Delta x_i}$$

Similarly from the expression for $S_{i-1}(x)$, one obtains

$$S_{i-1}''(x_i) = 2c_{i-1} + 6d_{i-1}\Delta x_{i-1} = \frac{2b_{i-1}}{\Delta x_{i-1}} + \frac{4b_i}{\Delta x_{i-1}} - \frac{6f[x_{i-1}, x_i]}{\Delta x_{i-1}}$$

Equating the above two expressions, one obtains the system of equations

$$(\Delta x_i)b_{i-1} + 2(\Delta x_{i-1} + \Delta x_i)b_i + (\Delta x_{i-1})b_{i+1}$$
$$= 3(\Delta x_{i-1} f[x_i, x_{i+1}] + \Delta x_i f[x_{i-1}, x_i]) \qquad (4.27)$$

for $i = 1, 2, \cdots, n-1$. Since there are $(n+1)$ unknowns $b_0, b_1, \cdots, b_n$, it is necessary to specify two more conditions at the *boundary points* x_0 and x_n ($i = 0, n$). One simple way to introduce boundary conditions is to take

$$b_0 = f'(x_0), \quad b_n = f'(x_n)$$

provided $f'(x_0)$ and $f'(x_n)$ are given. Alternatively, one may assume "natural" boundary conditions

$$S_0''(x_0) = 0, \quad S_{n-1}''(x_n) = 0$$

which means that the curvature at the boundary points is zero.. These conditions yield

$$2b_0 + b_1 = 3f[x_0, x_1], \quad b_{n-1} + 2b_n = 3f[x_{n-1}, x_n] \qquad (4.28)$$

Equations (4.27) and (4.28) form a row diagonally dominant tridiagonal system of equations, possessing unique solution. The system of equations can be solved by the Thomas method of Sect. 3.1.1, Chap. 3.

Subroutine SPLINE given below solves the system (4.27) and (4.28) by the above mentioned technique and computes S at m points $\bar{x}_1, \bar{x}_2, \cdots, \bar{x}_m$.

```
SUBROUTINE SPLINE(x,y,n,xbar,s,m)
!n=number of sub–intervals.     (Input)
! x(1),···,x(n+1)=abscissa of knots.     (Input)
! y(1),···,y(n+1)=ordinate of knots.     (Input)
! xbar(1),···,xbar(m)=abscissa of points for interpolation.     (Input)
! s(1),···,s(m)=spline computed function.     (Output)
!*********************************************
REAL :: x(n+1), y(n+1), xbar(m), s(m), dx(n+2),&
&f(n+2), diag(n+2), b(n+2), c(n+2), d(n+2)
dx(1)=1.0; dx(n+2)=1.0
DO i=2,n+1
dx(i)=x(i)−x(i−1); f(i)=(y(i)−y(i−1))/dx(i)
END DO
diag(1)=2.0; diag(n+1)=2.0
DO i=2,n
diag(i)=2.*(dx(i)+dx(i+1))
END DO
b(1)=3.0*f(2); b(n+1)=3.0*f(n+1)
DO i=2,n
b(i)=3.0*(dx(i)*f(i+1)+dx(i+1)*f(i))
END DO
! Solution of tridiagonal system begins
DO k=1,n
ratio=dx(k+2)/diag(k)
diag(k+1)=diag(k+1)−ratio*dx(k)
b(k+1)=b(k+1)−ratio*b(k)
END DO
b(n+1)=b(n+1)/diag(n+1)
do i=n,1,−1
b(i)=(b(i)−dx(i)*b(i+1))/diag(i)
END DO
! Computation of coefficients begins
DO i=1,n
d(i)=(b(i)+b(i+1)−2.0*f(i+1))/dx(i+1)**2
c(i)=(f(i+1)−b(i))/dx(i+1)−d(i)*dx(i+1)
END DO
! Interpolation begins
DO k=1,m
DO i=1,n
xbarx=xbar(k)−x(i)
IF(xbar(k)>=x(i).AND.xbar(k)<x(i+1)&
&s(k)=y(i)+xbarx*(b(i)+xbarx*(c(i)+xbarx*d(i)))
```

END DO; END DO
RETURN
END SUBROUTINE SPLINE

Example 1. Obtain the cubic spline fit for the data

x	-1	0	1	2
y	1	-2	-1	3

with the conditions $f'(-1) = -1$, $f'(2) = 5$.

Solution. For equally spaced data Eq. (4.27) becomes

$$b_{i-1} + 4b_i + b_{i+1} = \frac{3}{h}(y_{i+1} - y_{i-1})$$

where $h := \Delta x_i$, $(i = 1, 2, \cdots, n - 1)$. Here $n = 2$, $h = 1$, $b_0 = f'(-1) = -1$, $b_3 = f'(2) = 5$ and so we have the equations

$$4b_1 + b_2 = 3(y_2 - y_0) - b_0 = -5$$
$$b_1 + 4b_2 = 3(y_3 - y_1) - b_3 = 10$$

The solution of these equations is $b_1 = -2$, $b_2 = 3$.

For $-1 \le x \le 0$, in Eqs. (4.25), (4.26) $a_0 = f(-1) = 1$, $b_0 = -1$,

$$d_0 = b_0 + b_1 - 2(y_1 - y_0) = 3, \quad c_0 = (y_1 - y_0) - b_0 - d_0 = -5$$

Hence, $S_0(x) = 1 - 1 \cdot (x + 1) - 5 \cdot (x + 1)^2 + 3 \cdot (x + 1)^3 = -2 - 2x + 4x^2 + 3x^3$.

For $0 \le x \le 1$, $a_1 = f(0) = -2$, $b_1 = -2$,

$$d_1 = b_1 + b_2 - 2(y_2 - y_1) = -1, \quad c_1 = (y_2 - y_1) - b_1 - d_1 = 4$$

Hence, $S_1(x) = -2 - 2(x - 0) + 4(x - 0)^2 - 1 \cdot (x - 0)^3 = -2 - 2x + 4x^2 - x^3$.

For $1 \le x \le 2$, $a_2 = -1$, $b_2 = 3$

$$d_2 = b_2 + b_3 - 2(y_3 - y_2) = 0, \quad c_2 = (y_3 - y_2) - b_2 - d_2 = 1$$

Hence, $S_2(x) = -1 + 3(x - 1) + 1 \cdot (x - 1)^2 + 0 \cdot (x - 1)^3 = -3 + x + x^2$.

It may be verified that $S_0(0) = S_1(0)$, $S_0'(0) = S_1'(0)$, $S_0''(0) = S_1''(0)$ and $S_1(1)$ $= S_2(1)$, $S_1'(1) = S_2'(1)$, $S_1''(1) = S_2''(1)$. □

Example 2. Using subroutine SPLINE, compare the computed and exact value of $y = \sin x$ in $[0, \pi]$ at $x = 0.5$, 1.0, 1.5, 2.0, 2.5, 3.0, taking five data points at $x = 0$, $\pi/4$, $\pi/2$, $3\pi/4$ and π.

Solution. Writing the main program as

```
REAL :: x(5), y(5), xbar(6), s(6)
pi=3.141593
x(1)=0.; x(2)=pi/4; x(3)=pi/2; x(4)=3*pi/4; x(5)=pi
y(1)=0.; y(2)=sin(x(2)); y(3)=1.; y(4)=sin(x(4)); y(5)=0.
xbar(1)=.5; xbar(2)=1.; xbar(3)=1.5; xbar(4)=2.; xbar(5)=2.5; xbar(6)=3.
CALL SPLINE(x,y,4,xbar,s,6)
DO i=1,6
PRINT*, xbar(i), s(i), sin(xbar(i))
END DO
END
```

The output is tabulated below:

x	Computed value	Exact value
0.5	0.47912	0.47943
1.0	0.84073	0.84147
1.5	0.99739	0.99750
2.0	0.90824	0.90930
2.5	0.59843	0.59847
3.0	0.14082	0.14112

The answers are essentially in agreement . □

Exercises

1. Modify subroutine SPLINE that uses boundary conditions $b_0 = f'(x_0)$, $b_n = f'(x_n)$.

2. Obtain the cubic spline approximation for the data

x	0	1	2	3
y	1	4	10	8

given that $f'(0) = 5$, $f'(3) = -1$.

$[S_0(x) = 1 + 5x - 6x^2 + 4x^3$, $\quad 0 \le x \le 1$; $\quad S_1(x) = 10 - 22x + 21x^2 - 5x^3$, $1 \le x \le 2$; $S_2(x) = -70 + 98x - 39x^2 + 5x^3$, $\quad 2 \le x \le 3]$.

3. Obtain cubic spline approximation of Ex. 2, for the natural boundary conditions $f''(0) = f''(3) = 0$.

$[S_0(x) = 1 + \dfrac{5}{3}x + \dfrac{4}{3}x^3; \quad S_1(x) = 6 - \dfrac{40}{3}x + 15x^2 - \dfrac{11}{3}x^3; \quad S_2(x) = -42 +$
$\dfrac{176}{3}x - 21x^2 + \dfrac{7}{3}x^3].$

4. Obtain the cubic spline approximation for the data

x	-1	0	1	2
y	-1	1	3	35

given that $f''(0) = f''(3) = 0$.

$[S_0(x) = 1 - 2x - 6x^2; \quad S_1(x) = 1 - 2x - 6x^2 + 10x^3; S_2(x) = 19 - 56x + 48x^2 - 8x^3].$

Chapter 5
Differentiation and Integration

In the solution of practical problems, one often encounters derivatives and integral of a function f defined in a certain interval. In many cases, the function f is so complicated that even the powerful tools of calculus are either difficult to apply or become totally infructuous. In other cases, f may be defined in a *tabular form*, to which the results of calculus do not apply. In such difficulties, numerical answers may be sought by **approximating** the function f by an **interpolating polynomial** ϕ obtained from a table of values of $f(x)$, which if not given, may be constructed from the analytical definition of f. If L denotes the operation of differentiation or integration, then since $f(x) \approx \phi(x)$, we surmise that

$$L(f(x)) \approx L((\phi(x))$$

The right-hand side of the above relation is generally called a **rule** of differentiation or integration, as the case may be. Such rules as we shall see, are linear expressions in ordinate values at the points of discretisation.

The accuracy of a rule may be judged by using the property that L is a *linear operator*. This means that for two functions f, g and a constant c, L satisfies

$$L(f(x) + g(x)) = L(f(x)) + L(g(x))$$

$$L(c\,f(x)) = c\,L(f(x))$$

Thus, if $R(x)$ is the error in approximating $f(x)$ by $\phi(x)$, i.e. $f(x) = \phi(x) + R(x)$, then

$$L(f(x)) = L(\phi(x)) + L(R(x))$$

$L(R(x))$ is, therefore, the error committed in using the rule as an approximation. Its smallness will ensure accuracy.

© Springer Nature Singapore Pte Ltd. 2019
S. K. Bose, *Numerical Methods of Mathematics Implemented in Fortran*, Forum for Interdisciplinary Mathematics,
https://doi.org/10.1007/978-981-13-7114-1_5

We first consider Numerical Differentiation in Sect. 5.1, followed by Numerical Integration in Sect. 5.2.

5.1 Numerical Differentiation

At the outset, a few words of caution are necessary regarding numerical differentiation. First of all, even if the interpolating polynomial $\phi(x)$ is close to $f(x)$, it does not mean that the slope $\phi'(x)$ is necessarily close to $f'(x)$ or, the curvature proportional to $\phi''(x)$ is close to $f''(x)$. For accuracy, the interpolating points must be very close to each other, but then differencing of nearly equal function values is encountered. This fact endangers roundoff errors on the one hand, while in the case of experimental 'noisy' data for f, the result may be totally erroneous. A comparatively safe application is in the topic of numerical solution of differential equations, where the function values are noise free.

In developing the rules of numerical differentiation, we restrict to equally spaced points (nodes) x_i, ($i = 0, 1, \cdots, n$) of an interval $[a, b]$, so that $x_i - x_{i-1} = h$. Given the data $(x_i, y_i = f(x_i))$, we base our discussion on Gregory–Newton *forward difference formula* for the interpolation function $\phi(x)$ (Chap. 4, Eq. (4.13), Sect. 4.2.1):

$$\phi(x) = y_0 + t \, \Delta y_0 + \frac{t(t-1)}{2!} \Delta^2 y_0 + \frac{t(t-1)(t-2)}{3!} \Delta^3 y_0$$

$$+ \frac{t(t-1)(t-2)(t-3)}{4!} \Delta^4 y_0 + \cdots \text{ to } n+1 \text{ terms}$$

where $t = (x - x_0)/h$. Since $d\phi/dx = \frac{1}{h} d\phi/dt$, it follows that

$$f'(x) \approx \phi'(x) = \frac{1}{h} \left[\Delta y_0 + \frac{2t-1}{2} \Delta^2 y_0 + \frac{3t^2 - 6t + 2}{6} \Delta^3 y_0 \right.$$

$$\left. + \frac{2t^3 - 9t^2 + 11t - 3}{12} \Delta^4 y_0 + \cdots \text{ to } n+1 \text{ terms} \right] \qquad (5.1)$$

The error in formula (5.1) is $R'(x) = f'(x) - \phi'(x)$ where (see Chap. 4, Sect. 4.2.1),

$$R(x) = h^{n+1} \frac{t(t-1)\cdots(t-n)}{(n+1)!} f^{(n+1)}(\xi), \quad x_0 < \xi < x_n$$

Assuming $f \in C^{(n+2)}$, the above equation yields

$$R'(x) = \frac{dR}{dx} \frac{dt}{dx} = \frac{h^n}{(n+1)!} \left[\frac{d}{dt} \{ t(t-1) \cdots (t-n) \} f^{(n+1)}(\xi) \right.$$

$$+t(t-1)\cdots(t-n)\frac{d}{dt}\{f^{(n+1)}(\xi)\}\Bigg]$$

Often one is interested in evaluating the derivative at a nodal point x_k or $t = k$. In such a case, the above expression simplifies in to

$$R'(x_k) = (-1)^{n-k}\frac{k!(n-k)!}{(n+1)!}h^n f^{(n+1)}(\xi) \tag{5.2}$$

The error $R'(x)$ or $R'(x_k)$ is called **discritisation error** due to discritisation of continuous f.

Based on formulas (5.1) and (5.2), several widespread formulas can be derived. They are listed below:

$1^o.$ $n = 1$ (two points):

$$f_0' = f_1' = \frac{1}{h}(y_1 - y_0) - \frac{h}{2}f''(\xi)$$

$2^o.$ $n = 2$ (three points):

$$f_0' = \frac{1}{2h}(-3y_0 + 4y_1 - y_2) + \frac{h^2}{3}f'''(\xi)$$

$$f_1' = \frac{1}{2h}(y_2 - y_0) - \frac{h^2}{6}f'''(\xi)$$

$$f_2' = \frac{1}{2h}(y_0 - 4y_1 + 3y_2) + \frac{h^2}{3}f'''(\xi)$$

$3^o.$ $n = 3$ (four points):

$$f_0' = \frac{1}{6h}(-11y_0 + 18y_1 - 9y_2 + 2y_3) - \frac{h^3}{4}f^{iv}(\xi)$$

$$f_1' = \frac{1}{6h}(-2y_0 - 3y_1 + 6y_2 - y_3) + \frac{h^3}{12}f^{iv}(\xi)$$

$$f_2' = \frac{1}{6h}(y_0 - 6y_1 + 3y_2 + 2y_3) - \frac{h^3}{12}f^{iv}(\xi)$$

$$f_3' = \frac{1}{6h}(-2y_0 + 9y_1 - 18y_2 + 11y_3) + \frac{h^3}{4}f^{iv}(\xi)$$

etc. For any n, we have

$$f_0' = \frac{1}{h}\left[\Delta y_0 - \frac{1}{2}\Delta^2 y_0 + \frac{1}{3}\Delta^3 y_0 - \cdots n \text{ terms}\right] + (-1)^n\frac{h^n}{n+1}f^{(n+1)}(\xi)$$

Formulas for the second derivative $f''(x)$ can be obtained by differentiating Eq. (5.1). Since $\phi''(x) = \frac{1}{h^2}\frac{d^2\phi}{dt^2}$, it follows that

$$f''(x) \approx \phi''(x) = \frac{1}{h^2}\Big[\Delta^2 y_0 + (t-1)\,\Delta^3 y_0 + \frac{6t^2 - 18t + 11}{12}\,\Delta^4 y_0 + \cdots$$
$$+ \,(n-1)\text{ terms}\Big] \tag{5.3}$$

with an error term $R''(x)$. At the nodal points, the following formulas hold:

1°. $n = 2$ (three points):

$$f_0'' = \frac{1}{h^2}(y_0 - 2y_1 + y_2) - h\,f'''(\xi)$$
$$f_1'' = \frac{1}{h^2}(y_0 - 2y_1 + y_2) - \frac{h^2}{12}\,f^{(iv)}(\xi)$$
$$f_2'' = \frac{1}{h^2}(y_0 - 2y_1 + y_2) + h\,f'''(\xi)$$

2°. $n = 3$ (four points):

$$f_0'' = \frac{1}{h^2}(2y_0 - 5y_1 + 4y_2 - y_3) + \frac{11}{12}\,h^2\,f^{(iv)}(\xi)$$
$$f_1'' = \frac{1}{h^2}(y_0 - 2y_1 + y_2) - \frac{h^2}{12}\,f^{(iv)}(\xi)$$
$$f_2'' = \frac{1}{h^2}(y_1 - 2y_2 + y_3) - \frac{h^2}{12}\,f^{(iv)}(\xi)$$
$$f_3'' = \frac{1}{h^2}(-y_0 + 4y_1 - 5y_2 + 2y_3) + \frac{11}{12}\,h^2\,f^{(iv)}(\xi)$$

etc. More formulas of the above type may be added for increasing n and the order of the derivative. It is then revealed that the order of accuracy increases with n, while it decreases with respect to h for increasing order of the derivative. For an even n and even derivative, the formula at the middle point is a unity higher than at the remaining points. Therefore, as far as possible, one should perform numerical differentiation with points arranged symmetrically about the point and use central difference quotient formula. The remainders in these formulas are obtained by adopting the Taylor's expansion method, with the remainder in suitable from (as illustrated in Example 1). In treating the remainder term, one often requires the following property of continuous function:

Lemma 5.1 (Weighted Average of a Function). *Let $f \in C[a, b]$ and $\xi_i \in [a, b]$ be arbitrary points, $i = 1, 2, \cdots, n$. Then for constants $\lambda_i > 0$, there exists a point $\xi \in [a, b]$ such that*

$$\frac{\lambda_1 f(\xi_1) + \cdots + \lambda_n f(\xi_n)}{\lambda_1 + \cdots + \lambda_n} = f(\xi)$$

Proof $\min_{x \in [a, b]} f(x) \le f(\xi_i) \le \max_{x \in [a, b]} f(x)$

Hence, the weighted average lies between the maximum and minimum of $f(x)$. The lemma follows from the intermediate value of a continuous function.   □

Example 1. Prove the central difference quotient formula for the second derivative

$$f_1'' = \frac{1}{h^2}(y_0 - 2y_1 + y_2) - \frac{h^2}{12}f^{(iv)}(\xi)$$

using Taylor's expansion theorem.

Solution. Shifting the origin to the point x_1, the formula to be proved is

$$f''(0) = \frac{1}{h^2}[f(-h) - 2f(0) + f(h)] - \frac{h^2}{12}f^{(iv)}(\eta)$$

Let $F(h) = f(h) - 2f(0) + f(-h) - h^2 f''(0)$. Then $F(0) = 0$

$$F'(h) = f'(h) - f'(-h) - 2h f''(0), \quad F'(0) = 0$$

$$F''(h) = f''(h) + f''(-h) - 2 f''(0), \quad F''(0) = 0$$

$$F'''(h) = f'''(h) - f'''(-h), \qquad\qquad F'''(0) = 0$$

$$F^{(iv)}(h) = f^{(iv)}(h) + f^{(iv)}(-h), \qquad F^{(iv)}(0) \neq 0$$

Hence,

$$F(h) = \frac{h^4}{24}F^{(iv)}(\xi) = \frac{h^2}{12}\frac{f^{(iv)}(\xi) + f^{(iv)}(-\xi)}{2} = \frac{h^2}{12}f^{(iv)}(\eta)$$

by the lemma on weighted average. Here we assume that $f \in C^4[a, b]$. Thus

$$f(h) - 2f(0) + f(-h) - h^2 f''(0) = \frac{h^2}{12}f^{(iv)}(\eta)$$

which yields the formula   □

Example 2. The table below gives the value of distance travelled by a projectile at various time intervals:

Time T(s)	5	6	7	8	9
Distance Travelled s (km)	10.0	14.5	19.5	25.5	33.0

Estimate the velocity and acceleration at times $T = 5$, $T = 7$ and $T = 9$ s.

Solution. The data are equally spaced and so we form the finite difference table

T	y	Δ	Δ^2	Δ^3	Δ^4
5	10.0				
		4.5			
6	14.5		0.5		
		5.0		0.5	
7	19.5		1.0		0
		6.0		0.5	
8	25.5		1.5		
		7.5			
9	33.0				

Here $x_0 = 5$ and $h = 1$, with $x_2 = 7$ and $x_4 = 9$. For these abscissas, the value of the parameter t are, respectively, $t = 0$, $t = 2$ and $t = 4$.

Velocity. It is given by the first derivative. For $t = 0, 2, 4$, formula (5.1) yields the approximation

$$f'(x_0) \approx \frac{1}{h}\left[\Delta y_0 - \frac{1}{2}\Delta^2 y_0 + \frac{1}{3}\Delta^3 y_0 - \frac{1}{4}\Delta^4 y_0\right]$$

$$f'(x_2) \approx \frac{1}{h}\left[\Delta y_0 + \frac{3}{2}\Delta^2 y_0 + \frac{1}{3}\Delta^3 y_0 - \frac{1}{12}\Delta^4 y_0\right]$$

$$f'(x_4) \approx \frac{1}{h}\left[\Delta y_0 + \frac{7}{2}\Delta^2 y_0 + \frac{13}{3}\Delta^3 y_0 + \frac{25}{12}\Delta^4 y_0\right]$$

Thus, from the difference table, the velocities at $T = 5$, 7 and 9 s are

$$f'(5) \approx \frac{1}{1}\left[4.5 - \frac{1}{2} \times 0.5 + \frac{1}{3} \times 0.5 - \frac{1}{4} \times 0\right] = 4.4167 \text{ km/s}$$

$$f'(7) \approx \frac{1}{1}\left[4.5 + \frac{3}{2} \times 0.5 + \frac{1}{3} \times 0.5 - \frac{1}{12} \times 0\right] = 5.4167 \text{ km/s}$$

$$f'(9) \approx \frac{1}{1}\left[4.5 + \frac{7}{2} \times 0.5 + \frac{13}{3} \times 0.5 + \frac{25}{12} \times 0\right] = 8.4167 \text{ km.s}$$

Acceleration. It is given by the second derivative. For $t = 0, 2$ and 4, formula (5.3) similarly yields

$$f''(x_0) \approx \frac{1}{h^2}\left[\Delta^2 y_0 - \Delta^3 y_0 + \frac{11}{12}\Delta^4 y_0\right]$$

$$f''(x_2) \approx \frac{1}{h^2}\left[\Delta^2 y_0 + \Delta^3 y_0 - \frac{1}{12}\Delta^4 y_0\right]$$

$$f''(x_4) \approx \frac{1}{h^2}\left[\Delta^2 y_0 + 3\Delta^3 y_0 + \frac{35}{12}\Delta^4 y_0\right]$$

Hence, from the difference table, the accelerations at $T = 5$, 7 and 9 s are

$$f''(5) \approx \frac{1}{1^2}\left[0.5 - 0.5 + \frac{11}{12} \times 0\right] = 0 \text{ km/s}^2$$

$$f''(7) \approx \frac{1}{1^2}\left[0.5 + 0.5 - \frac{1}{12} \times 0\right] = 1 \text{ km/s}^2$$

$$f''(9) \approx \frac{1}{1^2}\left[0.5 + 3 \times 0.5 + \frac{35}{12} \times 0\right] = 2 \text{ km/s}^2 \qquad \square$$

Exercises

1. The distance travelled by a car in kilometres, at intervals of 2 min. are given as follows:

Time (m)	2	4	6	8	10
Distance (km)	0.75	2.00	3.50	5.35	8.00

Evaluate the velocity and acceleration at $T = 2$ and 7 min.

[Vel.: $v(2) = 32.12$ km/h, $v(7) = 54.84$ km/h. Accl.: $a(2) = 0.1177$ km/min^2, $a(7) = 0.1255$ km/min^2].

2. A table of complete elliptic integral $K(m)$ is given below:

m	0.1	0.2	0.3	0.4	0.5
$K(m)$	1.61244	1.65962	1.71389	1.77752	1.85407

Calculate $K'(m)$ for $m = 0.1$ and 0.3. Also calculate $K''(0.5)$.

[$K'(0.1) = 0.4407$, $K'(0.3) = 0.5847$, $K''(0.5) = 1.766$].

3. Given the table of Bessel function of order 1, denoted by $J_1(x)$:

x	0	0.1	0.2	0.3	0.4	0.5
$J_1(x)$	0.0	0.04994	0.09950	0.14832	0.19603	0.24227

Calculate $J_1'(x)$ at $x = 0$, 0.3 and 0.5.

[0.5001, 0.4833, 0.4541].

4. Using the backward difference interpolation formula, prove that

$$(i) \quad y_n' \approx \frac{1}{h}\left[\nabla + \frac{1}{2}\nabla^2 + \frac{1}{3}\nabla^3 + \frac{1}{4}\nabla^4 + \cdots\right] y_n$$

$$(ii) \quad y_n'' \approx \frac{1}{h^2}\left[\nabla^2 + \nabla^3 + \frac{11}{12}\nabla^4 + \cdots\right] y_n$$

Obtain the expression for the error in the case (i).

5. Using Lagrange's interpolation formula for three arbitrarily spaced points (x_0, y_0), (x_1, y_1) and (x_3, y_3), prove the differentiation formulas:

$$(i) \quad f'(x_0) \approx \frac{2x_0 - x_1 - x_2}{(x_0 - x_1)(x_0 - x_2)} y_0 + \frac{x_0 - x_2}{(x_1 - x_0)(x_1 - x_2)} y_1 + \frac{x_0 - x_1}{(x_2 - x_0)(x_2 - x_1)} y_2$$

$$(ii) \quad f''(x_0) \approx 2\left[\frac{y_0}{(x_0 - x_1)(x_0 - x_2)} + \frac{y_1}{(x_1 - x_0)(x_1 - x_2)} + \frac{y_2}{(x_2 - x_0)(x_2 - x_1)}\right]$$

Prove that the remainder in case(i) is $R'(x_0) = \dfrac{1}{6}(x_0 - x_1)(x_0 - x_2) f''(\xi)$.

5.1.1 Minimal Step Length h

In the formulas of numerical differentiation with constant h, the values of the function f are divided by h^m, where m is the order of the computed derivative. The values of f are subject to roundoff in a computer and if h is taken very small, the nearly equal values of f in a formula lead to loss of significance. To theoretically examine how small h can be taken, consider the central difference differentiation formulas

$$f_1' = \frac{1}{2h}(y_2 - y_0) - \frac{h^2}{6} f'''(\xi)$$

and

$$f_1'' = \frac{1}{h^2}(y_0 - 2y_1 + y_2) - \frac{h^2}{12} f^{(iv)}(\xi)$$

Let ϵ be the maximum roundoff error in y_0, y_1, y_2 and let $|f'''(\xi)| \le M_3, |f^{(iv)}(\xi)| \le M_4$, then the maximum computed error in f_1' and f_1'' are

$$\epsilon_1 = \frac{1}{2h}(\epsilon + \epsilon) + \frac{h^2}{6} M_3 = \frac{\epsilon}{h} + \frac{h^2}{6} M_3$$

and

$$\epsilon_2 = \frac{1}{h^2}(\epsilon + 2\epsilon + \epsilon) + \frac{h^2}{12} M_4 = \frac{4\epsilon}{h^2} + \frac{h^2}{12} M_4$$

These quantities are minimum when $d\epsilon_1/dh = 0$ and $d\epsilon_2/dh = 0$. These conditions yield

$$h = h_1 = \left(\frac{3\epsilon}{M_3}\right)^{1/3}$$

for minimum ϵ_1 and

$$h = h_2 = 2\left(\frac{3\epsilon}{M_4}\right)^{1/4}$$

for minimum ϵ_2. These expressions prove the need for lower limits h_1 and h_2 on h for the respective cases.

5.1.2 Richardson Extrapolation

The exact numerical differentiation formulas contain a discretised version of the derivative known as the **difference quotient** together with a discretisation error containing the information of its rate of decay as $h \to 0$. The general form of this error can be used to estimate its major contribution to the differentiation formulas. Consider, for example, the central difference formula for f_1' considered in the preceding section. Here, $R'(x_1) = -\frac{h^2}{6} f'''(\xi)$, where $\xi \to x_1$ and $f'''(\xi) \to f'''(x_1)$ as $h \to 0$. Hence, $\frac{h^2}{6}[f'''(\xi) - f'''(x_1)] \to 0$ faster than h^2. Using the order notation, we then have

$$f_1' = \frac{1}{2h}(y_2 - y_0) - \frac{h^2}{6} f'''(x_1) + o(h^2)$$

In general, therefore a pth-order differentiation formula can be written as

$$D(f) = D_h(f) + C h^p + o(h^p)$$

where D is a differentiation operator of some order and D_h is its discretised counterpart, and C is an unknown constant independent of h. D and D_h are linear operators.

The principal contribution of the error Ch^p can be estimated by choosing another step length αh, where $0 < \alpha < 1$. In particular, α can be chosen as $1/2$. If $D_{\alpha h}(f)$ denotes the computed discrete derivative for the step length αh, then as before

$$D(f) = D_{\alpha h}(f) + C (\alpha h)^p + o(h^p)$$

Eliminating C from the two equations, we obtain

$$D(f) = \frac{D_{\alpha h}(f) - \alpha^p D_h(f)}{1 - \alpha^p} + o(h^p) \tag{5.4}$$

$$= D_h(f) + \frac{D_{\alpha h}(f) - D_h(f)}{1 - \alpha^p} + o(h^p)$$

For computing the derivative from the first two terms of the right-hand side of (5.4), h is required to be sufficiently 'small'. To test the smallness of h suppose, we also compute $L_{h/\alpha}$ for the step length h/α. Then

$$\frac{D_{\alpha h}(f) - D_h(f)}{1 - \alpha^p} \approx C h^p \approx \frac{D_{h/\alpha}(f) - D_h(f)}{1 - 1/\alpha^p}$$

or,

$$R(h) := \frac{D_{\alpha h}(f) - D_h(f)}{D_h(f) - D_{h/\alpha}(f)} \approx \alpha^p \qquad (5.5)$$

Thus, if the expression in Eq. (5.4) approximately equals α^p, h is sufficiently small and can be accepted. In the contrary case, h is indicated to be not small enough and (5.4) may not be accurate.

The formula (5.4) is an extrapolation process, because let

$$D_h^1(f) := D_h(f) + \frac{D_{\alpha h}(f) - D_h(f)}{1 - \alpha^p} = D_{\alpha h}(f) + \frac{\alpha^p [D_{\alpha h}(f) - D_h(f)]}{1 - \alpha^p}$$

Hence, if $D_h(f) < D_{\alpha h}(f)$, then $D_h^1(f) < D_h(f) < D_{\alpha h}(f)$ and if $D_h(f) > D_{\alpha h}(f)$ then $D_h^1(f) < D_{\alpha h}(f) < D_h(f)$. This means that $D_h^1(f)$ always falls outside the interval $(D_h(f), D_{\alpha h}(f))$ or $(D_{\alpha h}(f), D_h(f))$.

The extrapolation process can be repeated a number of times if the discritisation error can be expressed as a series of powers of h. In this way, highly accurate value can be obtained *by computing a relatively small number of ordinates.* (see Exercise 4 below).

This process is due to Richardson and is also known as **deferred approach to the limit**.

Lewis Fry Richardson (1881–1953), English scientist. He was the first to apply finite differences to numerical weather prediction (1922). He also made contributions to the theory of eddy diffusion in the atmosphere, where *Richardson Number* is a fundamental quantity involving gradients of temperature and wind velocity. Beginning World War II in 1939, he made mathematical studies of causes of wars.

Example 1. Let $f(x) = \sin x$. Calculate $f'(x)$ for $x = \pi/3$, using central difference quotient formula with Richardson extrapolation. Try with $h = 1.6,\ 0.8,\ 0.4$ and 0.2 for calculation.

Solution. Let $x_1 = \pi/3 = 1.047197551 \approx 1.0472$. The central difference quotient formula

$$f'(x_1) = \frac{y_2 - y_0}{2h} - \frac{h^2}{6} f''(\xi)$$

leads to the form

$$D(f) = D_h(f) + C h^2 + o(h^2) = D_h^1(f) + o(h^2)$$

where according to formula (5.3) with $\alpha = 1/2$

$$D_h(f) = \frac{\sin(1.0472 + h) - \sin(1.0472 - h)}{2h}$$

$$D_h^1(f) = D_h(f) + \frac{D_{h/2}(f) - D_h(f)}{1 - 1/4} = \frac{4\,D_{h/2}(f) - D_h(f)}{3}$$

The accuracy testing function $R(h)$ given by Eq. (5.5) must satisfy

$$R(h) := \frac{D_{h/2}(f) - D_h(f)}{D_h(f) - D_{2h}(f)} \approx \left(\frac{1}{2}\right)^2 = 0.25$$

Calculating these functions for $h = 1.6,\ 0.8,\ 0.4,\ 0.2$, we obtain the table

h	$D_h(f)$	$D_h^1(f)$	$R(h)$
1.6	0.3123654	0.4936725	
0.8	0.4483457	0.4995793	0.2826
0.4	0.4867709	0.4999713	0.2577
0.2	0.4966712	0.4999963	0.2519
0.1	0.4991650		

Since $R(0.2) = 0.2519 \approx 0.25$, $f'(\pi/3)$ is best given by $D_h^1(f) = 0.4999963$ for $h = 0.2$. Of course, we know that the exact value is $f'(\pi/3) = \cos(\pi/3) = 0.5$. $\square$

Exercises

1. Calculate the value of $f'(0)$, using central difference quotient and Richardson extrapolation for the function tabulated below:

x	-4	-2	-1	0	1	2	4
$f(x)$	-8.46	-3.39	-2.03	0	1.98	4.64	7.65

Calculate $R(h)$ and draw conclusion about accuracy.

[$f'(0) = 2.00$, $R(h) = 0.38$. Not indicated to be accurate if the function values are accurate up to two decimal places].

2. A partial table of the error function $erf(x)$ is given below:

x	0.80	0.90	0.95	1.00	1.05	1.10	1.20
$erf(x)$	0.74210	0.79691	0.82089	0.84270	0.86244	0.88021	0.91031

Calculate $erf'(1)$ and $erf''(1)$ using central difference quotient and Richardson extrapolation. Comment on the accuracy of the two results.

$[erf'(1) = 0.41517, erf''(1) = -0.828.\ R(h) = 0.2484$ and 0, respectively, for the two answers. The first result is accurate but the second result may not be accurate].

3. By Taylor's expansion prove that

(i) $f'(x_1) = \dfrac{f(x_1+h) - f(x_1-h)}{2h} - \dfrac{h^2}{6} f'''(x_1) - \dfrac{h^4}{120} f^{(v)}(x_1) + o(h^4)$

(ii) $f''(x_1) = \dfrac{f(x_1-h) - 2f(x_1) + f(x_1+h)}{h^2} - \dfrac{h^2}{12} f^{(iv)}(x_1) - \dfrac{h^4}{360} f^{(vi)}(x_1) + o(h^4)$

4. By Exercise 3 above, a central difference derivative formula can be written as

$$D(f) = D_h(f) + C_1 h^2 + C_2 h^4 + o(h^4)$$

Prove that

$$D(f) = D_h^1(f) + o(h^2) = D_h^2(f) + o(h^4)$$

where

$$D_h^1(f) = \frac{1}{3}\left[4D_{h/2}(f) - D_h(f)\right], \quad D_h^2(f) = \frac{1}{15}\left[16D_{h/2}^1(f) - D_h^1(f)\right]$$

provided that

$$\frac{D_{h/2}(f) - D_h(f)}{D_h(f) - D_{2h}(f)} \approx \frac{1}{4}, \quad \frac{D_{h/2}^1(f) - D_h^1(f)}{D_h^1(f) - D_{2h}^1(f)} \approx \frac{1}{16}$$

5.2 Numerical Integration

In practice, many a time, even simple looking (definite) integrals defy analytical evaluation. For instance, the elliptic integral $K(m)$ (Exercise 1, Sect. 4.1.5, Chap. 4) and the error function $erf(x)$, are of this category. These two integrals, respectively, appear somehow in the study of large amplitude oscillations of a simple pendulum and in the probabilistic theory of errors. The difficulty of calculating an integral analytically is evident in another case, that of an integrand given as a table of numerical values. Thus, we consider the numerical version of the problem, viz. given numbers a, b and the function $f(x)$ (either in analytical or in tabular form), then estimate the number

$$I(f) = \int_a^b f(x)\,dx \tag{5.6}$$

Such a problem is called **numerical integration** on **numerical quadrature**.

The definite integral (5.6), possesses a number of simple properties. One of them is that I is a *linear operator*, in the sense explained at the beginning of this chapter. These properties fortunately enable, accurate numerical integration of a definite integral (in contrast to numerical differentiation). The basic procedure is to approximate $f(x)$ by a suitable interpolating $\phi(x)$ and take $I(f) \approx \int_a^b \phi(x)\,dx$. We begin by developing some basic rules of numerical integration based on equally spaced subdivision or **panels** of the interval $[a, b]$.

5.2.1 Basic Rules of Numerical Integration

Suppose $[a, b]$ is divided in to n panels each of length $h = (b - a)/n$, by the nodal points x_i, $(i = 0, 1, \cdots, n)$ where $x_0 := a$ and $x_n := b$, so that $x_i - x_{i-1} = h$. On $[a, b]$ suppose $f(x)$ is approximated by the Gregory–Newton forward difference formula for the interpolation function $\phi(x)$ (Chap. 4, Eq. (4.13), Sect. 4.2.1). If $R(x)$ is the error, then

$$f(x) = \phi(x) + R(x)$$
$$= y_0 + t\,\Delta y_0 + \frac{t(t-1)}{2!}\Delta^2 y_0 + \cdots n + 1 \text{ terms}$$
$$+ \frac{h^{n+1}}{(n+1)!} f^{(n+1)}(\xi)\, t(t-1)\cdots(t-n), \quad x_0 < \xi < x_n \tag{5.7}$$

where $t = (x - x_0)/h$. Integrating the above expression from a to b, we get an expression for $I(f)$. Giving n some low values, we obtain a set of formulas. The number of terms and complexity of the formulas increase with n.

The treatment of remainder term $R(x)$ requires the use of

(Generalised Mean Value Theorem). *Let* f, $g \in C[a, b]$ *and* $g(x) \geq 0$ *on* $[a, b]$, *then*

$$\int_a^b f(x)\,g(x)\,dx = f(\eta) \int_a^b g(x)\,dx, \quad a < \eta < b$$

The proof of the theorem can be found in a textbook of Integral Calculus.

We now develop the basic rules of numerical integration from Eq. (5.7).

1°. **Rectangle Rule** $(n = 0)$. In this very simple case, $h = b - a$ and $f(x) \approx \phi(x) = y_0$. Thus

$$I_R(f) \approx \int_a^b \phi(x)\, dx = h\, y_0$$
$$= \text{Area of the rectangle with base } h = b - a,$$
$$\text{and height } y_0 = f(x_0) = f(a) \tag{5.8}$$

The error of interpolation is $R(x) = h\, f'(\xi)\, t$, where $t = (x - x_0)/h$ according to Eq. (5.7). Hence, the error of this quadrature formula is

$$E(h) = \int_a^b h\, f'(\xi)\, t\, dx = h^2 \int_0^1 f'(\xi)\, t\, dt = h^2 f'(\eta) \int_0^1 t\, dt = \frac{h^2}{2} f'(\eta) \tag{5.9}$$

in which the generalised mean value theorem is applied.

2°. **Midpoint Rule** $(n = 1)$. With a change, take $x_0 = (a + b)/2$, the midpoint of $[a, b]$. Then as in 1°

$$f(x) \approx \phi(x) = y_0 + t\, \Delta y_0$$

where $t = (x - x_0)/h$. Hence

$$I_M(f) \approx \int_{x_0-h/2}^{x_0+h/2} [y_0 + t\, \Delta y_0]\, dx = \int_{-1/2}^{1/2} [y_0 + t\, \Delta y_0]\, h\, dt = h\, y_0$$
$$= \text{Area of the rectangle with base } h = b - a,$$
$$\text{and midpoint height } y_0 = f(x_0) = f\left(\frac{a+b}{2}\right) \tag{5.10}$$

For this case, the error $R(x) = \dfrac{h^2}{2} f''(\xi)\, t(t - 1)$ and the error of the quadrature formula (5.10) is

$$E(h) = \int_a^b R(x)\, dx = \frac{h^3}{2} \int_{-1/2}^{1/2} f''(\xi)\, t(t - 1)\, dt$$

We note that $t(t - 1)$ changes sign in $[-1/2, 1/2]$ and so the Generalised Mean Value Theorem cannot be applied to this case. Nevertheless, we have

$$E(h) = \int_{x_0-h/2}^{x_0+h/2} f(x)\, dx - h\, y_0, \quad E(0) = 0 - 0 = 0$$

Differentiating with respect to h,

$$E'(h) = \frac{1}{2} f\left(x_0 + \frac{h}{2}\right) - \left(-\frac{1}{2}\right) f\left(x_0 - \frac{h}{2}\right) - y_0, \quad E'(0) = \frac{1}{2} y_0 + \frac{1}{2} y_0 - y_0 = 0$$

Similarly

$$E''(h) = \frac{1}{4} f'\left(x_0 + \frac{h}{2}\right) - \frac{1}{4} f'\left(x_0 - \frac{h}{2}\right), \quad E''(0) = \frac{1}{4} f''(0) - \frac{1}{4} f''(0) = 0$$

$$= \frac{1}{4}\left[\frac{h}{2} - \left(-\frac{h}{2}\right)\right] f''(\xi) = \frac{1}{4} h f''(\xi)$$

Now integrating with respect to h, we obtain

$$E'(h) - E'(0) = E'(h) = \frac{1}{4} \int_0^h h f''(\xi) \, dh = \frac{h^2}{8} f''(\zeta)$$

by applying the Generalised Mean Value Theorem. Again integrating with respect to h, we get

$$E(h) - E(0) = \frac{1}{8} \int_0^h h^2 f''(\zeta) \, dh = \frac{1}{24} h^3 f''(\eta)$$

Thus,

$$E(h) = \frac{h^3}{24} f''(\eta) \tag{5.11}$$

Equations (5.10) and (5.11) show that for 'small' $h(< 1)$ and comparable values of $f'(\eta)$ and $f''(\eta)$, the midpoint rule (5.10) is more accurate than the rule (5.8) even though both represent area of rectangles.

$3°$. **Trapezoidal Rule** $(n = 1)$. Here $x_0 = a$, $x_1 = b$ and $f(x) \approx \phi(x) = y_0 + t \Delta y_0$ where $t = (x - x_0)/h$. Hence

$$I_T(f) \approx \int_a^b [y_0 + t \Delta y_0] \, dx = \int_0^1 [y_0 + t(y - y_0)] h \, dt$$

$$= h y_0 + h(y_1 - y_0) \cdot \frac{1}{2} = h\left(\frac{y_0 + y_1}{2}\right)$$

$$= \text{Area of the trapezium with base } h = b - a$$

$$\text{bounded by ordinates } y_0 \text{ and } y_1 \tag{5.12}$$

The error of the formula is

$$E(h) = \int_a^b R(x)\,dx = \int_0^1 \frac{h^2}{2} f''(\xi)\,t(t-1)\,h\,dt$$

$$= -\frac{h^3}{2} f''(\eta) \int_0^1 t(1-t)\,dt = -\frac{h^3}{12} f''(\eta) \qquad (5.13)$$

by the Generalised Mean Value Theorem. The expression for the error shows that the rule integrates a linear function exactly, because $f''(\eta) = 0$ for such a function.

4°. Simpson's Rule ($n = 2$). Here $x_0 = a$, $x_1 = \dfrac{a+b}{2}$ and $x_2 = b$. Then

$$I_S(f) \approx \int_a^b \left[y_0 + t\,\Delta y_0 + \frac{t(t-1)}{2}\,\Delta^2 y_0 \right] dx$$

$$= \int_0^2 \left[y_0 + t\,(y_1 - y_0) + \frac{t^2 - t}{2}\,(y_0 - 2\,y_1 + y_2) \right] h\,dt$$

$$= h \left[2\,y_0 + (y_1 - y_0)\cdot 2 + \frac{1}{2}(y_0 - 2\,y_1 + y_2)\cdot\left(\frac{8}{3} - 2\right) \right]$$

$$= \frac{h}{3}\,(y_0 + 4\,y_1 + y_2)$$

$$= \text{Area under the parabola through } (x_0,\,y_0),\ (x_1,\,y_1)$$

$$\text{and }\ (x_2,\,y_2) \text{ between the abscissas } x_0 \text{ and } x_2 \qquad (5.14)$$

The geometrical interpretation of the right-hand side follows from the fact that the formula is derived by integrating a second-degree (interpolation) polynomial, which is a *parabola*.

To obtain a suitable expression for the error $E(h)$ committed in (5.14), we have effectively

$$E(h) = \int_{x_0}^{x_0+2h} f(x)\,dx - \frac{h}{3}\,[f(x_0 + 4\,f(x_0 + h) + f(x_0 + 2h)]$$

$$= \int_{-h}^h f(x_1 + u)\,du - \frac{h}{3}\,[f(x_1 - h) + 4\,f(x_1) + f(x_1 + h)]$$

by putting $x = x_1 + u$. Differentiating the two sides with respect to h, we have

$$E'(h) = f(x_1 + h) + f(x_1 - h) - \frac{1}{3}\,[f(x_1 - h) + 4\,f(x_1) + f(x_1 + h)]$$

$$-\frac{h}{3}\,[-f'(x_1 - h) + f'(x_1 + h)]$$

$$= \frac{2}{3} [f(x_1 + h) + f(x_1 - h)] - \frac{4}{3} f(x_1) - \frac{h}{3} [f'(x_1 + h) - f'(x_1 - h)]$$

Thus $E'(0) = 0$. Again differentiating with respect to h, we obtain

$$E''(h) = \frac{2}{3} [f'(x_1 + h) - f'(x_1 - h)] - \frac{1}{3} [f'(x_1 + h) - f'(x_1 - h)]$$
$$- \frac{h}{3} [f''(x_1 + h) + f''(x_1 - h)]$$
$$= \frac{1}{3} [f'(x_1 + h) - f'(x_1 + h)] - \frac{h}{3} [f''(x_1 + h) + f''(x_1 - h)]$$

Consequently $E''(0) = 0$. A third differentiation simplifies in to

$$E'''(h) = -\frac{h}{3} [f'''(x_1 + h) - f'''(x_1 - h)] = -\frac{2h^2}{3} f^{(iv)}(\xi)$$

Integrating the above equation three times repeatedly and applying the Generalised Mean Value Theorem we get

$$E''(h) - E''(0) = E''(h) = -\frac{2}{3} f^{(iv)}(\xi_1) \int_0^h h^2 \, dh = -\frac{2}{9} f^{(iv)}(\xi_1) h^3$$

$$E'(h) - E'(0) = E'(h) = -\frac{2}{9} f^{(iv)}(\xi_2) \int_0^h h^3 \, dh = -\frac{1}{18} f^{(iv)}(\xi_2) h^4$$

$$E(h) - E(0) = E(h) = -\frac{1}{18} f^{(iv)}(\eta) \int_0^h h^4 \, dh = -\frac{h^5}{90} f^{(iv)}(\eta)$$

Thus, finally we obtain

$$E(h) = -\frac{h^5}{90} f^{(iv)}(\eta) \tag{5.15}$$

The above expression shows that Simpson's rule not only integrates a second-degree polynomial exactly, but it does so a third-degree cubic polynomial as well. This is true because $f^{(iv)}(\eta)$ vanishes for both the types of curve.

Thomas Simpson (1710–1761), English mathematician who lectured in coffee houses and reportedly kept low company, becoming editor of *Ladies Diary* in 1754. But earlier he published *The Nature and Laws of Chance* in 1740 and *The Doctrine and Application of Fluxions* in 1750. The latter book contained his work on interpolation and numerical integration together with the work of Cotes. His contributions to geometry, trigonometry and astronomy are equally important. The names of the trigonometric functions sine, cosine, tangent and cotangent are due to Simpson.

Example 1. Evaluate $\int_0^1 e^{-x^2} dx$ by the basic rules, estimating the error in each case.

Solution. Here $f(x) = e^{-x^2}$, $a = 0$ and $b = 1$. Therefore,

$$f(0) = 1, \quad f(1) = e^{-1} = 0.36788, \quad f(1/2) = e^{1/4} = 0.77880$$

Formulas (5.8), (5.10), (5.12) and (5.14), respectively, give

$I_R = 1 \cdot f(0) = 1$
$I_M = 1 \cdot f(1/2) = 0.77880$
$I_T = \frac{1}{2}[f(0) + f(1)] = \frac{1}{2}(1 + 0.36788) = 0.68394$
$I_S = \frac{1/2}{3}[f(0) + 4f(1/2) + f(1)] = \frac{1}{6}[1 + 4 \times 0.77880 + 0.36788] = 0.74718$

The exact value correct to five decimal places is $I = 0.74682$.

The errors are, respectively, given by formulas (5.9), (5.11), (5.13) and (5.15). To estimate the errors, we calculate the derivatives

$$f'(x) = e^{-x^2}(-2x), \quad f''(x) = e^{-x^2}(4x^2 - 2), \quad f'''(x) = e^{-x^2}(12x - 8x^3)$$

$$f^{(iv)}(x) = e^{-x^2}(12 - 48x^2 + 16x^4), \quad f^{(v)}(x) = e^{-x^2}(-120x + 160x^3 - 32x^5)$$

Therefore, from Eq. (5.9)

$$E_R = \frac{1}{2}|f'(\eta)| \leq \frac{1}{2} \max_{0 \leq x \leq 1} 2|x\, e^{-x^2}| = \frac{1}{\sqrt{2}} e^{-1/2} = 0.42888$$

since $|f'(x)|$ is maximum when $f''(x) = 0$ or, $x = 1/\sqrt{2}$. Similarly from Eq. (5.11)

$$|E_M| = \frac{1}{24}|f''(\eta)| \leq \frac{1}{24} \max_{0 \leq x \leq 1} 2|e^{-x^2}(2x^2 - 1)| = \frac{1}{12} \max\{1, e^{-1}\} = \frac{1}{12}$$

$$= 0.08333$$

where $|f''(x)|$ is maximum when $f'''(x) = 0$ i.e. $x = 0$ or $x = \sqrt{3/2} > 1$. So the maximum must occur at $x = 0$ or $x = 1$. Similarly, for the trapezoidal rule using Eq. (5.13)

$$E_T = \frac{1}{12}|f''(\eta)| \leq \frac{1}{6} \max\{1, e^{-1}\} = \frac{1}{6} = 0.16667$$

Finally, for the Simpson's rule in Eq. (5.15), $h = 1/2$ and

$$E_S = \frac{1}{90 \cdot 3}|f^{(iv)}(\eta)| \leq \frac{1}{2880} \max_{0 \leq x \leq 1} |e^{-x^2}(12 - 48x^2 + 16x^4)|$$

$$= \frac{1}{2880}\max\{12, 20e^{-1}\} = \frac{12}{2880} = 0.00417$$

In the above, $|f^{(iv)}(x)|$ is maximum when $f^{(v)}(x) = 0$, i.e. $x = 0$ or $4x^4 - 20x^2 + 15 = 0$. The admissible root of the second equation is $x^2 = 15/2 > 1$. Hence, the maximum is at either $x = 0$ or $x = 1$. $\qquad\qquad\square$

Example 2 (End-Corrected Trapezoidal Rule). Prove that

$$\int_a^b f(x)\,dx \approx \frac{h}{2}(y_0 + y_1) + \frac{h^2}{12}(y_0' - y_1'), \quad h = b - a$$

with error $\frac{h^5}{720}f^{(iv)}(\eta)$. Apply the rule to the problem of Example 1.

Solution. Let

$$\int_a^b f(x)\,dx \approx \frac{h}{2}(y_0 + y_1) + \frac{h^2}{12}(Ay_0' + By_1'), \quad h = b - a$$

We determine A and B so that the formula is exact for degree as high as possible. For this purpose, it is sufficient to test functions 1, x, x^2, etc. For $f(x) = 1$, the formula is obviously exact. If it is exact for $f(x) = x$, then

$$\frac{1}{2}(b^2 - a^2) = \frac{b - a}{2}(a + b) + A + B \Rightarrow B = -A$$

Similarly, if it is exact for $f(x) = x^2$, then

$$\frac{1}{3}(b^3 - a^3) = \frac{b - a}{2}(a^2 + b^2) + 2A(a - b) \Rightarrow A = \frac{1}{12}(b - a)^2$$

Hence the formula. It also integrates $f(x) = x^3$ exactly.

An alternative proof which also yields the expression for the error, is to consider the four-point Newton's divided difference formula for $x_0 = x_1 = a$ and $x_2 = x_3 = b$. The exact formula including the error term is

$$f(x) = f(a) + f[a, a](x - a) + f[a, a, b](x - a)^2 + f[a, a, b, b](x - a)^2(x - b) + R(x)$$

where

$$R(x) = \frac{f^{(iv)}(\xi)}{4!}(x-a)^2(x-b)^2$$

Integrating from a to b

$$\int_a^b f(x)\,dx \approx f(a)(b-a) + f[a,a]\frac{(b-a)^2}{2} + f[a,a,b]\frac{(b-a)^3}{3} + f[a,a,b,b]\Big[\frac{(b-a)^4}{4} - \frac{(b-a)^4}{3}\Big]$$

Here from Eq. (4.4) of Chap. 4, Sect. 4.1.2

$f[a,a] = f'(a)$
$f[a,a,b] = \{f[a,b] - f'(a)\}/(b-a), \quad f[a,b] = \{f(b) - f(a)\}/(b-a)$
$f[a,a,b,b] = \{f'(b) - 2f[a,b] + f'(a)\}/(b-a)^2$

Substituting in the integrated approximation and simplifying, we again obtain the required expression for the end-corrected trapezoidal rule. Moreover, the error of the approximation is

$$E = \int_a^b \frac{f^{(iv)}(\xi)}{24}(x-a)^2(x-b)^2\,dx = \frac{f^{(iv)}(\eta)}{24}\int_a^b (x-a)^2(x-b)^2\,dx$$

by application of generalised mean value theorem. Substituting $x = a + z$ and integrating, we obtain $E = \frac{h^5}{720}f^{(iv)}(\eta)$.

In particular, if $f(x) = e^{-x^2}$, with $a = 0,\ b = 1$ then $f'(x) = -2x\,e^{-x^2}$ and

$$\int_0^1 e^{-x^2}\,dx \approx \frac{1}{2}(e^0 + e^1) + \frac{1}{12}\cdot(-2)\cdot(0 - e^{-1}) = 0.74525 \qquad \Box$$

Exercises

1. Calculate the following integrals by the four basic rules, estimating the error in each case:

(i) $\displaystyle\int_0^1 \frac{dx}{1+x}$

(ii) $\displaystyle\int_0^1 \sin x^2\,dx$

[(i) $I_R = 1$, $I_M = 0.66667$, $I_T = 0.75$, $I_S = 0.69444$. Exact value $= 0.69315$.

$|E_R| \leq 0.5$, $|E_M| \leq 0.08333$, $|E_T| \leq 0.16667$, $|E_S| \leq 0.00833$.

(ii) $I_R = 0$, $I_M = 0.24740$, $I_T = 0.42074$, $I_S = 0.30518$,

$|E_R| \leq 1$, $|E_M| \leq 0.25$, $|E_T| \leq 0.5$, $|E_S| \leq 0.02639$].

2. Calculate $\int_0^1 f(x)\,dx$ for the test function

$$f(x) = x, \qquad 0 \leq x \leq 1$$
$$= 2 - x, \quad 1 \leq x \leq 2$$

by the trapezoidal and Simpson's rule. What is the exact value? Why cannot you estimate the errors in this case?

$[I_T = 0,\ I_S = 1.33333,$ Exact Value $= 1.$ The successive derivatives do not exist at $x = 1]$.

3. Apply Simpson's rule to estimate the value of the integrals

(i) $\displaystyle\int_0^1 \frac{x\,dx}{x^3 + 10}$ (ii) $\displaystyle\int_0^{1/2} \frac{\sin x}{x}\,dx$

$[(i)\ 0.04807,\ (ii)\ 0.47307]$.

4. Calculate the integral of Example 1 by the end-corrected trapezoidal rule. Estimate the error in the result.

$[I_{ECT} = 0.74525.\ |E_{ECT}| \leq 0.016666]$.

5. Prove the error formula Eq. (5.13) for the trapezoidal rule by the method employed in the case of Simpson's rule.

[Here $E(h) = \displaystyle\int_{x_0}^{x_0+h} f(x)\,dx - \frac{h}{2}[f(x_0) + f(x_0 + h)]$. Now show that $E''(h) = -\frac{h}{2} f''(x_0 + h)$ with $E(0) = E'(0) = 0$. Integrate twice using generalised mean value theorem, to obtain the form for $E(h)$].

6. End-Corected Simpson's Rule. Prove that

$$\int_a^b f(x)\,dx \approx \frac{h}{15}(7y_0 + 16y_1 + 7y_2) + \frac{h^2}{15}(y_0' - y_2'), \qquad h = \frac{b - a}{2}$$

[With respect to the midpoint x_1 assume

$$\int_{-h}^{h} f(x)\, dx = h[Af(-h) + Bf(0) + Af(h)] + C[f'(-h) - f'(h)]$$

For highest degree accuracy, determine A, B, C by applying the formula to $f(x) = 1$, x^2 and x^4. It can be proved that the error in the formula is $\frac{h^7}{4725} f^{(vi)}(\eta)$].

5.2.2 Newton–Cotes Formula

This formula is a generalisation of the formulas of the preceding subsection. Here, as before, the interval $[a, b]$ is divided in to n equal panels of length $h = (b - a)/n$ by the nodal points x_i ($i = 0, 1, \cdots, n$), with $x_0 = a$ and $x_n = b$. Since we want to derive a formula directly in terms of the ordinates $y_i (= f(x_i))$, we approximate $f(x)$ by the Lagrange interpolation polynomial $\phi(x)$ (Sect. 4.1.1, Chap. 4) on $[a, b]$, for the net of nodes . Setting $x = x_0 + ht$, $x_i = x_0 + ih$.

$$f(x) \approx \phi(x) = \sum_{i=0}^{n} \frac{(x - x_0)(x - x_1) \cdots (x - x_{i-1})(x - x_{i+1}) \cdots (x - x_n)}{(x_i - x_0)(x_i - x_1) \cdots (x_i - x_{i-1})(x_i - x_{i+1}) \cdots (x_i - x_n)} y_i$$

$$= \sum_{i=0}^{n} \frac{h^n t(t - 1) \cdots (t - i + 1)(t - i - 1) \cdots (t - n)}{h^n i(i - 1) \cdots 1 \cdot (-1) \cdots [-(n - 1)]} y_i$$

$$= \sum_{i=0}^{n} \frac{t(t - 1) \cdots (t - n)}{i!(-1)^{n-i}\, (n - i)!\, (t - i)} y_i$$

Hence

$$I_C(f) \approx \int_{x_0}^{x_0 + nh} \phi(x)\, dx = \sum_{i=0}^{n} \frac{(-1)^{n-i} y_i}{i!\, (n - i)!} \int_{0}^{n} \frac{t(t - 1) \cdots (t - n)}{t - i} \cdot \frac{(b - a)}{n}\, dt$$

$$=: (b - a) \sum_{i=0}^{n} H_i\, y_i \tag{5.16}$$

where the *Cotes' coefficients* H_i are given by

$$H_i = \frac{1}{n}\, \frac{(-1)^{n-i}}{i!\, (n - i)!} \int_{0}^{n} \frac{t(t - 1) \cdots (t - n)}{t - i}\, dt \tag{5.17}$$

Formulas (5.16) and (5.17) comprise Newton–Cotes' formula.

The Cotes' coefficients possess the following useful properties:

$1°$. *The sum of all the coefficients is equal to unity*:

$$\sum_{i=0}^{n} H_i = 1$$

Proof Let $f(x) = 1$, a polynomial of zero degree. Then $\phi(x) = 1$ by uniqueness of polynomials and formula (5.16) yields

$$\int_a^b 1 \cdot dx = (b - a) \sum_{i=0}^{n} H_i \cdot 1$$

Hence the property.   □

$2°$. *The symmetrical coefficients (the first and the nth, the second and the $(n - 1)$th, $\cdots$) are equal to one another*:

$$H_i = H_{n-i}$$

Proof Setting $t' = n - t$, we get from (5.17)

$$H_{n-i} = \frac{1}{n} \frac{(-1)^i}{(n-i)!\, i!} \int_0^n \frac{t(t-1)\cdots(t-n)}{t-n+i} dt$$

$$= \frac{1}{n} \frac{(-1)^i}{i!\,(n-i)!} \int_n^0 \frac{(n-t')(n-t'-1)\cdots(-t')}{-t'+1} (-dt')$$

$$= \frac{1}{n} \frac{(-1)^i}{i!\,(n-i)!} \int_0^n \frac{(-1)^n (t'-n)(t'-n+1)\cdots t'}{t'-i} dt' = H_i$$

The Cotes' coefficients H_i are compiled below for $n = 1 - 6$.

i	0	1	2	3	4	5	6
n							
1	1/2	1/2					
2	1/6	4/6	1/6				
3	1/8	3/8	3/8	1/8			
4	7/90	32/90	12/90	32/90	7/90		
5	19/288	75/288	50/288	50/288	75/288	19/288	
6	41/840	216/840	27/840	272/840	27/840	216/814	41/840

For $n = 7$ and $n \geq 10$, the coefficients are of *mixed sign*. For $n = 1$ and 2, the trapezoidal and the Simpson's rule are retrieved.

The analysis of the error in approximating $I_C(f)$ by the right-hand side of Eq. (5.16) is somewhat complicated. It was given by Steffensen in 1927, who proved that the error $E(x)$ is of the order of

$$O\left(h^{2[n/2]+3}\right)$$

where $[\cdot]$ denotes the integer part of the argument. From this, we see that quadrature rules for even n employing odd number of ordinates are more advantageous as to the degree of accuracy. This was the case with Simpson's rule.

Roger Cotes (1682–1716), English mathematician who edited the second edition of Newton's *Principia*. He made advances in the theory of logarithms, the integral calculus and numerical methods. He is also known for the study of spirals.

Johan Fredrik Steffensen (1873–1961), a Danish mathematician who made contributions in interpolation and concomitant errors.

5.2.3 Composite Rules

The expressions for the error of all the rules obtained in the previous two sections show that they are accurate only when h is 'small'. This is possible only when the length of the interval $b - a$ is 'small'. In the contrary case , which is more important in practice, one can divide the interval $[a, b]$ in to N panels of sufficiently small length, by the points

$$a = x_0 < x_1 < \cdots < x_N = b$$

and use a property of the definite integral to write

$$I(f) = \int_{x_0}^{x_N} f(x)\,dx = \int_{x_0}^{x_1} f(x)\,dx + \int_{x_1}^{x_2} f(x)\,dx + \cdots + \int_{x_{N-1}}^{x_N} f(x)\,dx$$

If $f(x)$ is now approximated by a certain interpolation polynomial $\phi(x)$, the above equation yields

$$I(f) \approx \sum_{i=1}^{N} \int_{x_{i-1}}^{x_i} \phi(x)\,dx \tag{5.18}$$

with an error in the calculated value equaling the difference of he two sides. We apply this technique to the basic rules of Sect. 5.2.1.

1°. Composite Rectangle Rule

Applying the rule (5.8) to Eq. (5.18) we obtain

$$I_{CR}(f) \approx \sum_{i=1}^{N} h\, y_{i-1} = h(y_0 + y_1 + \cdots + y_{N-1})$$

with an error

$$E(h) = \sum_{i=1}^{N} \left[\int_{x_{i-1}}^{x_i} f(x)\, dx - h\, y_{i-1} \right] = \frac{h^2}{2} \sum_{i=1}^{N} f'(\eta_i)$$

$$= \frac{(b-a)h}{2} \frac{1}{N} \sum_{i=1}^{N} f'(\eta_i) = \frac{(b-a)h}{2} f'(\eta)$$

by the weighted average lemma (see Sect. 5.1).

2°. Composite Midpoint Rule

In this case, as before

$$I_{CM}(f) = \sum_{i=1}^{N} h\, f\left(\frac{x_{i-1} + x_i}{2}\right) =: h\,(y_{1/2} + y_{3/2} + \cdots + y_{N-1/2})$$

with an error

$$E(h) = \sum_{i=1}^{N} \left[\int_{z_{i-1}}^{x_i} f(x)\, dx - h\, y_{i-1/2} \right] = \frac{h^3}{24} \sum_{i=1}^{N} f''(\eta_i) = \frac{(b-a)h^2}{24} f''(\eta)$$

where the weighted average lemma is applied.

3°. Composite Trapezoidal Rule

Using Eqs. (5.12), (5.18) yields

$$I_{CT}(f) \approx \sum_{i=1}^{N} \frac{h}{2} (y_{i-1} + y_i) = \frac{h}{2} \Big[(y_0 + y_1) + (y_1 + y_2) + \cdots + (y_{N-1} + y_N) \Big]$$

$$= h \left[\frac{y_0 + y_N}{2} + y_1 + \cdots + y_{N-1} \right] \qquad (5.19)$$

The error committed in this approximation, using Eq. (5.13) is

$$E(h) = \sum_{i=1}^{N} \left[\sum_{x_{i-1}}^{x_i} f(x)\,dx - \frac{h}{2} (y_{i-1} + y_i) \right] = -\frac{h^3}{12} \sum_{i=1}^{N} f''(\eta_i)$$

$$= -\frac{(b-a)h^2}{12} f''(\eta) \qquad (5.20)$$

using the lemma on weighted average.

4°. Composite Simpson's Rule

The basic Simpson's rule requires two equally spaced sum intervals each of length h. Hence we divide $[a, b]$ in ot an even number of $N = 2M$ panels. Following Eqs. (5.14) and (5.18), we get

$$I_{CS}(f) \approx \frac{h}{3}(y_0 + 4y_1 + y_2) + \frac{h}{3}(y_2 + 4y_3 + y_4) + \cdots + \frac{h}{3}(y_{2M-2} + 4y_{2M-1} + y_{2M})$$

$$= \frac{h}{3} \Big[y_0 + y_{2M} + 4 (y_1 + y_3 + \cdots + y_{2M-1}) + 2 (y_2 + y_4 + \cdots + y_{2M-2}) \Big]$$

$$(5.21)$$

The inherent error of this composite rule following Eq. (5.15) is

$$E(h) = -\frac{h^5}{90} \sum_{i=1}^{M} f^{(iv)}(\eta_i) = -\frac{(b-a)h^4}{180} f^{(iv)}(\eta) \qquad (5.22)$$

where $b - a = 2M\,h$ and the lemma on weighted average is used.

The composite trapezoidal and Simpson's rule are useful in practice and for that purpose we give below subroutines of the same names:

SUBROUTINE TRAPEZOIDAL(f,a,b,n,result)
! f=integrand. Must be declared external in the calling program.   (Input)
! a=lower limit of integration.   (Input)
! b=upper limit of integration.   (Input)

```
! n=number of panels of [a,b].     (Input)
! result=value of the integral.     (Output)
!*************************************
h=(b−a)/n
x=a; result=0.0
DO i=1,n
result=result+h/2.0*(f(x)+f(x+h))
x=x+h
END DO
RETURN
END SUBROUTINE TRAPEZOIDAL


SUBROUTINE SIMPSON(f,a,b,n,result)
! f=integrand. Must be declared external in the calling program.     (Input)
! a=lower limit of integration.     (Input)
! b=upper limit of integration.     (Input)
! n=number of panels of [a,b], must be an even number.     (Input)
! result=value of the integral.     (Output)
!*************************************
h=(b−a)/n
x=a; result=0.0
DO i=1,n,2
result=result+h/3.0*(f(x)+4*f(x+h)+f(x+2*h))
x=x+2*h
END DO
RETURN
END SUBROUTINE SIMPSON
```

The use of the subroutines is presented below.

Example 1. Calculate the value of the elliptic integral of the first kind

$$K(0.25) = \int_0^{\pi/2} \frac{dx}{\sqrt{1 - 0.25 \sin^2 x}}$$

by dividing the interval $[0, \pi/2]$ in to six equal parts and using the composite trapezoidal and the Simpson's rules.

Solution. We first tabulate the function $f(x) = 1/\sqrt{1 - 0.25 \sin^2 x}$ as

x	0	0.261799	0.523599	0.785398	1.047198	1.308997	1.570796
$f(x)$	1	1.008480	1.032796	1.069045	1.109400	1.142021	1.154701

By the composite trapezoidal rule ($h = 0.261799$)

$$K(0.5) \approx 0.261799 \times \left[\frac{1 + 1.154701}{2} + 1.008480 + 1.032796 + 1.06904 + 1.109400 \right.$$

$$\left. +1.142021 \right] = \underline{1.68575} \text{ (to five decimal places)}$$

Again, by the composite Simpson's rule

$$K(0.5) \approx \frac{0.261799}{3} \left[1 + 1.154701 + 4 \times (1.008480 + 1.069045 + 1.142021) \right.$$

$$\left. +2 \times (1.032796 + 1.109400) \right] = \underline{1.68575} \text{ (to five decimal places)}$$

Incidentally, the actual value agrees with the value.   □

Example 2. A rocket is launched from ground. Its acceleration is registered during the first 80 s and is given in the table below:

t	0	10	20	30	40	50	60	70	80
$a\ (m/s^2)$	30.00	31.63	33.44	35.47	37.75	40.33	43.29	46.69	50.67

Find the velocity and the height of the rocket at $t = 80$ s. Use Simpson's rule whenever you can otherwise use trapezoidal rule.

Solution. Here, the integrand is given as a table of values. If v is the velocity at time t, $a = dv/dt$. Hence by Simpson's rule

$$v_{80} = \int_0^{80} a\, dt = \frac{1}{3} [30.00 + 50.67 + 4 \times (31.63 + 35.47 + 40.33 + 46.69)$$

$$+ 2 \times (33..44 + 37.75 + 43.29)] = \underline{3420.36} \text{ m/s}$$

Again if h is the height at time t, $dh/dt = v$. Hence

$$h_{80} = \int_0^{80} v\, dt$$

To evaluate the integral we need a table of values of v. For this purpose we have

$$v_{10} = \frac{10}{2} \times (30.00 + 50.67) = 308.15$$

$$v_{20} = \frac{10}{3} \times (30.00 + 4 \times 31.63 + 33.44) = 633.20$$

$$v_{30} = 633.20 + \frac{10}{2} \times (33.44 + +35.47) = 977.75$$

$$v_{40} = \frac{10}{3} \times [30.00 + 37.75 + 4 \times (31.63 + 35.47) + 2 \times 33.44] = 1343.43$$

$$v_{50} = 1343.43 + \frac{10}{2} \times (37.75 + 40.33) = 1733.83$$

$$v_{60} = 1343.43 + \frac{10}{3} \times (37.75 + 4 \times 40.33 + 43.29) = 2151.30$$

$$v_{70} = 2151.30 + \frac{10}{2} \times (43.29 + 46.69) = 2601.20$$

$$v_{80} = 3420.36$$

Thus, $h_{80} = \dfrac{10}{3} [0 + 3420.36 + 4 \times (308.15 + 977.75 + 1733.83 + 2601.20)$

$$+ \ 2 \times (633.20 + 1343.43 + 2151.30)] \ = \underline{113.87} \ \text{km}$$

□

Example 3. Compute $\int_{0}^{1} e^{-x^2} dx$ by subroutines TRAPEZOIDAL and SIMPSON correct to five decimal places.

Solution. We write the main program as

EXTERNAL f
a =0.; b=1.
READ*, n
CALL TRAPEZOIDAL(f,a,b,n,result)
PRINT*, 'n=', n, ' result=', result
END

and append the subroutine TRAPEZOIDAL. The output for various values of n are obtained as

$n = 10, \quad \text{result} = 0.74621$

$n = 50, \quad \text{result} = 0.74680$

$n = 100, \text{result} = 0.74682$

$n = 120, \text{result} = 0.74682$

$n = 130, \text{result} = 0.74682$

Since stability is reached it is not expedient to further increase n, as roundoff errors will creep in.

In a similar manner calling subroutine SIMPSON, the following results are obtained:

$n = 6, \quad \text{result} = 0.74683$

$n = 10, \text{result} = 0.74683$

$n = 12, \text{result} = 0.74682$

$n = 14, \text{result} = 0.74682$

$n = 16, \text{result} = 0.74682$

Reaching stability we conclude that the result is 0.74682. Evidently subroutine SIMP-SON works much faster.

Exercises

1. Compute the value of π from the formulae

$(i) \ \dfrac{\pi}{4} = \displaystyle\int_0^1 \dfrac{dx}{1+x^2}$ $\qquad (ii) \ \dfrac{\pi}{2\sqrt{2}} = \displaystyle\int_0^1 \dfrac{1+x^2}{1+x^4}\,dx$

using 10 panel trapezoidal rule.

$[(i) \ \pi \approx 3.13993, \ (ii) \ \pi \approx 3.13923]$.

2. The prime number theorem state that the number of primes in the interval $a < x < b$ is approximately $\int_a^b dx/\ln x$. Use this for $a = 10$, $b = 100$ and compare with the exact value. Noting that the integrand varies slowly use trapezoidal rule with only 8 panels.

[24. Exact value = 26].

3. Calculate the Incomplete Gamma Function

$$\int_0^1 e^{-x} \sqrt{x}\, dx$$

to two decimal places by Simpson's rule with four panels.

[1.25]

4. The arc length L of an ellipse of semi-axes a and b is given by the formula $L = 4a\, E(m)$ where $m = 1 - b^2/a^2$ and

$$E(m) = \int_0^{\pi/2} \sqrt{1 - m\, \sin^2 \theta}\, d\theta$$

is the elliptic integral of the second kind. Calculate L for $a = 2$, $b = 1$ by six panel Simpson's rule.

[9.68862].

5. Calculate the value of the Fresnel integral

$$C(x) = \int_0^x \cos\left(\frac{\pi}{2} t^2\right) dt$$

for $x = 1$ by 8 panel Simpson's rule.

[0.77933].

6. Revisit Exercises 1 and 3 to compute the quantities correctly to five decimal places by using subroutine TRAPEZOIDAL, stating the number of panels used.

[(1) 3.14159 ((i) n =200 (ii) n =190), (3) 1.25559 (n = 300)].

7. Revisit Exercises 4 and 5 to compute the quantities to five decimal places by using subroutine SIMPSON, stating the number of panels used.

[(4) 9.68845 (n = 10) (5) 0.77989 (n = 20)].

8. Compute the following integrals by subroutine SIMPSON, correct to five decimals:

$$(i) \int_0^1 e^{x\sqrt{x}}dx \qquad (ii) \int_0^1 \sqrt{\frac{1-e^{-x}}{x}}\,dx \qquad (iii) \int_0^{\pi/2} \frac{\cos x}{1+x}\,dx$$

$$(iv) \int_0^2 \frac{\sin x}{x}\,dx \qquad (v) \int_0^1 \frac{dx}{\sqrt{e^x+x+1}}$$

[(i) 1.56240, (ii) 0.89057, (iii) 1.60541, (iv) 0.89057, (v) 0.57011].

9. A reservoir discharging through sluices at a depth h below the water surface has a surface area A for various values of h as given below:

h (ft)	10	1	12	13	14
A (sq. ft.)	950	1070	1200	1350	1530

If t denotes time in minutes, the rate of fall of the surface is given by $dh/dt = -48\sqrt{h}/A$. Estimate the time taken for the water level to fall from 14 to 10 ft. above the sluices.

[Required time $= \dfrac{1}{48}\displaystyle\int_{10}^{14} \frac{A}{\sqrt{h}}\,dh$; now tabulate $A/\sqrt{h}$. 29.10 min].

10. The velocity v of a particle at a distance s from a point on its path is given in the following data:

s (m)	0	2.5	5.0	7.5	10.0	12.5	15.0	17.5	20.0
v (m/s)	16	19	21	22	20	17	13	11	9

Estimate the time taken to traverse the distance of 20 m.

[Require time $= \displaystyle\int_0^{20} \frac{ds}{v} = 1.26\,\text{s}$].

11. The temperature θ at radius r in a finned radiator tube is given in the following table:

r (cm)	0.50	0.75	1.0	1.25	1.50	1.75	2.00
θ ($^\circ C$)	100	77.3	63.8	55.9	51.7	48.5	45.6

Calculate (i) average temperature $= \dfrac{1}{1.5} \displaystyle\int_{0.50}^{2.00} \theta\, dr$ and (ii) the heat flux $Q = -2\pi r k t \dfrac{d\theta}{dt}$ at inner radius $r = 0.50$ cm, given conductivity $k = 0.26$ c.g.s. units and thickness of a fin $t = 0.36$ cm.

[(i) $61.3^\circ C$ (ii) At $r = 0.5$, $d\theta/dr = -115.6$, $Q = 300.6$].

12. Prove the n-panel end-corrected trapezoidal and Simpson's rules:

(i) $\displaystyle\int_a^b f(x)\, dx = h\left(\frac{y_0 + y_n}{2} + y_1 + \cdots + y_{n-1}\right) + \frac{h^2}{12}(y_0' - y_n') + O(h^4)$

(ii) $\displaystyle\int_a^b f(x)\, dx = \frac{h}{15}(7y_0 + 16y_1 + 14y_2 + 16y_3 + \cdots + 7y_{2n}) + \frac{h^2}{15}(y_0' - y_{2n}') + O(h^6)$

5.2.4 *Gaussian Quadrature Formula*

For developing this formula, we first consider a function $y = f(t)$ specified on the standard interval $[-1, 1]$. The general case of an integral with respect to $x \in [a, b]$ can be reduced to this standard interval by the linear transformation

$$x = \frac{b+a}{2} + \frac{b-a}{2}t \tag{5.23}$$

If $t_1, t_2, \cdots, t_n$ are points in $[-1, 1]$, a quadrature formula is of the form

$$I_1(f) := \int_{-1}^{1} f(t)\, dt \approx \sum_{i=1}^{n} A_i\, f(t_i) \tag{5.24}$$

where $A_1, A_2, \cdots, A_n$ are certain coefficients called **weights** of the formula. Gauss posed the problem of determining the abscissas $t_1, t_2, \cdots, t_n$ and the coefficients $A_1, A_2, \cdots, A_n$ so that the quadrature formula (5.24) is exact for all polynomials $f(t)$ of degree N, *as high as possible*. Since we have at our disposal $2n$ constants t_i and A_i ($i = 1, 2, \cdots, n$) and a polynomial of degree $2n - 1$ is determined by $2n$ coefficients, this highest possible degree N in the general case should equal $2n - 1$.

To ensure exact validity of (5.24) for a polynomial of degree $2n - 1$, it is necessary and sufficient that it be valid for

$$f(t) = 1, t, t^2, \cdots, t^{2n-1} = t^k \quad (k = 0, 1, \cdots, 2n - 1) \qquad (5.25)$$

Indeed, let

$$\int_{-1}^{1} t^k dt = \sum_{i=1}^{n} A_i t_i^k, \quad (k = 0, 1, \cdots, 2n - 1)$$

Then if

$$f(t) = \sum_{k=0}^{2n-1} C_k t^k$$

we obtain

$$\int_{-1}^{1} f(t) \, dt = \sum_{k=0}^{2n-1} C_k \int_{-1}^{1} t^k dt = \sum_{k=0}^{2n-1} C_k \sum_{i=1}^{n} A_i t_i^k$$

$$= \sum_{i=1}^{n} A_i \sum_{k=0}^{2n-1} C_k t_i^k = \sum_{i=1}^{n} A_i f(t_i)$$

This proves the assertion.

The exact quadrature of t^k, $(k = 0, 1, 2, \cdots, 2n - 1)$ in (5.25) yields

$$\sum_{i=1}^{n} A_i t_i^k = \int_{-1}^{1} t^k dt = \frac{1 - (-1)^{k+1}}{k + 1} = \begin{cases} \frac{2}{k+1} & \text{for } k \text{ even} \\ 0 & \text{for } k \text{ odd} \end{cases}$$

We thus conclude that it is sufficient to determine t_i and A_i from the system of $2n$ equations

$$\sum_{i=1}^{n} A_i = 2$$

$$\sum_{i=1}^{n} A_i t_i = 0$$

$$\cdots\cdots\cdots\cdots\cdots \qquad (5.26)$$

$$\sum_{i=1}^{n} A_i t_i^{2n-2} = \frac{2}{2n - 1}$$

$$\sum_{i=1}^{n} A_i t_i^{2n-1} = 0$$

The system (5.26) is nonlinear and its solution in ordinary manner involves mathematical difficulties. However, the following device may be employed.

Instead of considering the power functions (5.25), consider the polynomial

$$f(t) = t^k P_n(t), \quad (k = 0, 1, \cdots, n - 1) \tag{5.27}$$

where $P_n(t)$ is the well-known **Legendre polynomial** of degree n. It satisfies the Rodrigue's formula

$$P_n(t) = \frac{1}{n! \, 2^n} \frac{d^n}{dt^n} [(t^2 - 1)^n]$$

and the **orthogonality property**

$$\int_{-1}^{1} P_m(t) P_n(t) \, dt = 0, \quad \text{if } m \neq n$$

(see Chap. 8, Sect. 8.4). The degree of $f(t)$ do not exceed $2n - 1$ and exact validity of (5.24) yields

$$\int_{-1}^{1} t^k P_n(t) \, dt = \sum_{i=1}^{n} A_i t_i^k P_n(t_i), \quad (k = 0, 1, \cdots, n - 1)$$

Now t^k can be expressed as a series of Legendre polynomials of various degrees, not exceeding k. The left-hand side of the above equation vanishes since by virtue of the orthogonality property,

$$\int_{-1}^{1} t^k P_n(t) \, dt = 0 \quad \text{for } k < n$$

Hence we obtain

$$\sum_{i=1}^{n} A_i t_i^k P_n(t_i) = 0, \quad (k = 0, 1, \cdots, n - 1) \tag{5.28}$$

Equation (5.28) are definitely satisfied for any value of A_i, if

$$P_n(t_i) = 0, \quad (i = 1, 2, \cdots, n) \tag{5.29}$$

Thus, to achieve the maximum accuracy of the quadrature formula (5.24), it is sufficient to take t_i as the zeros of the Legendre polynomial of degree n. As is known (see Chap. 8, Sect. 8.4, Theorem 9, *Remark*), these zeros are real and distinct and lie in the interval $(-1, 1)$. Knowing the values of t_i, the coefficients A_i $(i = 1, 2, \cdots, n)$ can be found from the *linear system* of the first n equations of (5.26). The determinant of this subsystem is the Vandermonde determinant equal to

$$\prod_{i>j} (t_i - t_j) \neq 0$$

since t_i are distinct. Hence A_i are determined uniquely.

Formula (5.24), where the t_i are zeros of the Legendre polynomial $P_n(t)$, Eq. (5.29), and the A_i $(i = 1, 2, \cdots, n)$ determined from n equations of the system (5.26), are called **Gaussian Quadrature Formula**.

Example 1. Derive the Gaussian quadrature formula for the case of three ordinates.

Solution. Here $n = 3$. The Legendre polynomial of degree 3 is

$$P_3(t) = \frac{1}{48} \frac{d^3}{dt^3}[(t^2 - 1)^3] = \frac{1}{2}(5t^3 - 3t)$$

whose zeros are

$$t_1 = -\sqrt{\frac{3}{5}} = -0.7745967, \quad t_2 = 0, \quad t_3 = \sqrt{\frac{3}{5}} = 0.7745967$$

To determine the coefficients A_1, A_2, A_3, we have from Eq. (5.26)

$$A_1 + A_2 + A_3 = 2$$
$$-\sqrt{\tfrac{3}{5}}A_1 + \sqrt{\tfrac{3}{5}}A_3 = 0$$
$$\tfrac{3}{5}A_1 + \tfrac{3}{5}A_3 = \tfrac{2}{3}$$

The solution of these equations is $A_1 = A_3 = 5/9$, $A_2 = 8/9$. Therefore

$$\int_{-1}^{1} f(t)\, dt \approx \frac{1}{9}\left[5 f\left(-\sqrt{\frac{3}{5}}\right) + 8 f(0) + 5 f\left(\sqrt{\frac{3}{5}}\right)\right] \qquad \square$$

The abscissas t_i and the weights A_i $(i = 1, 2, \cdots, n)$, calculated in this manner for $n = 1, 2, \cdots, 6$ are tabulated next page. For comprehensive tables, see the book by Stroud and Secrest listed in the bibliography at the end of this book.

To determine an expression for the error, let $f(x)$ be approximated by the Lagrange interpolation function $\phi(x)$ of degree $2n - 1$, that passes through the $2n$ nodes $t_1, t_2, \cdots, t_{2n}$ where $t_1, t_2, \cdots, t_n$ is the node set of the Gaussian formula (5.24). Thus $\phi(t_i) = f(t_i)$, $i = 1, 2, \cdots, 2n$. Taking in to account the error we have the exact relation

$$I_G^1(f) := \int_{-1}^{1} f(t)\, dt = \int_{-1}^{1} \phi(t)\, dt + \frac{1}{(2n+1)!} \int_{-1}^{1} f^{(2n)}(\xi)\,(t - t_1)\cdots(t - t_{2n})\, dt$$

Table of Gaussian Elements

n	i	t_i	A_i
1	1	0	2
2	1, 2	∓ 0.5773503	1
3	1, 3	∓ 0.7745967	0.5555556
	2	0	0.8888889
4	1, 4	∓ 0.8611363	0.3478548
	2, 3	∓ 0.3399810	0.6521452
5	1, 5	∓ 0.9061799	0.2369269
	2, 4	∓ 0.5384693	0.4786287
	3	0	0.5688889
6	1, 6	∓ 0.9324695	0.1713245
	2, 5	∓ 0.6612094	0.3607616
	3, 4	∓ 0.2386192	0.4679139

Since $\phi(t)$ is of degree $2n - 1$, the Gaussian formula (5.24) holds exactly and

$$\int_{-1}^{1} \phi(t)\, dt = \sum_{i=1}^{n} A_i\, \phi(t_i) = \sum_{i=1}^{n} A_i\, f(t_i)$$

Substituting this value in the preceding equation,

$$I_G^1(f) = \sum_{i=1}^{n} A_i\, f(t_i) + E_G^1$$

where E_G^1 is the error in Gaussian quadrature. Now letting $t_{n+1}, \cdots, t_{2n}$ tend to $t_1, \cdots, t_n$, respectively, the expression for E_G^1 becomes

$$E_G^1 = \frac{1}{(2n)!} \int_{-1}^{1} f^{(2n)}(\xi)[(t - t_1)(t - t_2) \cdots (t - t_n)]^2\, dt$$

$$= \frac{f^{(2n)}(\eta)}{(2n)!} \int_{-1}^{1} [(t - t_1)(t - t_2) \cdots (t - t_n)]^2\, dt$$

by using the generalised mean value theorem. Now, t_i $(i = 1, 2, \cdots, n)$ are the zeros of $P_n(t)$; hence $P_n(t) = K\, (t - t_1) \cdots (t - t_n)$, where K is a constant. Using Rodrigue's formula

$$K = \text{coefficient of } t^n \text{ in } \frac{1}{2^n n!} \frac{d^n}{dt^n}(t^2 - 1)^n$$

$$= \text{coefficient of } t^n \text{ in } \frac{1}{2^n n!} \frac{d^n}{dt^n}(t^{2n}) = \frac{1}{2^n n!} \frac{(2n)!}{n!} = \frac{(2n)!}{2^n (n!)^2}$$

Thus, $\quad E_G^1 = \dfrac{f^{(2n)}(\eta)}{(2n)!} \dfrac{1}{K^2} \displaystyle\int_{-1}^{1} [P_n(t)]^2 \, dt$

$$= \frac{f^{(2n)}(\eta)}{(2n)!} \cdot \frac{2^{2n}(n!)^4}{[(2n)!]^2} \cdot \frac{2}{2n+1} = \frac{2^{2n+1}(n!)^4}{[(2n)!]^3(2n+1)} \, f^{(2n)}(\eta) \qquad (5.30)$$

where the result

$$\int_{-1}^{1} [P_n(t)]^2 \, dt = \frac{2}{2n+1}$$

noted in Chap. 8, Sect. 8.4, is used. Equation (5.30) is the sought expression for the error.

Let us now return to the question of calculating

$$I_G(f) := \int_a^b f(x) \, dx \qquad (5.31)$$

Making the change of variable (5.23), we obtain by Eq. (5.24)

$$I_G(f) = \frac{b-a}{2} \int_{-1}^{1} f\left(\frac{b+a}{2} + \frac{b-a}{2} t\right) dt \approx \frac{b-a}{2} \sum_{i=1}^{n} A_i \, f(x_i); \qquad (5.32)$$

$$x_i = \frac{b+a}{2} + \frac{b-a}{2} t_i$$

Equation (5.32) is the required Gaussian formula. The error for this rule is

$$E_G = \frac{(b-a)^{2n+1}(n!)^4}{[(2n)!]^3(2n+1)} \, f^{(2n)}(\eta) \qquad (5.33)$$

Remark The Gaussian quadrature theory has also been treated in a general manner by integrating Lagarnge interpolation formula. The general approach has led to formulas that are particularly useful in computing *improper integrals* over an interval $[a, b]$. The technique uses the *orthogonal polynomial* over the given interval. Let

$$I(f) := \int_a^b f(x) \, w(x) \, dx \qquad (5.34)$$

where $w(x)$ is a certain *weight function* ≥ 0. If $f(x)$ is approximated by its Lagrange interpolation $\phi(x)$ over the points $a < x_1 < x_2 < \cdots < x_n < b$, that is

$$f(x) \approx \phi(x) = \sum_{i=1}^{n} l_i(x) f(x_i)$$

with $\qquad l_i(x) = \dfrac{\omega_n(x)}{(x - x_i)\,\omega'(x_i)}, \quad \omega_n(x) = (x - x_1)\cdots(x - x_n)$

then

$$I(f) \approx \sum_{i=1}^{n} A_i\, f(x_i) \tag{5.35}$$

where

$$A_i = \int_a^b l_i(x)\, w(x)\, dx = \frac{1}{\omega'(x_i)} \int_a^b \frac{\omega_n(x)\, w(x)}{(x - x_i)}\, dx \tag{5.36}$$

Selecting $\omega(x)$ as the *orthogonal polynomial* over $[a, b]$ with *weight function* $w(x)$, we obtain the corresponding Gauss quadrature formula (5.35) for the integral (5.34). However, evaluation of A_i from (5.36) requires advanced techniques (see the text of Stroud and Secrest). If $a = -1$, $b = 1$, $w(x) = 1$ and $\omega_n(x) = P_n(x)$ (Legendre polynomial) we obtain the formula (5.24).

If $a = -1$, $b = 1$, $w(x) = (1 - x^2)^{-1/2}$ and $\omega_n(x) = T_n(x)$ where $T_n(x)$ is the *Chebyshev polynomial* (Chap. 4, Sect. 4.3), then

$$x_i = \cos\left[\frac{(2i - 1)\pi}{2n}\right], \quad (i = 1, 2, \cdots, n)$$

and it can be proved that A_i are all equal to π/n (see Exercise 2, below for the case $n = 3$). Hence, we have the **Gauss–Chebyshev formula**

$$\int_{-1}^{1} f(x) \frac{dx}{\sqrt{1 - x^2}} \approx \frac{\pi}{n} [f(x_1) + f(x_2) + \cdots + f(x_n)] \tag{5.37}$$

If $a = 0$, $b = \infty$, $w(x) = e^{-x}$ and $\omega_n(x) = L_n(x)$ where $L_n(x)$ is the **Laguerre polynomial** (see Chap. 8, Sect. 8.4, Exercise 2), then

$$\int_0^{\infty} f(x)\, e^{-x}\, dx$$

can be successfully integrated by the **Gauss–Laguerre** formula. Similar formula exists for $a = -\infty$, $b = \infty$, $w(x) = e^{-x^2}$ and $\omega_n(x) = H_n(x)$, where $H_n(x)$ is the **Hermite polynomial** (see Chap. 8, Sect. 8.4, Exercise 3), to evaluate

$$\int_{-\infty}^{\infty} f(x)\, e^{-x^2}\, dx$$

The points x_i and the coefficients A_i, for the two cases are given in Stroud and Secrest (see Bibligraphy). This authoritative text also gives the error in such quadrature formulas.

Example 1. Find a quadrature formula

$$\int_0^1 \frac{f(x)\, dx}{\sqrt{x(1-x)}} \approx A_1\, f(0) + A_2\, f\left(\frac{1}{2}\right) + A_3\, f(1)$$

which is exact for all polynomials of degree ≤ 2. Then use the formula to compute

$$\int_0^1 \frac{dx}{\sqrt{x - x^3}}.$$

Solution. Since the formula is to be exact for all polynomials of degree ≤ 2, we obtain

$$\text{for } f(x) = 1, \quad I_1 = \int_0^1 \frac{dx}{\sqrt{x(1-x)}} = A_1 + A_2 + A_3$$

$$\text{for } f(x) = x, \quad I_2 = \int_0^1 \frac{x\, dx}{\sqrt{x(1-x)}} = \frac{1}{2} A_2 + A_3$$

$$\text{for } f(x) = x^2, \quad I_3 = \int_0^1 \frac{x^2 dx}{\sqrt{x(1-x)}} = \frac{1}{4} A_2 + A_3$$

$$\text{Now,} \quad I_1 = \int_0^1 \frac{dx}{\sqrt{x(1-x)}} = 2\int_0^1 \frac{dx}{\sqrt{1 - (2x-1)^2}} = \pi$$

$$I_2 = \int_0^1 \frac{x\, dx}{\sqrt{x(1-x)}} = 2\int_0^1 \frac{x\, dx}{\sqrt{1 - (2x-1)^2}} = \frac{\pi}{2}$$

$$I_3 = \int_0^1 \frac{x^2 dx}{\sqrt{x(1-x)}} = 2\int_0^1 \frac{x^2 dx}{\sqrt{1 - (2x-1)^2}} = \frac{3\pi}{8}$$

Hence we have the equations

$$A_1 + A_2 + A_3 = \pi, \quad \frac{1}{2} A_2 + A_3 = \frac{\pi}{2}, \quad \frac{1}{4} A_2 + A_3 = \frac{3\pi}{8}$$

whose solution is $A_1 = A_3 = \pi/4$, $A_2 = \pi/2$. Hence

$$\int_0^1 \frac{f(x)\,dx}{\sqrt{x(1-x)}} \approx \frac{\pi}{4}\left[f(0) + 2f\left(\frac{1}{2}\right) + f(1)\right]$$

For the second part

$$I = \int_0^1 \frac{dx}{\sqrt{x-x^3}} = \int_0^1 \frac{1}{\sqrt{1+x}}\frac{dx}{\sqrt{x(1-x)}} \approx \frac{\pi}{4}\left[1 + 2\sqrt{\frac{2}{3}} + \frac{1}{\sqrt{3}}\right] = 2.62331 \qquad \Box$$

Exercises

1. Construct a rule of the form

$$\int_{-1}^1 f(x)\,dx \approx A_1 f\left(-\frac{1}{2}\right) + A_2 f(0) + A_3 f\left(\frac{1}{2}\right)$$

which is exact for all polynomials of degree ≤ 2.

$[A_1 = A_3 = \frac{4}{3},\ A_2 = -\frac{2}{3}]$.

2. Prove the Gauss–Chebyshev formula for three points

$$\int_{-1}^1 f(x)\frac{dx}{\sqrt{1-x^2}} = \frac{\pi}{3}\left[f\left(\cos\frac{5\pi}{6}\right) + f\left(\cos\frac{3\pi}{6}\right) + f\left(\cos\frac{\pi}{6}\right)\right]$$

that is exact for all polynomials of degree ≤ 2.

3. Prove that Gaussian quadrature Eq. (5.35) is exact for polynomial of degree atmost $2n - 1$.

$[f(x) - \phi(x)$ has zeros at $x_1, x_2, \cdots, x_n$. If $f(x)$ is a polynomial of degree atmost $2n - 1$, $f(x) - \phi(x) = \omega_n(x) \cdot p_{n-1}(x)$ where $p_{n-1}(x)$ is a polynomial of degree atmost $n - 1$. Hence

$$\int_a^b f(x)w(x)\,dx - \int_a^b \phi(x)w(x)\,dx = \int_a^b p_{n-1}(x)\,\omega_n(x)\,dx = 0$$

if $\omega_n(x)$ is an orthogonal polynomial].

4. Prove that in Eq. (5.36) $A_i = \int_a^b [l_i(x)]^2 w(x)\,dx > 0$.

[Let $f(x) = [l_j(x)]^2$ (polynomial of degree $2n - 2$) in Exercise 2 above. Then since $l_j(x_i) = \delta_{ji}$, $\int_a^b [l_j(x)]^2 w(x)\,dx = \sum_{i=1}^n A_i [l_j(x_i)]^2 = A_j$].

5. Compute the value of the integral

$$\int_2^3 \frac{\cos 2x}{1 + \sin x}\,dx$$

by three point Gauss quadrature formula of Example 1.

[Set $x = (t + 5)/2$, then $I = \frac{1}{2}\int_{-1}^1 \frac{\cos(t + 5)}{1 + \sin((t + 5)/2)}\,dt \approx 0.20271$].

6. Integrate by four-point Gauss quadrature formula: $\int_1^2 \frac{dx}{1 + x^3}$.

[Set $x = (t + 3)/2$, then $I = \frac{1}{2}\int_{-1}^1 \frac{dt}{1 + [(t + 3)/2]^3} \approx 0.25435$].

7. Integrate by four-point Gauss–Chebyshev quadrature formula:

$$(i)\ \int_0^1 \cos 2x \frac{dx}{\sqrt{1 - x^2}} \qquad (ii)\ \int_0^1 \frac{\ln(x + 1)}{\sqrt{x(1 - x)}}\,dx$$

[(i) Set $I = \frac{1}{2}\int_{-1}^1 \cos 2x \frac{dx}{\sqrt{1 - x^2}}$. $x_i = \cos\left[\frac{(2i - 1)\pi}{8}\right]$, $i = 1, 2, 3, 4$, $A_i = \frac{\pi}{4}$. Hence $I \approx 0.35162$.

(ii) Set $x = (t + 1)/2$, then $I = \int_{-1}^1 \ln[(t + 3)/2] \frac{dt}{\sqrt{1 - t^2}}$. hence $I \approx 1.18266$].

8. (Lobatto's Rule). In the Gaussian quadrature for $\int_a^b f(x)\,dx$, if the points ± 1 are kept as fixed abscissas, then results the formula

$$\int_{-1}^1 f(x)\,dx \approx A_1 f(-1) + A_2 f(x_2) + \cdots + A_{n-1} f(x_{n-1}) + A_n f(1)$$

The rule is known after Lobatto. Show that for the case $n = 3$ that is exact for all polynomials of degree ≤ 3, the formula leads to Simpson's rule.

5.2.5 Romberg Integration

The method of Richardson extrapolation applied to the composite trapezoidal rule leads to a useful method known after W. Romberg.

Werner Romberg (1909–2003) is a German mathematician who gave this method in the year 1955.

The N-panel composite trapezoidal rule is of the form

$$I(f) = \int_a^b f(x)\,dx = T_h(f) + E(h) \tag{5.38}$$

where

$$T_h(f) = \frac{h}{2}[f(a) + f(b)] + h\,(y_1 + y_2 + \cdots + y_{N-1}) \tag{5.39}$$

and $E(h) = -[(b-a)h^2/12]f''(\eta)$. In these expressions $h = (b-a)/N$. The error $E(h)$ is certainly of the type $C_1 h^2 + o(h^2)$, but more than that, the Euler–Maclaurin sum formula (see Sect. 5.3 to follow) shows that it can be expressed as a power series in h^2, viz. $E(h) = C_1 h^2 + C_2 h^4 + \cdots + C_k h^{2k} + O(h^{2k+2})$, provided that $f \in C^{(2k+2)}[a, b]$. This means that Eq. (5.38), is actually of the type (suppressing the f in $T_h(f)$)

$$I(f) = T_h + C_1 h^2 + O(h^4)$$

For applying Richardson extrapolation to this equation, we consider $2N$ panels of $[a, b]$, each of length $h/2 = (b-a)/(2N)$. Then

$$I(f) = T_{h/2} + C_1\left(\frac{h}{2}\right)^2 + O(h^4)$$

Eliminating C_1 from the above two equations (see in this connection Eq. (5.4)), we obtain

$$I(f) = T_h^1 + O(h^4)$$

where

$$T_h^1 = T_h + \frac{T_{h/2} - T_h}{1 - 1/4} = T_{h/2} + \frac{1}{4-1}(T_{h/2} - T_h) \tag{5.40}$$

In the computation of $T_{h/2}$, it is advantageous to choose *even* N. For,

$$T_{h/2} = \frac{h}{4}[f(a) + f(b)] + \frac{h}{2}(y_1' + y_1 + y_2' + y_2 + \cdots + y_{N-1} + y_N')$$

$$= \frac{1}{2}T_h + \frac{h}{2} \qquad\qquad\qquad\qquad\qquad\qquad (5.41)$$

where $y_1', y_2', \cdots, y_N'$ are midpoints of the N panels $(x_0, x_1), (x_1, x_2), \cdots,$ (x_{N-1}, x_N) respectively, i.e. $y_i' = f\left(a + \dfrac{kh}{2}\right)$, $k = 1, 3, \cdots, 2N - 1$.

Continuing the extrapolation process, C_2 can similarly be eliminated to yield

$$I(f) = T_h^2 + O(h^6)$$

where

$$T_h^2 = T_{h/2}^1 + \frac{T_{h/2}^1 - T_h^1}{1 - 1/16} = T_{h/2}^1 + \frac{1}{4^2 - 1}(T_{h/2}^1 - T_h^1) \qquad (5.42)$$

Note that in the computation of $T_{h/2}^1$, the value of $T_h/4$ is required in virtue of Eq. (5.40). Similarly in general, if C_i is eliminated, then

$$I(f) = T_h^i + O\left(h^{2i+2}\right)$$

where T_h^i is given recursively by

$$T_h^i = T_{h/2}^{i-1} + \frac{1}{4^i - 1}\left(T_{h/2}^{i-1} - T_h^{i-1}\right) \qquad (5.43)$$

The computed estimates T_h^i of $I(f)$ can be conveniently arranged in tabular form like

i	$j = 1$	1	2
1	$T_h = T_{11}$		
2	$T_{h/2} = T_{21}$	$T_h^1 = T_{22}$	
3	$T_{h/4} = T_{31}$	$T_{h/2}^1 = T_{32}$	$T_h^2 = T_{33}$
.	$\cdots$	$\cdots$	$\cdots$

The acceptance criterion of an estimate T_{jj} can be based on stability and the following criteria. If T_{22} is assumed accurate enough, then

$$\frac{T_{h/2} - T_h}{3/4} \approx C h^2 = 4 C \left(\frac{h}{2}\right)^2 \approx 4 \frac{T_{h/4} - T_{h/2}}{3/4}$$

This means that the *difference ratios* must satisfy

$$\frac{T_{h/2} - T_h}{T_{h/4} - T_h} = \frac{T_{21} - T_{11}}{T_{31} - T_{21}} \approx 4$$

Similarly replacing h by $h/2$, $h/4$, $\cdots$, the other difference ratios of this column, $(T_{31} - T_{21})/(T_{41} - T_{31})$, $(T_{41} - T_{31})/(T_{51} - T_{41})$, $\cdots$ must approximate 4, in case T_{33} or T_{44}, $\cdots$ is accurate. The difference ratios of the second column, by similar arguments, is easily seen to approximate $4^2 = 16$, that of the third column to approximate $4^3 = 64$ and so on. Thus the diagonal element T_{jj} for which the ratios in the jth row are nearly 4, 16, 64,$\cdots$ respectively, is accepted as the estimate of $I(f)$.

There is, however, a computational difficulty in the above criterion, and the test may fail on occasions. As i and j increase, the differences in the ratios diminish and roundoff errors become prominent. A difference may also become zero, disabling the calculation of the ratio. Under such situation, one needs to consider values of the difference ratios only for early coarse division of $[a, b]$, that is , for low values of i and j.

It can be shown in general that $T_{22} = T_h^1$ is equivalent to Simpson's rule and similarly $T_{33} = T_h^2$ is of Newton–Cote type; but the rest like T_{44}, T_{55} etc. are not of this type.

```
SUBROUTINE ROMBERG(f,a,b,n,T,nrow)
! f=function to be integrated from a to b. Must be
! declared external in the calling program.     (Input)
! a=lower limit of integration.     (Input)
! b=upper limit of integration.     (Input)
! n=initial number of panels, must be even.     (Input)
! T=T-matrix of Romberg method.               (Output)
! nrow=number of rows of T-matrix to be computed,
!           not less than 3.     (Input)
! ratio=difference ratios overwritten on the T- matrix     (Output)
!***********************************************************
REAL :: T(nrow,nrow)
h=(b-a)/n
! Compute Romberg T- matrix
sum=0.5*(f(a)+f(b))
DO i=1,n-1
sum=sum+f(a+i*h)
```

```
END DO
T(1,1)=h*sum
PRINT 10, T(1,1)

DO i=2,nrow

    h=0.5*h; n=2*n
    sum=0.0
    do k=1,n-1,2
    sum=sum+f(a+k*h)
    END DO
    T(i,1)=0.5*T(i-1,1)+h*sum
    DO j=1,i-1
    T(i-1,j)=T(i,j)-T(i-1,j)
    T(i,j+1)=T(i,j)+T(i-1,j)/(4.0**j-1.)
    END DO
    PRINT 10, (T(i,j),j=1,i)
    10 FORMAT(7E15.7)

END DO
! Compute table of ratios
DO i=1, nrow-2
DO j=1,i
if(T(i+1,j)==0.0) THEN
ratio=0.0
ELSE
ratio=T(i,j)/T(i+1,j)
END IF
T(i,j)=ratio
END DO
PRINT 20, (T(i,j),j=1,i)
20 FORMAT(7F10.2)
END DO
RETURN
END SUBROUTINE ROMBERG
```

Example 1. compute $I = \int_0^1 e^{-x^2} dx$

by using subroutine ROMBERG.

Solution. Writing a suitable main program with n=2, nrow=6 and appending subroutine ROMBERG the following T table is obtained:

.7313702
.7429841 .7468553
.7458656 .7468261 .7468241
.7465845 .7468242 .7468241 .7468241
.7467642 .7468241 .7468241 .7468241 .7468241
.7468091 .7468241 .7468241 .7468241 .7468241 .7468241

The table of ratios is obtained as

4.03
4.01 15.31
4.00 16.00 .00
4.00   .00   .00 .00

The ratios converge nicely to 4 and 16 for $i = 3$. Hence, we conclude that $I = 0.7468241$. $\qquad\qquad\qquad\qquad\qquad\qquad\qquad\qquad\qquad\qquad\qquad\qquad\qquad$ □

Exercises

1. Compute the following integrals by subroutine ROMBERG:

(i) $\displaystyle\int_0^1 \frac{dx}{1+x}$ $\qquad$ (ii) $\displaystyle\int_0^2 \frac{\sin x}{x} dx$ $\qquad$ (iii) $\displaystyle\int_0^{1.25} \cos x^2\, dx$

(iv) $\displaystyle\int_0^2 (1 + 3e^{-x} \sin x^2)^{-1} dx$ $\qquad$ (v) $\displaystyle\int_0^2 x^{10} e^{4x^2 - 3x^4}\, dx$

$[(i)$ 0.6931472, $\quad(ii)$ 1.605413, $\quad(iii)$ 0.9774377, $\quad(iv)$ 1.498042 (ratio test fails), (v) 1.556354 (ratio test fails)]

5.2.6 *Adaptive Quadrature*

The composite rules and the Romberg integration are all based on panels of *equal size*. Such a choice of panels is a necessity, if the integrand is known only at a sequence of equally spaced points, e.g. when $f(x)$ is given as a table of function values. If on the other hand, $f(x)$ can be computed with equal ease for every point in the interval of integration, then it is usually economical to use panels determined by the local behaviour of the function. The integral $I(f)$ in this approach, can possibly be computed within prescribed accuracy, with fewer function evaluations. Integration schemes based on this principle are called **adaptive**.

In an adaptive quadrature scheme, the user specifies a finite interval $[a, b]$, provides a function subprogram which computes $f(x)$ for any x in the interval and an error tolerance ϵ. The routine attempts to compute an approximation Q, so that $|I - Q| < \epsilon$. The routine may decide that the prescribed accuracy is not attainable, do the best it can, and return an estimate of the accuracy actually achieved.

Routines of varying degree of complexities are reported in the literature. We explain the basic approach by adopting Simpson's rule. Suppose the interval $[a, b]$ is bisected to yield the two-panel nodes a, $a + h$, b where $h = (b - a)/2$. then Simpson's rule gives the approximation

$$I(f) \approx P_1 = \frac{h}{3} [f(a) + 4 f(a + h) + f(b)]$$

If the two panels are again bisected to yield a four panel, then the composite Simpson's rule yields another estimate

$$I(f) \approx Q_1 = \frac{h}{6} [f(a) + 4 f\left(a + \frac{h}{2}\right) + 2 f(a + h) + 4 f\left(a + \frac{3h}{4}\right) + f(b)]$$

If both P_1 and Q_1 pass the accuracy test $|P_1 - Q_1| < \epsilon$, where the absolute error ϵ is sufficiently small, then either P_1 or Q_1 can be accepted as the computed value of $I(f)$. Further refinement is possible by the Richardson extrapolation method. Considering the error term, we have as in Romberg integration

$$I(f) = P_1 + C h^5 + O(h^7)$$

as also
$$I(f) = Q_1 + C\left(\frac{h}{2}\right)^5 + O(h^7)$$

where C is a constant. Eliminating C, we obtain

$$I(f) = \frac{32 Q_1 - P_1}{31} + O(h^7)$$

and thus the result $Q = (32 Q_1 - P_1)/31$ is a better approximation of I(f).

If the accuracy test $|P_1 - Q_1| < \epsilon$ is not satisfied then for refinement, the interval $[a, b]$ is broken down in to two subintervals $[a, c]$ and $[c, b]$ where $c = (a + b)/2$. The P,Q – method is first applied to the interval $[a, c]$ and then to the interval $[c, b]$. The procedure is continued till the length of the last nested interval becomes zero. The 'result' of integration can be stored in a single location, initially its value being zero. On successful integration over any sub-interval, the value of 'result' is augmented by Q. Finally the accumulated value of the 'result' gives the estimate of the integral $I(f)$, to required tolerance of error. The following subroutine implements the above scheme:

SUBROUTINE ADAPTIVE_ SIMPSON(f,a,b,result,abserr)
! f=function name to be integrated from a to b, given as a function
! subprogram. Must be declared external in the calling program.   (Input)
! a=lower limit of integration.   (Input)
! b=upper limit of integration.   (Input)
! result=computed value of the integral of f.   (Output)
! abserr=absolute error tolerance.   (Input)
!*
result=0.0; xa=a; xb=b; kount=0**
10 h=(xb−xa)/2; kount=kount+1; PRINT*, kount
f0=f(xa); f1=f(xa+h/2); f2=f(xa+h); f3=f(xa+3*h/2); f4=f(xb)
P1=h/3*(f0+4*f2+f4);   Q1=h/6*(f0+f4+4*(f1+f3)+2*f2)
IF(ABS(P1−Q1)<h/(b−a)*abserr) THEN
result=result+(32.0*Q1−P1)/31.0
xa=xb; xb=b; if(xa>=xb) RETURN
GOTO 10
END IF
xb=xa+h
GOTO 10
END SUBROUTINE ADAPTIVE_ SIMPSON

The subroutine does not economise on computation of function value for keeping simplicity of the routine. The values of $f0$, $f2$, and $f4$ are potential candidates for use in later iterations, but all the five function values $f0, \cdots, f4$ are computed afresh in each iteration.

In routines of this type, multiplying abserr by $h/(b-a)$ ensures that the total accumulated error, in all the iterations, does not exceed abserr (see, for example, the text of Forsythe and Moler referred in the bibliography).

One of the advanced level adaptive schemes is QUANC8 given by Forsythe and Moler (see Bibliography). It is based on an 8-panel Newton–Cotes rule

$$\int_a^b f(x)\,dx \approx \sum_{k=0}^{8} A_k\, f(a+kh), \quad h = (b-a)/8$$

where $A_0 = 3956/14175$, $A_1 = 23552/14175$, $A_2 = -3712/14175$, $A_3 = 41984/14175$, $A_4 = -18160/14175$, $A_5 = A_3$, $A_6 = A_2$, $A_7 = A_1$, $A_8 = A_1$.

Exercise

1. Apply subroutine ADAPTIVE_SIMPSON to compute the integrals $(i) - (v)$ of
Exercise 1, Sect. 2.5.

5.3 Euler–Maclaurin Summation Formula

Integration, we know is the inverse operation of differentiation. *Operationally*, therefore, if D denotes differentiation, then $\frac{1}{D}$ would denote integration. To be precise
if

$$D f(x) = \frac{d f(x)}{dx}, \quad \text{then} \quad \frac{1}{D} f(x) = \int_{x_0}^{x} f(x)\, dx$$

where f is a function defined for $x \geq x_0$. Now, the discrete version of the differential
$df(x)$ is the *finite difference*

$$\Delta f(x) := f(x+h) - f(x) \tag{5.44}$$

where Δ is the finite difference operator. The inverse of this operation may be defined
as that operator, which yields $f(x)$ when operating on a certain function $F(x)$:

$$\frac{1}{\Delta} f(x) =: F(x), \quad \text{if} \quad \Delta F(x) = f(x) \tag{5.45}$$

The continuous and discrete versions of the integral viz. $\frac{1}{D} f(x)$ and $\frac{1}{\Delta} f(x)$ must
be connected in some way. This connection is provided by the Euler–Maclaurin
summation formula. To derive it, we require some facts about **Bernoulli Numbers**,
known after J. Bernoulli.

Jakob Bernoulli (1655–1705), eldest of the famous Swiss family of mathematicians. At his time
he greatly advanced algebra, differential and integral calculus, calculus of variations, mechanics,
theory of series and theory of probability. He discovered Bernoullian numbers for giving a rigid
derivation of the exponential series through complete induction. He is also famous for proving
Bernoulli's inequality $(1 + x)^n > 1 + nx$ for $x > 0$. In mechanics he investigated nonuniform and
elastic *catenaries*. He also treated the problem of *elastica* or a bent elastic beam. The term *integral*
was first used by him in its present mathematical sense. In probability theory he gave philosophical
thought on all the basic concepts, including the law of large numbers.

Lemma 5.2 (Bernoulli Numbers). *Let*

$$f(x) := \frac{x}{e^x - 1} =: \sum_{n=0}^{\infty} \frac{B_n}{n!} x^n$$

then the coefficients B_n are the Bernoulli numbers satisfying the recurrence relations

$$B_0 = 1, \quad \frac{B_n}{n!} + \frac{B_{n-1}}{(n-1)!} \frac{1}{2!} + \cdots + \frac{B_0}{0!} \frac{1}{(n+1)!} = 0$$

i.e. $B_0 = 1$, $B_1 = -\frac{1}{2}$, $B_2 = 6$, $B_4 = -\frac{1}{30}$, $B_6 = \frac{1}{42}$, $B_8 = -\frac{1}{30}$, $B_{10} = \frac{5}{66}, \cdots$,
$B_3 = B_5 = B_7 = \cdots = 0$.

Proof
$$f(x) = \frac{1}{1 + \dfrac{x}{2!} + \dfrac{x^2}{3!} + \cdots} = \sum_{n=0}^{\infty} \frac{B_n}{n!} x^n$$

Hence
$$\sum_{n=0}^{\infty} \frac{B_n}{n!} x^n \left(1 + \frac{x}{2!} + \frac{x^2}{3!} + \cdots \right) = 1$$

equating the coefficient of x^0, we have $B_0 = 1$. The coefficient of x^n similarly yields the desired recurrence relation. Successively setting $n = 1, 2, 3, 4, \cdots$ in the recurrence relation, we get the set of equations

$$B_1 + \frac{1}{2} = 0 \quad \text{or} \quad 2B_1 + 1 = 0$$

$$\frac{B_2}{2} + \frac{B_1}{2!} + \frac{B_0}{3!} = 0 \quad \text{or} \quad 3B_2 + 2B_1 + 1 = 0$$

$$\frac{B_3}{3!} + \frac{B_2}{2!\,2!} + \frac{B_1}{1!\,3!} + \frac{B_0}{4!} = 0 \quad \text{or} \quad 4B_3 + 6B_2 + 4B_1 + 1 = 0$$

$$\frac{B_4}{4!} + \frac{B_3}{3!\,2!} + \frac{B_2}{2!\,3!} + \frac{B_1}{1!\,4!} + \frac{B_0}{5!} = 0 \quad \text{or} \quad 5B_4 + 10B_3 + 10B_2 + 5B_1 + 1 = 0, \quad \text{etc.}$$

These relations yield the required coefficients.

It is noteworthy that except for B_1, the other odd-ordered numbers are equal to zero. This fact can be independently demonstrated as follows:

$$f(x) + \frac{x}{2} = \frac{x}{e^x - 1} + \frac{x}{2} = \frac{x}{2} \frac{e^x + 1}{e^x - 1} = \frac{x/2}{\tanh x/2} = 1 + \sum_{n=2}^{\infty} \frac{B_n}{n!} x^n$$

Now, $(x/2)/\tanh(x/2)$ is an even function, and so the odd powers in the series on the right-hand side must vanish. This means that $B_3 = B_5 = B_7 = \cdots = 0$.

Applying Taylor's series expansion to Eq. (5.40), we have *operationally*

$$\Delta f(x) = f(x+h) - f(x) = \sum_{k=1}^{\infty} \frac{h^k}{k!} D^k f(x) = \left(e^{hD} - 1\right) f(x)$$

This means that $\Delta \equiv e^{hD} - 1$ and so

$$hD \frac{1}{\Delta} = \frac{hD}{e^{hD} - 1} = \sum_{k=0}^{\infty} \frac{B_k}{k!} h^k D^k$$

by the lemma. Hence

$$\frac{d}{dx}\left[\frac{1}{\Delta} f(x)\right] = \sum_{k=0}^{\infty} \frac{B_k}{k!} h^{k-1} D^k f(x)$$

In the above relation, it is tacitly assumed that the nodes x_i $(i = 0, 1, 2, \cdots)$ of discritisation for the operator Δ are equally spaced with equal length h. Integrating the above relation from x_0 to x_n, we obtain

$$\frac{1}{\Delta} f(x_n) - \frac{1}{\Delta} f(x_0) = \frac{1}{h} \int_{x_0}^{x_n} f(x)\, dx - \frac{1}{2}[f(x_n) - f(x_0)]$$
$$+ \sum_{k=2}^{\infty} \frac{B_k}{k!} h^{k-1}\left[f^{(k-1)}(x_n) - f^{(k-1)}(x_0)\right] \qquad (5.46)$$

where the fact that $B_1 = -1/2$ has been used.

In the relation (5.46), the left hand side can be expressed as a summation over nodal function values. For this purpose, define the summation

$$S(x_i) = \sum_{j=0}^{i-1} f(x_j), \quad (i = 1, 2, \cdots) \quad \text{with} \quad S(x_0) = 0$$

It follows that

$$\Delta S(x_i) = S(x_{i+1}) - S(x_i) = f(x_i) = \Delta F(x_i)$$

according to Eq. (5.45). Hence $\Delta[S(x_i) - F(x_i)] = 0$. This means that $S(x_i) - F(x_i)$ is the same for all $i = 0, 1, 2, \cdots$, i.e. $S(x_i) - F(x_i) = S(x_0) - F(x_0) = -F(x_0)$. Thus

$$F(x_i) = \frac{1}{\Delta} f(x_i) = F(x_0) + S(x_0) \tag{5.47}$$

Using the expression (5.47) in Eq. (5.46), we obtain

$$F(x_0) + S(x_n) - F(x_0) = S(x_n) = \sum_{j=0}^{n-1} f(x_j)$$

$$= \frac{1}{h} \int_{x_0}^{x_n} f(x)\, dx - \frac{1}{2}[f(x_n) - f(x_0)] + \sum_{k=2}^{\infty} \frac{B_k}{k!} h^{k-1} \left[f^{(k-1)}(x_n) - f^{(k-1)}(x_0) \right]$$

In the infinite series, on the right-hand side, the odd order terms $k = 3, 5, \cdots$ vanish in virtue of vanishing of odd order Bernoulli numbers. Moreover, the series may diverge. If the series is terminated at $k = 2m$, we finally get the form

$$\int_{x_0}^{x_n} f(x)\, dx = h \left[\frac{1}{2} f(x_0) + f(x_1) + \cdots + f(x_{n-1}) + \frac{1}{2} f(x_n) \right]$$

$$- \sum_{k=1}^{m} \frac{B_{2k}}{(2k)!} h^{2k} \left[f^{(2k-1)}(x_n) - f^{(2k-1)}(x_0) \right] + R_{2m} \tag{5.48}$$

Equation (5.48) is the celebrated Euler–Maclaurin summation formula with a remainder that can be proved to be given by

$$R_{2m} = -n\, h^{2m+2} \frac{B_{2m+2}}{(2m+2)!} f^{(2m+2)}(\xi) \tag{5.49}$$

The first summation on the right-hand side is the composite trapezoidal rule. Its utility has been noticed in Romberg integration. It can be used for numerical integration, if the function f and its successive derivatives can be calculated at the nodal points. Conversely, if the integral of f can be analytically calculated, then it can be used for summation of series of f-values. However, it is hardly used for such purpose, because of better available methods.

Leonhard Euler (1707–1783), was the leading mathematician of the eighteenth century. Like Gauss, he contributed very significantly to the entire spectrum of mathematics as well as to numerous application areas such as mechanics, hydrodynamics, optics and astronomy. Concerning the depth and importance of his work, it was Gauss' opinion that 'studying Euler's papers remains the best way to learn about the various areas of mathematics and it can not be replaced by anything else'. He was born in Switzerland but in later life worked in St. Petersberg, Russia.

Colin Maclaurin (1698–1746), English mathematician, published the expansion formula in about 1737 independent from Euler, who had presented it in 1730.

5.4 Improper Integrals

Improper integrals are of two types: (i) when the integrand is singular (unbounded) in the interval of integration, and (ii) when the range of integration is infinite. Of course, there can be a case which is a combination of the two. This later case can be treated by separating the singular part and the infinite interval parts.

The integrable singularity of the integrand can be of two types: algebraic and logarithmic. In the former case, let the integral be

$$I(f) = \int_0^b \frac{f(x)}{x^\alpha}\,dx, \quad f(0) \neq 0, \quad 0 < \alpha < 1 \tag{5.50}$$

We can write

$$I(f) = \int_0^b \frac{f(x) - f(0)}{x^\alpha}\,dx + f(0) \int_0^b \frac{dx}{x^\alpha}$$
$$= \int_0^b \frac{f(x) - f(0)}{x^\alpha}\,dx + \frac{b^{1-\alpha} f(0)}{1 - \alpha} \tag{5.51}$$

The integral on the right-hand side is regular and can be integrated by methods of the preceding sections solving the problem. If $0 < \alpha \le 1/2$, the simple substitution $x = t^{1/\alpha}$ reduces the integral to

$$I(f) = \frac{1}{\alpha} \int_0^{b^\alpha} f\left(t^{1/\alpha}\right) t^{1/\alpha - 2}\,dt \tag{5.52}$$

which is a regular integral. Thus, Eqs. (5.51) or (5.52) solve the problem of Eq. (5.50).

An integral containing logarithmic singularity is

$$I(f) = \int_0^b g(x) \ln f(x)\,dx, \quad f(0) = 0 \tag{5.53}$$

Suppose that $f(x)/x^\alpha \to C$ as $x \to 0$ ($\alpha > 0$). Then

$$I(f) = \int_0^b g(x) \ln\left[\frac{f(x)}{x^\alpha}\right] dx + \alpha \int_0^b g(x) \ln x\,dx$$
$$= \int_0^b g(x) \ln\left[\frac{f(x)}{x^\alpha}\right] dx + \alpha g(b)\, b\, (\ln b - 1) - \alpha \int_0^b g'(x)\, x\, (\ln x - 1)\,dx \tag{5.54}$$

The two integrals in (5.54) are regular integrals that can be computed in the usual way.

A cauchy Principal Value singular integral can be written as

$$P \int_a^b \frac{f(x)}{x-c}\,dx = \int_a^b \frac{f(x)-f(c)}{x-c}\,dx + f(c)\,P \int_a^b \frac{dx}{x-c}, \quad a < c < b$$

$$= \int_a^b \frac{f(x)-f(c)}{x-c}\,dx + f(c)\ln\frac{b-c}{c-a} \qquad (5.55)$$

Equation (5.55) solves the problem of evaluation.

If the interval of integration is infinite as in

$$I = \int_1^\infty \frac{f(x)}{x^p}\,dx, \quad p > 1 \qquad (5.56)$$

with $\lim_{x\to\infty} f(x)$ nonzero and finite, then we may substitute $x = 1/u^\alpha$, $\alpha > 0$. This yields

$$I = \alpha \int_0^1 u^{p\alpha} f\left(\frac{1}{u^\alpha}\right)\frac{du}{u^{\alpha+1}} = \alpha \int_0^1 u^{(p-1)\alpha-1} f\left(\frac{1}{u^\alpha}\right) du$$

Maximising the smoothness of the integrand at $u = 0$, we pick α to yield a large value of the exponent $(p-1)\alpha - 1$. If we suppose that exponent value of 1 is sufficient, we take $\alpha = 2/(p-1)$ obtaining

$$I = \frac{2}{p-1} \int_0^1 u\, f\left(u^{-2/(p-1)}\right) du \qquad (5.57)$$

and the integral in (5.57) can be treated by usual methods.

A brute force method of computing an infinite integral is to replace the upper limit of integration ∞ by a very large number, say 1000 or 10000, so that the integrand becomes sufficiently small and apply subroutine QUANC8. It usually delivers the result without consuming too much time.

Gauss-type quadrature formulas are also applicable for evaluating improper integrals of the type

$$\int_{-1}^1 f(x)\frac{dx}{\sqrt{1-x^2}}, \quad \int_0^\infty e^x f(x)\,dx, \quad \int_{-\infty}^\infty e^{-x^2} f(x)\,dx$$

Evidently Gauss–Chebyshev, Gauss–Laguerre and Gauss–Hermite quadrature formulas are applicable in the respective cases.

Exercises

1. Compute the following improper integrals using ADAPTIVE_ SIMPSON or QUANC8:

$(i)\ \displaystyle\int_0^1 \frac{e^x}{\sqrt{x}}\,dx \quad (ii)\ \displaystyle\int_0^1 \frac{\sin x}{x^{3/2}}\,dx \quad (iii)\ \displaystyle\int_0^1 \cos x\,\ln x\,dx \quad (iv)\ \displaystyle\int_0^{\pi/2} \ln \sin x\,dx$

$(v) \displaystyle\int_0^1 \frac{\ln x}{1-x}\,dx$ $(vi) \displaystyle\int_0^\infty e^{-x}\sqrt{x}\,dx$ $(vii) \displaystyle\int_1^\infty \frac{\sin x}{x^2}\,dx$ $(viii) \displaystyle\int_{-1}^1 \frac{\sqrt{1-x^2}}{x-1/2}\,dx$

$[(i)\ 2.92530,\ (ii)\ 1.93526,\ (iii)\ -0.94606,\ (iv)\ -1.08879,\ (v)$ Divide the interval in to $[0,\ 1/2]$ and $[1/2,\ 1];\ -1.644934$ (exact value is $-\pi^2/6$), $(vi)\ 0.88623,\ (vii)$ $0.50404,\ (viii)\ -1.57078$ (exact value is $-\pi/2$)].

5.5 Double Integration

If the limits of integration are constants, a **double integral** is of the form

$$I(f) = \int_c^d \int_a^b f(x,\ y)\,dx\,dy$$

The trapezoidal and the Simpson's rules can be generalised by successive integrations in x and y in the following manner.

1°. **Trapezoidal Rule.** Integrating the inner integral by the trapezoidal rule

$$I_T(f) = \frac{b-a}{2} \int_c^d [f(a,\ y) + f(b,\ y)]\,dy$$

$$= \frac{(b-a)(d-c)}{4}[f(a,\ c) + f(a,\ d) + f(b,\ c) + f(b,\ d)] \qquad (5.58)$$

by repeating the rule over the variable y.

2°. **Simpson's Rule.** By application of Simpson's rule over x and then y, we obtain

$$I_S(f) = \frac{h}{3} \int_c^d [f(a,\ y) + 4\,f(a+h,\ y) + f(b,\ y)]\,dy$$

$$= \frac{hk}{9}\,[f(a,\ c) + 4\,f(a,\ c+k) + f(a,\ d)$$

$$+ 4\{f(a+h,\ c) + 4\,f(a+h,\ c+k) + f(a+h,\ d)\}$$

$$+ f(b,\ c) + 4\,f(b,\ c+k) + f(b,\ d)]$$

$$= \frac{hk}{9}\,[f(a,\ c) + f(a,\ d) + f(b,\ c) + f(b,\ d)$$

$$4\{f(a+h,\ c) + f(a+h,\ d) + f(a,\ c+k) + f(b,\ c+k)\} + 16\,f(a+h,\ c+k)]$$

$$\qquad (5.59)$$

Equations (5.58) and (5.59) give the required basic rules.

If the limits of the inner integral are functions of y then rules (5.58) and (5.59) can be modified suitably.

A practical way of computing a double integral is to successively apply Simpson's rule recursively. A *recursive* subroutine is given below:

RECURSIVE SUBROUTINE SIMPSON2D(f,a,b,result)

! Evaluates $\quad \int_a^b dx \int_c^d fn(x,\ y)dy.$

! f=Integrand function.
! a=Lower limit of integration.
! b=Upper limilt of integration
! result=Value of the Integral.
!**
```
h=0.1                                          ! h is step length
n=(b-a)/h; IF(n/2*2 /= n) n=n+1        ! n = Number of panels of [a, b].
                                               ! Must be an even number
IF(n /= 0) h=(b - a)/n; IF(n==0) h=0.0
x=a; result=0.0
DO i=1,n,2
result=result+h/3*(f(x)+4*f(x+h)+f(x+2*h))
x=x+2*h
END DO
RETURN
END SUBROUTINE SIMPSON2D
```
!*******************************
```
FUNCTION f(x)
COMMON /fxy/ x1
EXTERNAL g
c= · · · · · ·          ! Lower limit of y - integral, can be function of x
d= · · · · · ·          ! Upper limit of y - integral, can be function of x
x1=x
CALL SIMPSON2D(g,c,d,f1)
f=f1
RETURN
END FUNCTION f
```
!**************************
```
FUNCTION g(y)
COMMON /fxy/ x
g=fn(x,y)          ! fn(x,y) is the given integrand of the double integral
RETURN
END FUNCTION g
```
!********************

FUNCTION fn(x,y)
fn= Given function of x,y
RETURN
END FUNCTION fn

Example 1. Evaluate $\int_0^1 dx \int_0^x fn(x, y)\, dy$ for

$$fn(x, y) = x + y^2, \quad \text{and} \quad fn(x, y) = x + y^2/x$$

Solution. Select a = 0, b = 1 and c = 0, d = x for the recursive subroutine SIMP-SON2D, writing the program as

```
PROGRAM MAIN
EXTERNAL f
a=0.0; b=1.0
CALL SIMPSON2D(f,a,b,result)
PRINT*, result
END
!***********************************************************************
!*****
```

! Append function subprograms F(x) with c=0, d=x, and g(y). In the function subprogram for fn(x,y) write fn(x,y)=x+y**2 for the first case of the problem and two statements: IF(x==0.0) x=x+0.00000001 followed by fn=x+y**2/x for the second case.

On executing the program, the 'result' in the two cases are, respectively: 0.4444445 and 0.4166667.   □

One may adopt routines like ADAPTIVE_SIMPSON or QUANC8 recursively in a similar manner.

Chapter 6
Ordinary Differential Equations

Ordinary differential equations (or ODE in short) occur naturally in science and engineering whenever two quantities x and y are connected by some differential law. Historically, the motivation for building the early computers came from the need to compute ballistic trajectories accurately and quickly that required solution of certain ordinary differential equations. Today computers have become indispensable in similar situations in other subjects as well, and indeed go beyond the preliminary requirement.

An nth–order differential equation when resolved in terms of the highest order derivative can be written as

$$y^{(n)} = f(x, y, y', y'', \cdots, y^{(n-1)}) \tag{6.1}$$

If at a given point $x = x_0$, it is also given that

$$y = y_0, \quad y' = y_0', \quad y'' = y_0'', \cdots, \quad y^{(n-1)} = y_0^{(n-1)} \tag{6.2}$$

where $y_0, y_0', y_0'', \cdots, y_0^{(n-1)}$ are given numerical values, then Eqs. (6.1) and (6.2) constitute the *Initial Value Problem* (IVP) or the *Cauchy Problem*. In the particular case $n = 1$, the *first-order* problem becomes

$$y' = f(x, y), \quad y(x_0) = y_0 \tag{6.3}$$

In the theory of differential equations, it is proved that the IVP (6.3) has always a unique solution according to the following theorem.

© Springer Nature Singapore Pte Ltd. 2019
S. K. Bose, *Numerical Methods of Mathematics Implemented
in Fortran*, Forum for Interdisciplinary Mathematics,
https://doi.org/10.1007/978-981-13-7114-1_6

Theorem 6.1 *If $f \in C(R)$, $R = \{(x, y) \big| |x - x_0| < 0, \ |y - y_0| < b\}$ and satisfies in the region R the Lipschitz condition*

$$|f(x, \bar{y}) - f(x, y)| \le K |\bar{y} - y| \quad (K = Lipschitz \ \ constant)$$

then problem (6.3) has a unique solution.

The Lipschitz condition is satisfied by f if $\partial f / \partial y$ is bounded on R. For by the mean value theorem

$$|f(x, \bar{y}) - f(x, y)| = \left| (\bar{y} - y) \frac{\partial f}{\partial y}(x, \xi) \right| \le K |\bar{y} - y|$$

where

$$K = \frac{\max}{(x, y) \in R} \left| \frac{\partial f((x, y)}{\partial y} \right|$$

The existence theorem can be generalised for the nth-order equation.

When the order n is greater than or equal to 2, the initial conditions may be replaced by *boundary conditions* at the end points of the interval $[a, b]$ in which the solution is sought. The boundary conditions constitute functional relation between y and some of its successive derivatives at $x = a$ and b. Problems posed in this manner are called *Boundary Value Problems* (BVP).

The numerical solution of an initial or boundary value problem is sought in a discretised manner, that is to say, at discrete *grid points* $a = x_0, x_1, x_2, \cdots, x_n = b$. The points will be assumed to be equally spaced, but in general this may not be the case. We will find that while it is possible to design numerical methods for Initial Value Problems (IVPs) of great generality, Boundary Value Problems (BVPs) can be treated in a restricted manner.

We begin by considering the initial value problem for the first-order equation in Sect. 6.1, followed by the nth-order equation in Sect. 6.2. Boundary value problems are discussed in Sect. 6.4. In Sect. 6.3, 'stiff differential equations' are treated, when some IVPs become hard to treat.

6.1 Initial Value Problem for First-Order ODE

The differential equation and the initial condition are defined in Eq. (6.3). If the function f is differentiable sufficient number of times, the exact solution $y(x)$ can be expanded in a Taylor series about the point $x = x_0$:

$$y(x) = y_0 + (x - x_0) y'(x_0) + \frac{(x - x_0)^2}{2!} y''(x_0) + \cdots \tag{6.4}$$

The successive derivatives occurring in the series can be calculated by noting that

$$y' = f(x, y)$$

$$y'' = \frac{\partial}{\partial x}\{f(x, y)\} = f_x + f_y y' = f_x + f_y f$$

$$y''' = \frac{\partial}{\partial x}\{f_x + f_y f\} = f_{xx} + f_{xy}f + f_y(f_x + f_y f) + f(f_{yx} + f_{yy}f)$$

$$= f_{xx} + 2 f_{xy}f + f_x f_y + f_y^2 f + f_{yy}f^2$$

$$\cdots\cdots\cdots\cdots\cdots\cdots\cdots\cdots\cdots\cdots$$

It is clear that in general the difficulty of calculating the derivatives at x_0 becomes increasingly difficult with increasing order of the derivative and the series will have to be terminated at some stage. This is only possible when $x - x_0$ is suitably small.

For computation of the solution at grid points $x_0, x_1, x_2, \cdots$, the above consideration means that the grid size must be suitably small. If $h := x_1 - x_0$, we can calculate $y_1 := f(x_1)$ from Eq. (6.4) in the form

$$y_1 = y_0 + h\,\Phi(x_0, y_0;\ h)$$

where $\quad \Phi(x_0, y_0;\ h) := f(x_0, y_0) + \dfrac{h}{2!} f'(x_0, y_0) + \cdots + \dfrac{h^{k-1}}{k!} f^{(k-1)}(x_0, y_0)$

$$(6.5)$$

by retaining $k + 1$ terms in Eq. (6.4). It is clear that the process can be continued in **single steps** to compute $y_2 := y(x_2)$, $y_3 := y(x_3)$, $\cdots$, the recursive relation being

$$y_{n+1} = y_n + h\,\Phi(x_n, y_n;\ h), \quad n = 0, 1, 2, \cdots \qquad (6.6)$$

where Φ is defined by Eq. (6.5).

By tackling Eq. (6.5) in suitable ways, we obtain different methods of solving the first-order initial value problem. All these methods are *single-step* methods.

6.1.1 Euler's Method

Confining the expansion (6.5) only to the first term, that is taking $k = 1$, we obtain $\Phi(x_n, y_n;\ h) = f(x_n, y_n)$. Hence, from Eq. (6.6), we obtain the iteration

$$y_{n+1} = y_n + h\,f(x_n, y_n), \quad n = 0, 1, 2, \cdots \qquad (6.7)$$

Euler's formula (6.7) can also be viewed as application of the *rectangle rule* of numerical integration. For integrating the ODE of Eq. (6.3) and applying the rule,

$$y_{n+1} - y_n = h \int_{x_n}^{x_{n+1}} f(x, y) \, dx \approx h \, f[x_n, y(x_n)] = h \, f(x_n, y_n)$$

This equation evidently is of the form (6.7).

For accuracy, the step size h must be very small, because of the crude nature of approximation used. For increasing number of grid points n, the truncation error may also grow, that is to say, the method may become **unstable**. This is a potential hazard. As a result, the method is not a practical one for an arbitrarily defined function f.

It is customary in numerical solution of differential equations to call a method of **order** p if the error is $O(h^{p+1})$. The error in recangle rule is $O(h^2)$ and so *Euler's method is a first-order method*.

6.1.2 Modified Euler's Method

The numerical integration procedure for Eq. (6.7) is suggestive of employing the trapezoidal rule for greater accuracy. Thus, applying the rule to Eq. (6.3)

$$y_{n+1} - y_n = h \int_{x_n}^{x_{n+1}} f(x, y) \, dx \approx \frac{h}{2} [f(x_n, y_n) + f(x_{n+1}, y_{n+1})] \qquad (6.8)$$

for $n = 0, 1, 2, \cdots$. Equation (6.8) however is an implicit equation in y_{n+1}, because the quantity also occurs in the right-hand side of the equation. Noting that as h is small, solving the equation by iteration using Eq. (6.7), one obtains the scheme

$$\begin{aligned}
y_{n+1}^{(0)} &= y_n + hf(x_n, y_n) \\
y_{n+1}^{(k+1)} &= y_n + \frac{h}{2}[f(x_n, y_n) + f(x_n, y_{n+1}^{(k)})], \quad k = 0, 1, 2, \cdots
\end{aligned} \qquad (6.9)$$

This yields y_{n+1} at the point x_n. Equation (6.9) constitutes the *Modified Euler's Method*, also known as *Heun's Method*. The method is of the *second order* since the error in the trapezoidal rule is $O(h^3)$.

6.1.3 Runge–Kutta Methods

Here the objective is to increase the order of the general method (6.6) by somehow obviating the difficulty of computing the higher order derivatives y'', y''', $\cdots$ expressed in terms of the partial derivatives of f. The idea is to replace the derivatives by requisite number of *functional values* of f at intermediate points. We illustrate the procedure below:

1°. Second-Order Runge–Kutta Method

Let x be any grid point. If in the Taylor series (6.4) or (6.6), terms up to $y''(x)$ are retained then it will read

$$y(x + h) = y(x) + h\,y'(x) + \frac{h^2}{2}\,y''(x)$$

$$= y(x) + hf + \frac{h^2}{2}\,(f_x + ff_y) \tag{6.10}$$

We can also write by the Mean Value Theorem

$$y(x + h) = y(x) + h\,f(\xi,\,y(\xi))$$
$$= y(x) + A\,k_1 + B\,k_2, \quad \text{(say)}$$

where

$$k_1 = h\,f(x,\,y)$$
$$k_2 = h\,f(x + \alpha h,\,y + \beta k_1)$$

This means that $f(\xi,\,y(\xi))$ is replaced by the weighted average of $f(x,\,y)$ and $f(x + \alpha h,\,y + \beta k_1)$. Hence,

$$y(x + h) = y(x) + h[A\,f + B\,\{f + \alpha h f_x + \beta k_1 f_y + O(h^2)\}]$$
$$= y(x) + h(A + B)\,f + h^2\,B\,(\alpha f_x + \beta\,ff_y) + O(h^3) \tag{6.11}$$

by an application of Taylor's theorem for two variables. Comparing Eqs. (6.10) and (6.11), we find that

$$A + B = 1, \quad \alpha B = \frac{1}{2}, \quad \beta B = \frac{1}{2}$$

Since there are four unknowns, the above set of three equations do not determine A, B, α, β uniquely. An elegant particular solution is evidently

$$A = B = \frac{1}{2}, \quad \alpha = \beta = 1$$

yielding the solution

$$y(x + h) = y(x) + \frac{1}{2}(k_1 + k_2)$$
$$k_1 = h\,f(x,\,y)$$
$$k_2 = h\,f(x + h,\,y + k_1)$$

For a grid point x_{n+1}, $(n = 0, 1, 2, \cdots)$, we therefore compute

$$k_1 = h\, f(x_n, y_n) \tag{6.12a}$$
$$k_2 = h\, f(x_n + h, y_n + k_1) \tag{6.12b}$$

to yield

$$y_{n+1} = y_n + \frac{1}{2}(k_1 + k_2) \tag{6.12c}$$

Equation (6.12) constitutes the *Second-Order* Runge–Kutta. The method is of second order since the discretisation error is $O(h^3)$.

For higher order methods, a similar procedure is adopted by expanding $\Phi(x, y; h)$ in a Taylor series (like Eq. (6.10)) with higher degree terms in h and corresponding increase in the number of function values for weighted average of $h\, f(\xi, y(\xi))$. By equating the coefficients of various powers of h in the two equivalent quantities, a system of equations is obtained which is less than the number of unknowns. Suitable particular solution of the equations deliver the required formulae.

2°. Third-Order Runge–Kutta Method

By the procedure outlined above, it can be proved that with

$$
\begin{aligned}
k_1 &= h\, f(x_n, y_n) \\
k_2 &= h\, f\left(x_n + \frac{h}{2},\ y_n + \frac{k_1}{2}\right) \\
k_3 &= h\, f(x_n + h,\ y_n - k_1 + 2k_2)
\end{aligned}
\tag{6.13a}
$$

the value of y_{n+1} is given by

$$y_{n+1} = y_n + \frac{1}{6}(k_1 + 4k_2 + k_3) \tag{6.13b}$$

Stepping up to another order, we have the famous:

3°. Fourth-Order Runge–Kutta Method

For the grid point x_{n+1}, compute

$$
\begin{aligned}
k_1 &= h\, f(x_n, y_n) \\
k_2 &= h\, f\left(x_n + \frac{h}{2},\ y_n + \frac{k_1}{2}\right) \\
k_3 &= h\, f\left(x_n + \frac{h}{2},\ y_n + \frac{k_2}{2}\right) \\
k_4 &= h\, f(x_n + h,\ y_n + k_3)
\end{aligned}
\tag{6.14a}
$$

then the value of y_{n+1} is given by

$$y_{n+1} = y_n + \frac{1}{6}(k_1 + 2k_2 + 2k_3 + k_4) \tag{6.14b}$$

The formula after laborious calculations (see, for instance, Scheid's text noted in the bibliography) is remarkably simple in structure and is of fourth order with discretisation error $O(h^5)$.

If f is independent of y, we can demonstrate that the method is equivalent to Simpson's rule. In this case, the ODE is $y' = f(x)$, yielding

$$y_{n+1} = y_n + \int_{x_n}^{x_{n+1}} f(x)\,dx$$

$$= y_n + \frac{h}{6}\left[f(x_n) + 2f\left(x_n + \frac{h}{2}\right) + 2f\left(x_n + \frac{h}{2}\right) + f(x_n + h)\right]$$

which means that

$$\int_{x_n}^{x_{n+1}} f(x)\,dx = \frac{h}{6}\left[f(x_n) + 4f\left(x_n + \frac{h}{2}\right) + f(x_n + h)\right]$$

$\square$

A Fortran subroutine for the method can easily be coded. Those who want to use a ready-made one, may read Sect. 6.2, on system of ODEs. Experience with the method shows that it is *stable*, that is to say, the discretisation error does not grow with the number of grid points. The method has the main disadvantage that it requires four function evaluation per grid, which may be time-consuming for unwieldy functions. Nevertheless, the method is very popular with practitioners.

Carl David Tolme Runge (1856–1927), German mathematician and scientist. Initially, he studied literature but after attending Weierstrass's lectures turned to pure mathematics and in 1880 wrote a doctoral dissertation on differential geometry. In 1886, he obtained a chair at Hanover and within a year he moved away from pure mathematics to study the wavelengths of the spectral lines of elements other than hydrogen. He moved to a chair in applied mathematics at Götingen in 1904.

Martin Wilhelm Kutta (1867–1944), Polish mathematician. He is best known for the Runge–Kutta method (1901) and for the Kutta–Zhukovski aerofoil in hydro- and aerodynamics (1910). Runge presented Kutta's methods. He became a professor at Stuttgart, Germany in 1911.

Karl Heun (1859–1929), German mathematician, best known for the Heun differential equation which generalises the hypergeometric differential equation. His equation has four singularities compared to the latter equation which has three singularities. He became a professor at Karlsruhe, Germany in 1902, holding the chair of theoretical mechanics.

Example 1. Solve the ODE

$$\frac{dy}{dx} = \frac{2y}{x} + x^3, \quad y(1) = \frac{1}{2}$$

by Euler's method and Euler' modified method, at the two points $x = 1.1$ and 1.2, and compare the solutions with the exact solution.

Solution. Here $x_0 = 1$, $y_0 = 1/2$, $h = 0.1$

Euler's Method. At $x = 1.1$, $y_1 = y_0 + h\,f(x_0, y_0) = 0.5 + 0.1 \times \left(\frac{2 \times 0.5}{1} + 1^3\right)$

$= \underline{0.7}$

At $x = 1.2$, $y_2 = y_1 + h\,f(x_1, y_1) = 0.7 + 0.1 \times \left(\frac{2 \times 0.7}{1.1} + 1.1^3\right) = \underline{0.96037}$

Modified Euler's Method. At $x = x_1 = 1.1$, $y_1^0 = 0.7$

$$y_1^{k+1} = y_0 + \frac{h}{2}[f(x_0, y_0) + f(x_0, y_1^{(k)})], \quad k = 0, 1, 2, \cdots$$

The iterations yield

$$y_1^{(1)} = 0.5 \times \frac{0.1}{2}\left[\frac{2 \times 0.5}{1} + 1^3 + \frac{2 \times 0.7}{1} + 1^3\right] = 0.72200$$

$$y_1^{(2)} = 0.5 + \frac{0.1}{2}\left[\frac{2 \times 0.5}{1} + 1^3 + \frac{2 \times 0.722}{1} + 1^3\right] = 0.72220$$

$$y_1^{(3)} = 0.5 + \frac{0.1}{2}\left[\frac{2 \times 0.5}{1} + 1^3 + \frac{2 \times 0.7222}{1} + 1^3\right] = 0.72222$$

$$y_1^{(4)} = 0.5 + \frac{0.1}{2}\left[\frac{2 \times 0.5}{1} + 1^3 + \frac{2 \times 0.72222}{1} + 1^3\right] = 0.72222$$

Hence, $\underline{y_1 = 0.72222}$.

At $x = x_2 = 1.2$, $y_2^{(0)} = 0.96037$

$$y_2^{(k+1)} = y_1 + \frac{h}{2}[f(x_1, y_1) + f(x_1, y_2^{(k)})], \quad k = 0, 1, 2, \cdots$$

The iterations give

$$y_2^{(1)} = 0.7 + \frac{0.1}{2}\left[\frac{2 \times 0.7}{1.1} + 1.1^3 + \frac{2 \times 0.96037}{1.1} + 1.1^3\right] = 0.98404$$

$$y_2^{(2)} = 0.7 + \frac{0.1}{2} \left[\frac{2 \times 0.7}{1.1} + 1.1^3 + \frac{2 \times 0.98404}{1.1} + 1.1^3 \right] = 0.98619$$

$$y_2^{(3)} = 0.7 + \frac{0.1}{2} \left[\frac{2 \times 0.7}{1.1} + 1.1^3 + \frac{2 \times 0.98619}{1.1} + 1.1^3 \right] = 0.98639$$

$$y_2^{(4)} = 0.7 + \frac{0.1}{2} \left[\frac{2 \times 0.7}{1.1} + 1.1^3 + \frac{2 \times 0.98639}{1.1} + 1.1^3 \right] = 0.98640$$

$$y_2^{(5)} = 0.7 + \frac{0.1}{2} \left[\frac{2 \times 0.7}{1.1} + 1.1^3 + \frac{2 \times 0.98640}{1.1} + 1.1^3 \right] = 0.98641$$

The subsequent iterations remain at the same value. Hence, $y_2 = \underline{0.98641}$.

The given ODE is a first-order linear equation. When integrated by standard method, its solution is

$$y = \frac{1}{2} x^4$$

For $x = 1.1$, y_1(exact) = 0.73205 and at $x = 1.2$, y_2(exact) = 1.03680. This shows discrepancy at the second decimal place itself. □

Example 2. Solve Example 1 by the second- and fourth-order Runge–Kutta methods.

Solution. As before $x_0 = 1$, $y_0 = 1/2$ and $h = 0.1$

<u>Second-Order Runge–Kutta</u>. At $x = x_1 = 1.1$

$$k_1 = h f(x_0, y_0) = 0.1 \times \left[\frac{2 \times 0.5}{1} + 1^3 \right] = 0.2$$

$$k_2 = h f(x_0 + h, y_0 + k_1) = 0.1 \times \left[\frac{2 \times (0.5 + 0.2)}{1.1} + 1.1^3 \right] = 0.26037$$

So $y_1 = y_0 + \frac{1}{2}(k_1 + k_2) = 0.5 + \frac{1}{2} \times (0.2 + 0.26037) = \underline{0.73019}$.
At $x = x_2 = 1.2$

$$k_1 = h f(x_1, y_1) = 0.1 \times \left[\frac{2 \times 0.73019}{1.1} + 1.1^3 \right] = 0.26586$$

$$k_2 = h f(x_1 + h, y_1 + k_1) = 0.1 \times \left[\frac{2 \times (0.73019 + 0.26586)}{1.1} + 1.1^3 \right] = 0.33881$$

Hence, $y_2 = y_1 + \frac{1}{2}(k_1 + k_2) = 0.73019 + \frac{1}{2}(0.26586 + 0.33881) = \underline{1.03252}$.
<u>Fourth-Order Runge–Kutta</u>. At $x = x_1 = 1.1$

$$k_1 = h\,f(x_0,\ y_0) = 0.1 \times \left[\frac{2 \times 0.5}{1} + 1^3\right] = 0.2$$

$$k_2 = h\,f\left(x_0 + \frac{h}{2},\ y_0 + \frac{k_1}{2}\right) = 0.1 \times \left[\frac{2 \times (0.5 + 0.2/2)}{1 + 0.5} + (1 + 0.05)^3\right] = 0.23005$$

$$k_3 = h\,f\left(x_0 + \frac{h}{2},\ y_0 + \frac{k_2}{2}\right) = 0.1 \times \left[\frac{2 \times (0.5 + 0.23005/2)}{1 + 0.05} + (1 + 0.05)^3\right] = 0.23291$$

$$k_4 = h\,f(x_0 + h,\ y_0 + k_3) = 0.1 \times \left[\frac{2 \times (0.5 + 0.23291)}{1 + 0.1} + (1 + 0.1)^3\right] = 0.26636$$

Hence, $\qquad\qquad y_1 = y_0 + \dfrac{1}{6}(k_1 + 2k_2 + 2k_3 + k_4) = \underline{0.73205}$

At $x = x_2 = 1.2$

$$k_1 = h\,f(x_1,\ y_1) = 0.1 \times \left[\frac{2 \times 0.73205}{1.1} + 1.1^3\right] = 0.26620$$

$$k_2 = h\,f\left(x_1 + \frac{h}{2},\ y_1 + \frac{k_1}{2}\right) = 0.1 \times \left[\frac{2 \times (0.732205 + 0.26620/2)}{1.15} + 1.15^3\right] = 0.30255$$

$$k_3 = h\,f\left(x_1 + \frac{h}{2},\ y_1 + \frac{k_2}{2}\right) = 0.1 \times \left[\frac{2 \times (0.73205 + 0.30255/2)}{1.15} + 1.15^3\right] = 0.30571$$

$$k_4 = h\,f(x_1 + h,\ y_1 + k_3) = 0.1 \times \left[\frac{2 \times (0.73205 + 0.30571)}{1.2} + 1.2^3\right] = 0.34576$$

so that $\qquad\qquad y_2 = y_1 + \dfrac{1}{6}(k_1 + 2k_2 + 2k_3 + k_4) = \underline{1.03679}.$

The accuracy of these results, particularly that of the fourth-order formula is noteworthy. In the exercises, we consider only these two methods. □

Exercises

1. Apply second-order Runge–Kutta method to solve:

(i) $\dfrac{dy}{dx} = x + y,\quad y(0) = 1,\quad$ for $x = 0.1$ and 0.2

(ii) $\dfrac{dy}{dx} = xy,\qquad y(0) = 1,\quad$ for $x = 0.1$ and 0.2

(iii) $\dfrac{dy}{dx} = 3x^2 + y,\quad y(0) = 4,\quad$ for $x = 0.2$ and 0.3

taking steps of length $h = 0.1$. Compare the results with the exact solution of the problems.

[(i) 1.11000, 1.24205 (1.11034, 1.24281; $y = 2e^x - x - 1$) (ii) 1.0050, 1.02018 (1.00513, 1.02020; $y = e^{x^2/2}$). (iii) 4.89200, 5.42576 (4.89404, 5.42862; $y = 10e^x - 3x^2 - 6x - 6$)].

2. Apply fourth-order Runge–Kutta method to solve the ODEs:

(i) $\dfrac{dy}{dx} = x^2 + y^2$, $y(0) = 1$ for $x = 0.1$ and 0.2

(ii) $\dfrac{dy}{dx} = -x^2 y^2$, $y(0) = 2$ for $x = 0.1$ and 0.3

(iii) $\dfrac{dy}{dx} = y(x^2 - 1)$, $y(1) = 1$ for $x = 1.25$ and 1.5

(iv) $\dfrac{dy}{dx} = 1 + \sqrt{xy}$, $y(0) = 2$ for $x = 0.1$, 0.2 and 0.3

(v) $\dfrac{dy}{dx} = -\dfrac{y^2}{1+x}$, $y(0) = 1$ for $x = 0.5$ taking $h = 0.1$

[(i) 1.22175, 1,49477 (ii) 1.99867, 1.96464 (iii) 1.07005, 1.33863
(iv) 2.12908, 2.28638, 2.46333 (v) 0.71151].

3. Write a Fortran program for the fourth-order Runge–Kutta method.

4. The temperature θ of a well-stirred liquid by an isothermal heating coil is given by the equation

$$\frac{d\theta}{dt} = K(100 - \theta)$$

where K is a constant for the system. Solve the equation by Runge–Kutta fourth-order method to find θ at $t = 0.5$ and 1.0 s for $K = 2.5$ and initial condition $\theta = 25\,^\circ\mathrm{C}$ at $t = 0$ s.

[76.94 °C, 92.91 °C].

5. The magnetic flux ϕ in the iron core of a current containing a resistance is given by the differential equation

$$\frac{d\phi}{dt} + 1.8\,\phi + 0.01\,\phi^3 = 20, \quad \phi(0) = 0$$

Compute ϕ at $t = 1$ s in steps of $h = 0.2$ sec.

[3.33856, 5.51598, 6.79012, 7.46951, 7.81093].

6. The current in an L–R circuit with variable resistance is given by the differential equation

$$\frac{di}{dt} + (1 + 3i^2)i = e(t)$$

where

$$e(t) = 5t, \quad 0 \le t \le 0.2$$
$$= 1, \quad t > 0.2$$

Given initial condition $i(0) = 0$, find i at time 0.5 s in steps of 0.1 sec.

[0.02419, 0.09358, 0.17902, 0.31820].

6.1.4 Convergence of Single-Step Methods

A single-step method is defined by an equation of the form (6.6) which means that

$$\frac{y_{n+1} - y_n}{h} - \Phi(x_n, y_n; h) = 0 \tag{6.15}$$

Replacing $y'(x_n)$ by $[y(x_n + h) - y(x_n)]/h$ and $f(x_n, y(x_n))$ by $\Phi(x_n, y(x_n); h)$ in the ODE (6.3), the **local truncation error** at x_n is

$$t(x_n, h) := \frac{y(x_n + h) - y(x_n)}{h} - \Phi(x_n, y(x_n); h) \tag{6.16}$$

The **discretisation error** e_n at x_n in replacing $y(x_n)$ by y_n is $e_n = y(x_n) - y_n$.

Theorem 6.2 *In the single-step method (6.6) or (6.15), if*
(i) the method is **consistent** *of order p, i.e. $|t(x_n, h)| \le N h^p$ and*
(ii) Φ satisfies Lipschitz condition

$$|\Phi(x_n, y_n; h) - \Phi(x_n, y(x_n); h)| \le K_\phi |y_n - y(x_n)|$$

for all $x_n \in [x_0, b]$, $K_\phi > 0$, then

$$|e_n| \le \left[\frac{e^{(x_n - x_0)} - 1}{K_\phi} \right] N h^p$$

Proof Equations (6.15) and (6.16) give

$$y_{n+1} - y_n = h\, \Phi(x_n, y_n; h)$$
$$y(x_{n+1}) - y(x_n) = h\, \Phi(x_n, y(x_n); h) + h\, t(x_n, h)$$

On subtraction, we obtain

$$e_{n+1} = e_n + h\left[\Phi(x_n, y(x_n); h) - \Phi(x_n, y_n; h)\right] + h\, t(x_n, h)$$

and hence taking modulus and using conditions (i) and (ii)

$$|e_{n+1}| \leq |e_n| + h\,K_\phi|y(x_n) - y_n| + N\,h^{p+1}$$
$$= (1 + h\,K_\phi)|e_n| + N\,h^{p+1}$$

in which h, $K_\phi \geq 0$. The above is a *difference inequality*. Now consider the corresponding difference equation

$$\xi_{n+1} = (1 + h\,K_\phi)\,\xi_n + N\,h^{p+1}, \quad \xi_n = e_0 \tag{6.17}$$

We prove that $|e_n|$ is majored by ξ_n, i.e. $|e_n| \leq \xi_n$. This is evidently true for $n = 0$ by definition. The proof is completed by induction, i.e. if it is assumed that $|e_n| \leq \xi_n$, then

$$|e_{n+1}| \leq (1 + h\,K_\phi)\,\xi_n + N\,h^{p+1} = \xi_{n+1}$$

The solution of (6.17) consists of a complementary part which is a solution of $\xi_{n+1} = (1 + h\,K_\phi)\,\xi_n$. The solution $\xi_n \propto \rho^n$ gives rise to the characteristic equation $\rho = 1 + h\,K_\phi$ and the complementary solution is

$$\xi_n^C = C\,(1 + h\,K_\phi)^n$$

The inhomogeneous term in (6.17) is a constant and if the particular solution is k, then $k = (1 + h\,K_\phi)\,k + N\,h^{p+1}$, yielding

$$\xi_n^P = k = -\frac{N h^p}{K_\phi}$$

Hence, the complete solution of (6.17) using the initial condition $\xi_0 = e_0 = y(x_0) - y_0 = 0$ is

$$\xi_n = \xi_n^C + \xi_n^P = \frac{N h^p}{K_\phi}\left[(1 + h\,K_\phi)^n - 1\right]$$

Now, for any $x \geq 0$, $1 + x \leq e^x$ and so

$$|e_n| \leq \xi_n \leq \frac{N h^p}{K_\phi}\left(e^{nhK_\phi} - 1\right) = \frac{N h^p}{K_\phi}\left[e^{(x_n - x_0)K_\phi} - 1\right]$$

$\square$

The theorem means that $|e_n| = O(h^p)$ if x_n is fixed and $h \to 0$. The single-step method is thus convergent if $p > 0$, with conditions (i) and (ii) of the theorem holding true.

Remark The above analysis may suggest that consistency and convergence are synonymous, and by reducing h, an arbitrarily small error can be achieved. But one must bear in mind that round-off errors will inevitably creep—in the iteration of

$y_{n+1} = y_n + h\,\Phi(x_n, y_n; h)$. If a round-off error ϵ_n occurs in this iteration, then it can be shown that the result of Theorem 2 will be

$$|e_n| \le \frac{e^{(x_n - x_0)K_\phi} - 1}{K_\phi} \left(N\,h^p + \frac{\epsilon}{h} \right)$$

where $|\epsilon_n| \le \epsilon$. The contribution of ϵ is enormous when $h \to 0$, although $N\,h^p \to 0$. Thus, as in Sect. 1.1 of Chap. 5 on numerical differentiation, there will be an optimum value of h, say $h_{opt} > 0$ below which round-off errors will dominate. Hence, in the computation of y_{n+1}, h should be chosen small but greater than h_{opt}.

6.1.5 Adams–Bashforth–Moulton Predictor–Corrector Method

In contrast to a single-step method, a predictor–corrector method uses the previously determined function values of f at the equally spaced grid points x_n, x_{n-1}, x_{n-2}, $\cdots$ for computation of the solution y_{n+1} at the point x_{n+1}. A predictor formula is used to predict this value and then a corrector formula is used to improve the predicted value. Such methods are therefore **multistep**.

To develop such formulas, consider the Newton's backward difference formula applied to the function $f(x, y(x))$:

$$f(x, y(x)) = f_n + t\,\nabla f_n + \frac{t(t+1)}{2}\,\nabla^2 f_n + \frac{t(t+1)(t+2)}{6}\,\nabla^3 f_n + \cdots$$

where $t = (x - x_n)/h$ and $f_n = f(x_n, y_n)$. If the formula is substituted in

$$y_{n+1} = y_n + \int_{x_n}^{x_{n+1}} f(x, y)\,dx$$

we get

$$y_{n+1}^p = y_n + \int_{x_n}^{x_{n+1}} [f_n + t\,\nabla f_n + \frac{t(t+1)}{2}\,\nabla^2 f_n + \frac{t(t+1)(t+2)}{6}\,\nabla^3 f_n + \cdots]\,dx$$

$$= y_n + h \int_0^1 [f_n + t\,\nabla f_n + \frac{t(t+1)}{2}\,\nabla^2 f_n + \frac{t(t+1)(t+2)}{6}\,\nabla^3 f_n + \cdots]\,dt$$

$$= y_n + h \left[f_n + \frac{1}{2}\nabla f_n + \frac{5}{12}\nabla^2 f_n + \frac{3}{8}\nabla^3 f_n + \cdots \right] \qquad (6.18)$$

where the superscript p indicates that it is a predicted value. Formula (6.18) is called **Adams- Bashforth formula**.

A corrector formula is similarly obtained by using Newton's backward difference formula at f_{n+1}:

$$y_{n+1}^c = y_n + \int_{x_n}^{x_{n+1}} [f_{n+1} + t \nabla f_{n+1} + \frac{t(t+1)}{2} \nabla^2 f_{n+1}$$

$$+ \frac{t(t+1)(t+2)}{6} \nabla^3 f_{n+1} + \cdots] dx$$

$$= y_n + h \int_{-1}^{0} [f_{n+1} + t \nabla f_{n+1} + \frac{t(t+1)}{2} \nabla^2 f_{n+1}$$

$$+ \frac{t(t+1)(t+2)}{6} \nabla^3 f_{n+1} + \cdots] dt$$

$$= y_n + h \left[f_{n+1}^p - \frac{1}{2} \nabla f_{n+1}^p - \frac{1}{12} \nabla^2 f_{n+1}^p - \frac{1}{24} \nabla^3 f_{n+1}^p - \cdots \right] \quad (6.19)$$

The right-hand side of Eq. (6.19) contains y_{n+1} that can be replaced by the predicted value y_{n+1}^p given by Eq. (6.18). It supplies the corrector formula which is indicated by the superscript c. If wanted the formula can also be iterated by using the previously obtained corrected value in the right-hand side of Eq. (6.19). Formula (6.19) is known as **Adams–Moulton formula**.

In practice, it is convenient to use these formulas by ignoring the dotted terms and expressing the backward differences in terms of function values. For this purpose, we have

$$\nabla f_n = f_n - f_{n-1}, \quad \nabla^2 f_n = f_n - 2f_{n-1} + f_{n-2}$$

$$\nabla^3 f_n = f_n - 3f_{n-1} + 3f_{n-2} - f_{n-3}$$

and similarly for those of ∇f_{n+1}, $\nabla^2 f_{n+1}$ and $\nabla^3 f_{n+2}$. Inserting these expressions in Eqs. (6.18) and (6.19), the formulas become

$$y_{n+1}^p = y_n + \frac{h}{24} [55 f_n - 59 f_{n-1} + 37 f_{n-2} - 9 f_{n-3}] \quad (6.20)$$

and

$$y_{n+1}^c = y_n + \frac{h}{24} [9 f_{n+1}^p + 19 f_n - 5 f_{n-1} + f_{n-2}] \quad (6.21)$$

By iteration, the pair (6.20) and (6.21) yields y_{n+1} at the grid point x_{n+1} next to x_n provided three starting values y_1, y_2, y_3 are known at x_1, x_2, x_3; x_0, y_0 being already given by the initial condition. For this purpose, Runge–Kutta method may be employed to start the solution, replaced by the predictor–corrector iteration at subsequent grid points. The latter iteration is advantageous from the point of view of just two function evaluations of f, viz. f_n and f_{n+1}^p, since the rest are computed in the preceding stages of iteration.

The discretisation error of formulas (6.20) and (6.21) are, respectively,

$$\frac{251}{720}\,h^5\,f_n^{(iv)} \quad \text{and} \quad -\frac{19}{720}\,h^5\,f_n^{(iv)}$$

These errors are of the same order as that in the fourth-order Runge–Kutta method.

The following subroutine implements the scheme:

```
SUBROUTINE ABM(deriv,x0,y0,xlast,nsteps)
! Implements the Adams–Bashforth–Moulton method,
! started by Runge–Kutta Fourth Order Method.
! deriv=function f of Eq. (3); must be declared external
!            in the calling program.      (Input)
! x0, y0=initial point of Eq. (3).    (Input)
! xlast=abscissa of the last point of the entire grid.    (Input)
! nsteps=number of steps in the grid.    (Input)
! The output is generated by the routine as printed value of xn and yn.
!***********************************************************************
REAL :: f(4), k1, k2, k3, k4
f(1)=deriv(x0,y0); xn=x0; yn=y0; h=(xlast−x0)/nsteps
DO n=1,3
k1=h*f(n); k2=h*deriv(xn+h/2,yn+k1/2)
k3=h*deriv(xn+h/2,yn+k2/2); k4=h*deriv(xn+h,yn+k3)
f(n+1)=yn+1.0/6.0*(k1+2*(k2+k3)+k4)
xn=xn+h; yn=f(n+1); PRINT*, n, xn, yn
END DO
DO n=4,nsteps
ynp=yn+h/24*(55*f(4)−59*f(3)+37*f(2)−9*f(1))
xn=x0+n*h; fnp=deriv(xn,ynp)
yn=yn+h/24*(9*fnp+19*f(4)−5*f(3)+f(2))
f(1)=f(2); f(2)=f(3); f(3)=f(4); f(4)=deriv(xn,yn)
PRINT*, n, xn, yn
END DO
END SUBROUTINE ABM
```

John Couch Adams (1819–1892), British mathematician and astronomer. His famous achievement was predicting the existence and position of the planet Neptune using only mathematics. The calculations were made to explain discrepancies in Uranus's orbit. During the same period, but independently, Leverrier made the same discovery. The planet was actually observed in 1846 to within 10° from Adams' prediction and within 1° from Leverrier's prediction. He was a Professor at the University of Cambridge from 1859 till his death.

Francis Bashforth (1819–1912), British mathematician. His interest in ballistics led him to make a series of experiments upon which our present knowledge of air–resistance is founded. The Bashforth chronograph for recording the velocity of shot was invented by him. He was a Fellow of St. John's

College, Cambridge and for some time he was appointed Professor of applied mathematics to British army officers.

Forest Ray Moulton (1872–1952), U.S. astronomer. He proposed the hypothesis of *planetismals* in the formation of the solar system—that has now fallen out of favour. He discovered additional small satellites around Jupiter. The *Moulton plane* in geometry is named after him.

Exercises

[Use subroutine ABM to solve the problems given below:]
1. Given the ODE's, tabulate y in the indicated range:

(i) $\dfrac{dy}{dx} = xy, \quad y(1) = 1$ in the range $1(0.1)2$

(ii) $\dfrac{dy}{dx} = xy + y^2, \quad y(0) = 1$ in the range $0(0.1)0.8$

(iii) $\dfrac{dy}{dx} = (x + y)e^{-x}, \quad y(-1) = 1$ in the range $-1(0.2) + 1$

(iv) $\dfrac{dy}{dx} = \dfrac{\cos x}{1 + y^2}, \quad y(0) = 0$ in the range $0(0.5)5$

Take nsteps $=100$ per unit.
[(i) (1, 1),   (1.1, 1.11048),   (1.2, 1.24582),   (1.3, 1.41170),   (1.4, 1.61574),   (1.5, 1.86786),   (1.6, 2.18102),   (1.7, 2.57228),   (1.8, 3.06422),   (1.9, 3.68692),   (2, 4.48076);
(ii) (0, 1),   (0.1, 1.11638),   (0.2, 1.27673),   (0.3, 1.50324),   (0.4, 1.83769),   (0.5, 2.36677),   (0.6, 3.30375),   (0.7, 5.35050),   (0.8, 12.96553);
(iii) (−1,1), (−0.8, 1.06825), (−0.6, 1.24626), (−0.4, 1.53489), (−0.2, 1.91591), (0, 2.36420),   (0.2, 2.85267),   (0.4, 3.35614),   (0.6, 3.85365),   (0.8, 4.32944),   (1, 4.77280);
(iv) (0, 0), (0.5, 0.44242), (1, 0.71267), (1.5, 0.81433), (2, 0.75672), (2.5, 0.53833), (3, 0.13221), (3.5, −0.34521),   (4, −0.66633),   (4.5, −0.80911),   (5, −0.79780)].

6.1.6 Milne's Predictor–Corrector Method

In this method, Newton's forward difference formula is applied to the function $f(x, y(x))$ in the form

$$f(x, y(x)) = f_n + t \, \Delta f_n + \frac{t(t-1)}{2} \Delta^2 f_n + \frac{t(t-1)(t-2)}{6} \Delta^3 f_n + \cdots$$

where $t = (x - x_n)/h$ and $f_n = f(x_n, y_n)$. Substituting the expression with n replaced by $n - 3$ in the relation

$$y_{n+1} = y_{n-3} + \int_{x_{n-3}}^{x_{n+1}} f(x, y) \, dx$$

we obtain

$$
\begin{aligned}
y_{n+1}^p &= y_{n-3} + \int_{x_{n-3}}^{x_{n+1}} \left[f_{n-3} + t \, \Delta f_{n-3} + \frac{t(t-1)}{2} \Delta^2 f_{n-3} \right. \\
&\qquad \left. + \frac{t(t-1)(t-2)}{6} \Delta^3 f_{n-3} + \cdots \right] dx \\
&= y_{n-3} + h \int_0^4 \left[f_{n-3} + t \, \Delta f_{n-3} + \frac{t(t-1)}{2} \Delta^2 f_{n-3} \right. \\
&\qquad \left. + \frac{t(t-1)(t-2)}{6} \Delta^3 f_{n-3} + \cdots \right] dt \\
&= y_{n-3} + h \left[4 f_{n-3} + 8 \Delta f_{n-3} + \frac{20}{3} \Delta^2 f_{n-3} + \frac{8}{3} \Delta^3 f_{n-3} + \cdots \right] \\
&\approx y_{n-3} + \frac{4h}{3} [2 f_n - f_{n-1} + 2 f_{n-2}]
\end{aligned}
\tag{6.22}
$$

after neglecting the remainder in the forward difference formula. Formula (6.22) can be used to predict the value of y_{n+1} when y_{n-3}, y_{n-2}, y_{n-1} and y_n are known. This is indicated by superscript p in the left-hand side of (6.22). To obtain a *corrector* formula, we replace n by $n - 1$ in the forward difference formula and obtain

$$
\begin{aligned}
y_{n+1}^c &= y_{n-1} + \int_{x_{n-1}}^{x_{n+1}} f(x, y) \, dx \\
&= y_{n-1} + \int_{x_{n-1}}^{x_{n+1}} \left[f_{n-1} + t \, \Delta f_{n-1} + \frac{t(t-1)}{2} \Delta^2 f_{n-1} + \cdots \right] dx \\
&= y_{n-1} + h \int_0^2 \left[f_{n-1} + t \, \Delta f_{n-1} + \frac{t(t-1)}{2} \Delta^2 f_{n-1} + \cdots \right] dt \\
&= y_{n-1} + h \left[2 f_{n-1} + 2 \Delta f_{n-1} + \frac{1}{3} \Delta^2 f_{n-1} + \cdots \right] \\
&\approx y_{n-1} + \frac{h}{3} [f_{n+1}^p + 4 f_n + f_{n-1}]
\end{aligned}
\tag{6.23}
$$

where f_{n+1} is replaced by f_{n+1}^p obtained from Eq. (6.22). Formula (6.23) can be used as a corrector for y_{n+1}, indicated by superscript c. It can be even iterated by replacing f_{n+1}^p on the right-hand side by f_{n+1}^c obtained from a previous iteration. The corrector formula (6.23), it may be noted, is simply the Simpson's rule.

The discretisation error committed in (6.22) and (6.23) can be shown to be

$$\frac{28}{90} h^5 f_{n-3}^{(iv)} \quad \text{and} \quad -\frac{1}{90} h^5 f_{n-1}^{(iv)}$$

respectively.

The formulas of this method, viz. (6.22) and (6.23) are somewhat simpler than those in Adams–Bashforth–Moulton method. However, the method has the serious drawback that it can be *unstable* in some cases (see Sect. 6.1.7).

Edward Arthur Milne (1896–1950), British astrophysicist and applied mathematician. He is well known for his work on radiative equilibrium and mathematical theory of stellar structures. He held a Chair at Oxford University.

Exercises

1. Apply Milne's predictor–corrector method to the following problems:

(i) $\dfrac{dy}{dx} = \dfrac{1}{x+y}$, $y(0) = 2$, $y(0.2) = 2.0933$, $y(0.4) = 2.1755$,
$y(0.6) = 2.2493$

Calculate $y(0.8)$ and $y(1.0)$.

(ii) $\dfrac{dy}{dx} = \dfrac{1}{2} xy$, $y(0) = 1$, $y(0.1) = 1.0025$, $y(0.2) = 1.0101$,
$y(0.3) = 1.0228$

Calculate $y(0.4)$ and $y(0.5)$.

(iii) $\dfrac{dy}{dx} = \dfrac{1-xy}{x^2}$, $y(1) = 1$, $y(1.1) = 0.996$, $y(1.2) = 0.986$,
$y(1.3) = 0.972$

Calculate $y(1.4)$ and $y(1.5)$.

[(i) 2.3164, 2.4219 (ii) 1.0408, 1.0645 (iii) 0.955, 0.938].

6.1.7 Stability of the Methods

In the convergence of a method, we are concerned with the relationship between $y(x_n)$ and y_n as the grid size h tends to zero. A method however delivers a *difference equation* in y_n and the question arises as to whether y_n tends to $y(x_n)$ when n tends to ∞ as well. Historically, it was discovered that application of a formula like Milne's, sometimes led to errors unacceptably increasing (instead of decreasing) with increasing values of n. This led to the analysis of the difference equation of the method. If it can be proved that $y(x_n) \to y_n$ as $n \to \infty$, then we say that the method is asymptotically stable (A–stable).

Let us examine Milne's corrector formula

$$y_{n+1} = y_{n-1} + \frac{h}{3} \left(f_{n+1} + 4f_n + f_{n-1} \right) \tag{6.24}$$

The solution of this difference equation very much depends on f and for further progress, it is necessary to assume some form of f which will throw light on stability of the method. The original ODE is $y' = f(x, y)$ with initial condition $y = y_0$ at $x = x_0$. Intuitively, if we confine our attention to a neighbourhood of (x_0, y_0), we can expand $f(x, y)$ and write

$$y' \approx f(x_0, y_0) + (x - x_0) f_x(x_0, y_0) + (y - y_0) f_y(x_0, y_0)$$
$$=: \lambda(y - y_0) + g(x)$$

where $\lambda := f_y(x_0, y_0)$ and $g(x) := f(x_0, y_0) + f_x(x_0, y_0)(x - x_0)$. Introducing $v := (y - y_0)$, we get the simplified ODE

$$v' = \lambda v + g(x)$$

The inhomogeneous term will not contribute anything to the stability analysis as we are concerned with differences of solutions. Hence, for testing for stability of an equation like (6.24), we consider $f(x, y) = \lambda y$ or the ODE as

$$y' = \lambda y \quad \text{with} \quad y = y_0 \text{ at } x = x_0 \tag{6.25}$$

The solution of this equation is evidently $y = y_0 \, e^{\lambda(x - x_0)}$.

Thus, for Eq. (6.24), we take $f(x, y) = \lambda y$ and obtain

$$y_{n+1} - y_{n-1} - \frac{\lambda h}{3} \left(y_{n+1} + 4y_n + y_{n-1} \right) = 0$$

or

$$\left(1 - \frac{\lambda h}{3} \right) y_{n+1} - \frac{4}{3} \lambda h \, y_n - \left(1 + \frac{\lambda h}{3} \right) y_{n-1} = 0 \tag{6.26}$$

which is a *second-order liner difference equation*. Assuming $y_n \propto \rho^n$, we obtain the characteristic equation

$$\left(1 - \frac{\lambda h}{3} \right) \rho^2 - \frac{4}{3} \lambda h \, \rho - \left(1 + \frac{\lambda h}{3} \right) = 0$$

whose roots are

$$\rho = \frac{\frac{4}{3} \lambda h \pm \sqrt{\frac{16}{9} \lambda^2 h^2 + 4(1 + \frac{\lambda h}{3})}}{2(1 - \frac{\lambda h}{3})}$$

that is

$$\rho_1 = 1 + \lambda h + O(h^2), \quad \rho_2 = -1 + \frac{1}{3}\lambda h$$

Thus, the solution of Eq. (6.26) is approximately

$$y_1 = C_1(1 + \lambda h)^n + C_2\left(-1 + \frac{1}{3}\lambda h\right)^n$$

Taking $x_0 = 0, \; h = x_n/n$, we then obtain

$$y_1 = C_1\left(1 + \frac{\lambda x_n}{n}\right)^n + (-1)^n C_n\left(1 - \frac{1}{3}\frac{\lambda x_n}{n}\right)^n$$

$$\rightarrow C_1 e^{\lambda x_n} + (-1)^n C_2 e^{-\lambda x_n/3} \tag{6.27}$$

In the solution (6.27), the first term is consistent with the solution of (6.25). This shows that increase in the order of the difference equation has led to the induction of a spurious term represented by C_2. If $\lambda > 0$ this term will be exponentially decreasing, implying stability. But if $\lambda < 0$ the term will increase exponentially indicating instability. Methods of this type are conditionally stable and are called **weakly stable**.

Let us now similarly examine the Adams–Moulton corrector formula for stability. The formula (6.21) is

$$y_{n+1} = y_n + \frac{h}{24}[9f_{n+1} + 19f_n - 5f_{n-1} + f_{n-2}]$$

Setting $f(x, y) = \lambda y$, we obtain the difference equation

$$y_{n+1} - y_n - \frac{\lambda h}{24}[9y_{n+1} + 19y_n - 5y_{n-1} + y_{n-2}] = 0$$

The characteristic equation setting $y_n \propto \rho^n$ is

$$\rho^2(\rho - 1) - \frac{\lambda h}{24}(9\rho^3 + 19\rho^2 - 5\rho + 1) = 0$$

which is a cubic equation with three roots $\rho_1 = 1, \; \rho_2 = \rho_3 = 0$ when $h \rightarrow 0$. Solving the equation by iteration for small h,

$$\rho_1 \approx 1 + \frac{\lambda h}{24 \cdot 1^2} \cdot (9 + 19 - 5 + 1) = 1 + \lambda h$$

$$\rho_2, \; \rho_3 \approx \pm\sqrt{\frac{\lambda h}{24 \cdot (-1)} \cdot (9 \cdot 0 + 19 \cdot 0 - 5 \cdot 0 + 1)} = \pm i\sqrt{\frac{\lambda h}{24}}$$

Thus,

$$y_n = C_1(1 + \lambda h)^n + C_2\left(i\sqrt{\frac{\lambda h}{24}}\right)^n + C_3\left(-i\sqrt{\frac{\lambda h}{24}}\right)^n$$

$$= C_1\left(1 + \frac{\lambda x_n}{n}\right)^n + C_2\left(i\sqrt{\frac{\lambda x_n}{24n}}\right)^n + C_3\left(-i\sqrt{\frac{\lambda x_n}{24n}}\right)^n$$

$$\to C_1 e^{\lambda x_n} + C_2 \cdot 0 + C_3 \cdot 0 = C_1 e^{\lambda x_n} \quad \text{as} \quad n \to \infty$$

In the above relation x_0 is taken zero so that $h = x_n/n$. Thus, y_n tends to the correct solution $y_0 e^{\lambda x_n}$ as $n \to \infty$ and the method is unconditionally **strongly stable**. In general, if $\rho_1 = 1$, $|\rho_2|, |\rho_3|, \cdots, |\rho_k| < 1$ as $h \to 0$, then the solution $y_n = C_1\rho_1^n + C_2\rho_2^n + \cdots + C_k\rho^k$ is said to be **strongly stable**. The solution for the Adams–Moulton method has this very property.

An analysis of the Adams–Bashforth predictor formula (6.20) shows that this counterpart is also strongly stable, but the Milne predictor formula (6.22) is weakly stable. What about the stability of the single-step Euler and Runge–Kutta methods? In the first place, the difference equation for these methods are of first order and no extraneous solutions creep in. As regards stability, these methods can be easily proved to be strongly stable.

In the above conclusions, it should be noted that h must tend to zero. If this condition is violated and h becomes larger, then all the methods may turn out to be unstable in the end.

6.2 System of ODEs

If a differential equation of order $N > 1$, e.g. Eq. (6.1) is to be solved, then it can be converted into a system of N first-order equations by the substitution

$$y_1 = y, \; y_2 = y', \; y_3 = y'', \cdots, y_N = y^{(N-1)}$$

so that Eq. (6.1) becomes

$$y^{(N)} = f(x, \, y_1, \, y_2, \cdots, \, y_N)$$

this means that we have a system of N differential equations of the first order in terms of variables $y_1, y_2, \cdots, y_N$ that satisfy

$$\begin{aligned} y_1' &= y_2 \\ y_2' &= y_3 \\ &\cdots \quad \cdots \\ y_{N-1}' &= y_N \\ y_N' &= f(x, \, y_1, \, y_2, \cdots, \, y_N) \end{aligned}$$

subject to initial conditions (6.2) at $x = x_0$, that reads

$$y_1 = y_0, \ y_2 = y_0', \ y_3 = y_0'', \cdots, \ y_N = y_0^{(N-1)}$$

We are thus led to the consideration of a system of general form

$$\begin{aligned}
y_1' &= f_1(x, y_1, y_2, \cdots, y_N) \\
y_2' &= f_2(x, y_1, y_2, \cdots, y_N) \\
&\cdots\cdots\cdots\cdots\cdots\cdots \\
y_N' &= f_N(x, y_1, y_2, \cdots, y_N)
\end{aligned} \tag{6.28}$$

subject to initial conditions

$$y_1 = y_{10}, \ y_2 = y_{20}, \cdots, \ y_N = y_{N0} \quad \text{at} \quad x = x_0 \tag{6.29}$$

The form of Eq. (6.28) at once leads to generalisation of single- and multistep methods developed for first-order equations. To be precise, consider the case $N = 2$. Let the equations be

$$\begin{aligned}
y' &= f(x, y, z) \\
z' &= g(x, y, z)
\end{aligned} \tag{6.30}$$

with $y = y_0$, $z = z_0$ at $x = x_0$, then it can be proved that for the system (6.30):

1°. Fourth-Order Runge–Kutta.

Let at the discretised point (x_n, y_n, z_n)

$$\begin{aligned}
k_1 &= hf(x_n, y_n, z_n), & l_1 &= h\,g(x_n, y_n, z_n) \\
k_2 &= hf\left(x_n + \frac{h}{2}, y_n + \frac{k_1}{2}, z_n + \frac{l_1}{2}\right), & l_2 &= h\,g\left(x_n + \frac{h}{2}, y_n + \frac{k_1}{2}, z_n + \frac{l_1}{2}\right) \\
k_3 &= hf\left(x_n + \frac{h}{2}, y_n + \frac{k_2}{2}, z_n + \frac{l_2}{2}\right), & l_3 &= h\,g\left(x_n + \frac{h}{2}, y_n + \frac{k_2}{2}, z_n + \frac{l_2}{2}\right) \\
k_4 &= hf(x_n + h, y_n + k_3, z_n + l_3), & l_4 &= h\,g(x_n + h, y_n + k_3, z_n + l_3)
\end{aligned}$$

then

$$y_{n+1} = y_n + \frac{1}{6}(k_1 + 2k_2 + 2k_3 + k_4)$$

$$z_{n+1} = z_n + \frac{1}{6}(l_1 + 2l_2 + 2l_3 + l_4), \quad n = 0, 1, 2, \cdots \tag{6.31}$$

2°. Adams–Bashforth–Moulton Predictor–Corrector. For the system (6.30), the formulas (6.20) and (6.21) generalise into

$$y_{n+1}^p = y_n + \frac{h}{24}[55 f_n - 59 f_{n-1} + 37 f_{n-2} - 9 f_{n-3}]$$

$$z_{n+1}^p = z_n + \frac{h}{24}[55 g_n - 59 g_{n-1} + 37 g_{n-2} - 9 g_{n-3}]$$

$$y_{n+1}^c = y_n + \frac{h}{24}[9 f_{n+1}^p + 19 f_n - 5 f_{n-1} + f_{n-2}]$$

$$z_{n+1}^c = z_n + \frac{h}{24}[9 g_{n+1}^p + 19 g_n - 5 g_{n-1} + g_{n-2}], \quad n = 0, 1, 2, \cdots \qquad (6.32)$$

where f_n and g_n stand for $f(x_n, y_n, z_n)$ and $g(x_n, y_n, z_n)$, respectively. Similarly, f_{n+1}^p, g_{n+1}^p, respectively, stand for $f(x_{n+1}, y_{n+1}^p, z_{n+1}^p)$ and $g(x_{n+1}, y_{n+1}^p, z_{n+1}^p)$.

With some effort, schemes such as (6.31) and (6.32) can be coded for general-purpose computation of a system of N ODEs. The subroutine RK4 listed below implements **1°** in the general case. It uses another subroutine named 'derivatives' which computes the components of f.

```
SUBROUTINE derivatives(n,x,y,f)
! n=number of equations.    (Input)
! x= value of independent variable.    (Input)
! y=value of n–vector dependent variable.    (Input)
! f=right hand side n–vector function.    (Output)
!********************************************
REAL :: y(n), f(n)
f(1)=·············   ! First function on the R.H.S.
f(2)=·············   ! Second function on the R.H.S.
..............................................
f(n)=·············   ! nth function on the R.H.S.
RETURN
END SUBROUTINE derivatives
!********************************************
SUBROUTINE RK4(n,x,h,y)
! n=Number of equations.    (Input)
! x=Value of independent variable.    (Input)
!    It returns the value at the next grid point.    (Output)
! h=Step length.    (Input)
! y=Computed value of n–vector dependent variable
!       at the next grid point.    (Output)
!********************************************
real :: y(n), f(n), k1(n), k2(n), k3(n), k4(n), yt(n)
CALL derivatives(n,x,y,f)
```

```
DO i=1,n
k1(i)=h*f(i); yt(i)=y(i)+0.5*k1(i)
END DO
xh=x+0.5*h
CALL derivatives(n,xh,yt,f)
DO i=1,n
k2(i)=h*f(i); yt(i)=y(i)+0.5*k2(i)
END DO
CALL derivatives(n,xh,yt,f)
DO i=1,n
k3(i)=h*f(i); yt(i)=y(i)+k3(i)
END DO
x=x+h
CALL derivatives(n,x,yt,f)
DO i=1,n
k4(i)=h*f(i)
END DO
DO i=1,n
y(i)=y(i)+(k1(i)+2*k2(i)+2*k3(i)+k4(i))/6.0
END DO
RETURN
END SUBROUTINE RK4
```

Writing a subroutine for 2° is left to the reader.

Example 1. Solve Bessel's equation of order zero:

$$xy'' + y' + xy = 0$$

for initial conditions $y(0) = 1$, $y'(0) = 0$, by Runge–Kutta fourth-order method and calculate $y(0.1)$ and $y(0.2)$. Assume that $y'(x)/x \to -1/2$ as $x \to 0$.

Solution. The ODE is of second order. Inorder to convert it to a pair of first-order ODEs, set $z = y'$. Thus, we have the differential equations

$$y' = z =: f(x, y, z)$$
$$z' = -y - \frac{z}{x} =: g(x, y, z)$$

with initial conditions at $x = 0$, $y = 1$, $z = 0$.

For the point $x = 0.1$, by the formulas of subsection 1°, with $x_0 = 0$, $y_0 = 1$, $z_0 = 0$ (in the limit $z_0/x_0 = -1/2$),

$k_1 = hf(x_0, y_0, z_0) = 0.1 \times f(0, 1, 0) = 0$

$l_1 = h\,g(x_0, y_0, z_0) = 0.1 \times g(0, 1, 0) = 0.1 \times (-1 - 0.5) = -0.15$

$k_2 = hf(x_0 + \dfrac{h}{2}, y_0 + \dfrac{k_1}{2}, z_0 + \dfrac{l_1}{2}) = 0.1 \times \left(-\dfrac{0.15}{2}\right) = -7.5 \times 10^{-3}$

$l_2 = h\,g(x_0 + \dfrac{h}{2}, y_0 + \dfrac{k_1}{2}, z_0 + \dfrac{l_1}{2}) = 0.1 \times \left(-1 - \dfrac{-0.15/2}{0.1/2}\right) = 0.05$

$k_3 = hf(x_0 + \dfrac{h}{2}, y_0 + \dfrac{k_2}{2}, z_0 + \dfrac{l_2}{2}) = 0.1 \times \dfrac{0.05}{2} = 2.5 \times 10^{-3}$

$l_3 = h\,g(x_0 + \dfrac{h}{2}, y_0 + \dfrac{k_2}{2}, z_0 + \dfrac{l_2}{2}) = 0.1 \times \left[-\left(1 + \dfrac{-7.5 \times 10^{-3}}{2}\right) - \dfrac{0.05/2}{0.1/2}\right]$

$$= -0.149625$$

$k_4 = hf(x_0 + h, y_0 + k_3, z_0 + l_3) = 0.1 \times (-0.149625) = -1.49625 \times 10^{-2}$

$l_4 = h\,g(x_0 + h, y_0 + k_3, z_0 + l_3) = 0.1 \times \left[-(1 + 2.5 \times 10^{-3}) - \dfrac{-0.149625}{0.1}\right]$

$$= 4.93750 \times 10^{-2}$$

Hence,

$$y(0.1) = y_0 + \frac{1}{6}(k_1 + 2k_2 + 2k_3 + k)$$

$$= 1 + \frac{1}{6}(0 - 2 \times 7.5 \times 10^{-3} + 2 \times 2.5 \times 10^{-5} - 1.49625 \times 10^{-2})$$

$$= \underline{0.99584}$$

and

$$y'(0.1) = z(0.1) = z_0 + \frac{1}{6}(l_1 + 2l_2 + 2l_3 + l_4)$$

$$= 0 + \frac{1}{6}(-0.15 + 2 \times 0.05 - 2 \times 0.149625 + 4.93750 \times 10^{-3})$$

$$= \underline{-0.04998}$$

For the next point $x = 0.2$, we have $x_1 = 0.1$, $y_1 = 0.99584$, $z_1 = -0.04995$. As before, we obtain

$$k_1 = -4.99792 \times 10^{-3},\ l_1 = -4.96048 \times 10^{-2}$$
$$k_2 = -7.47816 \times 10^{-3},\ l_2 = -4.94797 \times 10^{-2}$$
$$k_3 = -7.47190 \times 10^{-3},\ l_3 = -4.93974 \times 10^{-2}$$
$$k_4 = -9.93765 \times 10^{-3},\ l_4 = -4.91485 \times 10^{-2}$$

and therefore

$$y(0.2) = \underline{0.98837}, \quad y'(0.2) = z(0.2) = \underline{-0.09940} \qquad \square$$

Exercises

1. Solve the following systems of ODE by the fourth-order Runge–Kutta method:

(i) $\quad \dfrac{dy}{dx} = x - z, \quad \dfrac{dz}{dx} = x + y, \quad y(0) = 1, \ z(0) = 0$

(ii) $\quad \dfrac{dy}{dx} = x - z, \quad \dfrac{dz}{dx} = \dfrac{xy}{z}, \quad y(1) = \dfrac{1}{2}, \ z(1) = 1$

(iii) $\quad \dfrac{dy}{dx} = x + y + z, \quad \dfrac{dz}{dx} = x^2 + z^2, \quad y(0) = 1, \ z(0) = -1$

to calculate $y(0.1)$, $z(0.1)$ and $y(0.2)$, $z(0.2)$.

[(i) $y(0.1) = 0.99983$, $z(0.1) = 0.10500$; $y(0.2) = 0.99867$, $z(0.2) = 0.21993$
(ii) $y(1.1) = 0.37093$, $z(1.1) = 1.03628$; $y(1.2) = 0.41890$, $z(1.2) = 1.07914$
(iii) $y(0.1) = 1.01003$, $z(0.1) = -0.90877$; $y(0.2) = 1.04049$,
$z(0.2) = -0.83088$].
2. Calculate $y(0.1)$, $z(0.1)$, $y(0.2)$, $z(0.2)$ and $y(0.3)$, $z(0.3)$ for the ODE system
(you may use subroutine RK4)

$$\frac{dy}{dx} = \sin x - z + 1, \quad \frac{dz}{dx} = \cos x - y, \quad y(0) = 1, \ z(0) = 2$$

[$y(0.1) = 0.90483$, $z(0.1) = 2.04670$; $y(0.2) = 0.81873$, $z(0.2) = 2.01740$;
$y(0.3) = 0.74082$, $z(0.3) = 2.03634$].
3. Use subroutine RK4 to compute $y(x)$ for $x \in [0, \ 0.5]$ in steps of 0.1:

(i) $\quad y'' + xy' + y = 0,$

(ii) $\quad y'' - xy'^2 + y^2 = 0,$

for initial conditions $y(0) = 1$, $y' = 0$.

[(i) $y(0.1) = 0.99501$, $y(0.2) = 0.98020$, $y(0.3) = 0.95600$, $y(0.4) = 0.92312$,
$y(0.5) = 0.88250$
(ii) $y(0.1) = 0.99501$, $y(0.2) = 0.98015$, $y(0.3) = 0.95578$, $y(0.4) = 0.92256$,
$y(0.5) = 0.88140$].
4. Consider a simple ecosystem consisting of deer and tigers. The deer have infinite
food supply, while tigers prey upon them. The mathematical model due to V. Volterra
is the classical predator–prey equation of the form

$$\frac{dr}{dt} = 2r - \alpha rf, \quad \frac{df}{dt} = -f + \alpha rf$$

where r denotes deer and f the tigers. Initially, $r(0) = r_0$, $f(0) = f_0$. Investigate the system for $\alpha = 0.01$, $r_0 = 300$ and $f_0 = 150$ using subroutine RK4 and prove that the system is periodic with time period that is approximately five units of time.

5. An *inverted* pendulum consists of a stiff bar of length l, which is supported at one end by a frictionless pin. The support pin is given a rapid up-and-down motion by an electric motor. A simple application of Newton's second law of motion yields the differential equation

$$\frac{d^2\theta}{dt^2} = \frac{3}{2l}(g - a\omega^2 \sin \omega t) \sin \theta$$

where θ is the deflection from the upward vertical and a, ω are, respectively, the amplitude and circular frequency imparted to the pin. The interesting aspect of the problem is that it is stable for certain values of a and ω corresponding to an inverted configuration. Examine the equation for $l = 10$ in., $a = 2$ in., $\omega = 10$ rad/s and $g = 386.09$ in/s^2, for initial conditions $\theta = 0.1$, $\frac{d\theta}{dt} = 0$ at $t = 0$. Use subroutine RK4, 500 times with step length $h = 0.1$ for integration of the equation, and show that the system is stable for these set of parameters.

6.2.1 Adaptive Step Size Runge–Kutta Methods

Because of stability of the single-step Runge–Kutta methods, one is inclined to apply straight away the fourth-order method with a *fixed* small step size. In a multiple order problem with complicated functions on the right-hand side, such a procedure may prove unacceptably slow. This raises the question of using proper step size in order to economise repeated function evaluation. There is no reason to keep it fixed and it is more appropriate to vary it according to the nature of the solution function. In this subsection, we consider the methodology of adaptive step size control, concluding with an adaptive RK4 that uses the technique developed.

Consider the case of a single ODE in y, e.g. Eq. (6.3). If y_n is the computed solution of $y(x_n)$ at the point x_n, a Runge–Kutta method of order p supplies the solution y_{n+1} at a distance of step length h (at the point $x_{n+1} = x_n + h$), and

$$y_{n+1} = y_n + C_n h^p + O(h^{p+1})$$

where C_n is independent of h, but in general depends on the point x_n and the function $f(x, y)$. Now, suppose the integration is carried out in two steps each of length $h/2$, then we obtain another estimate of the solution y_{n+1}^*, and as before

$$y_{n+1}^* = y_n + C_n \left(\frac{h}{2}\right)^p + O(h^{p+1})$$

Subtracting the second solution from the first, we obtain

$$C_n\left(\frac{h}{2}\right)^p \approx \frac{y_{n+1} - y_{n+1}^*}{2^p - 1} =: \delta_n \tag{6.33}$$

The quantity $|\delta_n|$ provides us a computable error estimate in the second value y_{n+1}^*, and it provides us a means of deciding on the next step length $\bar{h}$.

Let ϵ be the absolute *local error tolerance*, then for step size h, the error is

$$\delta_n^1 := \delta_n h = h\, C_n\left(\frac{h}{2}\right)^p$$

For the next step of size $\bar{h}$, we require that $\bar{\delta}_n \bar{h}$ complies with the tolerance $\epsilon \bar{h}$, i.e. $\bar{\delta}_n \bar{h} \leq \epsilon \bar{h}$, or

$$C_n\left(\frac{\bar{h}}{2}\right)^p \leq \epsilon$$

or

$$\delta_n^1 \left(\frac{\bar{h}}{2}\right)^p \leq \epsilon h \left(\frac{h}{2}\right)^p$$

or

$$\bar{h} \leq h \left(\frac{\epsilon h}{\delta_n^1}\right)^{1/p} = h \left(\frac{\epsilon}{\delta_n}\right)^{1/p} \tag{6.34}$$

Thus, if the preceding step h is successfully computed, complying the error test, the next step should preferably be slightly smaller than $h\,(\epsilon/\delta_n)^{1/p}$, say, $\phi h\,(\epsilon/\delta_n)^{1/p}$, where ϕ is a 'safety factor' usually taken as 0.9 or, in some case, 0.8. This of course is true if the preceeding step is covered by bisection into two substeps. Thus, every step of whatever appropriate size must be covered by two substeps. In the general case of a system of first-order ODEs, δ_n is chosen from among the maximum difference in the components $y(i)$.

The major disadvantage of the above methodology of applying say the fourth-order Runge–Kutta is eight function evaluation of f in covering the two halves of h. This was considered to be too much of an effort on the part of the CPU in large problems. Improvement in this direction was provided by E. Fehlberg in 1970 that resulted in the development of subroutine called RKF45. The method requires six function evaluations per step which provides an automatic error estimate and at the same time produces better accuracy than standard fourth-order method. The price paid is the complications of the formulas. In the following, we give a simple adaptive RK4 subroutine for a system of N first-order equations, based on the technique presented in this subsection using Eqs. (6.33) and (6.34). $p = 4$ for RK4. The subroutine is for single precision arithmetic.

```
SUBROUTINE ADAPTIVE_ RK4(n,x,xlast,y)
! n = Number of equations.    (Input)
! x = Value of independent variable.    (Input)
! xlast = Last value of x up to which the solution is required.    (Input)
! y = Computed value of n-vector dependent variable at adaptively
! determined grid points.    (Output)
!**************************************************************
REAL :: y(n), ystar(n)
IF(x>xlast) THEN ! Initial x must be less than xlast
PRINT*, 'Initial x must be less than xlast'; RETURN
END IF
PRINT*, x, y(1), y(2)    ! , y(3), y(4),······ y(n) and unlock the comment
                         ! if n > 2
h=0.1        ! Take small first step
10 DO i=1,n
ystar(i)=y(i)
END DO
CALL RK4(n,x,h,y)
PRINT*, x, y(1), y(2)    ! , y(3), y(4),······, y(n) and unlock the comment
                         ! if n > 2
CALL RK4(n,x,0.5*h,ystar)
CALL RK4(n,x+0.5*h,0.5*h,ystar)
x=x-0.5*h
deltan=0.0
deltan=MAX(deltan, ABS(y(i)-ystar(i)))
END DO
deltan=deltan/15.0; epsilon=0.000001    ! p = 4 for RK4
phi=0.9         ! phi is a "Safety Factor"
h=phi*h*(epsilon/delta)**0.25
! h=0.1        Unlocking this statement will yield RK4 solution with h=0.1
IF(x>xlast) RETURN
GOTO 10
END SUBROUTINE ADAPTIVE_ RK4
!**************************************************************
! Append SUBROUTINE RK4
! Append SUBROUTINE derivatives
```

In applying such subroutine, it must be kept in view that although effort is made
to reduce the truncation error, it does not automatically guaranty greater accuracy in
all cases.

Example 1. Solve Tchebyschev differential equation of order 2:

$$(1 - x^2)\,y' - x\,y' + 4\,y = 0$$

with initial conditions $y(0) = -1$, $y'(0) = 0$ in the approximate range $0\ (0.1)\ 0.9$ by using ADAPTIVE_RK4. Compare with the exact solution $y = 2x^2 - 1$.

Solution. Setting $y_1 := y$, the second-order equation is equivalent to the pair

$$y_1' = y_2$$
$$y_2' = (x\,y_2 - 4\,y_1)/(1 - x^2)$$

subject to initial conditions $y_1(0) = -1$, $y_2(0) = 0$.

x	y (computed)	y (exact)
0	−1.0	−1.0
0.1	−0.98000	−0.98000
0.20085	−0.91935	−0.91931
0.30018	−0.81979	−0.91978
0.40039	−0.67937	−0.67937
0.50066	−0.49869	−0.49869
0.60005	−0.27990	−0.27989
0.70064	−0.01820	−0.01820
0.80031	0.28097	0.28098
0.90018	0.62063	0.62064

The two solutions clearly agree to four decimal places.   □

Exercises

Apply subroutine ADAPTIVE_RK4 to solve the following IVP's:

1. $y'' + y = 0;\quad y(0) = 0,\ y'(0) = 1$ in the range $[0,\ 2\pi]$.
Compare with the exact solution $y = \sin x$.

2. Solve Bessel's equation of order 0:

$$x\,y'' + y' + x\,y = 0;\quad y(0) = 1,\ y'(0) = 0$$

in the range $[0,\ 1.5]$ assuming that $y'(x)/x \to -1/2$ as $x \to 0$. What is the value of y near $x = 1$?
$[y(1.00235) = 0.76416]$.

3. The non-dimensional hodograph equation of a projectile in a resisting medium is

$$\frac{du}{dx} = -u - \sin\psi/u,\quad \frac{d\psi}{dx} = -\cos\psi/u^2$$

where u = velocity, ψ = inclination of the path to the horizon, and x = arc length traversed. Initially, $u = 1$, $\psi = \pi/4$ at $x = 0$. Compute u and ψ when $x = 10$. At what value of x, the projectile reaches the highest point?
$[u = 0.1,\ \psi = -1.57077.$ Highest point at $x = 0.371]$.

4. The function $y = \sin^2 x$ satisfies two forms of ODE's:

$$(a) \ \ y'' + 4y - 2 = 0 \quad \text{and} \quad (b) \ \ 2yy'' - y'^2 + 4y^2 = 0$$

Given $y(0) = 0$, $y'(0) = 0$, solve (a) and (b) in $[0, 2\pi]$, taking $y'^2/y \to 4$ as $x \to 0$. Which of the two forms (a) or (b) would you choose for applying the subroutine ADAPTIVE_RK4?

[(a) For greater accuracy, Form (b) leads to a singular right-hand side for the pair of first-order ODE's, leading to loss of accuracy].

6.3 Stiff Differential Equations

As soon as one deals with more than one first-order differential equation, the *possibility* of a **stiff** set of equations arises. Stiffness occurs in a problem whenever there are two or more widely differing scales of the independent variable on which the dependent variable changes.

For understanding the phenomenon first consider the example

$$y'' - 9y' - 10y = 0, \quad y(0) = 1, \ y'(0) = -1 \tag{6.35}$$

The general solution of the ODE is

$$y = C_1 e^{-x} + C_2 e^{10x} \tag{6.36}$$

and the exact solution satisfying the initial conditions is $y = e^{-x}$. Now the numerical solution of (6.35) by any of the methods described so far would start off decaying as e^{-x}, but would 'explode' like e^{10x} as x becomes large. The reason is that truncation and round-off errors of the method used would cause, no matter how small the step length h is, the computation to follow the general solution (6.36) viz.

$$y_{\text{comp}} \approx e^{-x} + \epsilon e^{10x}$$

so that as x increases the second term explodes.

A slightly better situation is exemplified by the equation

$$y'' + 1001 y' + 1000 y = 0, \quad y(0) = 1, \ y'(0) = -1 \tag{6.37}$$

whose general solution is

$$y = C_1 e^{-x} + C_2 e^{-1000x} \tag{6.38}$$

whereas the exact solution is $y = e^{-x}$. Due to the presence of the component e^{-1000x} in (6.38), one would naturally take $h << 1/1000$ in a numerical method. Stability requirement of the RK4 method is that $1000h < 2.8$ (see the text by Gear). Hence, RK4 would be successful only if $h < 0.0028$, otherwise the numerical solution would explode. Such small step lengths are unreasonable for computing a function like e^{-x} ! This is the generic disease of stiff equations: we are required to follow the variation in the solution on the shortest length scale to maintain stability of integration, even though requirement of the computed function allows a much larger step size h.

As a sort of remedy, suppose we have a general *linear* N–equation system

$$\mathbf{y}' = -\mathbf{C}\mathbf{y} \qquad (6.39)$$

where $\mathbf{C}$ is a constant *positive definite matrix*. Here, $\mathbf{f}(x, \mathbf{y}) = -\mathbf{C}\mathbf{y}$ and Euler's method for the equation yields

$$\mathbf{y}_{n+1} = \mathbf{y}_n - h\mathbf{C}\mathbf{y}_n = (\mathbf{I} - h\mathbf{C})\mathbf{y}_n \qquad (6.40)$$

Iteration of the explicit scheme (6.40) will converge if the largest eigenvalue of the matrix $\mathbf{I} - h\mathbf{C}$ is less than 1 or, $h < 2/\lambda_{max}$, λ_{max} being the largest eigenvalue, which is a large number for stiff equations.

Let the Euler method be modified to the form

$$\mathbf{y}_{n+1} = \mathbf{y}_n + h\mathbf{f}(x_{n+1}, \mathbf{y}_{n+1}) \qquad (6.41)$$

This process is called *implicit differencing* against *explicit differencing* in Euler's method. Setting $\mathbf{f}(x, \mathbf{y}_{n+1}) = -\mathbf{C}\mathbf{y}_{n+1}$ and transposing, we get

$$\mathbf{y}_{n+1} = (\mathbf{I} + h\mathbf{C})^{-1}\mathbf{y}_n$$

If the eigenvalues of the matrix $\mathbf{C}$ are $\lambda \geq 0$, then the eigenvalues of $(I + h\mathbf{C})^{-1}$ are $(1 + h\lambda)^{-1}$, that are ≤ 1 for all $h > 0$. Thus, the method is stable for all step size h. The penalty we pay is the numerical inversion of $I + h\mathbf{C}$ at each step.

The above trick can be generalised for a general *non-linear* N–system

$$\mathbf{y}' = \mathbf{f}(x, \mathbf{y}) \qquad (6.42)$$

Its implicit differencing of Euler type is given in Eq. (6.41). The equation is non-linear but it can be approximately linearised by Taylor's theorem in the form

$$\mathbf{y}_{n+1} = \mathbf{y}_n + h\left[\mathbf{f}(x_{n+1}, \mathbf{y}_n) + \frac{\partial \mathbf{f}}{\partial \mathbf{y}}\bigg|_{\mathbf{y}_n}(\mathbf{y}_{n+1} - \mathbf{y}_n)\right] \qquad (6.42a)$$

where $\partial \mathbf{f}/\partial \mathbf{y}$ is the matrix of the partial derivatives of $\mathbf{f}$. Equation (6.42a) leads to the iteration

$$\mathbf{y}_{n+1} = \left(\mathbf{I} - h \left.\frac{\partial \mathbf{f}}{\partial \mathbf{y}}\right|_{\mathbf{y}_n}\right)^{-1} \left[\left(\mathbf{I} - h \left.\frac{\partial \mathbf{f}}{\partial \mathbf{y}}\right|_{\mathbf{y}_n}\right)\mathbf{y}_n + h\,\mathbf{f}(x_{n+1}, \mathbf{y}_n)\right] \qquad (6.43)$$

The method (6.43) is not guaranteed to be stable, and is called *semi-implicit*. It however often stabilises as in the case of Eq. (6.39), because of similar nature.

Euler's method, we know is a slow method. An improved method is the implicit trapezoidal, viz.

$$\mathbf{y}_{n+1} = \mathbf{y}_n + \frac{h}{2} \left[\mathbf{f}(x_n, \mathbf{y}_n) + \mathbf{f}(x_{n+1}, \mathbf{y}_{n+1})\right]$$

$$\approx \mathbf{y}_n + \frac{h}{2} \left[\mathbf{f}(x_n, \mathbf{y}_n) + \mathbf{f}(x_{n+1}, \mathbf{y}_n) + \left.\frac{\partial \mathbf{f}}{\partial \mathbf{y}}\right|_{\mathbf{y}_n} (\mathbf{y}_{n+1} - \mathbf{y}_n)\right]$$

or,

$$\mathbf{y}_{n+1} = \left(\mathbf{I} - \frac{h}{2} \left.\frac{\partial \mathbf{f}}{\partial \mathbf{y}}\right|_{\mathbf{y}_n}\right)^{-1} \left[\left(\mathbf{I} - \frac{h}{2} \left.\frac{\partial \mathbf{f}}{\partial \mathbf{y}}\right|_{\mathbf{y}_n}\right)\mathbf{y}_n + \frac{h}{2}\left\{\mathbf{f}(x_n\,\mathbf{y}_n) + \mathbf{f}(x_{n+1}, \mathbf{y}_n)\right\}\right]$$

$$(6.44)$$

It is quite complicated to design higher order implicit methods, especially with a good scheme for automatic step size adjustment. Higher order methods analogous to Runge–Kutta and predictor–corrector methods have been developed and Gear's book gives well-tested routine. For more details see also the text of Stoer and Bulirsch.

6.4 Boundary Value Problems

In the preceding section, we studied the initial value problem of solving ODEs, in which all the initial conditions are prescribed at a single point x_0. In the present section, we consider cases when some of the conditions are prescribed at a second point x_1. For instance, in the case of a second-order ODE in y, or a pair of first-order ODEs in y_1, y_2, the initial conditions $y_1(x_0) = y_{10}$, $y_2(x_0) = y_{20}$ be replaced by $y_1(x_0) = y_{10}$, $y_2(x_1) = y_{21}$. Supposing $x_1 > x_0$, we seek the solution in the interval $[x_0, x_1]$ and the problem thus becomes a **Boundary Value Problem** (BVP) as the conditions are prescribed at the boundary points x_0 and x_1.

The solution of BVPs in general is a more difficult problem as it requires some guesswork and improving the guess in order to obtain reasonably accurate result. In the case of highly non-linear equations, there may even be difficulty in obtaining necessary convergence.

There are essentially three different kinds of method for solving the boundary value problems. These are variously described as **shooting methods, finite differ-ence methods** and **collocation methods**. We proceed to treat these methods in the following subsections. For fuller treatment of the methods, see the texts of Conte and de Boor, Press et al., and Kantorovich and Krylov mentioned in the Bibliography.

6.4.1 Shooting Methods

Let the boundary value problem over the interval $[x_0, x_1]$, be such that *only one condition* is prescribed at x_1, the rest being given at the initial point x_0. To understand the procedure, let there be a a second-order ODE, or a pair of first-order ODEs

$$\frac{dy_1}{dx} = f_1(x, y_1, y_2), \quad \frac{dy_2}{dx} = f_2(x, y_1, y_2), \quad x \in [x_0, x_1] \tag{6.45}$$

with *boundary conditions*

$$y_1(x_0) = y_{10}, \quad y_2(x_1) = y_{21}$$

Here, $y_2(x_0)$ is unknown. If $y_2(x_0)$ were known, we would have numerically solved the initial value problem as in Sect. 6.2. Suppose we can *guess* $y_2(x_0)$ on any consideration such as a physical reason. Let two guesses α_0, α_1 be made about the value of $y_2(x_0)$, then a third guess by linear interpolation is

$$\alpha_2 := \alpha_0 + (\alpha_1 - \alpha_0) \frac{y_{21} - y_2(x_1; \alpha_0)}{y_2(x_1, \alpha_1) - y_2(x_1, \alpha_0)} \tag{6.46}$$

where $y_2(x_1; \alpha_0)$ and $y_2(x_1; \alpha_1)$, respectively, stand for the soution of (6.45) with initial conditions $y_1(x_0) = y_{10}$, $y_2(x_0) = \alpha_0$ or $y_2(x_0) = \alpha_1$. The procedure can be repeated with α_1, α_2 to generate α_3 as in the case of (6.46) and so on till convergence is attained, i.e. $y_2(x_1; \alpha_n) \rightarrow y_{21}$. We thus obtain the following algorithm:

1. Guess two approximations α_0 and α_1 of $y_1'(x_0)$.
2. Solve the IVP $y_1' = f_1(x, y_1, y_2)$, $y_2' = f_2(x, y_1, y_2)$ with $y_1(x_0) = y_{10}$, $y_2(x_0) = \alpha_0$ from x_0 to x_1 using a suitable method. Call the solution $y_2(x_1; \alpha_0)$.
3. Solve the IVP for initial conditions $y_1(x_0) = y_{10}$, $y_2(x_0) = \alpha_1$ from x_0 to x_1 and call the solution $y_2(x_1; \alpha_1)$.
4. Compute: $\quad \alpha_2 \leftarrow \alpha_0 + (\alpha_1 - \alpha_0) \dfrac{y_{21} - y_2(x_1; \alpha_0)}{y_2(x_1; \alpha_1) - y_2(x_1; \alpha_0)}$.
5. If $|\alpha_2 - \alpha_0| < \epsilon$, Stop (The value of $y_2(x_0)$ is α_0).
6. Set $\quad \alpha_0 \leftarrow \alpha_1$, $\alpha_1 \leftarrow \alpha_2$, $\quad$ Go to step 2.

The algorithm yields the initial value $y_2(x_0)$ and the initial value problem of Eq. (6.45) can now be solved for initial conditions $y_1(x_0) = y_{10}$, $y_2(x_0) = y_{20}$ (the computed value of α_k).

It should be borne in mind that convergence is not guaranteed in the method and necessary precaution must be taken in the use of the above algorithm.

Clearly, the method can be generalised to a system of N first-order ODEs when $N - 1$ initial conditions are prescribed at x_0 and one condition at x_1.

The generalisation of the method to the case of two or more conditions given at the end point x_1 is complicated and convergence even more difficult to attain. Press et al. (see Bibliography) have given some subroutines for the general case.

The next example illustrates the simpler case of a pair of first-order ODEs only.

Example 1. Let the ODE be

$$y'' + y = 0, \quad y(0) = 0, \quad y(1) = 1$$

Compute $y'(0)$ by the shooting method.

Solution. The given ODE is equivalent to the pair

$$y' = z, \quad z' = -y, \quad y(0) = 0, \quad y(1) = 1$$

To reduce the problem to an IVP, we require the value of $y'(0) = z(0) = \alpha$ (say). Adopting the algorithm given in the foregoing, let two initial guesses of α be $\alpha_0 = 0.5$ and $\alpha_1 = 1$. The algorithm leads to the following Fortran program, taking step length $h = 0.1$ for the subroutine RK4.

```
REAL :: y(2)
! Append subroutines RK4 and DERIVATIVES to this program
h=.1; x=0.; y(1)=0.; a0=.5; y(2)=a0
DO i=1,10
CALL RK4(2,x,h,y)
END DO
PRINT*, x, a0, y(1)
y1a0=y(1)
x=0.; y(1)=0.; a1=1.; y(2)=a1
DO i=1,10
CALL 1 RK4(2,x,h,y)
END DO
PRINT*, x, a1, y(1)
y1a1=y(1)
DO k=1,10
a2=a0+(a1-a0)*(1.-y1a0)/(y1a1-y1a0)
IF(ABS(a2-a0)<1.e-6) STOP
x=0.; y(1)=0.; y(2)=a2
DO i=1,10
CALL RK4(2,x,h,y)
END DO
PRINT*, x, a2, y(1)
a0=a1; a1=a2; y1a0=y1a1; y1a1=y(1)
END DO
END
```

Appending subroutines RK4 and DERIVATIVES to the above program, the executed result is

k	α_k	$y(1; \alpha_k)$
0	0.5	0.4207352
1	1.0	0.8414705
2	1.188396	1.0
3	1.188396	1.0

The algorithm quickly converges to the value $y'(0) = z(0) = \alpha_3 = 1.188396$. With this value of $z(0)$, RK4 can be used to compute $y(x)$ and $z(x)$ for $0 \le x \le 1$ as a usual IVP. □

Exercises

1. Given the BVP

$$y'' + 0.1\, y' + y = 0, \quad y(0) = 0, \quad y(1) = 1$$

Compute $y'(0)$ by the shooting method.

[1.24877].

2. Solve the equation

$$x^2 y'' - 2y + x = 0, \quad \text{with} \quad y(2) = y(3) = 0$$

to compute $y'(2)$.

[0.21053. Exact solution $y(x) = (19x^2 - 5x^3 - 36)/38x$].

3. For the BVP

$$2yy'' - y'^2 + 4y^2 = 0, \quad y(0) = 0, \quad y\left(\frac{\pi}{2}\right) = 1$$

Compute $y'(0)$.

[0. Exact solution $y = \sin^2 x$].

4. The equation for the common catenary is governed by the BVP

$$yy'' + 1 + y'^2 = 0, \quad y(0) = 1, \quad y(1) = 2$$

Compute $y'(0)$, taking $h = 0.1$. Is the result accurate enough? If not obtain it more accurately.

[2.00033. No. 2.0]

6.4.2   Finite Difference Methods

In a boundary value problem one or more ODEs are required to be solved in an interval $[a, b]$, with conditions specified at the boundary points a and b. We exclusively treat the case of one equation, but the methods can be generalised to the case of two or more equations. In *finite difference methods*, $[a, b]$ is divided into a grid or **mesh** by equally spaced points $a = x_0 < x_1 < \cdots < x_N = b$ where $x_n = x_0 + nh$, $(n = 1, 2, \cdots, N - 1)$ are the **interior mesh points**. The mesh is usually fine, i.e. h is small or N is large. The derivatives occurring in the ODE are next replaced by some finite difference formula, as described in Chap. 5, Sect. 5.1. Central difference quotient formulas are usually preferred on account of their greater accuracy. If $y(x_n)$ is simply denoted by y_n, the central difference quotient formulas for the first, second and fourth derivatives are, respectively,

$$y'(x_n) \approx \frac{y_{n+1} - y_{n-1}}{2h}$$

$$y''(x_n) \approx \frac{y_{n+1} - 2y_n + y_{n-1}}{h^2}$$

$$y^{(iv)}(x_n) \approx \frac{y_{n+2} - 4y_{n+1} + 6y_n - 4y_{n-1} + y_{n-2}}{h^2}$$

the error in each case being $O(h^2)$. Plugging the finite difference expansions for the derivatives in the ODE, one obtains a finite difference equation of certain order. It remains to solve the difference equation satisfying the boundary conditions in an appropriate manner. If the ODE is *linear*, the difference equation will also be linear and comparatively easier to treat. We illustrate the further procedure by treating a second-order *linear* ODE

$$y''(x) + f(x)\, y'(x) + g(x)\, y(x) = r(x) \tag{6.47}$$

in $[a, b]$, subject to boundary conditions $y(a) = \alpha$, $y(b) = \beta$. Substituting the finite difference quotients of the derivatives in (6.47), we obtain the equation

$$\frac{y_{n-1} - 2y_n + y_{n+1}}{h^2} + f(x_n) \frac{y_{n+1} - y_{n-1}}{2h} + g(x_n)\, y_n = r(x_n)$$

Setting $f_n = f(x_n)$, $g_n := (x_n)$, $r_n := r(x_n)$, the above equation reduces to the difference equation

$$\left(1 - \frac{h}{2} f_n\right) y_{n-1} + (-2 + h^2 g_n)\, y_n + \left(1 + \frac{h}{2} f_n\right) y_{n+1} = h^2 r_n, \quad n = 1, 2, \cdots, N - 1 \tag{6.48}$$

Equation (6.48) is evidently a *tridiagonal system*. In fact, setting $n = 1, 2, \cdots,$ $N - 1$ and noting that $y_0 = \alpha$ and $y_N = \beta$, we have the equations

$$(-2 + h^2 g_1) y_1 + \left(1 + \frac{h}{2} f_1\right) y_2 = h^2 r_1 - \left(1 - \frac{h}{2} f_1\right) \alpha$$

$$\left(1 - \frac{h}{2} f_2\right) y_1 + (-2 + h^2 g_2) y_2 + \left(1 + \frac{h}{2} f_2\right) y_3 = h^2 r_2$$

$$\cdots\cdots\cdots\cdots\cdots\cdots\cdots\cdots\cdots\cdots\cdots\cdots\cdots\cdots\cdots \quad (6.49)$$

$$\left(1 - \frac{h}{2} f_{N-2}\right) y_{N-3} + (-2 + h^2 g_{N-2}) y_{N-2} + \left(1 + \frac{h}{2} f_{N-2}\right) y_{N-1} = h^2 r_{N-2}$$

$$\left(1 - \frac{h}{2} f_{N-1}\right) y_{N-2} + (-2 + h^2 g_{N-1}) y_{N-1} = h^2 r_{N-1} - \left(1 + \frac{h}{2} f_{N-1}\right) \beta$$

Evidently, Eq. (6.49) is tridiagonal in $N - 1$ unknowns $y_1, y_2, \cdots, y_{N-1}$ and can be solved numerically by the methods of Sect. 1.1, Chap. 3. If the mesh is fine, we obtain discrete values of the solution $y(x)$ at sufficiently nearby points.

Boundary conditions can sometimes be more complicated then we have assumed in the foregoing. Let the condition at $x = a = x_0$ be

$$y'(x_0) + \gamma y(x_0) = \alpha \quad (6.50)$$

If $y'(x_0)$ is replaced by the finite difference quotient $(y_1 - y_0)/h$. Equation (6.50) reduces to

$$y_1 + (-1 + \gamma h) y_0 = \alpha h$$

This equation yields $y_0 = (y_1 - \alpha h)/(1 - \gamma h)$. The first equation of the set (6.49) for $n = 1$ thus becomes

$$\left(-2 + h^2 g_1 + \frac{1 - \frac{h}{2} f_1}{1 - \gamma h}\right) y_1 + \left(1 + \frac{h}{2} f_1\right) y_2 = h^2 r_1 + \frac{\alpha h (1 - \frac{h}{2} f_1)}{1 - \gamma h} \quad (6.51)$$

The first equation of the tridiagonal system (6.49) must now be replaced by Eq. (6.51), the other equations of the set remaining unchanged.

The accuracy in the above differencing of (6.50) is however $O(h)$. We can make it $O(h^2)$ by considering the central difference quotient of $y'(x_0)$ viz. $[y(x_0 + h) - y(x_0 - h)]/(2h)$ or $(y_1 - y_{-1})/(2h)$ where $y_{-1} := y(x_0 - h)$. Equation (6.50) then yields

$$y_1 - y_{-1} + 2h\gamma y_0 = 2h\alpha$$

or

$$y_{-1} = y_1 + 2h\gamma y_0 - 2h\alpha \quad (6.52)$$

There is now an extra unknown y_0. To take it in to account the system (6.48) is extended to $n = 0$, i.e.

$$\left(1 - \frac{h}{2} f_0\right) y_{-1} + (-2 + h^2 g_0) y_0 + \left(1 + \frac{h}{2} f_0\right) y_1 = h^2 r_0$$

Now, replacing y_{-1} by using Eq. (6.52), we get

$$\left[2h\gamma\left(1 - \frac{h}{2}f_0\right) + (-2 + h^2 g_0)\right] y_0 + 2y_1 = h^2 r_0 + 2h\alpha\left(1 - \frac{h}{2}f_0\right), \quad n = 0$$

Moreover, in this process for $n = 1$, we obtain in (6.48)

$$\left(1 - \frac{h}{2}f_1\right) y_0 + (-2 + h^2 g_1) y_1 + \left(1 + \frac{h}{2}f_1\right) y_2 = hr_1, \quad n = 1$$

The remaining equations for completion of the system will be the same as the last $N - 2$ equations of (6.49).

The accuracy of finite difference methods depends on the order of finite difference approximation of the derivatives and the fineness of the mesh. Higher order difference schemes are preferable unless the system becomes cumbersome to handle. On the other hand, computation with grid sizes h and $h/2$ is preferable for comparison of computed value of $y(x_n)$ at the grid points. Moreover, Richardson extrapolation can be resorted to for better approximation. Let $y_h(x_n)$ denote the computed value at the mesh point x_n, where $h = (b - a)/N$, $n = 1, 2, \cdots, N - 1$. Let $y_{h/2}(x_n)$ denote the computed value at the same point x_n, with $h/2$ mesh size, then

$$y(x_n) \approx \frac{4y_{h/2}(x_n) - y_h(x_n)}{3}$$

will usually yield significant improvement in the approximation.

Experience with finite difference methods is that they provide satisfactory numerical results. This is so because of convergence and stability of the methods (see Volkov's text cited in the Bibliography).

Example 1. Solve the BVP

$$y'' - xy = 0, \quad y(0) + y'(0) = 1, \quad y(1) = 1$$

dividing the interval $[0, 1]$ into a mesh of three subdivisions by the points 0, $1/3$, $2/3$, 1.

Solution. The finite difference equivalent of the given ODE is

$$\frac{y_{n-1} - 2y_n + y_{n+1}}{h^2} - x_n y_n = 0$$

Since $h = 1/3$, we get

$$y_{n-1} - \left(2 + \frac{x_n}{9}\right) y_n + y_{n+1} = 0$$

Setting $n = 0, 1, 2$, we obtain the system ($x_0 = 0$, $x_1 = \frac{1}{3}$, $x_2 = \frac{2}{3}$)

$$y_{-1} - 2y_0 + y_1 = 0$$

$$y_0 - \frac{55}{27} y_1 + y_2 = 0$$

$$y_1 - \frac{56}{27} y_2 + y_3 = 0$$

From the first boundary condition, we obtain

$$y_0 + \frac{y_1 - y_{-1}}{2h} = 1$$

With $h = 1/3$, the above equation yields

$$y_{-1} = \frac{1}{3}(2y_0 + 3y_1 - 2)$$

The second boundary condition evidently yields $y_3 = 1$. Hence, the system of equations reduces to

$$-2y_0 + 3y_1 = 1$$

$$y_0 - \frac{55}{27} y_1 + y_2 = 0$$

$$y_1 - \frac{56}{27} y_2 = -1$$

Solving the equations, we get

$$y_0 = y(0) = -\frac{82}{83} = -0.9880$$

$$y_1 = y\left(\tfrac{1}{3}\right) = -\frac{27}{83} = -0.3253$$

$$y_2 = y\left(\tfrac{2}{3}\right) = \frac{27}{83} = 0.3253 \qquad \square$$

Exercises

1. Solve the BVP

$$y'' + y = 0, \quad y(0) = 0, \quad y(1) = 1$$

taking $h = 1/4$.

$[y_1 = 0.2943, \quad y_2 = 0.5702, \quad y_3 = 0.8104]$.

2. Solve the ODE

$$(1+x^2) \, y'' - y = 1, \quad \text{with} \quad y'(0) = 0, \quad y(1) = 0$$

taking $h = 1/3$.

$[y_0 = -0.3204, \quad y_1 = -0.2826, \quad y_2 = -0.1731]$.

3. Solve the ODE

$$y'' = xy + 1, \quad \text{subject to} \quad y(0) + y'(0) = 1, \quad y(1) = 1$$

taking $h = 1/4$.

$[y_0 = -19.8523, \quad y_1 = -14.6080, \quad y_2 = -9.5294, \quad y_3 = -4.6861]$.

6.4.3 Collocation, Least Squares and Galerkin Methods

Long before the appearance of computers, approximate methods of solving boundary value problems of ODEs (and PDEs) were developed for problems arising in engineering. Beginning with simple ideas, the methods have been turned into powerful computational tools called *finite element methods* that have become fundamental to engineering analysis and design. In this subsection, we describe the basic ideas by considering a second-order *linear* ODE in $[a, b]$ viz.

$$L\,y = y'' + f(x)\,y' + g(x)\,y = r(x) \tag{6.53}$$

with boundary conditions

$$
\begin{aligned}
a_0\,y(a) + a_1\,y'(a) &= \alpha \\
b_0\,y(b) + b_1\,y'(b) &= \beta
\end{aligned}
\tag{6.54}
$$

where $f, g, r \in C[a, b]$, $a_0, \cdots, b_1$, α, β are constants such that $a_0^2 + b_0^2 > 0$. In general, the system may or may not have a unique solution, unlike the Cauchy problem. But it can be proved that a necessary and sufficient condition for the existence of a unique solution is that the corresponding homogeneous equation (with $r(x) \equiv 0$) has only the trivial solution $y(x) \equiv 0$. Such a test is hard to apply and we assume that Eqs. (6.53) and (6.54) do have a unique solution, which is to be approximated.

Let ϕ_0, ϕ_1, ϕ_2, $\cdots$, ϕ_N be a set of linearly independent **basis functions** $\in$ $C^2[a, b]$, where only ϕ_0 satisfies the inhomogeneous boundary conditions (6.54) (i.e. α, $\beta \neq 0$) while ϕ_1, ϕ_2, $\cdots$, ϕ_N satisfy corresponding homogeneous boundary conditions (with α, β set equal to zero). These functions are usually selected as polynomials of low degree or some simple trigonometric polynomials. An approximate solution of (6.53) is then sought as the linear combination

$$y_N(x) = \phi_0(x) + c_1 \phi_1(x) + \cdots + c_N \phi_N(x) \qquad (6.55)$$

$y_N(x)$ exactly satisfies the boundary conditions (6.54), but not (6.53). So, we construct the **discrepancy** or the **residual** defined by

$$\psi(x; c_1, \cdots, c_N) = L\, y_N(x) - r(x) = L\, \phi_0(x) - r(x) + \sum_{k=1}^{N} c_k L\, \phi_k(x) \quad (6.56)$$

The residual ψ equals the difference between the left-hand and the right-hand sides of (6.53) when in the left-hand side y is replaced by the approximation $y_N(x)$. Evidently, ψ depends linearly on the parameters c_1, c_2, $\cdots$, c_N.

For accuracy of the approximation, c_1, c_2, $\cdots$, c_N must be so determined that ψ is minimum over the interval $[a, b]$ in some sense. This approach gives rise to the methods of collocation, least squares and the **method of weighted residuals** due to B. G. Galerkin.

Boris Grigerievich Galerkin (1871–1945) was born in Belarus but was a Professor at St. Petersberg State Polytechnical University, Russia, in the structural mechanics and theory of elasticity department. After graduating from the same institute, he became a member of the Social–democratic party in 1905. He was jailed for more than a year in 1907, but prison conditions gave him an opportunity of studying science and engineering and lost interest in revolutionary activity. After release from jail in 1908, he became a teacher in the same institute and worked on structural mechanics and elasticity. He gave the method which bears his name in the year 1915 and later applied the method to a large number of problems. He became a Professor in the university in the year 1919.

1°. **Collocation Method.** Let x_1, x_2, $\cdots$, x_N be chosen at will in $[a, b]$ and let the residual be set equal to zero at these points. Then, we obtain the system of equations

$$\psi(x_i; c_1, c_2, \cdots, c_N) \equiv \sum_{k=1}^{N} c_k\, L\phi_k(x_i) + L\phi_0(x_i) - r(x_i) = 0, \quad (i = 1, 2, \cdots, N)$$

$$(6.57)$$

The system of equations (6.57) is linear in c_1, c_2, $\cdots$, c_N and can easily be solved. This yields the approximation y_N by collocation; the points x_1, x_2, $\cdots$, x_N being called the **collocation points**.

The method is illustrated in the following example.

Example 1. Solve by collocation method the BVP

$$y'' + y = 0, \quad 0 \le x \le 1$$
$$y(0) = 0, \quad y(1) = 1$$

using three functions.

Solution. The boundary conditions are inhomogeneous and so we take $\phi_0(x) = x$ that satisfies the boundary conditions and $\phi_1(x) = x(1-x)$, $\phi_2(x) = x^2(1-x)$. The latter two functions satisfy homogeneous boundary conditions. Hence, we consider the approximation

$$y_2(x) = x + c_1 x(1-x) + c_2 x^2(1-x)$$

Therefore,

$$y_2''(x) + y_2(x) = -2c_1 + c_2(2-6x) + x + c_1 x(1-x) + c_2 x^2(1-x)$$
$$= x + c_1[-2 + x(1-x)] + c_2[2 - 6x + x^2(1-x)]$$

For hopefully good approximation, we choose equally spaced points $x_1 = 1/3$, $x_2 = 2/3$. For these points, the above expression should vanish in the collocation method. This consideration leads to the equations

$$48 c_1 - 2 c_2 = 9 \quad \text{and} \quad 24 c_1 + 25 c_2 = 9$$

the solution of which is $c_1 = 81/416$, $c_2 = 9/52$. Thus, we obtain

$$y_2(x) = 1.19471 x - 0.02163 x^2 - 0.17308 x^3$$

The exact solution of the problem is $y(x) = \sin x / \sin 1$. The comparative values of $y(x)$ and $y_2(x)$ are tabulated below:

x	y	y_2
0.1	0.1186	0.1156
0.25	0.2940	0.2946
0.50	0.5698	0.5703
0.75	0.8101	0.8766
1	1	1

The values of $y(x)$ and $y_2(x)$ are remarkably matched. For greater accuracy, more collocation basis functions like $\phi_i(x) = x^i(1-x)$ should be included.   □

2°. **Least Squares Method.** In this method, the overall residual over the whole interval $[a, b]$ is minimised in the sense of least square. This means that

$$I = \int_a^b \psi^2(x; c_1, c_2, \cdots, c_N)\, dx$$

is sought to be minimum for certain values of c_i $(i = 1, 2, \cdots, N)$, so that from Eq. (6.56)

$$\frac{\partial I}{\partial c_i} = 2 \int_a^b [L\phi_0(x) - r(x) + \sum_{k=1}^N c_k\, L\phi_k(x)]\, L\phi_i(x)\, dx = 0, \quad i = 1, 2, \cdots, N$$

yielding the *normal equations* for the determination of c_i as

$$\sum_{k=1}^N c_k \int_a^b L\phi_k(x)\, L\phi_i(x)\, dx = \int_a^b [r(x) - L\phi_0(x)]\, L\phi_i(x)\, dx, \quad i = 1, 2, \cdots, N$$

If the set of functions $L\phi_1, \cdots, L\phi_N$ are linearly independent on the interval, the normal equations have a unique solution.

Example 2. Solve the problem of Example 1 by the least squares method.

Solution. Here, we have to minimise the integral

$$I = \int_0^1 [x + c_1(-2 + x(1-x)) + c_2(2 - 6x + x^2(1-x))]^2\, dx$$

Therefore,

$$\frac{\partial I}{\partial c_1} = 2\int_0^1 [x + c_1(-2 + x(1-x)) + c_2(2 - 6x + x^2(1-x))] \times (-2 + x(1-x))\, dx = 0$$

$$\frac{\partial I}{\partial c_2} = 2\int_0^1 [x + c_1(-2 + x(1-x)) + c_2(2 - 6x + x^2(1-x))] \times (2 - 6x + x^2(1-x))\, dx = 0$$

Evaluating the integrals, we obtain the normal equations as

$$2c_1 + c_2 = \frac{55}{101}, \quad \frac{101}{12}c_1 + \frac{131}{7}c_2 = \frac{19}{4}$$

whose solution is $c_1 = 0.18754$ and $c_2 = 0.16947$. So, the approximation of $y(x)$ is

$$y_2(x) = 1.18754\, x - 0.01807\, x^2 - 0.16947\, x^3$$

The following table gives a comparison of the computed values of $y(x) = \sin x / \sin 1$ and $y_2(x)$:

x	y	y_2
0.1	0.1186	0.1184
0.25	0.2940	0.2931
0.50	0.5698	0.5683
0.75	0.8101	0.8090
1	1	1

The two solutions clearly agree to two decimal places. $\square$

The popular Raylrigh–Ritz finite element method for elliptic partial differential equations closely follows this method.

$3°$. **Galerkin Method.** This method is based on the requirement that the basis functions $\phi_1, \phi_2, \cdots, \phi_N$ be orthogonal to (6.56), that is,

$$\int_a^b \phi_i(x)\, \psi(x; c_1, c_2, \cdots, c_N)\, dx = 0, \quad i = 1, 2, \cdots, N$$

This requirement leads to the following system of linear algebraic equations for the coefficients $c_1, c_2, \cdots, c_N$:

$$\sum_{k=1}^{N} c_k \int_a^b \phi_i(x)\, L\phi_k(x)\, dx = \int_a^b \phi_i(x)\, [r(x) - L\phi_0(x)]\, dx \tag{6.58}$$

Solving the linear system (6.58), we obtain the approximate solution in the form (6.55).

The method is also called the *Method of Weighted Residuals*.

Example 3. Solve the boundary value problem

$$y'' + y = -x, \quad 0 \le x \le 1, \quad y(0) = 0, \quad y(1) = 0$$

by the Galerkin method.

Solution. Since the boundary conditions are homogeneous we choose

$$\phi_0(x) = 0, \quad \phi_i(x) = x^i(1 - x), \quad i = 1, 2, \cdots$$

Let $\underline{N = 1}$. Then, the approximation is $y_1(x) = c_1 \phi_1(x) = c_1 x(1 - x)$ and

$$y_1''(x) + y_1(x) + x = c_1 (-2 + x(1 - x)) + x$$

By Galerkin method, the above expression is orthogonal with respect to $\phi_1(x)$ or

$$\int_0^1 x(1-x)[c_1(-2+x(1-x))+x]\,dx = 0$$

or

$$c_1 \int_0^1 x(1-x)(-2+x(1-x))\,dx = -\int_0^1 x^2(1-x)\,dx$$

Evaluating the integrals, we obtain $c_1 = 5/18$ so that

$$y_1(x) = \frac{5}{18}x(1-x)$$

Let $\underline{N=2}$. Then, the approximation is $y_2(x) = c_1\phi_1(x) + c_2\phi_2(x) = c_1 x(1-x) + c_2 x^2(1-x)$. This yields

$$y_2''(x) + y_2(x) + x = c_1(-2+x(1-x)) + c_2(2-6x+x^2(1-x)) + x$$

The orthogonality relations by Galerkin method yield

$$c_1\int_0^1 x(1-x)(-2+x(1-x))\,dx + c_2\int_0^1 x(1-x)(2-6x+x^2(1-x))\,dx = -\int_0^1 x^2(1-x)\,dx$$

$$c_1'\int_0^1 x^2(1-x)(-2+x(1-x))\,dx + c_2\int_0^1 x^2(1-x)(2-6x+x^2(1-x))\,dx = -\int_0^1 x^3(1-x)\,dx$$

Evaluating the integrals, the equations become

$$\frac{3}{10}c_1 + \frac{3}{20}c_2 = \frac{1}{12}, \quad \frac{3}{20}c_1 + \frac{13}{105}c_2 = \frac{1}{20}$$

whose solution is $c_1 = 71/369$ and $c_2 = 7/41$. Thus, the sought approximation is

$$y_2(x) = x(1-x)\left(\frac{71}{369} + \frac{7}{41}x\right)$$

The exact solution is $y(x) = \sin x/\sin 1 - x$. Comparison data for the solutions is given below:

x	y	y_1	y_2
0.25	0.0440	0.0521	0.0441
0.50	0.0697	0.0694	0.0694
0.75	0.0601	0.0521	0.0601

The accuracy of $y_2(x)$ is noteworthy.

Exercises

1. Solve by collocation method the BVP

$$y'' - y = x, \quad y(0) = 0, \quad y(1) = 1$$

using three functions. Compare with the exact solution $y(x) = \dfrac{\sinh x}{\sinh 1} - x$
$[y_2(x) = x + c_1 x(1-x) + c_2 x^2(1-x)$. Collocating at $x = 1/3$ and $2/3$, $c_1 = -279/840$, $c_2 = 27/28]$.

2. Solve by collocation

$$y'' + (1+x^2)\, y = -1, \quad y(-1) = 0 = y(1)$$

using approximation of the form $y_2(x) = c_1 (1-x^2) + c_2 x^2(1-x^2)$.

[Collocating at $x = 0, \pm 1/2$, $y_2(x) = 0.957(1-x^2) - 0.022\, x^2(1-x^2)]$.

3. Solve Exercise 2 by least square method.

$[y_2(x) = 0.985\,(1-x^2) - 0.078\, x^2(1-x^2)]$.

4. Solve Exercise 1 by Galerkin's method of weighted residuals.

$[c_1 = -1062/473, \quad c_2 = 18634/5203]$.

5. Integrate Bessel's equation of order 1

$$x^2 y'' + x y' + (x^2 - 1)\, y = 0, \quad y(1) = 1, \quad y(2) = 2$$

using approximation $y_1(x) = x + c_1(x-1)(2-x)$. Estimate c_1 using the least square and Galerkin methods.

$[c_1 = 0.28$ by least square and $c_1 = 1.49$ by Galerkin method].

Chapter 7
Partial Differential Equations

Partial Differential Equations or PDEs are often encountered in science and engineering. Such equations arise when a quantity of interest say u, is governed by a continuous differential law in more than one independent variable. Generally speaking a PDE is accompanied by additional equations that hold on the boundary of the region defined by the independent variables. PDE solution in practice therefore becomes a *boundary value problem* (BVP), in which time t may also be present when the problem is time dependent.

PDEs may be *linear* or *non-linear*. When they are linear, several analytical techniques have been developed in the literature for their solution. Even then, if the boundary of the region is not regular, it may become impossible to construct an analytical solution of a linear PDE. In such cases, numerical solution methods remain the only available tool for obtaining a numerical solution. In the case of non-linear PDEs, the scenario is quite different and requires special numerical tools for each generic case. Due to variety of applications, a vast literature exists on the subject of PDEs. The present chapter being an expository one, the focus is on some simple linear equations of mathematical physics in just two independent variables.

Just as the finite difference methods are very successful in the numerical solution of ODEs, these methods are equally prominent for PDEs. However, constructing a good stable and convergent scheme for a PDE is a nontrivial task and often the physical process being described by the differential equation has to be borne in mind. Finite differencing of linear PDEs leads to linear difference equations, and it will be seen in this chapter that such equations can quite easily be handled for computation. As alternative to differencing, *finite volume, finite element, spectral, variational,* and *Monte Carlo* methods are also very popular. These latter methods are however not considered here.

We begin with a linear first-order PDE, which has the look of a *conservation equation* for a continuum.

© Springer Nature Singapore Pte Ltd. 2019
S. K. Bose, *Numerical Methods of Mathematics Implemented in Fortran,* Forum for Interdisciplinary Mathematics,
https://doi.org/10.1007/978-981-13-7114-1_7

7.1 First-Order Equation

Consider the linear first-order PDE for a function $u(x, t)$, viz.

$$\frac{\partial u}{\partial t} + \frac{\partial u}{\partial x} = 0 \tag{7.1}$$

subject to initial condition $u(x, 0) = f(x)$. Here, we envisage t to represent time and x a one- dimensional space variable. The obvious solution of the problem is $u = f(x - t)$, but we seek the solution by finite differencing choosing *equally spaced points* along x- and t- axes $x_m = x_0 + mh$, $t_n = 0 + nk$, denoting $u(x_m, t_n)$ by u_m^n. As in Euler method for ODE, suppose we represent the time derivative by the difference quotient

$$\left.\frac{\partial u}{\partial t}\right|_{m,n} = \frac{u_m^{n+1} - u_m^n}{k} + O(k)$$

and the space derivative by the central difference quotient

$$\left.\frac{\partial u}{\partial x}\right|_{m,n} = \frac{u_{m+1}^n - u_{m-1}^n}{2h} + O(h^2)$$

then we obtain the difference equation

$$u_m^{n+1} = u_m^n - \frac{r}{2}(u_{m+1}^n - u_{m-1}^n) \tag{7.2}$$

where $r = k/h$. The finite difference scheme (7.2) is **explicit**. This means that knowing the unknown u at a given time step n (at space steps $m - 1$ and $m + 1$), the unknown can be computed at the next time step $n + 1$ (for different space steps m). Evidently, the process can be started at time step $n = 0$, since $u_m^0 = f(x_m)$, $u_{m+1}^0 = f(x_{m+1})$ and $u_{m-1}^0 = f(x_{m-1})$ are known. The formula (7.2) is $O(k + h^2)$ and is easy to apply, but it has serious drawback that it is *unstable* for increasing $n!$. This fact will be proved in Sect. 7.1.1 to follow. The scheme is therefore of little use.

We now derive some stable schemes.

7.1.1 Lax–Friedrichs Method

Let u_m^n be replaced by the average of u_{m+1}^n and u_{m-1}^n, then one obtains Lax's scheme

$$u_m^{n+1} = \frac{1}{2}(u_{m+1}^n + u_{m-1}^n) - \frac{r}{2}(u_{m+1}^n - u_{m-1}^n) \tag{7.3}$$

which again is an explicit formula for increasing time steps. The scheme is $O(h^2 + k)$.

7.1.2 Lax–Wendroff Method

Consider the Taylor expansion

$$u_m^{n+1} = u_m^n + k \left.\frac{\partial u}{\partial t}\right|_{m,n} + \frac{k^2}{2} \left.\frac{\partial^2 u}{\partial t^2}\right|_{m,n} + O(k^3)$$

$$= u_m^n - k \left.\frac{\partial u}{\partial x}\right|_{m,n} + \frac{k^2}{2} \left.\frac{\partial^2 u}{\partial x^2}\right|_{m,n} + O(k^3)$$

or

$$u_m^{n+1} = u_m^n - \frac{r}{2}(u_{m+1}^n - u_{m-1}^n) + \frac{r^2}{2}(u_{m+1}^n - 2u_m^n + u_{m-1}^n) \tag{7.4}$$

which too is an explicit formula, but fully of second order, i.e. $O(h^2 + k^2)$.

Remark 7.1 A major disadvantage of the Lax–Wendroff scheme (7.4), though of second order in both space x and time t, is that it is prone to produce oscillatory solution if the gradient of u has a sharp change at a given time t. Alternative schemes have also been considered without much success. Two such schemes are given as Exercises 2 and 3 of this section. A large body of study exists on the nature of finite difference schemes, an account of which is given in the text of Morton and Mayers [26] cited in the Bibliography.

Peter David Lax (1926-). Hungarian-born U.S. mathematician is famous for his contributions in partial differential equations, in particular, the methods that bear his name in Numerical Analysis. He proved the *equivalence theorem* (see Sect. 7.6) *in a seminar presentation* given in the year 1953. He is a Professor at the Courant Institute of Mathematical Sciences at New York University, U.S.A.

Burton Wendroff (1930-). U.S. mathematician. He is an Associate of the Los Alamos National Laboratory, and Adjunct Professor at University of New Mexico, U.S.A.

We now take up the question of stability of the three schemes (7.2), (7.3) and (7.4).

7.1.3 von Neumann Stability Analysis

If we are given a *linear difference equation* with variable coefficients, then the von Neumann stability analysis is of *local* nature. This means that we assume that the coefficients of the difference equation are so slowly varying as to be considered constant in space and time. In that case, the independent solutions or *eigenmodes* of the equation are of the form

$$u_m^n = A\xi^n e^{iKmh} \tag{7.5}$$

where $\xi(K)$ the *amplification factor* is a complex number and K is a *real* spatial *wave number* having any value. In this representation, the key fact is that the time dependence of a single eigenmode is nothing more than integer powers of ξ. Therefore, a difference equation is *unstable* if $|\xi(K)| > 1$ for some K, so that u_m^n grows with exponent n.

Let us check ξ for Eq. (7.2). Inserting (7.5) in (7.2), we get

$$\xi = 1 - \frac{r}{2}(e^{iKh} - e^{-iKh}) = 1 - ir \sin Kh$$

Hence, $|\xi|^2 = 1 + r^2 \sin^2 Kh > 1$. Thus, the scheme is unstable as was asserted in the case of (7.2).

For the case of Lax–Friesrichs scheme (7.3), we obtain

$$\xi = \frac{1}{2}(e^{iKh} - e^{-iKh}) - \frac{r}{2}(e^{iKh} - e^{-iKh})$$
$$= \cos Kh - ir \sin Kh$$

Therefore, $|\xi|^2 = \cos^2 Kh + r^2 \sin^2 Kh = 1 - (1 - r^2) \sin^2 Kh \le 1$ if

$$r = \frac{k}{h} \le 1 \tag{7.6}$$

Condition (7.6) is the famous **Courant–Friedrichs–Lewy (C–F–L) criterion**, often called the **Courant condition** in brief. Subject to (7.6) being satisfied, Lax's scheme is stable.

For the Lax–Wendroff scheme (7.4), we have

$$\xi = 1 - r^2 - \frac{r}{2}(e^{iKh} - e^{-iKh}) + \frac{r^2}{2}(e^{iKh} + e^{-iKh})$$
$$= 1 - r^2 - ir \sin Kh + r^2 \cos Kh$$

Hence,

$$|\xi|^2 = \left(1 - 2r^2 \sin^2 \frac{Kh}{2}\right)^2 + r^2 \sin^2 Kh$$
$$= 1 - 4r^2(1 - r^2) \sin^4 \frac{Kh}{2} \le 1$$

if $r \le 1$. Thus, the Courant condition is satisfied and the method is stable.

John von Neumann (1903–1957) is regarded as one of the foremost mathematicians of the twentieth century. He contributed to many areas, including quantum mechanics, operator theory, ergodic theory, game theory, computation theory, numerical analysis and meteorology.

Richard Courant (1888–1972), German–American mathematician. He was David Hilbert's assistant in Göttingen, Germany. He had to fight in 1915 during World War I, but was wounded and dismissed from military service. He returned to be associated with Göttingen, but later left Germany in 1933 and became a Professor at the New York University. The C–F–L criterion was given in 1927, during his tenure in Germany. In 1943, he discovered the *finite element method*, which was later reinvented by engineers. He is famous for his work in analysis.

Kurt Otto Friedrichs (1901–1982), German–American mathematician. He escaped from Germany in 1937 and joined the New York University. A prolific researcher, his main work is on PDEs that represent laws of physics and engineering sciences. In particular, he vastly contributed to compressible fluid flows and non-linear buckling of plates.

Hans Lewy (1904–1988), German/Polish-born American mathematician. He began researches at Göttingen but fled to U.S.A. in 1933. He is known for his extremely original and inventive work on PDEs. He was a Professor at University of California, Berkeley, U.S.A.

Example 1. Solve the first-order PDE $\dfrac{\partial u}{\partial t} + \dfrac{\partial u}{\partial x} = 0$, subject to the initial condition

$$
\begin{aligned}
u(x, 0) &= x, & 0 \leq x \leq 1 \\
&= 2 - x, & 1 \leq x \leq 2 \\
&= 0, & \text{elsewhere}
\end{aligned}
$$

by the Lax–Wendroff formula. Compare up to two time steps using $h = 1/2$ and $r = 1/2$.

Solution. For the given data, the Lax–Wendroff formula becomes

$$
\begin{aligned}
u_m^{n+1} &= u_m^n - \frac{1}{4}(u_{m+1}^n - u_{m-1}^n) + \frac{1}{8}(u_{m+1}^n - 2u_m^n + u_{m-1}^n) \\
&= \frac{3}{4}u_m^n - \frac{1}{8}u_{m+1}^n + \frac{3}{8}u_{m-1}^n
\end{aligned}
$$

For the <u>first time step</u> $(n = 0)$

$$
u_m^1 = \frac{3}{4}u_m^0 - \frac{1}{8}u_{m+1}^0 + \frac{3}{8}u_{m-1}^0
$$

The grid points along the x-axis in $[0, 2]$ are $x_m = mh = m/2$, $m = 0, 1, \cdots, 4$.

At $m = 0$, $x_m = 0$: $u_m^1 = \dfrac{3}{4} \cdot 0 - \dfrac{1}{8} \cdot \dfrac{1}{2} + \dfrac{3}{8} \cdot 0 = -\dfrac{1}{16}$

At $m = 1$, $x_m = \dfrac{1}{2}$: $u_m^1 = \dfrac{3}{4} \cdot \dfrac{1}{2} - \dfrac{1}{8} \cdot 1 + \dfrac{3}{8} \cdot 0 = \dfrac{1}{4}$

At $m = 2$, $x_m = 1$: $u_m^1 = \dfrac{3}{4} \cdot 1 - \dfrac{1}{8} \cdot \dfrac{1}{2} + \dfrac{3}{8} \cdot \dfrac{1}{2} = \dfrac{7}{8}$

At $m = 3$, $x_m = \dfrac{3}{2}$: $u_m^1 = \dfrac{3}{4} \cdot \dfrac{1}{2} - \dfrac{1}{8} \cdot 0 + \dfrac{3}{8} \cdot 1 = \dfrac{3}{4}$

At $m = 4$, $x_m = 2$: $u_m^1 = \dfrac{3}{4} \cdot 0 - \dfrac{1}{8} \cdot 0 + \dfrac{3}{8} \cdot \dfrac{1}{2} = \dfrac{3}{16}$

For the second time step $(n = 1)$, we need the values of u_m^1 at grid points $x_m = -1/2$ and $5/2$.

These are both calculated as 0. For this step

$$u_m^2 = \frac{3}{4} u_m^1 - \frac{1}{8} u_{m+1}^1 + \frac{3}{8} u_{m-1}^1$$

Proceeding as in the first step, we obtain

$$x_m : \quad 0 \quad \frac{1}{2} \quad 1 \quad \frac{3}{2} \quad 2$$

$$u_m^2 : -\frac{5}{64} \quad \frac{7}{128} \quad \frac{21}{32} \quad \frac{111}{128} \quad \frac{27}{64}$$

□

Exercises

1. Solve the problem of Example 1 for the initial conditions:

(i) $\quad \begin{aligned} u(x, 0) &= x, \quad 0 \le x \le 2 \\ &= 0, \quad \text{elsewhere} \end{aligned}$

(ii) $\quad \begin{aligned} u(x, 0) &= 2x - x^2, \quad 0 \le x \le 2 \\ &= 0, \quad \text{elsewhere} \end{aligned}$

the other data remaining the same.

$[(i) \quad \begin{array}{|ccccc|} \hline -\frac{1}{16} & \frac{1}{4} & \frac{3}{4} & \frac{1}{2} & \frac{15}{8} \\ -\frac{5}{64} & \frac{9}{18} & \frac{19}{32} & \frac{27}{64} & \frac{51}{32} \\ \hline \end{array} \qquad (ii) \quad \begin{array}{|ccccc|} \hline -\frac{3}{32} & \frac{7}{16} & \frac{15}{16} & \frac{15}{16} & \frac{9}{32} \\ -\frac{1}{8} & \frac{45}{256} & \frac{3}{4} & \frac{261}{256} & \frac{117}{25} \\ \hline \end{array}].$

2. Derive the *Upwind Scheme*

$$u_m^{n+1} = u_m^n - r\,(u_m^n - u_{m-1}^n) + O(h + k)$$

for Eq. (7.1) by taking the first-order backward finite difference formula for the space derivative $\partial u / \partial x$, and prove that the magnitude of the amplification factor for the scheme is given by

$$|\xi|^2 = 1 - 2r\,(1 - r)(1 - \cos Kh)$$

so that the scheme is stable if the C–F–L condition $r \le 1$ is satisfied.

3. Derive the *Leapfrog Scheme*

$$u_m^{n+1} = u_m^{n-1} - r\,(u_{m+1}^n - u_{m-1}^n) + O(h^2 + k^2)$$

by taking the second-order finite difference formula for the time derivative $\partial u/\partial t$ in Eq. (7.1). Prove that the amplification factor ξ for the scheme is given by

$$\xi = ir\,\sin kH \pm \sqrt{1 - r^2 \sin^2 kH}$$

so that $|\xi|^2 = 1$ for $r \le 1$ having locally oscillatory solution.

7.2 The Diffusion Equation

The equation is a *second-order* equation of the form

$$\frac{\partial u}{\partial t} = \frac{\partial^2 u}{\partial x^2} \tag{7.7}$$

where t usually refers to time and x a space variable in an interval $[a,\,b]$. We assume that Eq. (7.7) is subject to *initial condition* $u(x,\,0) = f(x)$ and *boundary conditions* $u(a,\,t) = \phi(t)$, $u(b,\,t) = \psi(t)$.

For finite differencing Eq. (7.7), we discretise the x, t-space by *equally spaced points* $x_m = a + mh$, $t_n = nk$, $m, n = 0, 1, 2, \cdots$ and denote $u_m^n := u(x_m, t_n)$.

7.2.1 Schmidt Method

Selecting the finite difference quotients for (7.7) as

$$\frac{\partial u}{\partial t} = \frac{u_m^{n+1} - u_m^n}{k} + O(k), \qquad \frac{\partial^2 u}{\partial x^2} = \frac{u_{m-1}^n - 2u_m^n + u_{m+1}^n}{h^2} + O(k^2)$$

we obtain the scheme

$$\begin{aligned} u_m^{n+1} &= u_m^n + r\,(u_{m-1}^n - 2u_m^n + u_{m+1}^n) \\ &= (1 - 2r)\,u_m^n + r\,(u_{m-1}^n + u_{m+1}^n) \end{aligned} \tag{7.8}$$

where $r = k/h^2$. Equation (7.8) is an *explicit scheme*, since it determines the unknown u at time step $n + 1$ from previously determined values at time step n at three grid points $m - 1$, m and $m + 1$ along the x-axis. The order of the scheme is $O(k + h^2)$.

Let us examine the stability of the scheme (7.8). Substituting the form (7.5) in it, we get for the amplification factor

$$\xi = 1 - 2r + r\,(e^{-iKh} + e^{iKh})$$
$$= 1 - 2r + 2r\cos Kh = 1 - 4r\,\sin^2\frac{Kh}{2}$$

For stability of the method, we require that $|\xi| \leq 1$, or

$$-1 \leq 1 - 4r\,\sin^2\frac{Kh}{2} \leq 1$$

or

$$r \leq \frac{1}{2\sin^2\dfrac{Kh}{2}}$$

which is always satisfied if $0 < r \leq 1/2$. The scheme (7.8) is therefore stable if the condition $0 < k/h^2 < 1/2$ is ensured. In this case, the round-off errors will not get out of bounds for increasing number of time steps.

We now consider two modifications of the Schmidt method that ensure *unconditional stability* for all values of r!

7.2.2   Laasonen Method

Here, we take the backward difference quotient for the time derivative

$$\frac{\partial u}{\partial t} = \frac{u_m^n - u_m^{n-1}}{k} + O(k)$$

and obtain as before

$$u_m^n = u_m^{n-1} + r\,(u_{m-1}^n - 2u_m^n + u_{m+1}^n)$$

Replacing n by $n + 1$, we obtain

$$u_m^{n+1} = u_m^n + r\,(u_{m-1}^{n+1} - 2u_m^{n+1} + u_{m+1}^{n+1}) \qquad (7.9)$$

or

$$-r\,u_{m-1}^{n+1} + (1 + 2r)\,u_m^{n+1} - r\,u_{m+1}^{n+1} = u_m^n$$

which forms a tridiagonal system of equations for the determination of the unknown at time step $n + 1$ in terms of that at time step n. The scheme is thus an *implicit scheme*, for it involves the solution of the tridiagonal system.

To investigate the amplification factor of the scheme, we substitute Eq. (7.5) in (7.9) obtaining

$$[-r\, e^{-iKh} + (1 + 2r) - r\, e^{iKh}]\xi = 1$$

or

$$[1 + 2r - 2r \cos Kh]\xi = 1$$

i.e.

$$\xi = \frac{1}{1 + 4r \sin^2 \dfrac{Kh}{2}}$$

for which $|\xi| \leq 1$, and so the scheme is unconditionally stable.

The scheme like that of Schmidt is however of the first order in time step k. The next scheme improves upon this aspect.

7.2.3 Crank–Nicolson Method

Here, we take the average of Eqs. (7.8) and (7.9), i.e.

$$u_m^{n+1} = u_m^n + \frac{r}{2}\left[(u_{m-1}^{n+1} - 2u_m^{n+1} + u_{m+1}^{n+1}) + (u_{m-1}^n - 2u_m^n + u_m^n)\right]$$

or

$$-r\, u_{m-1}^{n+1} + (2 + 2r)\, u_m^{n+1} - r\, u_{m+1}^{n+1} = r\, u_{m-1}^n + (2 - 2r)\, u_m^n + r\, u_{m+1}^n \quad (7.10)$$

which also *implicitly* forms a tridiagonal system for the determination of u at time step $n + 1$ in terms of those at step n. The solution to the problem is obtained by solving such a system of linear equations.

The advantage of taking the average of (7.8) and (7.9) is that both sides of the equation are centred at time step $n + 1/2$ and thus is of order k^2. It is evidently of order h^2 in space step m. These facts ensure faster convergence of the method.

Regarding stability of the method, the amplification factor ξ is given by

$$[-r\, e^{-iKh} + (2 + 2r) - r\, e^{iKh}]\xi = r\, e^{-iKh} + (2 - 2r) + r\, e^{iKh}$$

or

$$(1 + r - r \cos Kh)\xi = 1 - r + r \cos Kh$$

or

$$\xi = \frac{1 - 2r \, \sin^2 Kh/2}{1 + 2r \, \sin^2 Kh/2}$$

Hence, $|\xi| \leq 1$ and the method is unconditionally stable.

Erhard Schmidt (1876–1959), Estonia-born German mathematician. He is better known for orthogonalisation of polynomials (see Chap. 8) and as one of the founders of functional analysis. The method for parabolic equations was given in a festschrift article in (1924).

Pentti Laasonen (1928–2016), Finn mathematician, Professor at technical university of Helsinki. The method was presented in a paper in the year 1949.

John Crank (1916–2006), British mathematical Physicist. Head of Department of mathematics and later Vice-Principal of Brunel University. He is best known for his work on numerical solution of PDEs.

Phyllis Nicolson (1917–1968), British mathematician. During World War II, she worked on wartime problems at Manchester University, one being magnetron theory and practice. She is best known for her joint work with John Crank in 1947. Later she became a lecturer at Leeds University.

Example 1. Solve the diffusion equation $\dfrac{\partial u}{\partial t} = \dfrac{\partial^2 u}{\partial x^2}$ subject to the initial and boundary conditions

$$u(x, \, 0) = \sin \pi x, \quad 0 \leq x \leq 1$$
$$u(0, \, t) = u(1, \, t) = 0$$

using (i) the Schmidt method, and (ii) the Crank–Nicolson method, for $h = 1/3$ and $k = 1/36$. Calculate up to two time steps.

Solution. For the given values of h and k, the grid points are $x_m = mh$, $m = 0, 1, 2, 3$ and $t_n = nk$, $n = 0, 1, 2$. The boundary conditions give $u_0^n = u_3^n = 0$, $n = 0, 1, 2$. Also, $r = k/h^2 = 1/4$ and the Schmidt method is stable.
(i) <u>Schmidt Method</u>. By Eq. (7.8) with $r = 1/4$, the explicit scheme becomes

$$u_m^{n+1} = \frac{1}{4} \, (u_{m-1}^n + 2u_m^n + u_{m+1}^n)$$

For the first time step $(n = 0)$

At $m = 1$, $x_m = \dfrac{1}{3}$, $u_1^1 = \dfrac{1}{4} \, (u_0^0 + 2u_1^0 + u_2^0) = \dfrac{1}{4} \left(0 + 2 \sin \dfrac{\pi}{3} + \sin \dfrac{2\pi}{3} \right) = $
<u>0.6495</u>

At $m = 2$, $x_2 = \dfrac{2}{3}$, $u_2^1 = \dfrac{1}{4}(u_1^0 + 2u_2^0 + u_3^0) = \dfrac{1}{4}\left(\sin\dfrac{\pi}{3} + 2\sin\dfrac{2\pi}{3} + 0\right) =$
0.6495

For the second time step ($n = 1$)

At $m = 1$, $x_m = \dfrac{1}{3}$, $u_1^2 = \dfrac{1}{4}(u_0^1 + 2u_1^1 + u_2^1) = \dfrac{1}{4}(0 + 2 \times 0.6495 + 0.6495) =$
0.4871

At $m = 2$, $x_m = \dfrac{2}{3}$, $u_2^2 = \dfrac{1}{4}(u_1^1 + 2u_2^1 + u_3^1) = \dfrac{1}{4}(0.6495 + 2 \times 0.6495 + 0) =$
0.4871

(*ii*) <u>Crank–Nicolson Method.</u> By Eq. (7.10) with $r = 1/4$, the scheme becomes

$$-\frac{1}{4}u_{m-1}^{n+1} + \frac{5}{2}u_m^{n+1} - \frac{1}{4}u_{m+1}^{n+1} = \frac{1}{4}u_{m-1}^n + \frac{3}{2}u_m^n + \frac{1}{4}u_{m+1}^n$$

or

$$-u_{m-1}^{n+1} + 10\,u_m^{n+1} - u_{m+1}^{n+1} = u_{m-1}^n + 6\,u_m^n + u_{m+1}^n$$

For the first time step ($n = 0$)

At $m = 1$, $x_m = \dfrac{1}{3}$, $0 + 10u_1^1 - u_2^1 = 0 + 6u_1^0 + u_2^0 = 6\sin\dfrac{\pi}{3} + \sin\dfrac{2\pi}{3} =$
6.0622

At $m = 2$, $x_m = \dfrac{2}{3}$, $-u_1^1 + 10u_2^1 - 0 = u_1^0 + 6u_2^0 + 0 = \sin\dfrac{\pi}{3} + 6\sin\dfrac{2\pi}{3} =$
6.0622

The solution of the two equations yields $u_1^1 = u_2^1 = \underline{0.6736}$

For the second time step ($n = 1$)

At $m = 1$, $x_m = \dfrac{1}{3}$, $0 + 10u_1^2 - u_2^2 = 0 + 6u_1^1 + u_2^1 = 0.5894$

At $m = 2$, $x_m = \dfrac{2}{3}$, $-u_1^2 + 10,\, u_2^2 - 0 = u_1^1 + 6u_2^1 + 0 = 0.5894$

whose solution is $u_1^2 = u_2^2 = \underline{0.5239}$

The exact solution of the problem is $u(x, t) = e^{\pi^2 t}\sin\pi x$, which at the second time level yields $u(\frac{1}{3}, \frac{1}{18}) = u(\frac{2}{3}, \frac{1}{18}) = 0.5005$. $\square$

Exercises

1. Solve the equation $\dfrac{\partial u}{\partial t} = \dfrac{\partial^2 u}{\partial t^2}$, $0 < x < 1$, $t > 0$ with the conditions $u(0, t) = 0$, $u(1, t) = 1$ and $u(x, 0) = x(2 - x)$ by the Schmidt method, taking $h = 1/4$ and $k = 1/32$. Calculate up to two time steps.

[

x	1/4	1/2	3/4
t			
1/32	3/8	11/16	7/8
1/16	11/32	5/8	27/32

].

2. Solve by Crank–Nicolson method, the equation $\dfrac{\partial u}{\partial t} = \dfrac{\partial^2 u}{\partial x^2}$, $0 < x < 1$, $t > 0$, satisfying the conditions $u(0, t) = 0$, $u(1, t) = 0$ and $u(x, 0) = 100x(1 - x)$. Compute u for two time steps with $h = 1/4$ and $k = 1/4$.

[

x	1/4	1/2	3/4
t			
1/4	14.2857	17.1429	14.2857
1/2	6.9388	10.6122	6.9388

].

3. Solve the equation $\dfrac{\partial u}{\partial t} = \dfrac{\partial^2 u}{\partial x^2}$, $0 < x < 1$, $t > 0$ by Crank–Nicolson method, given that $u(0, t) = 0$, $u(1, t) = 200t$ and $u(x, 0) = 0$. Calculate u for two time steps with $h = 1/4$ and $k = 1/8$.

[

x	1/4	1/2	3/4
t			
1/8	1.1905	3.5714	9.5238
1/4	5.2154	13.2653	27.4376

].

4. Show that the solution of the Schrödinger equation

$$i \frac{\partial \psi}{\partial t} = - \frac{\partial^2 \psi}{\partial x^2} + V(x)\, \psi$$

where ψ is complex valued and *unitary*, i.e. $\displaystyle\int_{-\infty}^{\infty} |\psi|^2 dx = 1$ can be symbolically written as $\psi(x, t) = e^{-iHt}\, \psi(x, 0)$, $H := -\dfrac{\partial^2}{\partial x^2} + V(x)$. Prove by using the Cayley approximation $e^{-iH\Delta t} \approx (1 - \frac{1}{2} iH\Delta t)/(1 + \frac{1}{2} iH\Delta t)$ that the equation can be solved by the Crank–Nicolson scheme

$$\left(1 + \frac{1}{2} iH\Delta t\right) \psi_j^{n+1} = \left(1 - \frac{1}{2} iH\Delta t\right) \psi_j^{n}$$

which is unconditionally stable and unitary.

7.3 The Wave Equation

The wave equation in one space dimension x is of the form

$$\frac{\partial^2 u}{\partial t^2} = \frac{\partial^2 u}{\partial x^2} \tag{7.11}$$

It may be subject to initial conditions at time $t = 0$ of the form

$$u(x, 0) = f(x), \quad \frac{\partial u}{\partial t}(x, 0) = g(x)$$

and boundary conditions at $x = a$ and $x = b$ of the form

$$u(a, t) = \phi(t), \quad u(b, t) = \psi(t)$$

where f, g, ϕ and ψ are continuous functions. In the theory of wave equations, it is proved that the exact solution of the IVP due to D'Alembert is

$$u(x, t) = \frac{1}{2}[f(x - t) + f(x + t)] + \frac{1}{2}\int_{x-t}^{x+t} g(\tau)\, d\tau$$

But that is not our point of view in finite difference solution.

Let the x, t space be discretised by *equally spaced points* $x_m = a + mh$, $t_n = nk$, $m, n = 0, 1, 2, \cdots$ and let $u(x_m, t_n)$ be denoted by u_m^n as usual. The central difference quotients of the two partial derivatives in (7.11) are

$$\frac{\partial^2 u}{\partial t^2} = \frac{u_m^{n-1} - 2u_m^n + u_m^{n+1}}{k^2} + O(k^2)$$

$$\frac{\partial^2 u}{\partial x^2} = \frac{u_{m-1}^n - 2u_m^n + u_{m+1}^n}{h^2} + O(h^2)$$

Substituting in Eq. (7.11) and neglecting the small quantities, we obtain the scheme

$$u_m^{n+1} = 2(1 - r^2)\, u_m^n + r^2(u_{m-1}^n + u_{m+1}^n) - u_m^{n-1} \tag{7.12}$$

where $r = k/h$. The scheme is explicit, determining u_m^{n+1} at level $n + 1$ in terms of the previously determined values at the preceding two levels n and $n - 1$. But, when the calculation is started at the level $n = 0$, the values of u_m^0 and u_m^{-1} are required in Eq. (7.12). The former is given by $f(x_m)$ according to the first of the boundary conditions, while the latter has to be computed from the second boundary condition. Using central difference quotient for $\partial u/\partial t$, we have

$$\frac{\partial u}{\partial t}\bigg|_{t=0} = \frac{u_m^1 - u_m^{-1}}{2k} = g(x_m)$$

This equation yields

$$u_m^{-1} = u_m^1 - 2k\, g(x_m) \tag{7.13}$$

Equation (7.13) helps in eliminating u_m^{-1} from the first equation of (7.12) for $n = 0$ and yields u_m^1 explicitly. The computation of u_m^n at the succeeding levels $n \geq 2$ is completed from (7.12). The method is evidently of second order throughout.

In order to examine the stability of the method, we substitute Eq. (7.5) in (7.12), giving

$$\xi = 2(1 - r^2) + r^2(e^{-iKh} + e^{iKh}) - \xi^{-1}$$

This yields the quadratic equation for the amplification factor ξ as

$$\xi^2 - \left(2 - 4r^2 \sin^2 \frac{Kh}{2}\right)\xi + 1 = 0$$

which has the form $\xi^2 - 2B\xi + 1 = 0$ with roots $\xi = B \pm \sqrt{B^2 - 1}$. For stability, we require that $|\xi| \leq 1$ or $-1 \leq B \pm \sqrt{B^2 - 1} \leq 1$ which means that we require $|B| \leq 1$. Thus, for stability

$$-1 \leq 1 - 2r^2 \sin^2 \frac{Kh}{2} \leq 1$$

or

$$r^2 \leq \frac{1}{\sin^2 \dfrac{Kh}{2}}$$

This condition is always satisfied if $r^2 \leq 1$ or $r \leq 1$, which again is the *Courant Condition*. Thus, for stability, we require that $k \leq h$.

Example 1. Solve the initial boundary value problem

$$\frac{\partial^2 u}{\partial t^2} = \frac{\partial^2 u}{\partial x^2}, \quad 0 < x < 1$$

subject to the initial conditions

$$u(x, 0) = \sin \pi x, \quad \frac{\partial u}{\partial t}(x, 0) = 0, \quad 0 \leq x \leq 1$$

and the boundary conditions

$$u(0, t) = 0, \quad u(1, t) = 0, \quad t > 0$$

Assume $h = 1/4$ and $k = 3/16$ and compute up to three time steps. Compare with the exact solution $u(x, t) = \sin \pi x \cos \pi t$.

Solution. For the given values of h and k, $r = 3/4$ and the computations are stable. The grid points are

$$x_m = mh, \quad m = 0, 1, 2, 3, 4 \quad \text{and} \quad t_n = nk, \quad k = 0, 1, 2$$

The initial conditions, using Eq. (7.13), give

$$u_m^0 = \sin\left(\frac{m\pi}{4}\right) \quad \text{and} \quad u_m^{-1} = u_m^1, \quad m = 0, 1, 2, 3, 4$$

while the boundary conditions give $u_m^n = 0$, $m = 0, 4$.
  The explicit finite difference solution (7.12) becomes in this case

$$u_m^{n+1} = \frac{7}{8} u_m^n + \frac{9}{16} (u_{m-1}^n + u_{m+1}^n) - u_m^{n-1}, \quad m = 1, 2, 3$$

For $n = 0$, we obtain

$$u_m^1 = \frac{7}{8} u_m^0 + \frac{9}{16} (u_{m-1}^0 + u_{m+1}^0) - u_m^{-1}$$

or since $\quad u_m^{-1} = u_m^1, \quad u_m^1 = \frac{7}{16} u_m^0 + \frac{9}{32} (u_{m-1}^0 + u_{m+1}^0)$

At $m = 1$, $\quad u_1^1 = \frac{7}{16} u_1^0 + \frac{9}{32} (u_0^0 + u_2^0) = 0.59061$

At $m = 2$, $\quad u_2^1 = \frac{7}{16} u_2^0 + \frac{9}{32} (u_1^0 + u_3^0) = 0.83525$

At $m = 3$, $\quad u_3^1 = \frac{7}{16} u_3^0 + \frac{9}{32} (u_2^0 + u_4^0) = 0.59061$

From these solutions, we have for $n = 1$,

$$u_1^2 = \underline{0.27951}, \quad u_2^2 = \underline{0.39528}, \quad u_3^2 = \underline{0.27951}$$

and for $n = 2$, $\quad u_1^3 = \underline{-0.12369}, \quad u_2^3 = \underline{-0.17493}, \quad u_3^3 = \underline{-0.12369}$

The exact solution is

$$u_1^3 = u\left(\frac{1}{4}, \frac{9}{16}\right) = \sin\frac{\pi}{4} \cos\frac{9\pi}{16} = -0.13795 = u_3^3$$

$$u_2^3 = u\left(\frac{1}{2}, \frac{9}{16}\right) = \sin\frac{\pi}{2} \cos\frac{9\pi}{16} = -0.19509 \qquad \Box$$

When the space and time steps are equal $h = k$ and $r = 1$. In this case, Eq. (7.12) simplifies into

$$u_m^{n+1} = u_{m-1}^n + u_{m+1}^n - u_m^{n-1} \tag{7.14}$$

and the method remains stable. The following exercises assume this simplified procedure.

Exercises

1. Solve the equation $\dfrac{\partial^2 u}{\partial t^2} = \dfrac{\partial^2 u}{\partial x^2}$, $0 < x < 1$, $t > 0$ satisfying the initial conditions $u(x, 0) = 0 = \dfrac{\partial u}{\partial t}(x, 0)$ and boundary conditions $u(0, t) = 0$, $u(1, t) = \frac{1}{2} \sin \pi t$. Assume $h = k = 1/4$ and compute up to four time steps.

[
x t	1/4	1/2	3/4	1
1/4	0	0	0	0.3536
1/2	0	0	0.3536	0.5
3/4	0	0.3536	0.5	0.3536
1	0.3536	0.5	0.3536	0
].

2. Solve the equation $\dfrac{\partial^2 u}{\partial t^2} = \dfrac{\partial^2 u}{\partial x^2}$, $0 < x < 1$, $t > 0$ up to $t = 1$, satisfying the conditions $u(0, t) = u(1, t) = 0$, $u(x, 0) = 0$, $\dfrac{\partial u}{\partial t}(x, 0) = \sin \pi x$. Assume $h = k = 0.2$.

[
x t	1/5	2/5	3/5	4/5
1/5	0.1176	0.1902	0.1902	0.1176
2/5	0.1902	0.3078	0.3078	0.1902
3/5	0.1902	0.3078	0.3078	0.1902
4/5	0.1176	0.1902	0.1902	0.1176
1	0	0	0	0
].

3. Solve the equation $\dfrac{\partial^2 u}{\partial t^2} = \dfrac{\partial^2 u}{\partial x^2}$ for $x = 0(0.2)1$ and $t = 0(0.2)1$, given that

$u(0, t) = u(1, t)$, $u(x, 0) = x(1 - x)$, $\dfrac{\partial u}{\partial t}(x, 0) = x(1 - x)$.

[

x t	0.2	0.4	0.6	0.8
0.2	0.32	0.40	0.40	0.32
0.4	0.24	0.48	0.48	0.24
0.6	0.16	0.32	0.32	0.16
0.8	0.08	0	0	0.80
1.0	−0.16	−0.24	−0.24	−0.16

].

7.4 Poisson Equation

The *Poisson equation* is a prominent equation in mathematical physics. Its form is

$$\frac{\partial^2 u}{\partial x^2} + \frac{\partial^2 u}{\partial y^2} = \rho(x, y) \tag{7.15}$$

where $\rho(x, y)$ is a given function of x and y in a region R. When $\rho(x, y) = 0$ the equation is called **Laplace's equation**. On the boundary ∂R, we assume that u is given, say $u = f(x, y)$.

For finite differencing, let the (x, y) plane be divided into a network of squares of side h by the mesh or grid points $x = x_0 + mh$, $y = y_0 + nh$, $m, n = 0, 1, 2, \cdots$. Writing $u_{m,n} = u(x_m, y_n)$, the central difference quotient for the second-order derivatives are

$$\frac{\partial^2 u}{\partial x^2} = \frac{u_{m-1,n} - 2 u_{m,n} + u_{m+1,n}}{h^2} + O(h^2)$$

$$\frac{\partial^2 u}{\partial y^2} = \frac{u_{m,n-1} - 2 u_{m,n} + u_{m,n+1}}{h^2} + O(h^2)$$

Substituting these expressions in (7.8), we obtain the finite difference scheme

$$u_{m-1,n} + u_{m+1,n} + u_{m,n-1} + u_{m,n+1} - 4 u_{m,n} = h^2 \rho_{m,n} \tag{7.16}$$

where $\rho_{m,n} = \rho(x_m, y_n)$. The grid points in (7.16) are schematically shown in Fig. 7.1.

Fig. 7.1 Schematic of nodes
in Eq. (7.16)

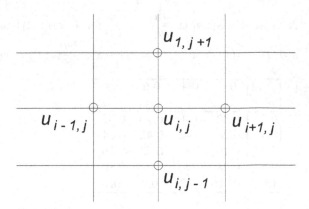

Fig. 7.2 Nodes of a
rectangular R

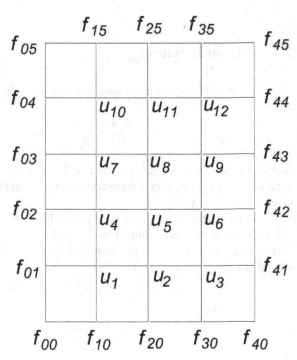

The difference scheme (7.16) leads to a system of linear equations for the unknown u at the *interior grid points* of R. In order to understand this, suppose that R is a rectangle as shown in Fig. 7.2.

Let the unknown at the interior points be denoted by a vector having components $u_1, u_2, \cdots, u_{12}$ running along horizontals as shown. At the boundary grid points, the quantity is however known from the boundary condition. Considering the nodes 1, 2, 3, $\cdots$ in succession and applying Eq. (7.16), we obtain the equations

$$
\begin{aligned}
-4u_1 +u_2 \quad\quad +u_4 \quad\quad\quad\quad\quad\quad &= h^2\rho_1 - f_{10} - f_{01}\\
u_1 -4u_2 +u_3 \quad\quad +u_5 \quad\quad\quad\quad &= h^2\rho_2 - f_{20}\\
u_2 -4u_3 \quad\quad\quad +u_6 \quad\quad\quad &= h^2\rho_3 - f_{30} - f_{41}\\
u_1 \quad\quad\quad -4u_4 +u_5 \quad\quad +u_7 \quad &= h^2\rho_4 - f_{02}\\
u_2 \quad\quad +u_4 -4u_5 +u_6 \quad\quad +u_8 \quad &= h^2\rho_5\\
u_3 \quad\quad +u_5 -4u_6 \quad\quad\quad +u_9 &= h^2\rho_6 - f_{42}\\
\cdots \quad \cdots \quad \cdots \quad \cdots \quad\quad \cdots \quad\quad &\cdots = \cdots\cdots\\
u_9 + u_{11} - 4u_{12} &= h^2\rho_{12} - f_{35} - f_{44}
\end{aligned}
$$

The above system is diagonally dominant and therefore has a unique solution. Because of sparsity of the matrix of the coefficients on the left-hand side, relaxation method such as the Seidel method is well suited for computer solution.

In the case of a boundary ∂R of arbitrary shape, there is no difficulty in writing down an equation for a grid point well inside ∂R, but there is some difficulty at points close to the boundary. This is illustrated in Fig. 7.3.

These latter nodes are indicated by small triangles. For these nodes, instead of (7.16), we linearly interpolate the value at the grid point. For instance, consider the grid point n which lies between grid points $n + 1$ and n' (outside R). The boundary crosses the abscissa at the point n'. Let the distance of this point from n be ρ, then linear interpolation means

$$
u_n = \frac{1}{h + \rho} (h\, f_{n'} + \rho\, u_{n+1})
$$

This relation introduces a *linear equation* for the grid point n. Similar other grid points contribute similar equations to the system. Taking into account all the nodes,

Fig. 7.3 Grid for a curved boundary

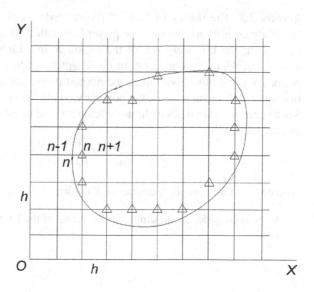

the full system for the solution of the problem is obtained.

Siméon Denis Poisson (1781–1840), French mathematician and physicist. He held a number of important positions since he was in his 20s. He made remarkable application of mathematics to physical subjects with his memoirs on theory of electricity and magnetism. In celestial mechanics, he made advances in planetary theory over those of Lagrange.

Pierre–Simon Laplace (1749–1827), French mathematician and astronomer. He made deep investigations in celestial mechanics, proving that the solar system was stable. In these investigations, he introduced *spherical harmonics* and the well-known equation that bears his name. He founded the *probability theory* and discovered the famous *Laplace transform* in his researches in differential equations.

Remark 7.2 The Laplace's equation and the wave equation differ in sign in one of the two second derivatives, yet the finite difference schemes in the two cases are somewhat different. The schemes adhere in some sense to the nature of these two equations. In the theory of linear partial differential equations of second order of the form

$$A \frac{\partial^2 u}{\partial x^2} + B \frac{\partial^2 u}{\partial x \partial y} + C \frac{\partial^2 u}{\partial y^2} + D \frac{\partial u}{\partial x} + E \frac{\partial u}{\partial y} + Fu = G$$

where $A, B, \cdots, G$ are functions of x and y, the equation is classified depending on the sign of the discriminant $B^2 - 4AC$. If $B^2 - 4AC < 0$, the equation is called *elliptic* and if $B^2 - 4AC > 0$, it is called *hyperbolic*. If $B^2 = 4AC$ it is called *parabolic*. The theory of each of these equations has a special feature and the numerical schemes are in tune with such features. Evidently, the Laplace's equation is elliptic while the wave equation is hyperbolic. The diffusion equation on the other hand is parabolic.

Remark 7.3 The theory of PDE of second order is concomitant with the type of boundary conditions that must be prescribed with it. If u is prescribed as a function of x, y on the boundary ∂R of the region R in which the PDE holds, then it is called *Dirichlet* type condition. In this chapter, we have exclusively assumed such boundary condition. Sometimes the normal derivative $\partial u / \partial \nu$ is prescribed on the boundary ∂R. In such cases, the boundary condition is said to be of *Neumann* type. Such conditions are dealt by finite differencing the first-order partial derivatives with respect to x, y in

$$\frac{\partial u}{\partial \nu} = l \frac{\partial u}{\partial x} + m \frac{\partial u}{\partial y}$$

where (l, m) are direction cosines of the normal.

We now consider some simple applications of the finite difference scheme (7.16).

Fig. 7.4 Solution grid

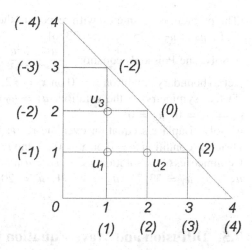

Example 1. Find the solution of the Laplace's equation $\dfrac{\partial^2 u}{\partial x^2} + \dfrac{\partial^2 u}{\partial y^2} = 0$ subject to Dirichlet condition $u(x, y) = x - y$ on ∂R, where R is the right-angled triangle with vertices $(0, 0)$, $(4, 0)$ and $(0, 4)$.

Solution. The region is shown in Fig. 7.4, together with the grid in which the solution is sought. The value of u at the boundary nodes is shown within parentheses. There are only three nodes at which the values u_1, u_2, u_3 are to be determined. Following Eq. (7.16), we have for the three nodes

$$-4u_1 + u_2 + u_3 - 1 + 1 = 0$$
$$u_1 - 4u_2 + 2 + 0 = 0$$
$$u_1 - 4u_3 - 2 + 0 - 2 = 0$$

The solution of the equations is easily found to be $u_1 = \frac{2}{7}$, $u_2 = \frac{15}{14}$, $u_3 = \frac{1}{14}$. □

Exercises

1. Solve Laplace's equation $\dfrac{\partial^2 u}{\partial x^2} + \dfrac{\partial^2 u}{\partial y^2} = 0$ over the square mesh bounded by the sides $x = 0$, $y = 0$, $x = 3$ and $y = 3$, satisfying the boundary conditions $u = 2$ at the mesh points $(2, 0)$, $(3, 1)$, $(1, 3)$, $(0, 2)$ and $u = 3$ at $(1, 0)$, $(3, 2)$, $(2, 3)$, $(0, 1)$, respectively.
[From symmetry about the centre $u_1 = u_4$ and $u_2 = u_3$. $u_1 = u_4 = 8/3$ and $u_2 = u_3 = 7/3$].

2. Solve the Poisson's equation $\dfrac{\partial^2 u}{\partial x^2} + \dfrac{\partial^2 u}{\partial y^2} = -4(x^2 + y^2)$ over the square mesh bounded but the sides $x = 0$, $y = 0$, $x = 3$ and $y = 3$, subject to Dirichlet condition $u = 0$ on the boundary of the region.

[The problem is symmetric with respect to the line $x = 0$, and hence $u_2 = u_3$. $u_1 = 37/6$, $u_2 = u_3 = 25/3$, $u_4 = 73/6$].

3. Solve the Poisson's equation $\dfrac{\partial^2 u}{\partial x^2} + \dfrac{\partial^2 u}{\partial y^2} = xy$, $-2 \le x \le 2$, $-2 \le y \le 2$ subject to boundary condition $u = 0$ on $x = \pm 2$, $y = \pm 2$.
[Detect symmetry of the solution. $u_1 = u_9 = -1/4$, $u_3 = u_7 = 1/4$, $u_2 = u_4 = u_6 = u_8 = 0$, $u_5 = 0$].

4. Solve Laplace's equation over the rectangle $0 \le x \le 4$, $0 \le y \le 5$, subject to Dirichlet condition $u = 0$ on $x = 0$, 4 and $u = 1$ on $y = 0$, 5.
[Symmetries of the solution: $u_1 = u_3 = u_{10} = u_{12}$, $u_4 = u_6 = u_7 = u_9$, $u_2 = u_{11}$, $u_5 = u_8$. $u_1 = 33/71$, $u_2 = 41/71$, $u_4 = 20/71$, $u_5 = 27/71$].

7.5 Diffusion and Wave Equation in Two Dimensions

The diffusion equation in two dimensions is extension of Eq. (7.7) in the form

$$\frac{\partial u}{\partial t} = \frac{\partial^2 u}{\partial x^2} + \frac{\partial^2 u}{\partial y^2} \tag{7.17}$$

Let the equation be subject to initial condition $u(x, y, 0) = f(x, y)$ and boundary conditions $u(a, y, t) = \phi_1(y, t)$, $u(b, y, t) = \psi_1(y, t)$, $u(x, c, t) = \phi_2(x, t)$, $u(x, d, t) = \psi_2(x, t)$. Thus, the solution is sought over a domain $R = [a \le x \le b] \times [c \le y \le d] \times [0, T]$.

An explicit method such as the Schmidt method, Eq. (7.8), can be generalised in the obvious way. However, the Crank–Nicolson scheme (7.10) is the best way to treat the problem in one dimension. Its extension in two dimensions in central difference notation is evidently

$$u_{l,m}^{n+1} = u_{l,m}^{n} + \frac{r}{2}\left(\delta_x^2 u_{l,m}^{n+1} + \delta_x^2 u_{l,m}^{n} + \delta_y^2 u_{l,m}^{n+1} + \delta_y^2 u_{l,m}^{n}\right) \tag{7.18}$$

where

$$\delta_x^2 u_{l,m}^{n} := u_{l-1,m}^{n} - 2u_{l,m}^{n} + u_{l+1,m}^{n}$$

and similarly for $\delta_y^2 u_{l,m}^{n}$. Here, it is assumed that discretisations in the x- and y-directions are of equal length h by the points $x_l = a + lh$ and $y_m = b + mh$. The discretisation in t is by the points $t_n = nk$. The scheme (7.18) is certainly viable, but it no longer leads to a tridiagonal system of equations, though it is still sparse.

Another possibility, which we prefer, is a slightly different way of generalising the Crank–Nicolson method. It is called Peaceman–Rachford Alternate Direction Implicit (ADI) method. This method consists of two steps: in the first step, we advance from t_n to $t_{n+1/2}$ and use *implicit differences* for $\partial^2 u/\partial x^2$ and *explicit differences* for

$\partial^2 u/\partial y^2$. In the second step, we advance from $t_{n+1/2}$ to t_{n+1} using *explicit differences* for $\partial^2 u/\partial x^2$ and *implicit differences* for $\partial^2 u/\partial y^2$. Thus, the ADI method is

$$
\begin{aligned}
u_{l,m}^{n+1/2} &= u_{l,m}^{n} + \frac{r}{2}\left(\delta_x^2 u_{l,m}^{n+1/2} + \delta_y^2 u_{l,m}^{n}\right) \\
u_{l,m}^{n+1} &= u_{l,m}^{n+1/2} + \frac{r}{2}\left(\delta_x^2 u_{l,m}^{n+1/2} + \delta_y^2 u_{l,m}^{n+1}\right)
\end{aligned}
\tag{7.19}
$$

The advantage of this method is that each substep requires only the solution of a simple tridiagonal system. The method is second-order accurate in time and space and unconditionally stable.

We now turn to the wave equation in two dimensions, which is

$$
\frac{\partial^2 u}{\partial t^2} = \frac{\partial^2 u}{\partial x^2} + \frac{\partial^2 u}{\partial y^2}
\tag{7.20}
$$

subject to the initial conditions

$$
\begin{aligned}
u(x,\,y,\,0) &= f(x,\,y), \quad a \le x \le b, \quad c \le y \le d \\
\frac{\partial u}{\partial t}(x,\,y,\,0) &= g(x,\,y), \quad a \le x \le b, \quad c \le y \le d
\end{aligned}
$$

and boundary conditions as in the case of Eq. (7.17).

The central difference scheme that can be written for Eq. (7.20) is

$$
\delta_t^2 u_{l,m}^{n} = r^2(\delta_x^2 + \delta_y^2)\,u_{l,m}^{n}
$$

where $r^2 = k^2/h^2$ and it is assumed that the grid lengths in the x- and y-directions are both equal to h. The above equation leads to the three-level *explicit* scheme

$$
u_{l,m}^{n+1} = 2(1 - 2r^2)\,u_{l,m}^{n} + r^2\left(u_{l-1,m}^{n} + u_{l,m-1}^{n} + u_{l+1,m}^{n} + u_{l,m+1}^{n}\right) - u_{l,m}^{n-1}
\tag{7.21}
$$

and can be tackled as in the case of one dimension (see Eq. 7.12). The method is obviously of second order in both space and time.

We now study the stability of the scheme. Substituting

$$
u_{l,m}^{n} = A\xi^{n} e^{iK_1 lh} e^{iK_2 mh}
$$

in Eq. (7.21), we obtain

$$
\xi^2 = 2(1 - 2r^2)\,\xi + r^2(e^{-iK_1 H} + e^{-iK_2 h} + e^{iK_1 h} + e^{iK_2 h})\,\xi - 1
$$

or

$$
\xi^2 - 2[1 - 2r^2(\sin^2\theta_1 + \sin^2\theta_2)]\,\xi + 1 = 0
$$

where $\theta_1 := K_1 h/2$, $\theta_2 := K_2 h/2$. The equation is of the form $\xi^2 - 2B\xi + 1 = 0$ and $|\xi| \leq 1$ if $|B| < 1$ (see Sect. 7.3) or

$$-1 \leq 1 - 2r^2(\sin^2 \theta_1 + \sin^2 \theta_2) \leq 1$$

or

$$r^2 \leq \frac{1}{\sin^2 \theta_1 + \sin^2 \theta_2}$$

which is always satisfied if $r^2 \leq 2$ or $r \leq 1/\sqrt{2} = 0.707$. Thus, for stability, we require that $k \leq 0.707\,h$.

7.6 Convergence: Lax's Equivalence Theorem

In the preceding sections, we have developed finite difference approximations of some linear PDEs with constant coefficients, laying emphasis on the stability of the corresponding linear difference equations. The finite difference approximations bore to the respective PDEs **consistency** in the sense that the *local truncation error* of the approximation was of the order of some positive power of h or k. For instance, we came across first-order approximation, second- order approximation, etc.

The question of **convergence** of these methods was settled by P.D. Lax (1953) in a very general manner. The gist of the result is that *consistency* and *stability* together imply convergence of the method. This means that all the derived methods that were stable are in fact also convergent.

Lax in fact treated the problem abstractly using Banach space terminology, considering a one-parameter family of vectors $\mathbf{u}(t)$ of elements of $\mathcal{B}$ with real parameter t, such that

$$\frac{d\mathbf{u}(t)}{dt} = L\,\mathbf{u}(t), \quad 0 \leq t \leq T \tag{7.22}$$

and $\mathbf{u}(0) = \mathbf{u}_0$. Here, L is a linear operator that may contain inhomgeneous term, and $\mathbf{u}_0$ is a given element of $\mathcal{B}$. In our application, L is a differential operator for which boundary conditions are assumed to be linear and are taken care of by assuming that the domain of L is restricted to those functions satisfying the boundary conditions. Equation (7.22) is thus a system of linear PDEs of *evolution*.

Extending the notion of consistency, stability and convergence for the discretisation of the above-stated problem, it was proved that

Lax's Equivalence Theorem. *Given a properly posed initial value problem and a finite difference approximation to it that satisfies the consistency condition, stability is the necessary and sufficient condition for convergence.*

Proof The proof will be restricted to proving the sufficiency of the conditions that stability and consistency imply convergence. This part is significant for our purpose because adopting a stable scheme which is consistent as well will ensure convergence. The proof of the necessity of the conditions requires use of deeper concepts of Functional Analysis in Banach spaces.

Discretisation of Eq. (7.22) with respect to t leads to a linear difference equation of the form

$$\tilde{\mathbf{u}}^{n+1} = \tilde{\mathbf{u}}^n + k L \tilde{\mathbf{u}}^n \tag{7.23}$$

where $\tilde{\mathbf{u}}$ is an approximation of the exact $\mathbf{u}$, distinction that was ignored in the development of the difference scheme, tacitly assuming convergence.

The stability of the system is theoretically studied in a different manner, where a varied general linear form

$$A_1 \tilde{\mathbf{u}}^{n+1} = A_0 \tilde{\mathbf{u}}^n + F^n \tag{7.24}$$

is considered where F^n consists of not only the data arising from the inhomogeneous term of $L \mathbf{u}$, but also from the contribution of inhomogeneous boundary conditions, if present. The matrix A_1 (as also A_0) must be such that $||A_1|| = O(1/k)$ for (7.24) to be compatible with (7.23), where $|| \cdot ||$ is any norm. Hence, if A_1^{-1} exists

$$||A_1^{-1}|| \le K_1 k \tag{7.25}$$

where K_1 is a constant. The finite difference system (7.24) is defined as *stable* when given two initial data $\mathbf{u}_0$ and $\mathbf{v}_0$, then

$$||\tilde{\mathbf{u}}^n - \tilde{\mathbf{v}}^n|| \le K ||\mathbf{u}_0 - \mathbf{v}_0|| \tag{7.26}$$

Hence,

$$||\tilde{\mathbf{u}}^n - \tilde{\mathbf{v}}^n|| = ||A_1^{-1} A_0 (\tilde{\mathbf{u}}^{n-1} - \tilde{\mathbf{v}}^{n-1})|| = \cdots = ||(A_1^{-1} A_0)^n (\mathbf{u}_0 - \mathbf{v}_0)|| \le K ||\mathbf{u}_0 - \mathbf{v}_0||$$

or

$$||(A_1^{-1} A_0)^n)|| \le K \tag{7.27}$$

Hence, for stable solution, we must have

$$||(A_1^{-1} A_0)^n A_1^{-1}|| \le (K K_1) k \tag{7.28}$$

If a finite difference relation like (7.24) for the exact values $\mathbf{u}^n$ and $\mathbf{u}^{n+1}$ is considered, then

$$T_n := A_1 \mathbf{u}^{n+1} - (A_0 \mathbf{u}^n + F^n) \tag{7.29}$$

is the *truncation error* occurring in the exact system defined by (7.24). If T_m^n represents the truncation error for the element $(m-1) h \leq x \leq mh$, then the varied linear system is said to be *consistent* if $T_m^n \to 0$ as $h^2 + k^2 \to 0$ (compare with Sect. 6.14, Chap. 6).

From Eqs. (7.24) and (7.29), we have by subtraction

$$A_1 (\tilde{\mathbf{u}}^{n+1} - \mathbf{u}^{n+1}) = A_0 (\tilde{\mathbf{u}}^n - \mathbf{u}^n) - T^n$$

or

$$(\tilde{\mathbf{u}}^{n+1} - \mathbf{u}^{n+1}) = (A_1^{-1} A_0) (\tilde{\mathbf{u}}^n - \mathbf{u}^n) - A_1^{-1} T^n$$

Replacing $n+1$ by $n, n-1, n-2, \cdots, 2, 1$, we obtain the recurring system of equations

$$\tilde{\mathbf{u}}^n - \mathbf{u}^n = (A_1^{-1} A_0) (\tilde{\mathbf{u}}^{n-1} - \mathbf{u}^{n-1}) - A_1^{-1} T^{n-1}$$
$$\tilde{\mathbf{u}}^{n-1} - \mathbf{u}^{n-1} = (A_1^{-1} A_0) (\tilde{\mathbf{u}}^{n-2} - \mathbf{u}^{n-2}) - A_1^{-1} T^{n-2}$$
$$\cdots\cdots\cdots\cdots\cdots\cdots\cdots\cdots\cdots\cdots\cdots\cdots\cdots\cdots\cdots$$
$$\tilde{\mathbf{u}}^2 - \mathbf{u}^2 = (A_1^{-1} A_0) (\tilde{\mathbf{u}}^1 - \mathbf{u}^1) - A_1^{-1} T^1$$
$$\tilde{\mathbf{u}}^1 - \mathbf{u}^1 = (A_1^{-1} A_0) (\tilde{\mathbf{u}}^0 - \mathbf{u}^0) - A_1^{-1} T^0$$

Hence, using inequality (7.28), and the consistency conditions

$$||\tilde{\mathbf{u}}^n - \mathbf{u}^n|| \leq (K K_1) k \sum_{l=0}^{n-1} ||T^l|| \to 0, \quad \text{as} \quad k \to 0$$

Thus, $\tilde{\mathbf{u}}^n \to \mathbf{u}^n$ as $k \to 0$, proving convergence of the solution in the limit. $\qquad\square$

Remark 7.4 For a discussion on the relation of the definition of stability adopted here to von Neumann stability, see the text of Morton and Mayers [26].

Remark 7.5 Conceptually, a properly posed IVP like (7.22) means that the solution **u** depends continuously on the initial data $\mathbf{u}_0$ as t increases during the evolution process. Mathematically, it means that if $\mathbf{u}^n$ and $\mathbf{v}^n$ are two solutions at the same time step $t = nk$, subject to initial conditions $\mathbf{u}^0$ and $\mathbf{v}^0$, respectively, then $||\mathbf{u}^n - \mathbf{v}^n|| \leq ||\mathbf{u}^0 - \mathbf{v}^0||$.

Chapter 8
Approximation

In this chapter, we study approximation of a function $f \in C[a, b]$ by another simpler function $\phi \in C[a, b]$ in a general manner. The function ϕ is mostly considered a polynomial of certain degree n. In some cases, ϕ as a *rational function* is also considered. Interestingly, this deeply theoretical development originated in Chebyshev's work on mechanisms of machines (see the text of Steffens cited in the Bibliography); but we confine to a few significant results. These theoretical results have contemporarily been put to good use in computer software for efficiently computing transcendental mathematical functions, and obtaining easy to evaluate functional forms that approximate complicated expressions appearing in theoretical and empirical experimental studies.

We begin with the definition of **norm** of f, denoted by $\| f \|$. As in the case of (finite dimensional) vectors (Chap. 3, Sect. 3.2), the norm can be defined in one of several ways. The useful ones are

(a) The 2–norm: $\| f \|_2 = \left(\int_a^b f^2(x)\,dx \right)^{1/2}$.

(b) Laplace's 1–norm: $\| f \|_1 = \int_a^b |f(x)|\,dx$.

(c) Chebyshev's maximum norm: $\| f \|_\infty = \max\limits_{x \,\in\, [a,\,b]} |f(x)|$.

That the three quantities on the right-hand side of (a), (b) and (c) indeed define norm is a matter of detail, and is left to the reader as an exercise in analysis. The *Chebyshev norm* (c) appeared informally in Sects. 4.1.4 and 4.3 of Chap. 4. It forms the basis of the theory of **uniform approximation**, an outline of which is given in the ensuing subsection. The *2–norm* (a) forms the basis of **least squares approximation** and is easiest to develop. It is discussed in Sect. 2.2. The *Laplace norm* has also been explored in theory but its practical utility is very limited (see the text of J. R. Rice, cited in the Bibliography).

© Springer Nature Singapore Pte Ltd. 2019
S. K. Bose, *Numerical Methods of Mathematics Implemented in Fortran*, Forum for Interdisciplinary Mathematics,
https://doi.org/10.1007/978-981-13-7114-1_8

8.1 Uniform Approximation by Polynomials

The earliest result concerning the degree to which a polynomial can approximate a function is **Weierstrass's Theorem**:

Let $f \in [a, b]$, then for every $\epsilon > 0$ there exists a polynomial ϕ of degree n, such that $\|f - \phi\|_\infty < \epsilon$.

Karl Weierstrass (1815–1897), German mathematician. He is famous for his fundamental contributions in analysis. He is considered to be one of the founders of function theory; the starting point of his work being power series. He fully understood the importance of applications of mathematics for studying problems of physics and astronomy. He gave a nonconstructive proof of the above theorem in 1885.

Sergei Natanovich Bernstein (1880–1968), Ukrainian mathematician. He solved Hilbert's nineteenth problem in his doctoral dissertation of 1904, submitted at Sorbonne, France. But on his return to Kharkov, he had to again work for a doctoral dissertation as foreign qualification for university posts were not recognised. In the second dissertation, he solved Hilbert's twentieth problem. He reproved Weierstrass's theorem in 1912 by a constructive method. Probability theory and application to genetics were his other areas of research.

We shall give here Bernstein's proof in a form that explicitly expresses the order of dependence of ϵ on n. For this purpose, we require the following.

Definition 8.1 The *modulus of continuity* of $f \in C[a, b]$ is the function

$$\omega(\delta) := \max_{x, y \in [a, b]} |f(x) - f(y)| \tag{8.1}$$

where $|x - y| \leq \delta$.

Evidently, since f is continuous, $\omega(\delta) \to 0$ as $\delta \to 0$. Further, if f has continuous first derivative f' on $[a, b]$, then

$$\omega(\delta) \leq \max_{x \in [a, b]} |f'(x)| \cdot \delta$$

Definition 8.2 The Bernstein polynomial $B_n(x)$ is the polynomial

$$B_n(x) := \sum_{k=0}^{n} \binom{n}{k} f\left(\frac{k}{n}\right) x^k (1 - x)^{n-k} \tag{8.2}$$

where $\binom{n}{k} = \dfrac{n!}{k!(n - k)!}$.

Clearly, $B_n(x)$ is a polynomial of degree n. However, it is not an interpolating polynomial because it does not equal $f(x)$ if $f(x)$ is a polynomial of degree n.

The proof of Weierstrass's theorem requires two simple lemmas.

Lemma 8.1 $\displaystyle\sum_{k=0}^{n} \binom{n}{k} x^k (1-x)^{n-k} = 1$

$$\sum_{k=0}^{n} \left(\frac{k}{n} - x\right)^2 \binom{n}{k} x^k (1-x)^{n-k} = \frac{x(1-x)}{n}$$

Proof By binomial theorem

$$\sum_{k=0}^{n} \binom{n}{k} p^k q^{n-k} = (p+q)^n$$

Setting $p = x$, $q = 1 - x$ we obtain the first identity. Next, if we differentiate the binomial identity twice in succession with respect to p, we have

$$\sum_{k=0}^{n} \binom{n}{k} k p^{k-1} q^{n-k} = n(p+q)^{n-1}$$

$$\sum_{k=0}^{n} \binom{n}{k} k(k-1) p^{k-2} q^{n-k} = n(n-1)(p+q)^{n-2}$$

Thus, setting $p = x$, $q = 1 - x$,

$$\sum_{k=0}^{n} \binom{n}{k} \frac{k}{n} x^k (1-x)^{n-k} = x, \quad \sum_{k=0}^{n} \binom{n}{k} \frac{k^2}{n^2} x^k (1-x)^{n-k} = \left(1 - \frac{1}{n}\right) x^2 + \frac{1}{n} x$$

Hence

$$\sum_{k=0}^{n} \left(\frac{k}{n} - x\right)^2 \binom{n}{k} x^k (1-x)^{n-k} = \left(1 - \frac{1}{n}\right) x^2 + \frac{x}{n} - 2x \cdot x + x^2 \cdot 1 = \frac{x - x^2}{n}$$

$\square$

Lemma 8.2 *Let x, $y \in [0, 1]$ and let $\delta > 0$. Let ν be the integral part of $|x - y|/\delta$, then*

$$f(x) - f(y) \le (\nu + 1)\,\omega(\delta)$$

Proof Without loss of generality assume that $x < y$, then

$$\nu \le \frac{y - x}{\delta} \le \nu + 1$$

Let $h = (y - x)/(\nu + 1)$, i.e., divide $[x, y]$ in to $\nu + 1$ equal parts each of length h by points $x_0 = x$, $x_1 = x_0 + h, \cdots , x_{\nu+1} = x_\nu + h$, then

$$|f(x) - f(y)| = \left| \sum_{i=0}^{\nu} [f(x_{i+1}) - f(x_i)] \right| \le \sum_{i=0}^{\nu} |f(x_{i+1}) - f(x_i)|$$

$$\le (\nu + 1) \max_{i = 0, \cdots , \nu} |f(x_{i+1} - f(x_i)| \le (\nu + 1)\omega(h) \le (\nu + 1)\omega(\delta)$$

since $h \le \delta$.   □

Theorem 8.1 *If $f \in C[a, b]$ with modulus of continuity $\omega(\delta)$ then*

$$\|f - B_n\| \le \frac{5}{4}\omega(n^{-1/2})$$

Proof There is no loss of generality in assuming an interval $[0, 1]$, because if $a \le x \le b$, then the transformation $z = (x - a)/(b - a)$ lies in $[0, 1]$. Now by Lemma 1

$$f(x) - B_n(x) = \sum_{k=0}^{n} \left[f(x) - f\left(\frac{k}{n}\right) \right] \binom{n}{k} x^k (1 - x)^{n-k}$$

Hence, $|f(x) - B_n(x)| \le \sum_{k=0}^{n} \left| f(x) - f\left(\frac{k}{n}\right) \right| \phi_k(x)$

where $\phi_k(x) = \binom{n}{k} x^k (1 - x)^{n-k}$. Using Lemma 8.2, the above inequality yields

$$|f(x) - B_n(x)| \le \omega(\delta) \sum_{k=0}^{n} (1 + \nu(k)) \phi_k(x) = \omega(\delta) \left[1 + \sum_{k=0}^{n} \nu(k) \phi_k(x) \right]$$

by the first result of Lemma 8.1. For $\nu(k) \ne 0$, i.e., $\nu(k) \ge 1$,

$$\nu(k) \le \frac{|x - (k/n)|}{\delta} \le \frac{(x - k/n)^2}{\delta^2}$$

Evidently, the inequality between the right-hand and left-hand members remain valid even for $\nu(k) = 0$. Hence, we can write

$$|f(x) - B_n(x)| \le \omega(\delta) \left[1 + \frac{1}{\delta^2} \sum_{k=0}^{n} \left(x - \frac{k}{n} \right)^2 \phi_k(x) \right]$$

$$= \omega(\delta) \left[1 + \frac{1}{\delta^2} \frac{x(1 - x)}{n} \right] \le \omega(\delta) \left(1 + \frac{1}{4\delta^2 n} \right)$$

since $x(1 - x)$ is maximum at $x = 1/2$. Setting $\delta^2 = 1/n$ and taking the maximum over $x \in [0, 1]$, we obtain the theorem. $\qquad\qquad\square$

The Bernstein polynomials approximate $f(x)$ for increasing n for all $x \in [0, 1]$. In other words, the approximation is **uniform** on $[0, 1]$.

The *smoothness* of a function $f \in [0, 1]$ is expressed by the Hölder condition $|f(x) - f(y)| \leq K |x - y|^\alpha$, where constant $K > 0$ and exponent α is also greater than 0. In this case, $\omega(\delta) \leq K \delta^\alpha$ and

$$|f(x) - B_n(x)| \leq \frac{5}{4} K n^{-\alpha/2}$$

Thus if $\alpha \leq 1$ (for $\alpha = 1$, f is said to be Lipschitz bounded), the convergence of the Bernstein polynomials for increasing n is extremely slow. In general, therefore, Bernstein polynomials are of limited practical use even though they are easy to handle. For smoother functions, possessing successive derivatives, the rate of convergence for increasing n becomes much faster according Jackson's theorem. The theorem is of the nature of existence theorem:

Theorem 8.2 *Let $f \in C^p[0, 1]$ and let $\omega_p(\delta)$ be the modulus of continuity of $f^{(p)}$; then for every $n \geq p$, there exists a polynomial $J a_n(x)$ of degree n, such that*

$$\| f - J a_n \|_\infty \leq A_p n^{-p} \omega_p(n^{-1})$$

where A_p is a constant (independent of f).

The proof of this theorem is omitted here.

Dunham Jackson (1888–1946), U.S. mathematician, worked in approximation theory, and trigonometrical and orthogonal polynomials

8.2 Best Uniform (Minimax) Approximation

Consider the set Φ of all polynomials of degree at most n and let Q_n be an arbitrary member of the set.

Definition 8.3 Let $f \in C[a, b]$. If there exists a polynomial ϕ_n^* of degree at most n, such that

$$\| f - \phi_n^* \|_\infty \leq \| f - Q_n \|_\infty$$

then ϕ_n^* is said to be the *best uniform approximation* of f on $[a, b]$.

In respect of the above definition, it can be proved that the best uniform approximation ϕ_n^* certainly exists and is unique. This means that the unique ϕ_n^* satisfies the relation

$$\|f - \phi_n^*\|_\infty = \min_{Q_n \in \Phi} \|f - Q_n\|_\infty$$

For proof, the interested reader may consult Wendroff's book cited in the Bibliography. Best uniform approximation is thus also called Minimax approximation. For different values of n, the function f possesses different minimax approximations ϕ_1^*, ϕ_2^*, $\cdots$. In general, the task of constructing these approximations is not simple, and moreover the errors need not be arbitrarily small for all n.

Theorem 8.3 *If x_0, x_1, $\cdots$, x_{n+1} are arbitrary $n + 2$ points $\in [a, b]$, then for a function f approximated by a polynomial Q_n at most of degree n,*

$$\|f - Q_n\|_\infty \geq |f[x_0, x_1, \cdots, x_{n+1}]|/W(x_0, x_1, \cdots, x_{n+1}) \qquad (8.3)$$

where

$$W(x_0, x_1, \cdots, x_{n+1}) = \sum_{i=0}^{n+1} 1/|(x_i - x_0) \cdots (x_i - x_{i-1})(x_i - x_{i+1}) \cdots (x_i - x_{n+1})|$$

in the notation of Newton's divided differences defined in Chap. 4, Sect. 4.1.2.

Proof From Example 2 of Sect. 4.1.2, Chap. 4, for any function g

$$g[x_0, x_1, \cdots, x_{n+1}] = \sum_{i=0}^{n+1} g(x_i)/\omega_{n+1}'(x_i)$$

where $\omega_{n+1}'(x_i) = (x_i - x_0) \cdots (x_i - x_{i-1})(x_i - x_{i+1}) \cdots (x_i - x_{n+1})$. Setting $g = f - Q_n$, where Q_n is at most of degree n, $Q_n[x_0, x_1, \cdots, x_{n+1}] = 0$ and therefore

$$f[x_0, x_1, \cdots, x_{n+1}] = \sum_{i=0}^{n+1} [f(x_i) - Q_n(x_i)]/\omega_{n+1}'(x_i)$$

Hence,

$$|f[x_0, x_1, \cdots, x_{n+1}]| \leq \sum_{i=0}^{n+1} |f(x_i) - Q_n(x_i)|/|\omega_{n+1}'(x_i)|$$

$$\leq \|f - Q_n\|_\infty \cdot \sum_{i=0}^{n+1} 1/|\omega_{n+1}'(x_i)|$$

$$= \|f - Q_n\|_\infty \cdot W(x_0, x_1, \cdots, x_{n+1})$$

which proves the theorem. $\square$

Example 1. Find lower bound (8.3) for $Q_1(x)$ and $Q_2(x)$ as approximations of e^x on $[-1, 1]$, for $x_0 = -1$, $x_1 = 0$, $x_2 = +1$ and $x_3 = 0$.

Solution. $f(x) = e^x$, $x \in [-1, 1]$. For $Q_1(x)$,

$$f[x_0, x_1, x_2] = \frac{f(x_0)}{(x_0 - x_1)(x_0 - x_2)} + \frac{f(x_1)}{(x_1 - x_0)(x_1 - x_2)} + \frac{f(x_2)}{(x_2 - x_0)(x_2 - x_1)}$$

$$= \frac{e^{-1}}{2} + \frac{e^0}{-1} + \frac{e^1}{2} = 0.54308$$

and $W(x_0, x_1, x_2) = 1/2 + 1/1 + 1/2 = 2$. Hence $\|f - Q_1\|_\infty \geq 0.54308/2 = 0.27154$.

For $Q_2(x)$, similarly let $x_3 = \epsilon$, $\epsilon \to 0$, then

$$f[x_0, x_1, x_2, x_3] = \frac{e^{-1}}{(-1)(-2)(-1)} + \frac{e^0}{(1)(-1)(-\epsilon)} + \frac{e^1}{(2)(1)(1 - \epsilon)} + \frac{e^\epsilon}{(\epsilon + 1)\epsilon(\epsilon - 1)}$$

$$= \frac{e^{-1}}{-2} + \frac{e^0}{\epsilon} + \frac{e^1}{2} - \frac{e^0}{\epsilon} = 1.17520$$

Hence $W(x_0, x_1, x_2, x_3) = 1/2 + 1/2 = 1$ and $\|f - Q_2\|_\infty \geq 1.17520/1 = 1.17520$. □

Example 2. Let $f(x) = x^{n+1}$. Find the nth degree polynomial of best uniform approximation for f in $[-1, 1]$.

Solution. Consider the polynomial $\bar{T}_{n+1}(x) = 2^{-n} T_{n+1}(x)$, where $T_{n+1}(x)$ is the Chebyshev polynomial of degree $n + 1$. According to properties **1** and **2** of Chebyshev polynomials (see Sect. 4.3, Chap. 4), one can write

$$\bar{T}_{n+1}(x) = x^{n+1} - \alpha x^{n-1} - \beta x^{n-3} - \cdots$$

where α, β, $\cdots$ are definite constants. From Theorem 4.3 of Chap. 4 (Sect. 4.3)

$$\|\bar{T}_{n+1}\|_\infty \leq \|\bar{P}_{n+1}\|_\infty$$

where $\bar{P}_{n+1}$ is a polynomial of degree $n + 1$, with leading coefficient 1. Thus if $\phi^*(x) = \alpha x^{n-1} + \beta x^{n-3} + \cdots$, then the above estimate yields

$$\|x^{n+1} - \phi^*\|_\infty \leq \|x^{n+1} - \bar{P}_n(x)\|_\infty$$

where $\bar{P}_n(x)$ is an arbitrary polynomial of degree $\leq n$. Thus $\phi^*(x)$ is the best uniform estimate of x^{n+1}. Moreover,

$$\|x^{n+1} - \phi^*\|_\infty = \|\bar{T}_{n+1}\|_\infty = 2^{-n} \|T_{n+1}\|_\infty = 2^{-n}$$

Hence, ϕ^* approximates f rapidly with increasing n.   □

In the above example, we note that $f(x) - \phi^*(x) = \bar{T}_{n+1}(x)$ attains extreme values at $x_m = \cos(m\pi/(n+1))$, $m = 0, 1, \cdots, n+1$ in the interval $[-1, 1]$. Thus the difference $f(x) - \phi^*(x)$ attains maximum modulus value at $n+2$ points with alternate sign. This is true in general; but toward preparation we have the result due to C. J. de la Vallée–Poussin.

Theorem 8.4 *Let there be a polynomial $P_n \in \Phi$ of at most degree n such that $f(x) - P_n(x)$ alternates in sign at the $n+2$ points $a \le x_0 < x_1 < \cdots < x_{n+1} \le b$, that is*

$$f(x_i) - P_n(x_i) = \sigma(-1)^i \lambda_i, \quad (i = 0, 1, \cdots, n+1) \tag{8.4}$$

where σ is $+1$ or -1 and $\lambda_i > 0$ $(i = 0, 1, \cdots, n+1)$, then the best polynomial approximation ϕ_n^* of degree at most n satisfies

$$min\{\lambda_0, \lambda_1, \cdots, \lambda_{n+1}\} \le \|f - \phi_n^*\|_\infty \le \|f - P_n\|_\infty \tag{8.5}$$

Proof The right-hand inequality is obvious. To prove the left-hand inequality suppose it is hypothesised that

$$\|f - \phi_n^*\|_\infty < min\{\lambda_0, \lambda_1, \cdots, \lambda_{n+1}\} =: \mu \quad (say)$$

then we show that there is a contradiction. By definition of ϕ_n^*, there exists a polynomial $Q_n \in \Phi$ of degree $\le n$, such that

$$\|f - \phi_n^*\|_\infty \le \|f - Q_n\|_\infty < \mu$$

in which $\mu < \lambda_i$ $(i = 0, 1, \cdots, n+1)$ or,

$$|f(x_i) - Q_n(x_i) < \mu \quad \text{i.e.,} \quad -\mu < f(x_i) - Q_n(x_i) < \mu$$

Now consider the difference of the polynomials P_n and Q_n, viz., $\Delta(x) = P_n(x) - Q_n(x)$. Evaluating $\Delta(x)$ at the points $x_0, x_1, \cdots, x_{n+1}$, we have assuming $\sigma = +1$ (an analogous argument would work for $\sigma = -1$):

$$\Delta(x_0) = P_n(x_0) - Q_n(x_0) = [f(x_0) - Q_n(x_0)] + [P_n(x_0) - f(x_0)] < \mu - \lambda_0 < 0$$

$$\Delta(x_1) = P_n(x_1) - Q_n(x_1) = [f(x_1) - Q_n(x_1)] + [P_n(x_1) - f(x_1)] > -\mu + \lambda_1 > 0$$

etc. In general, $\Delta(x_i) = (-1)^{i+1}$. Therefore, $\Delta(x_i)$ changes sign $(n+2) - 1 = n + 1$ times. This means that $P_n(x) - Q_n(x)$ is a polynomial of degree n vanishing $n+1$ times. By the fundamental theorem of algebra, this cannot happen unless $Q_n \equiv P_n$ and the starting hypothesis $|f(x) - \phi_n^*(x)| < \mu$, $x \in [a, b]$ yields for the $n+2$ points x_i, $\lambda_i < \mu$, which is impossible.   □

Charles Joseph de la Vallée Poussin (1866–1962), Belgian mathematician, is best known for the proof of the prime number theorem in 1896. He obtained the above theorem in 1908. His other area of interest was analysis.

For the polynomial P_n satisfying (8.4), Eq. (8.5) gives an estimate of the error from the minimax approximation ϕ_n^*.

Another conclusion that can be drawn is that if $\lambda_i = \|f - P_n\|_\infty$ for $i = 0, 1, \cdots$, $n + 1$, then

$$\|f - P_n\|_\infty \geq \|f - \phi_n^*\|_\infty \geq \min_i \{\lambda_i\} = \|f - P_n\|_\infty$$

which implies that

$$\|f - \phi_n^*\|_\infty = \|f - P_n\|_\infty = |f(x_i) - P_n(x_i)| \quad (i = 0, 1, \cdots, n + 1)$$

This suggests the following theorem of far-reaching importance originating from the works of Chebyshev (1859), A. A. Markov and V. A. Markov (1892).

Theorem 8.5 *For a polynomial ϕ_n^* of degree at most n to be a polynomial of best uniform approximation to a function $f \in C[a, b]$, it is necessary and sufficient that there exists at least $n + 2$ points $a \leq x_0 < x_1 < \cdots < x_{n+1} \leq b$, such that*

$$f(x_i) - \phi^*(x_i) = \sigma(-1)^i \|f - \phi_n^*\|_\infty \qquad (8.6)$$

where $i = 0, 1, \cdots, n + 1$ and $\sigma = 1$ or -1 simultaneously for all i.

The sufficiency part of the theorem for exactly $n + 2$ points of sign alterations is easily proved. For, if the condition (8.6) is satisfied

$$\|f - \phi_n^*\|_\infty = |f(x_i) - \phi^*(x_i)| \geq \min_i |f(x_i) - \phi^*(x_i)|$$

which is true according to (8.5). It is more difficult to prove the necessary part of the theorem, as it is based on several constructs and contradiction. The interested reader may consult the texts of Wendroff or Hammerlein (see Bibliography).

The points $x_0, x_1, \cdots, x_{n+1}$ which satisfy the conditions (8.6) of the theorem are called Chebyshev **alternants**. At such points $f(x) - \phi_n^*(x)$ attains extremum values and hence $f'(x) - \phi_n^*(x) = 0$ at these points. The extremum values are of equal magnitude, alternately changing in sign. The theorem is therefore also called *equioscillation theorem*.

From the definition and the above theorem, E. Y. Remez (1934) developed the *Exchange Algorithm* for constructing best uniform polynomial approximation of a given function $f \in C[a, b]$. But near-best uniform approximation can be obtained in much simpler ways.

Remez Exchange Algorithm. Let $\phi_n^*(x) = a_0^* + a_1^* x + \cdots + a_n^* x^n$ and let $\mu = -\sigma \| f - \phi_n^* \|_\infty$, then according to Eq. (8.6)

$$a_0^* + a_1^* x_i + \cdots + a_n^* x_i^n - (-1)^i \mu = f(x_i), \quad i = 0, 1, \cdots, n+1$$

Had the Chebyshev alternants x_i been known, the above set of $n + 2$ *linear equations* could have been solved for the $n + 2$ unknowns $a_0, a_1, \cdots, a_n$ and μ. So Remez's method proceeds iteratively, starting with points motivated by Chebyshev interpolation nodes (see Eq. (4.23), Chap. 4):

Step 1. Set $x_i = \dfrac{b+a}{2} + \dfrac{b-a}{2} \cos \dfrac{(n-i+1)\pi}{n+1}, \quad i = 0, 1, \cdots, n+1$

(Evidently $x_0 = a$, $x_{n+1} = b$).

Step 2. Compute (say by Gaussian elimination) $a_0^*, a_1^*, \cdots, a_n^*, \mu$ from the set of equations

$$a_0^* + a_1^* x_i + \cdots + a_n^* x_i^n - (-1)^i \mu = f(x_i), \quad i = 0, 1, \cdots, n+1$$

Step 3. Locate the extreme points $a \le y_0 < y_1 < \cdots < y_{n+1} \le b$ of the function $f(x) - \phi_n^*(x)$, by computing the zeros of $f'(x) - \phi_n^{*\prime}(x)$, by the bisection method. High accuracy in this computation is not required since the coefficients $a_0^*, \cdots, a_n^*$ are insensitive to small changes in x_i.

Step 4. If $|y_i - x_i| < \epsilon$ for $i = 0, 1, \cdots, n+1$, STOP. Else set $x_i = y_i$ and Go To **Step 2**.

It can be proved that the process does converge for every choice of starting values in **Step 1**, provided that the computed value of μ does not vanish. With mild additional assumptions regarding differentiability of f, it can also be proved that the *convergence is quadratic*. Hence, in practice, only a small number of iterations suffice. We, however, do not pursue the matter because near-best uniform approximation is easily constructed.

Evgeny Yacovlevich Remez (1896–1975). Ukrainian mathematician. His main work has been on constructive theory of functions and approximation theory.

Fike (see Bibliography) quotes some best uniform approximation results that are essentially based on the above theory:

(1) $\sin\left(\dfrac{\pi}{4} x\right) \approx 0.7853981609\, x - 0.0807454325\, x^3 + 0.0024900010\, x^5$

$\qquad\qquad - 0.0000359505\, x^7, \quad -1 \le x \le 1, \ \epsilon_1 = 3.20 \times 10^{-7}$

(2) $\cos\left(\dfrac{\pi}{4} x\right) \approx 0.9999999724 - 0.3084242536\, x^2 + 0.0158499153\mathbf{x}^4$

$\qquad\qquad -0.0003188805\, x^6, \quad -1 \le x \le 1, \ \epsilon = 2.76 \times 10^{-6}$

(3) $\ln(1+x) \approx 0.0000607 + 0.9965407\,x - 0.4678348\mathbf{x}^2 + 0.2208915\,x^3$

$\qquad -0.0565718\,x^4, \quad 0 \le x \le 1, \ \epsilon = 6.07 \times 10^{-5}$

(4) $\tan^{-1} x \approx 0.99538\,x - 0.288690\,x^3 + 0.079339\,x^5, \quad -1 \le x \le 1,$
$\epsilon = 6.09 \times 10^{-4}$

where ϵ, ϵ_1 are, respectively, maximum absolute and relative errors.

In the following, we give some simple exercises.

Example 3. Determine the linear minimax approximation $A + Bx$ of $\sqrt{x}$ in $\left[\dfrac{1}{16}, 1\right]$.

Solution. The number of alternation points is at least $1 + 2 = 3$. Let us assume the points to be $1/16$, α, 1. As in Remez's method, Eq. (8.6) yields the equations

$$A + \frac{B}{16} - \mu = \sqrt{\frac{1}{16}} = \frac{1}{4}, \quad A + B\alpha + \mu = \sqrt{\alpha}, \quad A + B \cdot 1 - \mu = \sqrt{1} = 1$$

The first and the third equations yield $B = 4/5$. At the interval point $\alpha, d(A + Bx - \sqrt{x})/dx = 0$ yields $\alpha = 1/4B^2 = 25/64$. Hence, the last two equations become

$$A + \mu = \sqrt{\alpha} - B\alpha = \frac{5}{16}, \quad A - \mu = 1 - B = \frac{1}{5}$$

Solving these equations we get $A = 41/160$. Hence, the minimax approximation is

$$\sqrt{x} \approx \frac{41}{160} + \frac{4}{5}\,x$$

$\qquad\qquad\qquad\qquad\qquad\qquad\qquad\qquad\qquad\qquad\qquad\qquad\qquad\qquad$ □

Exercises

1. Find a lower bound for the maximum absolute error (following Theorem 3), when $\sqrt{x}$ is approximated by a quadratic polynomial $Q_2(x)$ in the interval $[0, 2]$. [Choosing $x_0 = 0$, $x_1 = 1/2$, $x_2 = 1$, $x_3 = 2$, $f[x_0, \cdots, x_3] = 0.35702$, $W(x_0, \cdots, x_3) = 6$, $\max|\sqrt{x} - Q_2(x)| \ge 0.05950$].

2. As in Ex. 1 above find a lower bound when e^x is approximated by a cubic in $[-1, 1]$. [Choosing $x_0 = -1$, $x_1 = -1/2$, $x_2 = 0$, $x_3 = 1/2$, $x_4 = 1$, $f[x_0, \cdots, x_4] = 0.04343$, $W(x_0, \cdots, x_4) = 32/3$. $\max|e^x - Q_3(x)| \ge 0.00408$].

3. Determine the best uniform linear approximation $A + Bx$ of e^x in $[0, 1]$.

[See Example 3. $A = \frac{1}{2}[\alpha + (1 - \alpha)e]$, $B = e - 1$ where $\alpha = \ln(e - 1)$. $e^x \approx$
$0.89407 + 1.71828x$].

4. Determine the linear approximation $A + Bx$ of $\sqrt{x}$ with minimax *relative error*
in $[1/16, 1]$.
[Apply Theorem 8.4 and Remez's method to $E(x) = (\sqrt{x} - A - Bx)/\sqrt{x} = 1 -$
$(A + Bx)/\sqrt{x}$. As in Example 3, $(A + B/16)/(1/4) - \mu = 1$, $(A + B\alpha)/\sqrt{\alpha} +$
$\mu = 1$, $(A + B) - \mu = 1$ and $E'(\alpha) = 0$. Therefore, $B = 4A$, $\alpha = A/B = 1/2$,
$A = 2/9$, $B = 8/9$. Hence, $\sqrt{x} \approx 2/9 + 8x/9$].

5. If $f(x)$ is an even function in $[-a, a]$, then prove that the minimax polynomial
approximation $\phi_n^*(x)$ of degree at most n is also an even function.
[Apply Theorem 8.5].

8.2.1 Near-Best Uniform Approximation

Owing to some difficulty in obtaining best uniform approximation by the Remez
exchange algorithm, we now consider procedures that yield near-best uniform
approximations.

8.2.1.1 Polynomial Interpolation with Chebyshev Interpolation Points

If $f \in C^{(n+1)}[a, b]$ and $f^{(n+1)}$ varies but little in $[a, b]$, then the remainder in
Lagrange interpolation, denoted by L_n, becomes minimal when the interpolation
points are taken as the Chebyshev points (4.23) of Sect. 4.3, Chap. 4. In this respect
from Eq. (4.24) of that chapter

$$\|f - L_n\|_\infty \leq \frac{M_{n+1}}{(n+1)!} \frac{(b-a)^{n+1}}{2^{2n+1}}$$

where $|f^{(n+1)}(\xi)| \leq M_{n+1}$, $\xi \in (a, b)$. The maximum norm on the left-hand side
therefore vanishes as $n \to \infty$ if M_{n+1} remains bounded. Near-best uniform approximation can therefore be obtained by employing interpolation polynomial of suitable
degree n over the Chebyshev points.

The above-stated scheme is implemented in the following subroutine named
CHEBYSHEV for the standard interval $[-1, 1]$ obtained by the linear transformation
$x \leftarrow (2x - b - a)/(b - a)$, $x \in [a, b]$. Moreover, the interpolation polynomial is
written as $a_1^* + a_2^* x + \cdots + a_{n+1}^* x^n$ in keeping with the subscript notation used in
the case of matrices. Accordingly, the Chebyshev and the Remez interpolation points
(Eq. (4.23), Chap. 4 and Step 1 of the Remez algorithm), respectively take the form

$$x_i = \cos \frac{(2i-1)\pi}{2(n+1)} \quad \text{and} \quad x_i = \cos \frac{(n-i+2)\pi}{n+1}$$

Hence, these points must satisfy the system of linear equations

$$a_1^* + a_2^* x_i + a_3^* x_i^2 \cdots + a_{n+1}^* x_i^n = f(x_i), \quad i = 1, 2, \cdots, n + 1 \qquad (8.7)$$

Since the points x_i are distinct, the system of equations has always a solution, which can be determined by the Gauss elimination method (Chap. 3, Sect. 3.1.2). The solution yields the coefficients of the interpolation polynomial $L_n(x)$:

$$f(x) \approx L_n(x) = a_1^* + a_2^* x + a_3^* x^2 + \cdots + a_{n+1}^* x^n \qquad (8.8)$$

```
SUBROUTINE CHEBYSHEV(n,astar)
! Determines the coefficients of the interpolated polynomial
! on the interval [−1, 1] of a function f(x) choosing either.
! Chebyshev Interpolation Point or Remez Points.
! Uses function subprogram fn.
! n=degree of the interpolation polynomial.    (Input)
! astar(1), astar(2),···, astar(n+1) = coefficients of the
! nth degree ploynomial    (Output)
!***********************************************
REAL :: x(n+1), fn(n+1), astar(n+1), A(n+1,n+1)
DO i=1,n+1
x(i)=cos((2*i-1)*1.570796/(n+1))          ! (Chebyshev Points)
! x(i)=cos((n-i+2)*3.141592653/(n+1))    ! (Remez Points)
fn(i)=f(x(i))
END DO
DO i=1,n+1
DO j=1,n+1
A(i,j)=x(i)**(j-1)
END DO
END DO
CALL GAUSS(n+1,A,fn)          ! (The solution is returned in fn)
DO i=1,n+1
astar(i)=fn(i)
END DO
RETURN
END SUBROUTINE CHEBYSHEV
!***************************************************************
FUNCTION f(x)
f=···········
RETURN
END FUNCTION f(x)
!***************************************************************
! Append SUBROUTINE GAUSS
```

Applying the above subroutine CHEBYSHEV, for the functions $\sin(\frac{\pi}{4}x)$ and $\cos\left(\frac{\pi}{4}x\right)$, the following interpolated approximations are obtained:

$$\sin\left(\frac{\pi}{4}x\right) \approx 0.78540\,x - 0.08075\,x^3 + 0.00249\,x^5 - 0.00004\,x^7$$

$$\cos\left(\frac{\pi}{4}\right) \approx 1 - 0.30842\,x^2 + 0.01582\,x^4 - 0.00004\,x^6$$

The right-hand side expressions in both the cases are the same irrespective of the choice of Chebyshev or Remez interpolation points. The two expressions are remarkably accurate to single precision arithmetic, when they are compared to the accurate expressions obtained by the Exchange Algorithm quoted in the preceding section.

8.2.2 Chebyshev Series and Economisation of Polynomials

The Chebyshev polynomials $T_n(x)$ introduced in Sect. 4.3 of Chap. 4 satisfy the property that

$$\int_{-1}^{1} \frac{T_i(x)\,T_j(x)}{\sqrt{1-x^2}}\,dx = \int_{0}^{\pi} \cos(i\theta)\,\cos(j\theta)\,d\theta = \begin{cases} \pi, & i = j = 0 \\ \pi/2, & i = j \neq 0 \\ 0, & i \neq j \end{cases}$$

In Sect. 8.4 of this chapter, on *orthogonal polynomials*, such property will be called **orthogonality**. Following this property, one can contemplate expansion of a given function $f \in C[-1, 1]$ in a Fourier type *Chebyshev series*

$$f(x) = \frac{c_0}{2} + \sum_{j=1}^{\infty} c_j\,T_j(x) \tag{8.9}$$

Naively multiplying the two sides by $T_i(x)/\sqrt{1-x^2}$ and integrating from -1 to 1, we get from the orthogonality property

$$c_i = \frac{2}{\pi} \int_{-1}^{1} \frac{f(x)\,T_i(x)}{\sqrt{1-x^2}}\,dx = \frac{2}{\pi} \int_{0}^{\pi} f(\cos\theta)\,\cos i\theta\,d\theta, \quad i = 0,\,1,\,2,\,\cdots \tag{8.10}$$

However importantly, according to a theorem in analysis that when $f \in C[-1, 1]$ the infinite series in Eq. (8.9) indeed converges to $f(x)$. This means that the nth partial sum

$$c_n(x) = \frac{c_0}{2} + \sum_{j=1}^{n} c_j\,T_j(x) \tag{8.11}$$

approximates $f(x)$, which improves with increasing n. In fact it can be proved that if $f \in C^p[-1, 1], \quad p \geq 1$, then

$$\|f - c_n\|_\infty \leq \frac{A_p \ln n}{n^p}, \quad n \geq 2$$

where A_p is dependent on f. For proof, see the texts of Rivlin and Meinardus cited in the Bibliography.

The exact determination of the coefficients c_i is often difficult except in simple cases. One useful simple case is the following:

$$x^k = \frac{\gamma_0}{2} + \sum_{j=1}^{k} \gamma_j T_j(x), \quad \gamma_j = \frac{2}{\pi} \int_{-1}^{1} \frac{x^k T_j(x)}{\sqrt{1-x^2}} dx = \frac{2}{\pi} \int_0^\pi \cos^k \theta \cos j\theta \, d\theta$$

(8.12)

The definite integral for γ_j can be evaluated by noting that when k is odd, j takes the values 1, 3, 5, $\cdots$, k and when k is even, j takes the values 0, 2, 4, $\cdots$, k. Now the integrand for γ_j can be written as

$$\frac{1}{2^{k+1}} (e^{i\theta} + e^{-i\theta})^k (e^{ij\theta} + e^{-ij\theta}), \quad i = \sqrt{-1}$$

$$= \frac{1}{2^{k+1}} \left[e^{ik\theta} + \binom{k}{1} e^{i(k-2)\theta} + \binom{k}{2} e^{i(k-4)\theta} + \cdots + e^{-ik\theta} \right] (e^{ij\theta} + e^{-ij\theta})$$

The term $e^{i(k-2r)\theta} \cdot e^{ij\theta}$ becomes 1 if $r = (j+k)/2$, and analogously, we get 1 also from the term $e^{i(2r-k)\theta} \cdot e^{-ij\theta}$ with the same value of r. When we integrate over $[0, \pi]$, all other terms vanish and we are left with

$$\gamma_j = \frac{2}{\pi} \cdot \frac{1}{2^{k+1}} \cdot \binom{k}{(j+k)/2} \cdot 2\pi = \frac{1}{2^{k-1}} \binom{k}{(j+k)/2}$$

(8.13)

For particular values of $k = 0, 1, 2, \cdots, 10$, the formulae (8.12) and (8.13) yield the following useful expressions:

$$1 = T_0(x), \qquad\qquad\qquad x = T_1(x)$$

$$x^2 = \tfrac{1}{2}[T_0(x) + T_2(x)], \qquad\qquad x^3 = \tfrac{1}{4}[3\, T_1(x) + T_3(x)],$$

$$x^4 = \tfrac{1}{8}[3\, T_0(x) + 4\, T_2(x) + T_4(x)], \quad x^5 = \tfrac{1}{16}[10\, T_1(x) + 5\, T_3(x) + T_5(x)]$$

$$x^6 = \tfrac{1}{32}[5\, T_0 + 15\, T_2 + 6\, T_4 + T_6]$$

$$x^7 = \tfrac{1}{64}[35\, T_1(x) + 21\, T_3(x) + 7\, T_5(x) + T_7(x)]$$

$$x^8 = \tfrac{1}{128}[70\, T_0 + 56\, T_2 + 28\, T_4 + 8\, T_6 + T_8]$$

(8.14)

$$x^9 = \tfrac{1}{256}[126\, T_1(x) + 84\, T_3(x) + 36\, T_5(x) + 9\, T_7(x) + T_9(x)]$$

$$x^{10} = \tfrac{1}{512}[252\, T_0 + 210\, T_2 + 40\, T_4 + 45\, T_6 + 10\, T_8 + T_{10}], \quad etc.$$

The first few in the above set of relations can of course be derived from Eq. (4.18) of Chap. 4.

The fact that the Chebyshev series expression on the right-hand side of Eq. (8.13) decreases by a factor $1/2^{k-1}$ as k increases, and that $|T_n(x)| \le 1$, one can economise a certain polynomial $Q_n(x) = \sum_{k=0}^{n} \alpha_k x^k$ to the same order of accuracy by a polynomial of degree less than n. The procedure is examplified below.

Example 1. Let the slowly converging Madhav–Gregory infinite series

$$\tan^{-1} x = x - \frac{x^3}{3} + \frac{x^5}{5} - \frac{x^7}{7} + \frac{x^9}{9} - \cdots, \quad -1 \le x \le 1$$

be truncated to the 9-th power of x for approximation of $\tan^{-1} x$, in the range $[-\alpha, \alpha]$, where $\alpha = \tan \frac{\pi}{8}$. Economise the polynomial to the same order of accuracy.

Solution. $\alpha = \tan(\pi/8) = 0.4142136$. Hence if

$$f(x) := \tan^{-1} x, \quad Q_9(x) = x - \frac{x^3}{3} + \frac{x^5}{5} - \frac{x^7}{7} + \frac{x^9}{9}$$

then $f(\alpha) = \pi/8 = 0.3926991$ and $Q_9(\alpha) = 0.3927040$. Since the function is monotonic increasing

$$\|f - Q_9\|_\infty = |f(\alpha) - Q_9(\alpha)| = 0.49 \times 10^{-5}$$

Via the substitution $x = \alpha t$ we pass to the interval $[-1, 1]$ so that

$$S_9(t) := Q_9(\alpha t) = 0.4142136\, t - 0.0236893\, t^3 - 0.0024387\, t^5$$
$$- 0.0002989\, t^7 + 0.0000399\, t^9$$

Expressing t, t^3, t^5, t^7, t^9 in terms of $T_1(t)$, $T_3(t)$, $T_5(t)$, $T_7(t)$, $T_9(t)$ from Eq. (8.14), we obtain

$$S_9(t) = 0.3978269\, T_1(t) - 0.0053112\, T_3(t) + 0.0001253\, T_5(t)$$

$$-0.0000033\, T_7(t) + 0.0000002\, T_9(t)$$

The error due to the last two terms is less than 0.49×10^{-5}. Hence, to the same accuracy

$$Q_9(t) = 0.4143870\, t - 0.0237508\, t^3 + 0.0020048\, t^5$$

$$= 1.0004186\, x - 0.3341992\, x^3 + 0.1644182\, x^5$$

Thus $\tan^{-1} x$ can be approximated by the above quintic in x to the same order of accuracy in the interval $[-\alpha, \alpha]$.   □

Exercises

1. Prove that the Chebyshev series expansion of an even (odd) function is a series of even (odd) order Chebyshev polynomials.

2. Applying the Trapezoidal Rule, prove that the Chebyshev coefficients can be approximately computed from the expression

$$c_i \approx \frac{1}{n}\left[f(0) + (-1)^i f(-1) + 2 \sum_{k=1}^{n-1} f\left(\cos \frac{\pi k}{n} \right) \cos \frac{\pi k i}{n} \right]$$

3. Prove that the Chebyshev series expansion of $\sin^{-1} x$, $x \in [-1, 1]$ is

$$\sin^{-1} x = \frac{4}{\pi}\left[T_1(x) + \frac{1}{9} T_3(x) + \frac{1}{25} T_5(x) + \cdots \right]$$

Hence, show that $\dfrac{\pi^2}{8} = 1 + \dfrac{1}{9} + \dfrac{1}{25} + \cdots$

[Since this is a case of odd function, $c_{2i} = 0$. The odd coefficients are

$$c_{2i+1} = \frac{2}{\pi} \int_{-1}^{1} \sin^{-1} x \cdot \cos((2i + 1) \cos^{-1} x) \frac{dx}{\sqrt{1 - x^2}}, \quad i = 0, 1, 2, \cdots$$

or, setting $x = \cos\theta$, $c_{2i+1} = \dfrac{2}{\pi} \displaystyle\int_0^{\pi} \left(\dfrac{\pi}{2} - \theta \right) \cos(2i + 1)\theta\, d\theta = \dfrac{4}{\pi} \dfrac{1}{(2i + 1)^2}$.

Finally, set $x = 1$, and $T_1(1) = 1$].

4. Prove the Chebyshev series expansion

$$\ln \frac{1+x}{1-x} = 4\left[T_1(x) + \frac{1}{3} T_3(x) + \frac{1}{5} T_5(x) + \cdots \right]$$

[From Eq. (8.10), $c_i = \dfrac{2}{\pi} \displaystyle\int_{-1}^{1} \ln \dfrac{1+x}{1-x} \dfrac{T_i(x)}{\sqrt{1-x^2}}\, dx = \dfrac{4}{\pi} \displaystyle\int_0^{\pi} \ln\left(\cot \dfrac{\theta}{2} \right) \cos i\theta\, d\theta$.

Hence, $c_0 = 0$ and $c_i = \dfrac{2}{\pi} \displaystyle\int_0^{\pi} \dfrac{\sin i\theta}{\sin\theta}\, d\theta =: \dfrac{2}{\pi i} \cdot I_i$. Now prove that

$I_i = \frac{1}{2} I_i + \frac{1}{2} I_{i-2}$ or, $I_i = I_{i-2}$].

5. The polynomial $P_3(x) := 5 - 20x - 9x^2 + x^3$ is given in $[0, 4]$. Find the second-degree polynomial $P_2(x)$ such that $\delta = \max\limits_{x \in [0, 4]} |P_3(x) - P_2(x)|$ becomes the least possible. What is the value of δ?

[Set $x = 2(t + 1)$, $-1 \le t \le 1$. Then $P_3(t) = -16 T_0 - 82 T_1 - 6 T_2 + 2 T_3$. For P_2 truncate at T_2. Hence, $|P_3(t) - P_2(t)| = 2|T_3| \le 2$. Thus $\delta = 2$ and $P_2(t) = -16 - 82 t - 6(2t^2 - 1)$. Therefore $P_2(x) = 7 - 29 x - 3 x^2$].

6. Find a polynomial $P(x)$ of degree as low as possible such that

$$\max_{|x| \le 1} |e^x - P(x)| \le 0.05$$

$[e^x = 1 + x + \dfrac{x^2}{2} + \dfrac{x^3}{6} + \dfrac{x^4}{24} + \dfrac{x^5}{120} + \cdots\cdots$. For $|\text{error}| \le 0.05$ truncate at x^4. Hence, $e^x \approx 1.5156\, T_0 + 1.1250\, T_1 + 0.5208\, T_2 + 0.0417\, T_3 + 0.0052\, T_4$. For $|\text{error}| \le 0.05$, neglect T_3, T_4. Hence, $P(x) = 0.995 + 1.125\, x + 1.042\, x^2]$.

7. Taking the Chebyshev approximation of $\cos\left(\dfrac{\pi}{4} x\right)$, $x \in [-1, 1]$ as

$$P(x) := 1.00000 - 0.30842\, x^2 + 0.01585\, x^4 - 0.00032\, x^6$$

Economise the polynomial for $|\text{error}| < 10^{-5}$.
$[P(x) = 0.85163\, T_0 - 0.14644\, T_2 + 0.00192\, T_4 - 0.00001\, T_6$. Neglect T_6 to obtain $P(x) = 0.99615 - 0.30824\, x^2 + 0.10536\, x^4]$.

8. The *error function* is defined by

$$\text{erf}(x) = \frac{2}{\sqrt{\pi}} \int_0^x e^{-t^2/2}\, dt$$

Determine a low-degree polynomial $P(x)$ in $[-1, 1]$ for which $|f(x) - P(x)| \le 10^{-4}$.
[Integrating the series expansion of $e^{-t^2/2}$

$$\frac{\sqrt{\pi}}{2}\, \text{erf}(x) = x - \frac{x^3}{6} + \frac{x^5}{40} - \frac{x^7}{336} + \frac{x^9}{3456} - \frac{x^{11}}{42240} + \cdots$$

For $|\text{error}| \le 10^{-4}$ truncate at x^9. Hence, $\dfrac{\sqrt{\pi}}{2}\text{erf}(x) \approx 0.889116\, T_1 - 0.034736\, T_3 + 0.001278\, T_5 - 0.000036\, T_7 + 0.000000\, T_9$. For $|\text{error}| \le 10^{-4}$ neglect T_7 and T_9. Hence, $P(x) = 1.1280\, x - 0.1856\, x^3 + 0.0230\, x^5]$.

8.3   Least Squares Approximation

We now consider the approximation of $f \in C[a, b]$ by a function $\phi \in C[a, b]$ based on the $\|\cdot\|_2$–norm. The 2–norm is taken here in a slightly more general form according to

$$\|f\|_2 = \left(\int_a^b [f(x)]^2 w(x)\, dx\right)^{1/2} \tag{8.15a}$$

or, alternatively in a discrete version

$$\|f\|_2 = \left(\sum_{\nu=0}^{N} [f(x_\nu)]^2 w(x_\nu) \right)^{1/2} \tag{8.15b}$$

Here the **weight function** $w(x) > 0$, $x \in (a, b)$ and in Eq. (8.15b) $x_0, x_1, \cdots, x_N$ are certain fixed points in (a, b). In the space of continuous functions, the norms are induced by the **scalar** (or **inner**) **product** $\langle \cdot \rangle$ of two elements f, g defined by

$$\langle f, g \rangle = \int_a^b f(x) g(x) w(x) \, dx \tag{8.16a}$$

or

$$\langle f, g \rangle = \sum_{\nu=0}^{N} f(x_\nu) g(x_\nu) w(x_\nu) \tag{8.16b}$$

in the discrete version $w(x)$ is usually taken as unity, if nothing is stated specially to the contrary.

Let the approximating function ϕ be a **generalised polynomial** of the form

$$\phi = a_0 \phi_0 + a_1 \phi_1 + \cdots + a_n \phi_n \tag{8.17}$$

where $\phi_0, \phi_1, \cdots, \phi_n$ are linearly independent functions continuous on $[a, b]$ and $a_0, a_1, \cdots, a_n$ are constants. When $\phi_i(x) = x^i$, ϕ is a polynomial of degree n.

Definition 8.4 The best generalised polynomial approximation ϕ of f on $[a, b]$ is that polynomial which makes $\| f - \phi\|_2$ the least. For this reason, the best approximation is called **least squares approximation**.

The original idea is due to Gauss, who considered smoothing of experimental data containing errors. He proposed that the sum of the squares of the errors be minimised, in contrast to Laplace's proposal of minimising the sum of the absolute values of the errors which tantamount to minimising the 1-norm. It is easy to see that if there are N measurements $y_1, y_2, \cdots, y_N$ of a quantity y, then the least squares method produces the average value $\bar{y}$ of these measurements as the approximation to y. Indeed, the sum of squares of the errors is $(y - y_1)^2 + \cdots + (y - y_N)^2$, which is minimum for $(y - y_1) + \cdots + (y - y_N) = 0$, yielding the solution $\bar{y} = (y_1 + \cdots + y_N)/N$.

To obtain the least squares approximation, we have

$$\|f - \phi\|_2^2 = \left\langle f - \sum_{i=0}^{n} a_i \phi_i, \ f - \sum_{j=0}^{n} a_j \phi_j \right\rangle$$

$$= \langle f, f \rangle + \sum_{i=0}^{n} \sum_{j=0}^{n} a_i a_j \langle \phi_i, \phi_j \rangle - 2 \sum_{i=0}^{n} a_i \langle f, \phi_i \rangle$$

The right-hand side is a quadratic expression in $a_0, \cdots, a_n$. It is known from the theory of such expressions that it possesses a unique minimum or least value. For the least value, differentiating the above expression with respect to a_i $(i = 0, 1, \cdots, n)$ and equating to zero, one obtains the so-called **normal equations**

$$\sum_{j=0}^{n} a_j \langle \phi_i, \phi_j \rangle = \langle f, \phi_i \rangle \quad (i = 0, 1, \cdots, n) \tag{8.18}$$

for the best values of $a_0, a_1, \cdots, a_n$. The matrix $\langle \phi_i, \phi_j \rangle$ $(i, j = 0, 1, \cdots, n)$ is called the **Gram matrix**.

For the scalar product (8.13a), the Gram matrix

$$\langle \phi_i, \phi_j \rangle = \int_a^b \phi_i(x) \, \phi_j(x) \, w(x) \, dx \tag{8.19}$$

is invertible since $\phi_0, \cdots, \phi_n$ are assumed to be linearly independent. This conclusion follows from the contrary result as follows.

Theorem 8.6 *The system of functions $\phi_0, \phi_1, \cdots, \phi_n$ is linearly dependent if and only if the Gram determinant vanishes.*

Proof If $\phi_0, \cdots, \phi_n$ are linearly dependent, constants $c_0, c_1, \cdots, c_n$, not all zero, can be found such that

$$\sum_{j=0}^{n} c_j \, \phi_j(x) = 0$$

Multiplying the equality by $\phi_i(x) \, w(x)$, $i = 0, 1, \cdots, n$ and integrating, one obtains the relationships

$$\sum_{j=0}^{n} c_j \langle \phi_i, \phi_j \rangle = 0, \quad i = 0, 1, \cdots, n \tag{8.20}$$

which may be treated as a homogeneous system of linear equations having nonzero solution $c_0, c_1, \cdots, c_n$. Eliminating $c_0, \cdots, c_n$ we obtain $\det[\langle \phi_i, \phi_j \rangle] = 0$.

Conversely, suppose that the Gram determinant is equal to zero. This means that the system of of equations (8.20) has a nonzero solution c_j, $(j = 0, c_1, \cdots, c_n)$. This means that

$$\left\langle \phi_i, \sum_{j=0}^{n} c_j \, \phi_j \right\rangle = \sum_{j=0}^{n} c_j \langle \phi_i, \phi_j \rangle = 0, \quad i = 0, 1, \cdots, n$$

Now multiplying by c_i and summing over i, one has

$$\left\langle \sum_{i=0}^{n} c_i \phi_i, \sum_{j=0}^{n} c_j \phi_j \right\rangle = \left\| \sum_{j=0}^{n} c_j \phi_j \right\|_2^2 = 0$$

Consequently, $c_0 \phi_0 + \cdots + c_n \phi_n = 0$, i.e. $\phi_0, \phi_1, \cdots, \phi_n$ are linearly dependent. $\qquad \square$

Thus for the scalar product (8.19), the normal Eq. (8.18) possess a unique solution, if $\phi_0, \phi_1, \cdots, \phi_n$ are linearly independent.

Jørgen Pedersen Gram (1850–1960), Danish mathematician. His main area of work was number theory, viz., investigation of number of primes less than a given number. He was an amateur mathematician as he worked in an insurance company in more and more senior roles.

Next consider the scalar product (8.16b). In this discrete case setting $x = x_\nu$, $(\nu = 0, 1, 2, \cdots, N)$ in Eq. (8.18), one can construct an N–vector $\bar{\phi} = \sum_{j=0}^{n} a_j \bar{\phi}_j$ where

$\bar{\phi} = [\phi(x_0), \cdots, \phi(x_N)]^T$, $\bar{\phi}_j = [\phi_j(x_0), \cdots, \phi_j(x_N)]^T$, $(j = 0, 1, \cdots, n)$. If $n > N$, then in the vector space of N dimensions, the vectors $\bar{\phi}_0, \bar{\phi}_1, \cdots, \bar{\phi}_n$ are always linearly dependent. Hence, it is assumed that $n \leq N$. Moreover, it is assumed that $x_0, x_1, \cdots, x_N$ are distinct data points. The *normal equations* now become

$$\sum_{j=0}^{n} a_j \langle \bar{\phi}_i, \bar{\phi}_j \rangle = \langle \bar{f}, \bar{\phi}_i \rangle, \quad (i = 0, 1, \cdots, n) \qquad (8.21)$$

where $\bar{f} = [f(x_0), \cdots, f(x_N)]^T$. The Gram matrix becomes

$$\langle \bar{\phi}_i, \bar{\phi}_j \rangle = \sum_{\nu=0}^{N} \phi_i(x_\nu) \phi_j(x_\nu) w(x_\nu), \quad (i, j = 0, 1, \cdots, n)$$

and the normal equations (8.21) have a unique solution if and only if the vectors $\bar{\phi}_0, \bar{\phi}_1, \cdots, \bar{\phi}_n$ are linearly independent as in the preceding case. This requires more than linear independence of the functions $\phi_0, \phi_1, \cdots, \phi_n$.

Theorem 8.7 *The vectors* $\bar{\phi}_0, \bar{\phi}_1, \cdots, \bar{\phi}_n$ *are linearly independent if and only if functions* $\phi_0, \phi_1, \cdots, \phi_n \in$ *Haar space* U *on* $[a, b]$.

Proof The *Haar space* is defined in Chap. 4. Let $\phi_0, \phi_1, \cdots, \phi_n \in U$, then $\phi_0, \phi_1, \cdots, \phi_n$ are linearly independent on $[a, b]$, each function possessing at most n zeros in the interval. If possible let $\bar{\phi}_0, \bar{\phi}_1, \cdots, \bar{\phi}_n$ be linearly dependent. Then constants $c_0, c_1, \cdots, c_n$, not all zero, can be found such that

$$\sum_{j=0}^{n} c_j \bar{\phi}_j = 0, \quad \text{or} \quad \sum_{j=0}^{n} c_j \phi_j(x_\nu) = 0, \quad (\nu = 0, 1, \cdots, n) \qquad (8.22)$$

This must hold for all choice of distinct x_ν. Hence, $x_0, \cdots, x_N$ form zeros of $\phi_j(x)$. Since their number is $N + 1 > n$, $\phi_j \notin U$, which is a contradiction.

Conversely, let $\bar{\phi}_0, \bar{\phi}_1, \cdots, \bar{\phi}_n$ be linearly independent, then constants $c_0, c_1, \cdots, c_n \neq 0$ *can not be found* such that (8.22) holds. This means that the functions ϕ_j are linearly independent and cannot have $N + 1 > n$ zeros. This means that $\phi_j \in$ Haar space U. $\square$

Example 1. Construct the unit weight least squares approximation of $f(x) = \sqrt{x}$ on the interval $[0, 1]$ by a linear function. Calculate the root-mean-square deviation of the approximation required.

Solution. The required approximating function is $\phi(x) := a_0 + a_1 x$. Hence, $\phi_0(x) = 1$, $\phi_1(x) = x$. With $w(x) = 1$ (unit weight),

$$\langle \phi_0, \phi_0 \rangle = \int_0^1 1^2\, dx = 1, \quad \langle \phi_0, \phi_1 \rangle = \int_0^1 1 \cdot x\, dx = \frac{1}{2}, \quad \langle \phi_1, \phi_1 \rangle = \int_0^1 x^2\, dx = \frac{1}{3}$$

$$\langle f, \phi_0 \rangle = \int_0^1 \sqrt{x} \cdot 1\, dx = \frac{2}{3}, \quad \langle f, \phi_1 \rangle = \int_0^1 \sqrt{x} \cdot x\, dx = \frac{2}{5}$$

Hence, the normal equations are

$$1 \cdot a_0 + \frac{1}{2} a_1 = \frac{2}{3}, \quad \frac{1}{2} a_0 + \frac{1}{3} a_1 = \frac{2}{5}$$

Solving the two equations, $a_0 = 4/15$, $a_1 = 4/5$. Hence, $\phi(x) = 4/15 + (4/5)\, x$.

The root-mean-square error of the approximation is

$$\sqrt{\int_0^1 \left(\sqrt{x} - \frac{4}{15} - \frac{4}{5} x \right)^2 dx} = \frac{\sqrt{2}}{30}$$

$\square$

Exercises

1. Prove that the linear least squares approximation of $x^{1/3}$ in $[0, 1]$ is $\dfrac{3}{7} + \dfrac{9}{14}\, x$, with root-mean-square error $\sqrt{\dfrac{561}{980}}$.

2. Prove that the second-degree least squares approximation of $\sqrt{x}$ in $[0, 1]$ is $\dfrac{1}{35}(6 + 48\, x - 20\, x^2)$. What is the root-mean-square error?

3. Prove that the least squares approximation of degree 2 of the function $1/(1 + x^2)$ in $[-1, 1]$ is $\dfrac{1}{4}[3(2\pi - 5) + 15(3 - \pi)\, x^2]$.

8.3.1 *Least Squares Polynomial Approximation over Given Data Set*

Suppose a set of data (x_0, y_0), (x_1, y_1), $\cdots$, (x_n, y_n) is given for a function f and we want to obtain the least squares fit by a polynomial $\phi(x) := a_0 + a_1 x + \cdots + a_n x^n$, $(n \leq N)$. We note that the polynomial is constituted of the monomials 1, x, $x^2, \cdots$, x^n that belong to Haar space U. Hence, f can be least squares approximated by ϕ by minimising

$$S := \|f - \phi\|_2^2 = \sum_{j=0}^{N} (y_j - a_0 - a_1 x_j - \cdots - a_n x_j^n)^2$$

Hence,

$$\frac{\partial S}{\partial a_i} = -2 \sum_{j=0}^{N} (y_j - a_0 - a_1 x_j - \cdots - a_n x_j^n) \cdot x_j^i = 0$$

for the minimum, or

$$a_0 \sum_{j=0}^{N} x_j^i + a_1 \sum_{j=0}^{N} x_j^{i+1} + \cdots + a_n \sum_{j=0}^{N} x_j^{i+n} = \sum_{j=0}^{N} x_j^i y_j \tag{8.22a}$$

for $i = 0, 1, \cdots, n$. These are the *normal equations* for the problem.

Setting $s_i = \sum_j x_j$, $v_i = \sum_j x_j^i y_j$, the equations take the form

$$\begin{aligned}
s_0 a_0 + s_1 a_1 + \cdots + s_n a_n &= v_0 \\
s_1 a_0 + s_2 a_1 + \cdots + s_{n+1} a_n &= v_1 \\
\cdots\cdots\cdots\cdots\cdots\cdots\cdots\cdots \quad \cdots \\
s_n a_0 + s_{n+1} a_1 + \cdots + s_{2n} a_n &= v_n
\end{aligned} \tag{8.22b}$$

where, evidently $s_0 = N + 1$.

In the case of a linear fit (or *regression*) $n = 1$, the normal equations become

$$(N + 1) a_0 + s_1 a_1 = v_0, \quad s_1 a_0 + s_2 a_1 = v_1$$

whose solution is

$$a_1 = \frac{(N+1) v_1 - s_1 v_0}{(N+1) s_2 - s_1^2}, \quad a_0 = \frac{v_0}{N+1} - a_1 \frac{s_1}{N+1}$$

The degree n of the polynomial in any practical application of the above procedure should be selected on the basis of the trend shown by the data points. If the trend is linear, a linear fit should be used; if it is parabolic a second-degree fit is appropriate.

High degree fits are seldom attempted and in fact it has been noticed that the normal equations become *ill-conditioned* for degree exceeding 5. A good statistical rule is to begin with the first degree and continue fitting higher degree polynomials of degree $\leq n(<5)$, until

$$\frac{S_k}{n-k-1} > \frac{S_{k-1}}{n-k}$$

where S_k is the sum of squares of errors of polynomial of degree k.

The difficulty of ill-condition can be gauged from the fact that the matrix of the coefficients of the normal equations is

$$\sigma_{ij} := \sum_{k=0}^{N} x_k^{i+j}, \quad (i, j = 0, 1, \cdots, n)$$

If the points x_k are uniformly distributed (in the probabilistic sense) over $[0, 1]$ and N is large

$$\sigma_{ij} = N \int_0^1 x^{i+j}\, dx$$

$$= \frac{N}{i+j+1}$$

Thus the matrix of the coefficients is the Hilbert matrix

$$\begin{bmatrix} 1/1 & 1/2 & 1/3 & \cdots & 1/n+1 \\ 1/2 & 1/3 & 1/4 & \cdots & 1/n+2 \\ \cdots & \cdots & \cdots & \cdots & \cdots \\ 1/(n+1) & 1/(n+2) & 1/(n+3) & \cdots & 1/(2n) \end{bmatrix}$$

whose determinant tends to 0 as n increases, making it ill-conditioned.

The problem of ill-condition and solution in general can be altogether avoided if we choose the functions $\phi_0, \phi_1, \cdots, \phi_n$ to be **orthogonal**, that is if they satisfy the condition

$$\langle \phi_i, \phi_j \rangle = 0 \quad \text{for} \quad i \neq j$$

In this case the solution of the normal equations (8.18) becomes

$$a_i = \frac{\langle f, \phi_i \rangle}{\langle \phi_i, \phi_i \rangle} \tag{8.23}$$

This prospect is explored in the next section.

Exercises

1. The following table shows height (h) and weight (w) of eight persons:

h (cm)	150	155	160	165	170	175	180	185
w (kg)	51	53	59	58	62	68	71	68

Assuming a linear relationship between the height and weight, find the regression line and estimate the weight of two persons with height 140 and 172 cm.
[$w = -34.4644 + 0.5714\,h$. 45.536 and 63.821 kg].

2. Experimental data on constant filtration of a $CaCO_3$ slurry through a canvas medium is given in the following table:

x	5	10	15	20	25	30	35	40
y	0.530	0.716	0.806	0.869	0.943	1.013	1.096	1.160

x in this case is the total volume (ml) of filtrate collected till time t (s) and y is t/x. A chemical engineer knows that this process is described by the equation $y = a_0 + a_1 x$. Obtain least squares estimate of a_0 and a_1.
[$a_0 = 0.57614$, $a_1 = 0.01469$].

3. Experiments with a periodic process gave the following data:

t^o	0	50	100	150	200	250	300	350
y	0.754	1.762	2.041	1.412	0.303	−0.484	−0.380	0.520

If the process is modelled as $y = a_0 + a_1 \sin t$, estimate a_0 and a_1 by the least squares method.
[Let $x = \sin t$ and form the data for x. $a_0 = 0.75258$, $a_1 = 1.31281$].

4. Show that the following regression models are linearisable:
(i) $y = a + \dfrac{b}{x^n}$ (ii) $y = \dfrac{x^m}{a + bx^n}$ (iii) $y = ax^b$ (iv) $y = \exp[-(ax)^b]$.
Obtain the linearised forms.
[(i) $y = a + b(x^{-n})$ (ii) $x^m/y = a + bx^n$ (iii) $\ln y = \ln a + b(\ln x)$
(iv) $\ln \ln(1/y) = b \ln a + b(\ln x)$].

5. Fit the saturation growth rate model $y = ax/(b + x)$ to the data given below:

x	2	4	6	8	10
y	1.4	2.0	2.4	2.6	2.7

[$1/y = 1/a + (b/a)/x$; $a = 2.80948$, $b = 1.49190$].

6. In an experiment to determine the specific heat γ for a certain gas from the ideal gas law $pV^\gamma = C$, the following values of pressure p were obtained for predetermined values of volume V:

V (in³)	50	30	20	15	10
p (lb/in²)	16.6	39.7	78.5	115.5	195.3

Estimate γ by the least squares method.

$[\gamma = 1.54]$.

7. The dynamic viscosity μ (poise) as a function of the temperature T ($^\circ C$) for water is well modelled by the Walther formula $\mu = \mu_0 \exp(T_0/T)^n$. Assuming $n = 1$, estimate the unknown parameters by least squares, given the data:

T	10	20	30	40	50	60	70	80	90	100
μ	1.310	1.011	0.803	0.659	0.555	0.478	0.416	0.367	0.328	0.296

$[\ln \mu = (\ln \mu_0) + T_0 (1/T); \mu_0 = 0.350, \quad T_0 = 15.72]$.

8. A quadratic function $y = a_0 + a_1 x + a_2 x^2$ should be fitted with the following data:

x	8	10	12	16	20	30
y	4.41	8.12	12.40	23.60	37.92	88.31

Determine least squares estimate of a_0, a_1, a_2.

$[a_0 = -1.56277, \quad a_1 = -0.05820, \quad a_2 = 0.1018]$.

9. The variation of refractive index μ (unitless) of a polished plane metal surface can be modelled by the Cauchy equation $\mu = a + b/\lambda^2 + c/\lambda^4$, where λ (10^{-9}m) is the wavelength of light. Calculate a, b, and c for gold given the data:

λ	400	450	500	550	600	650	700
μ	0.360	0.385	0.415	0.740	0.845	0.890	0.925

$[$Put $x = 1/\lambda^2$. $a = 2.137, \quad b = -6.633 \times 10^{-3}, \quad c = 6.151 \times 10^{-6}]$.

10. A bivariate function z depends linearly on independent variables as $z = a_0 + a_1 x + a_2 y$. Show that the normal equations for estimating a_0, a_1, a_2 from $n + 1$ given data are

$$\begin{bmatrix} n+1 & \sum x_i & \sum y_i \\ \sum x_i & \sum x_i^2 & \sum x_i y_i \\ \sum y_i & \sum x_i y_i & \sum y_i^2 \end{bmatrix} \begin{bmatrix} a_0 \\ a_1 \\ a_2 \end{bmatrix} = \begin{bmatrix} \sum z_i \\ \sum x_i z_i \\ \sum y_i z_i \end{bmatrix}$$

11. The density of population of fish in a river depends on the rise/width x and the amount of calcium in water y. If the square root of the number of fish per hectare is $z = a_0 + a_1 x + a_2 y$, estimate a_0, a_1, a_2 by least squares, given the data

x	25	30	45	60	75	90	100	110
y	3	4	5	8	5	8	4	5
z	5	6	6	6	5	5	2	2

$[$Use Ex. 10. $a_0 = 5.204, \quad a_1 = -0.051, \quad a_2 = 0.541]$.

8.4 Orthogonal Polynomials

In the preceding section, it was shown that an arbitrary function $f \in C[a, b]$ could be compactly approximated by a generalised polynomial consisting of *orthogonal functions* (Eqs. (8.17), (8.23)). Pursuing this idea, we consider the latter functions to be polynomials of certain degree. We have already come across two special functions of this type, viz., the *Legendre* and the *Chebyshev* polynomials in Chaps. 5 and 4, respectively. These functions figure prominently in Function Theory also, but in the following we give a straightforward development.

Let the *orthogonal polynomials* be denoted as $p_0(x), p_1(x), \cdots, p_n(x), \cdots$ where it is assumed that (1) the *exact degree* of p_n is n, and (2) $\langle p_i, p_j \rangle = 0$ for $i \neq j$ (orthogonality condition). Further if $\langle p_i, p_i \rangle = 1$, the orthogonal polynomials are called *orthonormal*. We first show the following.

Theorem 8.8 (Gram–Schmidt Orthogonalisation). *Let* $p_0(x) = 1$, *then given a scalar product* $\langle \cdot, \cdot \rangle$, *the set of orthogonal polynomials* $p_1(x), p_2(x), \cdots, p_n(x), \cdots$ *is recursively given by the equation*

$$p_n(x) = x^n - \sum_{i=0}^{n-1} \frac{\langle x^n, p_i \rangle}{\langle p_i, p_i \rangle} p_i(x), \quad n = 1, 2, \cdots \tag{8.24}$$

Proof Suppose that $p_0, p_1, \cdots, p_{n-1}$ have been constructed such that $\langle p_i, p_j \rangle = 0$ for $i \neq j, (j = 0, 1, \cdots, n - 1)$. We can then set

$$p_n(x) = x^n - \alpha_0 \, p_0(x) - \cdots - \alpha_{n-1} \, p_{n-1}(x)$$

and determine α_i by putting $\langle p_n, p_i \rangle = 0$. for $i = 0, 1, \cdots, n - 1$. Thus

$$0 = \langle p_n, p_i \rangle = \langle x^n, p_i \rangle - \sum_{j=0}^{n-1} \alpha_j \langle p_j, p_i \rangle = \langle x^n, p_i \rangle - \alpha_i \langle p_i, p_i \rangle$$

or, $$\alpha_i = \frac{\langle x^n, p_i \rangle}{\langle p_i, p_i \rangle}, \quad i = 0, 1, \cdots, n - 1$$

Hence, we obtain Eq. (8.24), satisfyng the orthogonality condition. $\qquad \square$

In practice, one could use the Gram–Schmidt orthogonalisation procedure to actually construct the orthogonal polynomials for given interval and weight function, but there is a better way to do this in view of the following.

Theorem 8.9 *Let* $p_0, p_1, \cdots, p_n, \cdots$ *be orthogonal polynomials. Then there exist constants* $A_n \neq 0$, B_n, $C_n \neq 0$ *such that*

$$p_n(x) = (A_n x + B_n) \, p_{n-1}(x) - C_n \, p_{n-2}(x), \quad \text{for } n = 1, 2, \cdots \tag{8.25}$$

where $p_{-1}(x) \equiv 0$.

Proof Let the leading coefficient of p_n be α_n, i.e. $p_n(x) = \alpha_n x^n + \cdots$ for $n = 0, 1, 2, \cdots$, then by definition $\alpha_n \neq 0$. Let

$$A_n = \frac{\alpha_n}{\alpha_{n-1}} \neq 0$$

Then $p_n(x) - A_n x\, p_{n-1}(x)$ is a polynomial of degree $n - 1$. Hence, we can write

$$p_n(x) - A_n x\, p_{n-1}(x) = b_0\, p_0(x) + \cdots + b_{n-1}\, p_{n-1}(x)$$

Taking scalar product with $p_j(x)$, $j \le n - 1$ we get by orthogonality

$$b_j \langle p_j,\ p_j \rangle = \langle p_n,\ p_j \rangle - A_n \langle x\, p_{n-1},\ p_j \rangle$$
$$= -A_n \langle p_{n-1},\ x\, p_j \rangle$$

since by definition of scalar product $\langle xp, q \rangle = \langle p, xq \rangle$. Now $x\, p_j(x)$ is a polynomial of degree $j + 1$. Hence, if $j + 1 < n - 1$ or $j < n - 2$, $\langle p_{n-1},\ x\, p_j \rangle = 0$ and therefore $b_j = 0$ for $j < n - 2$. Setting $B_n := b_{n-1}$ and $C_n := -b_{n-2}$, we obtain the recurrence relation (8.25).

If we take scalar product of Eq. (8.25) with p_{n-1}, we easily obtain

$$B_n = -A_n \frac{\langle x\, p_{n-1},\ p_{n-1} \rangle}{\langle p_{n-1},\ p_{n-1} \rangle}$$

Similarly if we take scalar product with p_{n-2}, we obtain

$$C_n = A_n \frac{\langle x\, p_{n-1},\ p_{n-2} \rangle}{\langle p_{n-2},\ p_{n-2} \rangle}$$

But following Eq. (8.25)

$$p_{n-1}(x) = (A_{n-1}x + B_{n-1})\, p_{n-2}(x) - C_{n-1} p_{n-3}(x)$$

Taking scalar product with p_{n-1},

$$\langle p_{n-1},\ p_{n-1} \rangle = A_{n-1} \langle x\, p_{n-2},\ p_{n-1} \rangle = A_{n-1} \langle x\, p_{n-1},\ p_{n-2} \rangle$$

by orthogonality and definition. Hence

$$C_n = \frac{A_n}{A_{n-1}} \frac{\langle p_{n-1},\ p_{n-1} \rangle}{\langle p_{n-2},\ p_{n-2} \rangle} \neq 0$$ $\square$

For the purpose of constructing orthogonal polynomials by the application of the above theorem, one usually selects the leading coefficient α_n or equivalently the ratio A_n, so that the resulting sequence is simple enough in some sense.

Remark. The orthogonal polynomial $p_n(x)$ possesses the remarkable property that it has exactly n simple real zeros in $[a, b]$. For, let $p_n(x)$ change sign at the distinct points $\xi_1, \xi_2, \cdots, \xi_m \in [a, b]$, then $m \leq n$. Suppose $m < n$, then with $\bar{x} \in (\max\{\xi_1, \cdots, \xi_m\}, b)$, the polynomial

$$q(x) := p_n(\bar{x})(x - \xi_1) \cdots (x - \xi_m)$$

has a degree $m < n$. Hence

$$\langle p_n, q \rangle = \int_a^b p_n(x)\, q(x)\, w(x)\, dx = 0$$

But for any $x \in [a, b]$, $q(x)$ has the same sign as $p_n(x)$ except at the m zeros ($p_n(\bar{x})$ in the definition of $q(x)$ captures the sign of the non-fluctuating part of $p_n(x)$). This means that $p(x) \cdot q(x) > 0$, $x \neq \xi_i$ ($i = 1, 2, \cdots, m$) and the above integral must be positive and cannot vanish, which is a contradiction. Hence $m = n$.

We now treat some important special cases.

1°. **Legendre Polynomials.** Let the interval be $[-1, 1]$ with weight function $w(x) = 1$ for orthogonal polynomials denoted by $P_n(x)$. Let the polynomial be normalised by the condition $P_n(1) = 1$ and so

$$A_1 + B_1 = 1, \quad A_n + B_n - C_n = 1 \quad (n \geq 2)$$

For $n = 0$, $P_0(x) = \text{constant} = P_0(1) = 1$. For $n = 1$

$$B_1 = -A_1 \frac{\int_{-1}^1 x \cdot 1 \cdot 1\, dx}{\int_{-1}^1 1 \cdot dx} = 0, \quad A_1 = 1$$

Hence $P_1(x) = x$. For $n = 2$,

$$B_2 = -A_2 \frac{\int_{-1}^1 x \cdot x \cdot x\, dx}{\int_{-1}^1 x \cdot x\, dx} = 0, \quad C_2 = \frac{A_2}{A_1} \frac{\int_{-1}^1 x \cdot x\, dx}{\int_{-1}^1 1 \cdot 1\, dx} = \frac{A_2}{3}$$

and therefore, $A_2 = 3/2$, $C_2 = 1/2$.

Hence $P_2(x) = \frac{1}{2}(3x^2 - 1)$. For $n = 3$,

$$B_3 = -A_3 \frac{\int_{-1}^1 x \cdot \frac{1}{4}(3x^2 - 1)^2\, dx}{\int_{-1}^1 \frac{1}{4}(3x^2 - 1)^2\, dx} = 0, \quad C_3 = \frac{A_3}{A_2} \frac{\int_{-1}^1 \frac{1}{4}(3x^2 - 1)^2\, dx}{\int_{-1}^1 x^2\, dx} = \frac{2}{5} A_3$$

and so, $A_3 = 5/3, \quad C_3 = 2/3.$

Hence $P_3(x) = \dfrac{1}{2}(5x^3 - 3x)$, etc. In general, one identifies that

$$B_n = 0, \quad C_n = \frac{n-1}{n}, \quad A_n = 1 - B_n + C_n = \frac{2n-1}{n}$$

and so the Legendre polynomials satisfy the recurrence relations

$$n\, P_n(x) = (2n-1)\, x\, P_{n-1}(x) - (n-1)\, P_{n-2}(x)$$

In the theory of differential equations, it is shown that

$$P_n(x) = \frac{1}{n!\, 2^n} \frac{d^n}{dx^n} [(x^2 - 1)^n]$$

known as *Rodrigue's Formula* that satisfies the second-order differential equation

$$(1 - x^2)\, P_n''(x) - 2x\, P_n'(x) + n(n+1)\, P_n(x) = 0$$

satisfying the orthogonality relations

$$\int_{-1}^{1} P_i(x)\, P_j(x)\, dx = \begin{cases} \dfrac{2}{2i+1}, & i = j \\ 0, & i \neq j \end{cases}$$

These polynomials appeared in the derivation of the Gauss quadrature formula (Chap. 5, Sect. 5.2.4).

2°. Chebyshev Polynomials. In Sect. 4.3 of Chap. 4, it was shown that the Chebyshev Polynomials $T_n(x) = \cos(n\, \cos^{-1} x)$ satisfy the recurrence relation

$$T_n(x) = 2x\, T_{n-1}(x) - T_{n-2}(x)$$

which is of the form (8.25) with $A_n = 2, \; B_n = 0, \; C_n = -1$. The orthogonality relations are

$$\int_{-1}^{1} \frac{T_i(x)\, T_j(x)}{\sqrt{1 - x^2}}\, dx = \int_{0}^{\pi} \cos(i\theta)\, \cos(j\theta)\, d\theta = \begin{cases} \pi, & i = j = 0 \\ \pi/2, & i = j \neq 0 \\ 0, & i \neq 0 \end{cases}$$

Thus Chebyshev Polynomials are orthogonal on $[-1, 1]$ with weight function $w(x) = (1 - x^2)^{-1/2}$. Also $T_n(1) = 1$.

Exercises

1. Calculate the Legendre Polynomials of order 0–4 using the recurrence relation:

$$[P_0(x) = 1, \qquad P_3(x) = \frac{1}{2}(5x^3 - 3x)$$

$$P_1(x) = x, \qquad P_4(x) = \frac{1}{8}(35x^4 - 30x^2 + 3)$$

$$P_2(x) = \frac{1}{2}(3x^2 - 1)].$$

2. **(Laguerre Polynomials).** The Laguerre Polynomials $L_n(x)$ are orthogonal in $[0, \infty]$ with weight function $w(x) = e^{-x}$ and satisfy the recurrence relations

$$n L_n(x) = (2n - 1 - x) L_{n-1}(x) - (n - 1) L_{n-2}(x), \quad (n = 1, 2, \cdots)$$

Calculate $L_n(x)$ for $n = 0, 1, 2, 3, 4$.
$[L_0(x) = 1, \quad L_1(x) = -x + 1, \quad L_2(x) = \frac{1}{2}(x^2 - 4x + 2),$
$L_3(x) = \frac{1}{6}(-x^3 + 9x^2 - 18x + 6), \; L_4(x) = \frac{1}{24}(x^4 - 16x^3 + 72x^2 - 96x + 24)].$

3. **(Hermite Polynomials).** The Hermite Polynomials $H_n(x)$ are orthogonal in $(-\infty, \infty)$ with weight function $w(x) = e^{-x^2}$. They satisfy the recurrence relations

$$H_n(x) = 2x H_{n-1}(x) - 2(n - 1) H_{n-2}(x), \quad (n = 1, 2, \cdots)$$

Calculate $H_n(x)$ for $n = 0, 1, 2, 3, 4$.
$[H_0(x) = 1, H_1(x) = 2x, H_2(x) = 4x^2 - 2, H_3(x) = 8x^3 - 12x, H_4(x) = 16x^4 - 48x^2 + 12].$

8.5 Orthogonal Polynomials over Discrete Set of Points: Smoothing of Data

We now touch upon the topic of least squares approximation by polynomials when the scalar product is defined over a discrete set of $N + 1$ points as in Eq. (8.16b). For simplicity, we assume that the weight function is unity and the set of points is equally spaced. Using the common equal space as a unit of scale, the points can be taken as $0, 1, 2, \cdots, N$. A polynomial of degree $n \leq N$ ($n \geq 1$) which passes through the first $n + 1$ of these points is

$$p_{nN}(x) := 1 + \sum_{j=1}^{n} (-1)^j \binom{m}{j} \binom{n+j}{j} \frac{x(x - 1) \cdots (x - j + 1)}{N(N - 1) \cdots (N - j + 1)} \qquad (8.26)$$

For $n = 0$, we define $p_{0N} := 1$. $p_{nN}(x)$ so defined is known as *Gram Polynomial* or sometimes as *Chebyshev Polynomial*. In particular, for $n = 1, 2, 3$, the summation in (8.26) reduces to

$$p_{1N}(x) = 1 - 2\frac{x}{N}$$

$$p_{2N}(x) = 1 - 6\frac{x}{N} + 6\frac{x(x-1)}{N(N-1)} \tag{8.27}$$

$$p_{3N}(x) = 1 - 12\frac{x}{N} + 30\frac{x(x-1)}{N(N-1)} - 20\frac{x(x-1)(x-2)}{N(N-1)(N-2)}$$

The polynomials $p_{nN}(x)$ are *orthogonal with respect to the discrete scalar product* (8.16b). In fact, it can be proved that

$$\langle p_{mN},\, p_{nN}\rangle = \begin{cases} 0, & m \neq n \\[4pt] N+1, & m = n = 0 \\[6pt] \dfrac{(n+N+1)(n+N)\cdots(N+1)}{N(N-1)\cdots(N-n+1)\cdot(2n+1)}, & m = n \geq 1 \end{cases} \tag{8.28}$$

Thus *ab initio*, for example,

$$\langle p_{0N},\, p_{0N}\rangle = \sum_{\nu=0}^{N} 1 \cdot 1 = N+1$$

$$\langle p_{0N},\, p_{1N}\rangle = \sum_{\nu=0}^{N} 1 \cdot \left(1 - \frac{2\nu}{N}\right) = N+1 - \frac{2}{N}\cdot\frac{N(N+1)}{2} = 0$$

$$\langle p_{1N},\, p_{1N}\rangle = \sum_{\nu=0}^{N} \left(1 - \frac{2\nu}{N}\right)^2 = N+1 - \frac{4}{N}\cdot\frac{N(N+1)}{2}$$

$$+ \frac{4}{N^2}\cdot\frac{N(N+1)(2N+1)}{6} = \frac{(N+2)(N+1)}{3N}$$

which agree with the values obtained from Eq. (8.28).

Consider now the least squares approximation of a function f by Gram polynomials defined on the set of points $\{0, 1, \cdots, N\}$. If the approximating polynomial ϕ is of degree $n \leq N$, then by the method of least squares, Eqs. (8.17) and (8.23) become

$$\phi(x) = \sum_{j=0}^{n} a_j\, p_{jN}(x), \quad a_j = \frac{\langle f,\, p_{jN}\rangle}{\langle p_{jN},\, p_{jN}\rangle} \tag{8.29}$$

The formulae (8.29) have a useful application in **smoothing** of a table of values of $f(x_i)$ for *equally spaced* x_i. The values $f(x_i)$ often contain error, specially when they are results of experimental observations. Joining such data by lines results in a crude zig-zag curve. In order to recover the trend of the function by a *smooth curve*, one can replace $f(x_i)$ by $\phi(x_i)$—the least squares polynomial approximation over some neighbouring data points.

To illustrate, consider three points say $i-1$, i, $i+1$, so that $N=2$, and $n=1$ (linear polynomial). Using Eqs. (8.29), (8.27) and (8.28),

$$\phi(x) = a_0 \, p_{02}(x) + a_1 \, p_{12}(x) = a_0 + a_1\left(1 - 2 \cdot \frac{x}{2}\right)$$

where

$$a_0 = \frac{\langle f, \, p_{02}\rangle}{\langle p_{02}, \, p_{02}\rangle} = \frac{1}{3}\,\langle f, \, 1\rangle = \frac{1}{3}\,(f_{i-1} + f_i + f_{i+1})$$

$$a_1 = \frac{\langle f, \, p_{12}\rangle}{\langle p_{12}, \, p_{12}\rangle} = \frac{1}{2}\,[f_{i-1}\, p_{12}(0) + f_i \, p_{12}(1) + f_{i+1}\, p_{12}(2)]$$

$$= \frac{1}{2}\,[f_{i-1}\cdot(1-0) + f_i\cdot(1-1) + f_{i+1}\cdot(1-2)] = \frac{1}{2}[f_{i-1} - f_{i+1}]$$

Thus for the three points $i-1$, i, $i+1$, i.e., $x = 0,\ 1,\ 2$, we obtain

$$\bar{f}_i = \phi(1) = a_0 + a_1 \cdot 0 = \frac{1}{3}\,(f_{i-1} + f_i + f_{i+1}) \tag{8.30}$$

$$\bar{f}_{i-1} = \phi(0) = a_0 + a_1 \cdot (1-0) = \frac{1}{6}\,(5\,f_{i-1} + 2\,f_i - f_{i+1}) \tag{8.30a}$$

$$\bar{f}_{i+1} = \phi(2) = a_0 + a_1 \cdot (1-2) = \frac{1}{6}\,(-f_{i-1} + 2\,f_i + 5\,f_{i+1}) \tag{8.30b}$$

In the middle portion of the table f_i may be modified by $\bar{f}_i$ (Eq. (8.30)), while at the beginning and end of the table, formulae (8.30a) and (8.30b) may, respectively, be used for smoothing.

Other formulae such as *five point third-degree* least squares smoothing can be similarly found. Here $N=4$ with designated points $i-2$, $i-1$, i, $i+1$, $i+2$ and $n=3$. From Eqs. (8.29), (8.27) and (8.28), we obtain

$$\phi(x) = a_0 \, p_{04}(x) + a_1 \, p_{14}(x) + a_2 \, p_{24}(x) + a_3 \, p_{34}(x)$$

$$= a_0 + a_1\left(1 - \frac{x}{2}\right) + a_2\left(1 - \frac{3}{2}x + \frac{1}{2}x\,(x-1)\right)$$

$$+ a_3\left(1 - 3x + \frac{5}{2}x(x-1) - \frac{5}{6}x(x-1)(x-2)\right)$$

where

$$a_0 = \frac{\langle f, p_{04} \rangle}{\langle p_{04}, p_{04} \rangle} = \frac{1}{5} (f_{i-2} + f_{i-1} + f_i + f_{i+1} + f_{i+2})$$

$$a_1 = \frac{\langle f, p_{14} \rangle}{\langle p_{14}, p_{14} \rangle} = \frac{2}{5} \left(f_{i-2} + \frac{1}{2} f_{i-1} - \frac{1}{2} f_{i+1} - f_{i+2} \right)$$

$$a_2 = \frac{\langle f, p_{24} \rangle}{\langle p_{24}, p_{24} \rangle} = \frac{2}{7} \left(f_{i-2} - \frac{1}{2} f_{i-1} - f_i - \frac{1}{2} f_{i+1} + f_{i+2} \right)$$

$$a_3 = \frac{\langle f, p_{34} \rangle}{\langle p_{34}, p_{34} \rangle} = \frac{1}{10} (f_{i-2} - 2 f_{i-1} + 2 f_{i+1} - f_{i+2})$$

Hence, the smoothing values (for $x = 0, 1, 2, 3, 4$) are obtained as

$$\bar{f}_{i-2} = \phi(0) = \frac{1}{70} (69 f_{i-2} + 4 f_{i-1} - 6 f_i + 4 f_{i+1} - f_{i+2})$$

$$\bar{f}_{i-1} = \phi(1) = \frac{1}{35} (2 f_{i-2} + 27 f_{i-1} + 12 f_i - 8 f_{i+1} + 2 f_{i+2})$$

$$\bar{f}_i = \phi(2) = \frac{1}{35} (-3 f_{i-2} + 12 f_{i-1} + 17 f_i + 12 f_{i+1} - 3 f_{i+2}) \qquad (8.31)$$

$$\bar{f}_{i+1} = \phi(3) = \frac{1}{35} (2 f_{i-2} - 8 f_{i-1} + 12 f_i + 27 f_{i+1} + 2 f_{i+2})$$

$$\bar{f}_{i+2} = \phi(4) = \frac{1}{70} (-f_{i-2} + 4 f_{i-1} - 6 f_i + 4 f_{i+1} + 69 f_{i+2})$$

For other (higher order) formulae, see Hildebrand cited in the Bibliography. A guideline for selecting a particular formula is based on the observation that the amount of smoothing *increases* with N and *decreases* with increasing values of n. In practice, however, it is convenient to employ a smoothing formula involving a small number of points together with a low-degree polynomial, e.g., Eqs. (8.30) and (8.31), and repeat the smoothing process a few number of times. Repeated smoothing is, however, not recommended.

The smoothing process by least squares polynomial approximation can also be substantiated from the point of probability theory. If $f(x_i)$ is an observed value, the actual value is of the form

$$f_i^*(x_i) = f(x_i) + \eta_i$$

where η_i is the error with zero mean value and variance σ^2 independent of i. If the errors in the sequence of observations are assumed independent, then

$$E[\eta_i] = 0, \quad E[\eta_i \eta_j] = \begin{cases} 0, & i \neq j \\ \sigma^2, & i = j \end{cases}$$

where $E[\cdot]$ denotes expectation value. Hence, the coefficient a_j^* in the polynomial approximation (8.29) is of the form

$$a_j^* = \frac{\langle f^*, p_{jn} \rangle}{\langle p_{jN}, p_{jN} \rangle} = \frac{\langle f + \eta, p_{jN} \rangle}{\langle p_{jN}, p_{jN} \rangle} = \frac{\langle f, p_{jN} \rangle}{\langle p_{jN}, p_{jN} \rangle} + \frac{\langle \eta, p_{jN} \rangle}{\langle p_{jN}, p_{jN} \rangle} =: a_j + \gamma_j$$

Now,

$$E[\gamma_j] = \frac{1}{\langle p_{jN}, p_{jN} \rangle} E[\langle \eta, p_{jN} \rangle] = \frac{1}{\langle p_{jN}, p_{jN} \rangle} E\left[\sum_{i=0}^{\infty} \eta_i \, p_{jN}(x_i) \right]$$

$$= \frac{1}{\langle p_{jN}, p_{jN} \rangle} \sum_{i=0}^{N} p_{jN}(x_i) \, E[\eta_i] = 0$$

Thus $E[a_j^*] = a_j$, i.e., the expected value of a_j^* equals a_j. Moreover, it can similarly be proved that

$$E[\gamma_j \, \gamma_k] = \begin{cases} 0, & j \neq k \\ \dfrac{\sigma^2}{\langle p_{jN}, p_{jN} \rangle}, & j = k \end{cases}$$

Now in the smoothing formulas we find that $\langle p_{jN}, p_{jN} \rangle > 1$. For instance, in the formulas (30) and (31), the variance of the error in the smoothed values of $\bar{f}_i = a_0^*$ and $\bar{f}_i = a_0^* - a_2^*$ are, respectively,

$$\frac{\sigma^2}{3} \quad \text{and} \quad \frac{\sigma^2}{5} + \frac{\sigma^2}{7/2} = \frac{17}{35} \sigma^2$$

Thus, *the variance of the error in the smoothed values decreases*, justifying the terminology **smoothing of data**.

Example 1. The time-averaged velocity of flow $\bar{u}$ in an open channel with sandy bed, as measured by Acoustic Doppler Velocitimeter is tabulated below:

y (cm)	0.5	0.75	1.0	1.25	1.5	2.0	2.5	3.0
$\bar{u}$ (cm/s)	12.96	22.79	24.17	25.59	28.70	32.63	35.03	33.95

3.5	4.0	4.5	5.0	6.0	7.0	8.0	9.0
36.03	36.43	36.19	36.44	38.17	38.80	38.47	39.32

where y is the height above the bed level. Smooth the data using three-point linear formulae (30, a, b).

Solution. We can apply Eqs. (8.30), (8.30a) and (8.30b) to three equally spaced ranges (0.5, 1.5), (2, 5) and (6, 9). We thus obtain the smoothed data

y	0.5	0.75	1.0	1.25	1.5	2.0	2.5	3.0
$\bar{u}$	14.37	19.97	24.18	26.15	28.42	32.12	33.87	35.00

3.5	4.0	4.5	5.0	6.0	7.0	8.0	9.0
35.47	36.22	36.35	36.36	37.80	38.48	38.86	39.12

Note that the variation of $\bar{u}$ with y is now more gradual.   □

8.6 Trigonometric Approximation

Many natural phenomena, such as light and sound, have *periodic character*. If such a phenomenon is described by a periodic function f, then $f(x + \tau) = f(x)$ for all real x, where τ—a fixed number is called the **period** of the function. Now, the trigonometric functions sine and cosine are known to be periodic with period 2π as well as bounded. Hence, if $\tau = 2\pi$, we may consider approximation of f by a *trigonometric polynomial* ϕ, where

$$\phi(x) = \frac{a_0}{2} + \sum_{j=1}^{n} (a_j \cos jx + b_j \sin jx) \tag{8.32}$$

in which a_j, b_j are constants and $x \in [0, 2\pi]$. If $\tau \neq 2\pi$, we can consider the function $g(x) = f(\tau x/(2\pi))$, where

$$g(x) = f\left(\frac{\tau x}{2\pi}\right) = f\left(\frac{\tau x}{2\pi} + \tau\right) = f\left[\frac{\tau}{2\pi}(x + 2\pi)\right] = g(x + 2\pi)$$

The function g is therefore periodic with period 2π. Thus without loss of generality we can take $\tau = 2\pi$.

On the interval $[0, 2\pi]$, the functions $\cos jx$ and $\sin jx$ are orthogonal for $j = 0, 1, 2, \cdots, , n$ since

$$\int_0^{2\pi} \cos jx \sin kx \, dx = \frac{1}{2} \int_0^{2\pi} [\sin(j+k)x - \sin(j-k)x] \, dx = 0$$

$$\int_0^{2\pi} \cos jx \cos kx \, dx = \frac{1}{2} \int_0^{2\pi} [\cos(j+k)x + \cos(j-k)x] \, dx = \begin{cases} 0 & \text{for } j \neq k \\ \pi & \text{for } j = k \end{cases}$$

and likewise for $\int_0^{2\pi} \sin jx \sin kx \, dx$. Thus as the solution of normal equation (8.18) we obtain the **Fourier Coefficients**

$$a_j = \frac{1}{\pi} \int_0^{2\pi} f(x) \cos jx \, dx, \quad b_j = \frac{1}{\pi} \int_0^{2\pi} f(x) \sin jx \, dx \tag{8.33}$$

for $j = 0, 1, 2, \cdots, n$.

If the function f is given, the Fourier coefficients can be calculated in many cases, but that is not our point of view. We rather assume that $f(x)$ is given at a discrete set of data points. Dividing the interval $[0, 2\pi]$ into N equal parts by division points $x_\nu = 2\nu\pi/N$ $(\nu = 0, 1, 2, \cdots, N)$, the corresponding values of $f(x)$ can be designated as y_ν $(\nu = 0, 1, 2, \cdots, N)$ where $y_N = y_0$ due to periodicity of $f(x)$. If the integrals appearing in the Fourier coefficients (8.33) are numerically evaluated by the Rectangle Rule, we get

$$a_j \approx \frac{2}{N} \sum_{\nu=0}^{N-1} y_\nu \cos\left(j\frac{2\nu\pi}{N}\right), \quad b_j \approx \frac{2}{N} \sum_{\nu=0}^{N-1} y_\nu \sin\left(j\frac{2\nu\pi}{N}\right) \tag{8.34}$$

for $j = 0, 1, 2, \cdots, n$. The least squares approximation is then given by Eqs. (8.32) and (8.34).

We can show that formulas (8.32) and (8.34) are least squares approximation under the *discrete inner product* (8.16b), when the discrete points of division are $x_\nu = 2\nu\pi/N$ $(\nu = 0, 1, \cdots, N-1)$ where $N > 2n$. To show this, we note that

$$\sum_{\nu=0}^{N-1} [\cos(jx_\nu) + i \sin(jx_\nu)] = \sum_{\nu=0}^{N-1} e^{ijx_\nu} = \sum_{\nu=0}^{N-1} e^{2\pi i\nu j/N}$$

$$= \frac{1 - \left(e^{2\pi ij/N}\right)^N}{1 - e^{2\pi ij/N}} = 0$$

where $i = \sqrt{-1}$ and $j = 0, 1, 2, \cdots, n$. If the order of the polynomial $n < N/2$, it follows that the Gram matrix

$$\langle \cos jx, \sin kx \rangle = \sum_{\nu=0}^{N-1} \cos jx_\nu \sin kx_\nu = \frac{1}{2}\sum_{\nu=0}^{N-1}[\sin(j+k)x_\nu - \sin(j-k)x_\nu] = 0$$

$$\langle \cos jx, \cos kx \rangle = \sum_{\nu=0}^{N-1} \cos jx_\nu \cos kx_\nu = \frac{1}{2}\sum_{\nu=0}^{N-1}[\cos(j+k)x_\nu + \cos(j-k)x_\nu]$$

$$= \begin{cases} 0 & \text{for } j \neq k \\ N/2 & \text{for } j = k \end{cases}$$

and similarly for $\langle \sin jx, \sin kx \rangle$. Due to orthogonality of cosine and sine functions on the discrete set of points x_ν, the solution of the normal equations (8.18) is

$$a_j = \frac{\langle f, \cos jx \rangle}{\langle \cos jx, \cos jx \rangle}, \quad b_j = \frac{\langle f, \sin jx \rangle}{\langle \sin jx, \sin jx \rangle}$$

that are nothing but Eq. (8.34).

Exercises

1. Set up a trigonometric interpolating polynomial of order 4 for $y = f(x)$ in $[0, \pi]$
given by the table

x	0	$\pi/4$	$\pi/2$	$3\pi/4$	π
y	1	2	2.4	2.6	1

[Make a data set for $y - 1$ by extending the table to 2π symmetrically for negative
values of $y - 1$, and use Eq. (8.34), $N = 8, n = 4$. $\phi(x) = 1 + \cos x - 0.9 \cos 2x +$
$0.1 \cos 3x - 0.1 \cos 4x + 1.3 \sin x - 0.2 \sin 2x + 0.2 \sin 4x$].

2. Set up a trigonometric polynomial of order 3 for $y = f(x)$ in $[0, 2\pi]$ given the
data

x	0	$\pi/6$	$\pi/3$	$\pi/2$	$2\pi/3$	$5\pi/6$	π
y	1.21	1.32	1.46	1.40	1.34	1.18	1.07

$7\pi/6$	$4\pi/3$	$3\pi/2$	$5\pi/3$	$11\pi/6$	2π
1.01	1.05	1.10	1.14	1.17	1.21

[Use Eq. (8.34), $N = 12$, $n = 3$, $a_0 = 2.408$, $a_1 = 0.237$, $a_2 = 0.172$, $a_3 = 0$,
$b_1 = 1.440$, $b_2 = 0.158$, $b_3 = 0.561$].

8.7 Rational Approximations

As before suppose that we have a function $f \in C[a, b]$. Instead of a polynomial, we
now contemplate approximating it by a *rational integral function* $R_{m,n}$, that is to say
by the quotient of two polynomials P_m and Q_n. Let

$$R_{m,n}(x) := \frac{P_m(x)}{Q_n(x)} := \frac{a_0 + a_1 x + \cdots + a_m x^m}{b_0 + b_1 x + \cdots + a_n x^n} \tag{8.35}$$

where it is assumed that *the two polynomials $P_m(x)$ and $Q_n(x)$ possess no common
factors. $R_{m,n}(x)$* is then said to be *irreducible*.

Rational approximations are superior to polynomial approximations in the sense
that they account for possible *poles* of the function f not lying in $[a, b]$. Such poles
are represented approximately by the zeros of the denominator of Eq. (8.35). In this
respect, polynomials fail completely.

A complete theory of best uniform (minimax) rational approximation exists in the
literature (see Achieser cited in the bibliography) where existence of such approximation is proved. Remez exchange algorithm has been extended to this case also,

but it suffers from the difficulty of solving nonlinear equations. Following the development, near-best uniform approximations have been developed for many transcendental functions in computer function libraries.

Fröberg (see Bibliography) quotes highly accurate rational approximations for the functions $\sin x$, $\cos x$, e^{-x}, and $\ln((1 + x)/2)$, in which $m = n = 4$ for the first two functions and $m = n = 3$ for the other two functions.

Apparently, it is tedious work to obtain such highly accurate approximations. So we proceed to describe some other rational approximation methods that do not produce minimax approximations, but are easier to apply. In these methods, it is difficult to estimate the error in as much as accurate evaluation of function by some other method is required.

8.7.1 Padé Approximations

This method due to H. E. Padé (1892) is a classical method for deriving a rational approximation from a power series expansion. Let $f(x)$ possess the Taylor series expansion $f(x) = \sum_{j=0}^{\infty} c_j x^j$ valid within the 'circle of convergence' $|x| < r$, where $x = r$ is the nearest singularity of $f(x)$ defined by the limiting ratio $r = \lim_{n \to \infty} = \frac{c_{n-1}}{c_n}$. If $f(x)$ is approximated by the Padé approximation of *order* (m, n) viz. $R_{m,n} = P_m(x)/Q_n(x)$ defined in Eq. (8.35), then

$$
\begin{aligned}
P_m(x) - Q_n(x)\, f(x) = {} & (a_0 + a_1 x + \cdots + a_m x^m) \\
& - [c_0 b_0 + (c_1 b_0 + c_0 b_1) x + \cdots \\
& + (c_m b_0 + c_{m-1} b_1 + \cdots + c_{m-n} b_n) x^m + \cdots]
\end{aligned}
$$

where for convenience, we let $c_j = 0$ for $j < 0$. This expression must equal zero for holding exactly. Thus, as approximation, equating the coefficients of all the powers upto x^{m+n} to zero, we obtain the $m + n + 1$ *linear homogeneous equations*

$$
a_0 = c_0 b_0
$$

$$
a_1 = c_1 b_0 + c_0 b_1
$$
$$
\cdots\cdots\cdots\cdots\cdots\cdots\cdots
$$
$$
a_m = c_m b_0 + c_{m-1} b_1 + \cdots + c_{m-n} b_n
$$

(8.36a)

$$c_{m+1}b_0 + c_m b_1 + \cdots + c_{m-n+1}b_n = 0$$

$$c_{m+2}\,b_0 + c_{m+1}\,b_1 + \cdots + c_{m-n+2}\,b_n = 0 \qquad (8.36b)$$

$$\cdots\cdots\cdots\cdots\cdots\cdots\cdots\cdots\cdots\cdots\cdots$$

$$c_{m+n}b_0 + c_{m+n-1}b_1 + \cdots + c_m b_n = 0$$

Setting $b_0 = 1$ without loss of generality, we can proceed to solve these equations for b_1, b_2, $\cdots$, b_n from Eq. (8.36b) and then a_0, a_1, $\cdots$, a_m from Eq. (8.36a).

A Padé approximation to $f(x)$ is not an approximation designed for a specific interval like $[-1, 1]$; it is merely an approximation near $x = 0$. In general, the nearer x is to the origin, the more accurate the approximation. One may estimate the error with the aid of the first nonzero term in the Taylor series expansion for $P_m(x) - Q_n(x)\,f(x)$.

The two-dimensional array of the approximations

$$R_{0,0}(x)\ R_{0,1}(x)\ R_{0,2}(x)\ \cdots$$

$$R_{1,0}(x)\ R_{1,1}(x)\ R_{1,2}(x)\ \cdots$$

$$R_{2,0}(x)\ R_{2,1}(x)\ R_{2,2}(x)\ \cdots$$

$$\cdots\qquad\cdots\qquad\cdots\qquad\cdots$$

is called a *Padé table* for $f(x)$. A practical observation based on experience is that for constant value of $m + n$, the most accurate approximation to $f(x)$ in the Padé table for $f(x)$ lies *on or next to the diagonal of the table*. The approximation is often found to be *remarkably good*, extending beyond the circle of convergence of the original Taylor series of $f(x)$.

Henri Eugène Padé (1863–1953), French mathematician who is remembered for his development of the above-described method in a systematic manner. The Padé approximations were part of his doctoral dissertation.

Example 1. Obtain Padé approximation of

$$\frac{\sin^{-1} x}{x} = 1 + \frac{x^2}{2\cdot 3} + \frac{1\cdot 3\, x^4}{2\cdot 4\cdot 5} + \frac{1\cdot 3\cdot 5\, x^6}{2\cdot 4\cdot 6\cdot 7} + \frac{1\cdot 3\cdot 5\cdot 7\, x^8}{2\cdot 4\cdot 6\cdot 8\cdot 9} + \cdots$$

in the form $\qquad \dfrac{a_0 + a_1 x^2 + a_2 x^4}{b_0 + b_1 x^2 + b_2 x^4}$

Solution. Here $c_0 = 1$, $c_1 = 1/6$, $c_2 = 3/40$, $c_3 = 5/112$ and $c_4 = 35/1152$. The coefficients a_0, a_1, a_2, b_0, b_1, b_2 are given by the system of Eq. (8.36b), which in this case is

$$a_0 = c_0 b_0, \quad a_1 = c_1 b_0 + c_0 b_1, \quad a_2 = c_2 b_0 + c_1 b_1 + c_0 b_2$$

$$c_3 b_0 + c_2 b_1 + c_1 b_2 = 0, \quad c_4 b_0 + c_3 b_1 + c_2 b_2 = 0$$

Hence, with $b_0 = 1$, $a_0 = 1$, $a_1 = \dfrac{1}{6} + b_1$, $a_2 = \dfrac{3}{40} + \dfrac{1}{6} b_1 + b_2$ where

$$\frac{3}{40} b_1 + \frac{1}{6} b_2 = -\frac{5}{112}, \quad \frac{5}{112} b_1 + \frac{3}{40} b_2 = -\frac{35}{1152}$$

The solution of these equations is

$$b_1 = -0.94490, \quad b_2 = 0.15735, \quad a_1 = -0.77824, \quad a_2 = 0.07487$$

Hence $$\frac{\sin^{-1} x}{x} \approx \frac{1 - 0.77824\, x^2 + 0.07487\, x^4}{1 - 0.94490\, x^2 + 0.15735\, x^4}$$ □

Based on Eqs. (8.36a, b), a subroutine named PADE is presented below, assuming without loss of generality that $c_0 = 1$, so that $a_0 = 1$ with $b_0 = 1$:

```
SUBROUTINE PADE(m,n,c,a,b)
! Determines the coefficients of polynomials
! of PADE approximations of a function expressed
! as a truncated power series.
! m=degree of the polynomial in the numerator.        (Input)
! n=degree of the polynomial in the denominatot.      (Input)
! c(1), c(2), · · · , c(m+n)=coeffs. of the truncated power series.   (Output)
! a(1), a(2), · · · , a(m)=coeffis. of the polyl. in the numerator.    (Output)
! b(1), b(2), · · · , b(n)=coeffs. of the polyl. in the denominator.   (Output)
! It is assumed that c(0)=b(0)=a(0)=1.
REAL :: a(m), b(n), c(m+n), PA(n,n)
DO i=1,n
b(i)=-c(m+i)
DO j=1,n
PA(i,j)=c(m+i-j)
IF(m+i-j==0) PA(i,j)=1.0
END DO
END DO
CALL GAUSS(n,PA,b)
a(1)=b(1)+c(1)
DO i=2,m
a(i)=c(i)+b(i)
DO k=1,i-1
a(i)=a(i)+c(i-k)*b(k)
END DO
END DO
RETURN
END SUBROUTINE PADE
!*****************************************
! Append SUBROUTINE GAUSS
```

Exercises

Prove the following Padé approximations:

1. $\sqrt{1-x} \approx \dfrac{4-3x}{4-x}$

2. $e^x \approx \dfrac{2+x}{2-x}$, $e^x \approx \dfrac{6+2x}{6-4x+x^2}$

3. $\ln(1+x) \approx \dfrac{6x+3x^2}{6+6x+x^2}$, $\ln(1+x) \approx \dfrac{30x+21x^2+x^3}{30+36x+9x^2}$. [May use subroutine PADE].

4. $\cos(x) \approx \dfrac{1-0.45635\,x^2+0.02070\,x^4}{1+0.04365\,x^2+0.00086\,x^4}$ [Use subroutine PADE].

8.7.2 Rational Function Interpolation

Suppose a table of data $(x_i,\ y_i = f(x_i))$, $(i = 0, 1, 2, \cdots)$ of a function $f(x)$ is given, and a **rational interpolation** $f(x) \approx R_{m,n}(x)$ is attempted, where $R_{m,n}(x)$ is given by Eq. (8.35); then

$$f(x_i) = R_{m,n}(x_i) = \frac{P_m(x_i)}{Q_n(x_i)} \quad i = 0, 1, 2, \cdots \cdots$$

or, taking $b_0 = 1$,

$$a_0 + a_1 x_i + a_2 x_i^2 + \cdots + a_m x_i^m - f(x_i)\,[b_1 x_i + b_2 x_i^2 + \cdots + b_n x_i^n]$$
$$= f(x_i), \quad i = 0, 1, 2, \cdots, m+n \quad (8.37)$$

Equation (8.37) forms a system of linear equations in $m+n+1$ unknowns a_0, a_1, $a_2 \cdots$, a_m; $b_1, b_2, \cdots, b_n$. Hence, it is required that $i = 0, 1, 2, \cdots, m+n$. The linear system of equations posses a solution, since $Q_n(x_i) = 0$, $n \geq 1$ implies that $a_0 + a_1 x + a_2 x^2 + \cdots + a_m x^m$ has $m+n > m$ zeros, which is not true. A rational interpolation may, however, fail for some some specific values of m and n. For example, consider the following small set of data:

x	0	1	2
y	1	2	2

Since the number of data points is 3, we take $m = n = 1$, that is, $P_1(x) = a_0 + a_1 x$, and $Q_1(x) = 1 + b_1 x$. The determinant of the matrix of the coefficients of the left-hand side of Eq. (8.37) for this example is

$$\begin{vmatrix} 1 & 0 & 0 \\ 1 & 1 & -2 \\ 1 & 2 & -4 \end{vmatrix} = 0$$

Thus rational approximation of the form $P_1(x)/Q_1(x)$ does not exist for this example. However, if we take $m = 2$, $n = 0$, the approximation $P_2(x)/Q_1(x) = P_2(x)/1 = P_2(x)$ does exist, since it is a case of polynomial approximation. In fact $P_2(x) = 1 + \frac{3}{2}x - \frac{1}{2}x^2$.

A useful alternative to rational function interpolation is to employ Chebyshev polynomials $T_j(x)$ instead of integral powers x^j, to represent $f(x) \approx R_{m,n}(x)$ as

$$R_{m,n}(x) = \frac{a_0' T_0(x) + a_1' T_1(x) + a_2' T_2(x) + \cdots + a_m' T_m(x)}{b_0' T_0(x) + b_1' T_1(x) + b_2' T_2(x) + \cdots + b_n' T_n(x)} \tag{8.38}$$

with $b_0' = 1$ as before. Assuming without loss of generality that $x \in [-1, 1]$, the Chebyshev nodes are (see Chap. 4, Sect. 4.3):

$$x_i = \cos \frac{(2i - 1)\pi}{2(n + 1)}, \quad i = 1, 2, 3, \cdots, (m + n + 1) \tag{8.39}$$

for determining the $m + n + 1$ unknown coefficients $a_0', a_1', a_2', \cdots, a_m', b_1', b_2', \cdots, b_n'$, provided that $f(x_i)$ are also known. The condition for interpolation $R_{m,n}(x_i) = f(x_i)$ now yield the system of linear equations

$$a_0' + a_1' T_1(x_i) + a_2' T_2(x_i) + \cdots + a_m' T_m(x_i) - f(x_i) [b_1' T_1(x_i) + b_2' T_2'(x_i) + \cdots + b_n' T_n(x_i)]$$
$$= f(x_i), \quad i = 0, 1, 2, \cdots, m + n \tag{8.40}$$

The Chebyshev interpolation points (8.39) can also be employed for rational interpolation (8.37). The following subroutine RATIONAL_APPROX determines the coefficients of the interpolation formulae for the two cases, using the subroutine GAUSS. The subroutine uses an integer parameter named *select* for solving either of the two systems of Eq. (8.37) or (8.40). The step select $\leftarrow$ 1 solves the rational function approximation (8.37), while select $\leftarrow$ 2 does the same for the Chebyshev polynomial formulation (8.40):

```
SUBROUTINE RATIONAL_APPROX(m,n,x,f,select)
! Determines the coefficients of the polynomials appearing
! in rational approximations of the forms (35) and (38).
! select=1 for form (8.35), or 2 for form (8.38).          (Input)
! m = degree of the polynomial in the numerator.                 (Input)
! n = degree of the plynomial in the denominator.                (Input)
! x(1),x(2),···,x(m+n+1)=abscissa of nodal points.               (Input)
! f(1),f(2),···,f(m+n+1)=function values at the nodal points.    (Input)
! The solution is returned in the vector f.        (Output)
! a0=f(1), a1=f(2),···, am=f(m+1); b0=1, b1=f(m+2),···, bn=f(m+n+1), OR
! a'0=f(1), a'1=f(2),···, a'm=f(m+1); b'0=1, b'1=f(m+2),···, b'n=f(m+n+1)
!***********************************************************************
INTEGER :: select
```

```
REAL :: x(m+n+1), f(m+n+1), A(m+n+1,m+n+1)
DO i=1,m+n+1
A(i,1)=1.0
DO j=2,m+n+1
IF(select==1) THEN
IF(j<=m+1) A(i,j)=x(i)**(j-1)
IF(j>=m+2) A(i,j)=- x(i)**(j-m-1)*f(i)
ELSE IF(select==2) THEN
IF(j<=m+1) A(i,j)=cos((j-1)*acos(x(i)))
IF(j>=m+2) A(i,j)=- cos((j-m-1)*acos(x(i)))*f(i)
END IF
END DO
END DO
CALL GAUSS(m+n+1,A,f)
RETURN
END SUBROUTINE RATIONAL_APPROX
!****************************************
! Append subroutine GAUSS
```

Remark. In case the data are drawn from experimental observations, then the least squares fit of the data by a rational polynomial expression can be obtained by application of the method of least squares to Eq. (8.37). The estimate of the coefficients $a_0, a_1, a_2, \cdots, a_m$ and $b_1, b_2, \cdots, b_n$ determined by the method yields the rational approximation fit.

Exercises

[Use subroutine RATIONAL_APPROX].
1. Show that for the data

x	1	2	3	4
y	1	2	3	3

$R_{2,1}(x)$ fails, but $R_{1,2}(x) = (0.42857 + 0.21429\,x)/(1 - 0.42857\,x + 0.07142\,x^2)$.
2. For the data

x	0	1	2	3
y	1	0	1/3	1

show that $R_{2,1}(x) = (1 - 2\,x + x^2)/(1 + x)$ and $R_{1,2}(x) = (1 - x)/(1 - 4\,x + x^2)$.
3. For the data

(a)

x	0	1	2	3	4
y	2	0	0	2/13	2/7

and (b)

x	0	1	2	3	4
y	2	1	1	0	−4/5

show that (a) $R_{2,2}(x) = (2 - 3\,x + x^2)/(1 + x + x^2)$ and (b) $R_{2,2} = (12 - 13\,x + 3\,x^2)/(6 - 4\,x)$.

4. Find $R_{2,2}(x)$ for $(\sin^{-1} x)/x$, using Chebyshev interpolation points adopting the procedures given by Eqs. (8.39) and (8.40).
$[R_{2,2}(x) = (1 - 1.08239\, x^2 + 0.19026\, x^4)/(1 - 1.24889\, x^2 + 0.32146\, x^4)$, and $R_{2,2}(x) = (1 - 1.08244\, x^2 + 0.19030\, x^4)/(1 - 1.24895\, x^2 + 0.32151\, x^4)$ (Chebyshev)].

8.7.3 Near-Best Uniform Rational Approximation: Maehly's Method

The theorem on best uniform rational approximation is an extension of that for polynomials (Theorem 8.5)—a result originally due to Chebyshev. In this extension ϕ_n^* is simply replaced by the best uniform rational $R_{m,n}^*$. The REMEZ Exchange Algorithm also finds extension for finding $R_{m,n}^*$, but that algorithm becomes far more elaborate [see Ralston and Wilf (1967)] for presentation here. In that development, it is kept in view that the maximum error of the approximation is less than a desired value for which purpose a *starting solution is required*. A good starting solution is provided by the rational Chebyshev interpolation Eqs. (8.38)–(8.40) of the preceding section.

As a useful alternative, Padé's method of rational approximation was generalised by Hans J. Maehly (1960), (1963). Expanding $f(x)$ in Chebyshev series, let $f(x) = \sum_0^\infty c_i T_i(x), x \in [-1, 1]$ be approximated by $P_m(x)/Q_n(x)$ where

$$c_i = \frac{2}{\pi} \int_{-1}^{1} \frac{f(x)\, T_i(x)}{\sqrt{1 - x^2}}\, dx = \frac{2}{\pi} \int_0^\pi f(\cos\theta)\, \cos i\theta\, d\theta, \qquad i = 0, 1, 2, \cdots\cdots$$

and $P_m(x) = a_0 T_0(x) + a_1 T_1(x) + \cdots + a_m T_m(x)$; $Q_n(x) = b_0 T_0(x) + b_1 T_1(x) + \cdots + b_n T_n(x)$. Suppressing the argument x in $T_i(x)$, we have

$$P_m(x) - Q_n(x)\, f(x) = \sum_{i=0}^{m} a_i T_i - \sum_{j=0}^{n} b_j \left(\sum_{i=0}^{\infty} c_i T_i T_j \right)$$

Now it is elementary to prove that $T_i T_j = \dfrac{1}{2}(T_{i+j} + T_{|i-j|})$ (see Ex. 2b, Sect. 4.3, Chap. 4). Hence,

$$2 \sum_{i=0}^{\infty} c_i T_i T_j = \sum_{i=0}^{\infty} c_i\, (T_{i+j} + T_{|i-j|})$$

$$= \sum_{i=0}^{\infty} c_i T_{i+j} + \sum_{i=1}^{j-1} c_i T_{j-i} + \sum_{i=j+1}^{\infty} c_i T_{i-j} + c_0 T_j + c_j T_0$$

$$= \sum_{i=j}^{\infty} c_{i-j} T_i + \sum_{i=1}^{j-1} c_{j-i} T_i + \sum_{i=1}^{\infty} c_{i+j} T_i + c_0 T_j + c_j T_0$$

$$= \sum_{i=1}^{\infty} (c_{i+j} + c_{|i-j|}) T_i + c_0 T_j + c_j T_0$$

Substituting the above expression and setting for notational convenience $a_i = 0$ for $i > m$ and $b_i = 0$ for $i > n$, we obtain on simplification

$$P_m(x) - Q_n(x) f(x) = \sum_{i=0}^{\infty} a_i T_i - \left(b_0 c_0 + \frac{1}{2} \sum_{j=1}^{n} b_j c_j \right) T_0$$

$$- \sum_{i=1}^{\infty} \left[b_0 c_i + \frac{1}{2} b_i c_0 + \frac{1}{2} \sum_{j=1}^{n} b_j (c_{i+j} + c_{|i-j|}) \right] T_i$$

Equating to zero the first $m + n + 1$ coefficients of the above Chebyshev series yield the homogeneous system of equations

$$a_0 = b_0 c_0 + \frac{1}{2} \sum_{j=1}^{n} b_j c_j$$

$$a_i = b_0 c_i + \frac{1}{2} b_i c_0 + \frac{1}{2} \sum_{j=1}^{n} b_j (c_{i+j} + c_{|i-j|}), \quad i = 1, 2, \cdots, m+n$$

(8.41)

Setting $b_0 = 1$, Eq. (8.41) yields the coefficients $a_0, \cdots, a_m, b_1, \cdots, b_n$.

In practice, Maehly's method yields a near-minimax rational approximation. It is particularly suitable for slowly converging Taylor series for $f(x)$, when its Chebyshev series converges rapidly.

Hans Jakob Maehly (1920–1961) was a Swiss physicist, who contributed to numerical computing.

The following subroutine implements the method:

SUBROUTINE MAEHLY(m,n,c0,c,a,b)
! **Computes the coefficients of the Rational Chebyshev Approximation**
! **by Maehly's Method.**
! **m=degree of the Chebyshev polynomial in the numerator.**   (Input)
! **n=degree of the Chebyshev polynomial in the denominator.**   (Input)
! **c0, c(1), c(2), $\cdots$, c(m+n) are the coefficints of the Chebyshev**
! **polynomial approximation of the given function.**   (Input)
! **a(1), a(2), $\cdots$, a(m+1) are the coefficients of the Chebyshev polynomials**
! **appearing in the numerator of the rational approximation.**   (Output)
! **b(1), b(2), $\cdots$, b(n) are the coefficients of the Chebyshev polynomial**
! **appearing in the denominator.**   (Output)
!**

```
REAL :: c(m+n), a(m+1), b(n), MA(n,n)
DO i=1,n
b(i)=-2*c(i+m)
DO j=1,n
delta=0.0
IF(i+m==j) delta=1.0
MA(i,j)=delta*c0+c(i+j+m)+c(ABS(i-j+m))
END DO;   END DO
CALL GAUSS(n,MA,b)
a0=c0
DO i=1,n
a0=a0+0.5*b(i)*c(i)
END DO
DO i=1,m
a(i)=c(i)+b(i)*c0
DO j=1,n
a(i)=a(i)+0.5*b(j)*(c(i+j)+c(ABS(i-j)))
END DO;   END DO
DO i=m+1,2,-1
a(i)=a(i-1)
END DO
a(1)=a0
RETURN
END SUBROUTINE MAEHLY
!*****************************
! Append SUBROUTINE GAUSS
```

Example 1. Obtain Maehly approximation from Fraser (1965) series

$$\frac{1}{2} \tan^{-1} x = \sum_{k=0}^{\infty} \frac{(-1)^k (\sqrt{2} - 1)^{2k+1}}{2k+1} T_{2k+1}(x)$$

in the form $\dfrac{a_0 T_0(x) + a_1 T_1(x) + \cdots + a_5 T_5(x)}{b_0 T_0(x) + b_1 T_1(x) + \cdots + b_4 T_4(x)}$

Hence, obtain a rational approximation of $\tan^{-1} x$, and calculate π from the approximation.

Solution. Here $c_0 = c_2 = c_4 = c_6 = c_8 = 0$ and $c_1 = 0.414214$, $c_3 = -0.023689$, $c_5 = 0.002439$, $c_7 = -0.000299$ and $c_9 = 0.000040$. The higher order coefficients are approximately taken as zero.. Consistent with the vanishing even coefficients c_{2k}, we must have from Eq. (8.41), $a_0 = a_2 = a_4 = b_1 = b_3 = 0$. The remaining required coefficients given by Eq. (8.41) yield

$$a_1 = b_0 c_1 + \frac{1}{2} b_2 (c_3 + c_1) + \frac{1}{2} b_4 (c_5 + c_3)$$

$$a_3 = b_0 c_3 + \frac{1}{2} b_2 (c_5 + c_1) + \frac{1}{2} b_4 (c_7 + c_1)$$

$$a_5 = b_0 c_5 + \frac{1}{2} b_2 (c_7 + c_3) + \frac{1}{2} b_4 (c_9 + c_1)$$

$$0 = a_7 = b_0 c_7 + \frac{1}{2} b_2 (c_9 + c_5) + \frac{1}{2} b_4 c_3$$

$$0 = a_9 = b_0 c_9 + \frac{1}{2} b_2 c_7 + \frac{1}{2} b_4 c_5$$

Taking $b_0 = 1$ as usual, the last two equations can be solved for b_2 and b_4. The solution is $b_2 = 0.369573$, $b_4 = 0.013432$. With these values of b_2, b_4 the first three equations yield $a_1 = 0.486234$, $a_3 = 0.056082$ and $a_5 = 0.000788$. Hence

$$\frac{1}{2} \tan^{-1} x \approx \frac{0.486234 \, T_1(x) + 0.056082 \, T_3(x) + 0.000788 \, T_5(x)}{T_0(x) + 0.369573 \, T_2(x) + 0.013432 \, T_4(x)}$$

Now, $T_0(x) = 1$, $T_1(x) = x$, $T_2(x) = 2x^2 - 1$, $T_3(x) = 4x^3 - 3x$, $T_4(x) = 8x^4 - 8x^2 + 1$ and $T_5(x) = 16x^5 - 20x^3 + 5x$. Inserting these expressions in the approximation of $\tan^{-1} x$, we finally obtain

$$\tan^{-1} x \approx \frac{x + 0.64787 \, x^3 + 0.03916 \, x^5}{1 + 0.98110 \, x^2 + 0.16689 \, x^4}$$

The approximation yields $\pi \approx 3.14160$ □

Exercises

[Use subroutine MAEHLY].

1. It is known that $\tan^{-1} x$ can be represented as a continued fraction:

$$\tan^{-1} x = \frac{x}{1+} \frac{x^2}{3+} \frac{4x^2}{5+} \frac{9x^2}{7+} \frac{16x^2}{9+} \cdots \cdots \infty$$

Show that taking terms upto the fifth convergent as written above

$$\tan^{-1} x \approx \frac{x + 0.52063 \, x^3 + 0.06772 \, x^5}{1 + 0.85397 \, x^2 + 0.15238 \, x^4}$$

Show from the approximation that $\pi \approx 3.15696$.

2. It is known that for $|x| \leq 1$

$$\sin \left(\frac{\pi}{2} x \right) = x \, [1.276279 - 0.285262 \, T_1^*(x^2) + 0.009118 \, T_2^*(x^2) - 0.000137 T_3^*(x^2) + 0.000001 \, T_4^*(x^2)$$

where $T_n^*(x)$ is the *shifted Chebyshev polynomial* $T_n^*(x) := T_n(2x - 1)$. Show that

$$\sin\left(\frac{\pi}{2}x\right) \approx \frac{1.57209 - 0.54606\,x^2 + 0.03163\,x^4}{1 + 0.08286\,x^2 + 0.00321\,x^4}, \qquad |x| \le 1$$

Verify that $\sin(\pi/2) \approx 0.97384$.

3. As in Exercise 2 above, it is known that

$$\cos\left(\frac{\pi}{2}x\right) = 0.472001 - 0.499403\,T_1^*(x^2) + 0.027992\,T_2^*(x^2) - 0.000597\,T_3^*(x^2) + 0.000007\,T_4^*(x^2)$$

Show that

$$\cos\left(\frac{\pi}{2}x\right) \approx \frac{1.00149 - 1.07597\,x^2 + 0.13828\,x^4}{1 + 0.10488\,x^2 + 0.00508\,x^4}, \qquad |x| \le 1$$

Hence, verify that $\cos\dfrac{\pi}{4} = 0.72198$, whose exact value is $1/\sqrt{2} = 0.70711$.

4. Assuming that

$$\frac{1}{\Gamma(x+1)} = 0.731219\,T_0(x) + 0.523174\,T_1(x) - 0.249637\,T_2(x) - 0.021396\,T_3(x) + 0.018788\,T_4(x)$$
$$- 0.001881\,T_5(x)$$

Show that

$$\Gamma(x+1) \approx \frac{0.96082 + 0.03361\,x + 0.20898\,x^2 + 0.08711\,x^3}{1 + 0.57403\,x - 0.49549x^3}, \qquad |x| \le 1$$

Hence, calculate $\Gamma(1)$, $\Gamma(1/2)$ and $\Gamma(-1/2)$.

[0.96082, 1.69561, −3.39122. (Exact values: 1, 1.77245, −3.54490 respectively)].

Chapter 9
Matrix Eigenvalues

Given an $n \times n$ (real or complex) matrix A and an n-vector $\mathbf{x}$, suppose we construct the n-vector $A\mathbf{x} =: \mathbf{y}$, then A can be interpreted as a *linear transformation* which carries $\mathbf{x}$ to $\mathbf{y}$ in an n-dimensional space. In particular, if a non-zero vector $\mathbf{x}$ can be found such that A carries $\mathbf{x}$ to a collinear vector $\lambda\mathbf{x}$, i.e.

$$A\mathbf{x} = \lambda\mathbf{x}$$

then $\mathbf{x}$ is called an **eigenvector** of A and λ its corresponding **eigenvalue**. From the above definition, it follows that $\mathbf{x}$ satisfies the system of n linear equations

$$(A - \lambda I)\mathbf{x} = 0 \tag{9.1}$$

where I is the unit diagonal matrix. Since $\mathbf{x}$ is non-null

$$\det(A - \lambda I) = 0$$

or

$$\begin{vmatrix} a_{11} - \lambda & a_{12} & a_{13} & \cdots & a_{1n} \\ a_{21} & a_{22} - \lambda & a_{23} & \cdots & a_{2n} \\ \cdots & \cdots & \cdots & \cdots & \cdots \\ a_{n1} & a_{n2} & a_{n3} & \cdots & a_{nn} - \lambda \end{vmatrix} = 0 \tag{9.2}$$

The determinant is called the **characteristic** or **secular determinant** of A. When expanded, it yields the **characteristic polynomial** equation of degree n that has exactly n roots real or complex. The matrix A has then n eigenvalues all of which need not be distinct.

© Springer Nature Singapore Pte Ltd. 2019
S. K. Bose, *Numerical Methods of Mathematics Implemented in Fortran*, Forum for Interdisciplinary Mathematics,
https://doi.org/10.1007/978-981-13-7114-1_9

Corresponding to an eigenvalue λ, there exists at least one eigenvector obtained by solving Eq. (9.1). Such an eigenvector is obtained only up to a certain scalar multiple. For, if $\mathbf{x}$ is an eigenvector, $\alpha\mathbf{x}$ (α a scalar) is also evidently a solution of Eq. (9.1). One may normalise the eigenvectors by dividing them by the magnitude of the largest eigenvector of the system. If λ is *distinct*, then in view of Eq. (9.2), we ignore one equation, say the nth one, and solve the remaining $n - 1$ equations for say, x_1, x_2, $\cdots$, x_{n-1} in terms of x_n. Distinct eigenvalues yield linearly independent eigenvectors according to Theorem 9.1 of the following section. If λ is *repeated r* times, we ignore r linearly dependent equations from (1), and treating the remaining $n - r$ equations we may be able to construct r linearly independent eigenvectors. Thus, the matrix A may not always have n *linearly independent* eigenvectors.

Example 1. Let

$$A = \begin{bmatrix} 2 & 1 & 1 \\ 1 & 2 & 1 \\ 1 & 1 & 2 \end{bmatrix}$$

Find the eigenvalues and eigenvectors of A.

Solution. The characteristic equation is

$$\begin{vmatrix} 2-\lambda & 1 & 1 \\ 1 & 2-\lambda & 1 \\ 1 & 1 & 2-\lambda \end{vmatrix} = 0$$

or $\quad 4 - 9\lambda + 6\lambda^2 - \lambda^3 = (\lambda - 1)^2(\lambda - 4) = 0$. Hence, $\lambda_1 = 4$, $\lambda_2 = \lambda_3 = 1$, adopting the convention of writing the eigenvalues in descending order of magnitude.
The eigenvectors are given by

$$\begin{aligned} (2-\lambda)x_1 + \qquad x_2 + \qquad x_3 &= 0 \\ x_1 + (2-\lambda)x_2 + \qquad x_3 &= 0 \\ x_1 + \qquad x_2 + (2-\lambda)x_3 &= 0 \end{aligned}$$

For $\lambda_1 = 4$, deleting the third equation and solving the first two, we obtain

$$\frac{x_1}{\begin{vmatrix} 1 & 1 \\ -2 & 1 \end{vmatrix}} = \frac{-x_2}{\begin{vmatrix} -2 & 1 \\ 1 & 1 \end{vmatrix}} = \frac{x_3}{\begin{vmatrix} -2 & 1 \\ 1 & -2 \end{vmatrix}}$$

or $x_1/3 = x_2/3 = x_3/3$, that is, $x_1 = x_2 = x_3 = \alpha$ (say). Selecting $\alpha = 1$, the eigenvector corresponding to λ_1 is $[1, 1, 1]^T$.
For $\lambda_2 = \lambda_3 = 1$, deleting the last two dependent equations, we obtain

$$x_1 = -(x_2 + x_3)$$

Selecting $x_2 = 1$, $x_3 = 0$ and $x_2 = 0$, $x_3 = 1$, we get two linearly independent eigenvectors $[-1, 1, 0]^T$ and $[-1, 0, 1]^T$. Other vectors constructed from the equation $x_1 = -(x_2 + x_3)$ will be linearly dependent on these two vectors.   □

Example 2. Find the eigenvalues and eigenvectors of the matrix

$$\begin{bmatrix} 2 & 1 & 1 \\ 0 & 2 & 1 \\ 0 & 0 & 2 \end{bmatrix}$$

Solution. The characteristic equation of the matrix reduces to the equation $(2 - \lambda)^3 = 0$. Hence, $\lambda_1 = \lambda_2 = \lambda_3 = 2$. The eigenvectors are given by

$$
\begin{aligned}
(2 - \lambda) x_1 + \quad & x_2 + \quad & x_3 &= 0 \\
& (2 - \lambda) x_2 + \quad & x_3 &= 0 \\
& & (2 - \lambda) x_3 &= 0
\end{aligned}
$$

which yields $x_2 = x_3 = 0$. Thus, the matrix has only one linearly independent eigenvector $[1, 0, 0]^T$.

Matrix eigenvalues enter the study of *small oscillations* of a mechanical system about position of stable equilibrium in a straight forward way. Let $q_1, q_2, \cdots, q_n$ be the *generalised coordinates* of the system, then the dynamical equations of motion reduce to the form

$$\ddot{\mathbf{q}} + K\,\mathbf{q} = 0 \tag{9.3}$$

where $\mathbf{q} =:= \lfloor q_1, q_2, \cdots, q_n \rfloor^T$ and K is a *symmetric stiffness matrix*. The nomenclature is derived from the vibration of a mass m suspended by a spring of 'stiffness' k, so that $K = k/m$ in a single differential equation. The proof of Eq. (9.3) follows from Lagrange's equations of motion

$$\frac{d}{dt}\left(\frac{\partial T}{\partial \dot{q}_i}\right) + \frac{\partial V}{\partial q_i} = 0 \tag{9.4}$$

for the motion of the system, where T and V are, respectively, the kinetic and potential energies of the system given by the expressions

$$T = \frac{1}{2} \sum_{i=1}^{n} \sum_{j=1}^{n} a_{ij}\, \dot{q}_i\, \dot{q}_j, \quad V = \frac{1}{2} \sum_{i=1}^{n} \sum_{j=1}^{n} b_{ij}\, q_i\, q_j \tag{9.5}$$

where $a_{ij} = a_{ji}$ and $b_{ij} = b_{ji}$. The above expressions are a consequence of the fact that q_i is small during the oscillations. Inserting the expressions (9.5) in Eq. (9.4), yield Eq. (9.3) with $K = A^{-1}B$, where $A := [a_{ij}]$, and $B := [b_{ij}]$. For oscillations of period $2\pi/\omega$, we can consider the solution as $\mathbf{q} = \mathbf{q}_0\, e^{i\omega t}$ and Eq. (9.3) yields

$$(K - \omega^2 I)\, \mathbf{q}_0 = 0$$

so that the angular frequencies ω are given by the *secular* equation

$$\det (K - \omega^2 I) = 0$$

Since K is real symmetric matrix, the roots ω^2, and hence, ω are always real according to Theorem 9.3 of the next section.

There are several other applications of eigenvalues. Thus, computation of eigenvalues is an important problem which by no means is easy. The computation of eigenvectors is of somewhat less importance and degree of difficulty. Hence, we shall pay attention mostly to the former.

9.1   General Theorems

In the following we prove some basic theorems regarding eigenvalues and eigenvectors of matrices.

Theorem 9.1 *If the eigenvalues λ_1, $\lambda_2, \cdots, \lambda_m$ ($m \leq n$) are distinct, then the corresponding eigenvectors $\mathbf{x}_1$, $\mathbf{x}_2, \cdots, \mathbf{x}_m$ are linearly independent.*

Proof The proof is by contradiction. If possible let the eigenvectors $\mathbf{x}_1$, $\mathbf{x}_2$ be linearly dependent, that is, constants c_1, c_2 both not zero can be found such that

$$c_1\, \mathbf{x}_1 + c_2\, \mathbf{x}_2 = 0$$

Hence, $A\, (c_1\, \mathbf{x}_1 + c_2\, \mathbf{x}_2) = c_1\, \lambda_1\mathbf{x}_1 + c_2\, \lambda_2\mathbf{x}_2 = 0$
Multiplying the first equation by λ_1 and subtracting from the second equation we get $c_2\, (\lambda_1 - \lambda_2)\, \mathbf{x}_2 = 0$. Since $\lambda_1 \neq \lambda_2$, $c_2 = 0$ and hence from the first equation above $c_1 = 0$, which is contradiction The general case follows from induction. $\square$

Theorem 9.2 *If A is real with complex eigenvalue λ and complex eigenvector $\mathbf{x}$, then the complex conjugate λ^* is also a complex eigenvalue with complex eigenvector $\mathbf{x}^*$. If λ is repeated r times so is λ^*.*

Proof Since $A\mathbf{x} = \lambda\mathbf{x}$, $(A\mathbf{x})^* = (\lambda\mathbf{x})^*$ or $A^*\mathbf{x}^* = \lambda^*\mathbf{x}^*$. But A is real, so $A^* = A$. Hence, $A\mathbf{x}^* = \lambda^*\mathbf{x}^*$. $\square$

Theorem 9.3 *If A is real and symmetric, then all the eigenvalues are real.*

Proof Let λ be an eigenvalue of A with eigenvector $\mathbf{x}$; then $A\mathbf{x} = \lambda\mathbf{x}$. Forming the *scalar product* of the vector $A\mathbf{x}$ with $\mathbf{x}$

$$(A\mathbf{x}, \mathbf{x}) = (\lambda\mathbf{x}, \mathbf{x}) = \lambda(\mathbf{x}, \mathbf{x}) = \lambda\|\mathbf{x}\|^2$$

Similarly
$$(\mathbf{x}, A\mathbf{x}) = (\mathbf{x}, \lambda\mathbf{x}) = \lambda^*(\mathbf{x}, \mathbf{x}) = \lambda^*\|\mathbf{x}\|^2$$

Now

$$A\mathbf{x} = \sum_{j=1}^{n} a_{ij} x_j, \quad (A\mathbf{x}, \mathbf{x}) = \sum_{i=1}^{n}\sum_{j=1}^{n} a_{ij} x_j x_i^*, \quad (\mathbf{x}, A\mathbf{x}) = \sum_{i=1}^{n}\sum_{j=1}^{n} x_i a_{ij}^* x_j^*$$

Since A is real symmetric $a_{ij}^* = a_{ij} = a_{ji}$ and therefore $(A\mathbf{x}, \mathbf{x}) = (\mathbf{x}, A\mathbf{x})$. Hence, $\lambda\|\mathbf{x}\|^2 = \lambda^*\|\mathbf{x}\|^2$. Since $\|\mathbf{x}\| \neq 0$, $\lambda = \lambda^*$. $\square$

Theorem 9.4 *If A is complex Hermitian, then all eigenvalues are real.*

Proof Recall that for a Hermitian A, $a_{ji} = a_{ij}^*$ and the proof is as in Theorem 9.3. $\square$

Theorem 9.5 *If $\lambda_1, \lambda_2, \cdots, \lambda_n$ are eigenvalues of A; A^k ($k = positive\ integer$) has eigenvalues $\lambda 1^k, \lambda_2^k, \cdots, \lambda_n^k$.*

Proof If λ is an eigenvalue of A with eigenvector $\mathbf{x}$, $A\mathbf{x} = \lambda\mathbf{x}$. Hence, $A(A\mathbf{x}) = \lambda A\mathbf{x} = \lambda^2\mathbf{x}$ or $A^2\mathbf{x} = \lambda^2\mathbf{x}$. Similarly, $A^3\mathbf{x} = \lambda^3\mathbf{x}$ etc. and $A^k\mathbf{x} = \lambda^k\mathbf{x}$, i.e. λ^k is an eigenvalue of A^k. $\square$

Theorem 9.6 *If $\lambda_1, \lambda_2, \cdots, \lambda_n$ are eigenvalues of A, then A^{-1} has eigenvalues $\lambda_1^{-1}, \lambda_2^{-1}, \cdots, \lambda_n^{-1}$.*

Proof As in Theorem 9.5, $A^{-1}(A\mathbf{x}) = \lambda A^{-1}\mathbf{x}$ or $A^{-1}\mathbf{x} = \lambda^{-1}\mathbf{x}$. $\square$

The next theorem requires the extremely useful notion of similar matrices, which is required for the computation of eigenvalues. Let the matrix A represent a *linear transformation* $A\mathbf{x} = \mathbf{y}$, where $\mathbf{x}$ and $\mathbf{y}$ are expressed with respect to some *basis vector* $\mathbf{e}$ by the transformation $\mathbf{e}' = S^T\mathbf{e}$, where S *is the change of basis matrix* such that $\det(S) \neq 0$; then in the new basis, $\mathbf{x}, \mathbf{y}$ become $\mathbf{x}', \mathbf{y}'$ where

$$\mathbf{x} = \sum_{i=1}^{n} x_i e_i = \sum_{j=1}^{n} x_j' e_j' = \sum_{j=1}^{n} x_j' \sum_{i=1}^{n} s_{ij} e_i = \sum_{i=1}^{n}\left(\sum_{j=1}^{n} s_{ij} x_j'\right) e_i$$

Since e_i are linearly independent, we obtain

$$x_i = \sum_{j=1}^{n} s_{ij} x_j'$$

or $\mathbf{x} = S\mathbf{x}'$. Similarly $\mathbf{y} = S\mathbf{y}'$. Therefore $S\mathbf{y}' = \mathbf{y} = A\mathbf{x} = AS\mathbf{x}'$, that is, $\mathbf{y}' = (S^{-1}AS)\mathbf{x}'$. Hence in the changed basis $\mathbf{y}' = A'\mathbf{x}'$ where $A' = S^{-1}AS$ is called matrix **similar** to A.

Theorem 9.7 *Similar matrices have the same characteristic polynomial and have the same eigenvalues.*

Proof Let B be *similar* to A: $B = S^{-1}AS$, $\det(S) \neq 0$. Then

$$\det(B - \lambda I) = \det(S^{-1}AS - \lambda I) = \det\{S^{-1}(A - \lambda I)S\}$$
$$= \det(S^{-1}) \times \det(A - \lambda I) \times \det(S) = \det(A - \lambda I)$$

as $\det(S^{-1}) = 1/\det(S)$. $\qquad\qquad\qquad\qquad\qquad\qquad\qquad\qquad\qquad\qquad\square$

Theorem 9.8 *If the matrix A has n linearly independent eigenvectors then taking them as basis vectors, a diagonal matrix Λ similar to A is obtained whose diagonal entries are the eigenvalues of A.*

Proof Denote the eigenvectors by $\mathbf{e}_1, \mathbf{e}_2, \cdots, \mathbf{e}_n$, then $A\mathbf{e}_i = \lambda_i \mathbf{e}_i$, $i = 1, 2, \cdots, n$. Let A represent the linear transformation $A\mathbf{x} = \mathbf{y}$ in some basis. In the eigenvector basis, $\mathbf{x} = x_1\mathbf{e}_1 + \cdots + x_n\mathbf{e}_n$ and the transformation becomes

$$\mathbf{y} = \sum_{i=1}^{n} x_i A\mathbf{e}_i = \sum_{i=1}^{n} \lambda_i x_i \mathbf{e}_i = \begin{bmatrix} \lambda_1 & 0 & \cdots & 0 \\ 0 & \lambda_2 & \cdots & 0 \\ & & \cdots & \\ 0 & 0 & \cdots & \lambda_n \end{bmatrix} \begin{bmatrix} x_1 \\ x_2 \\ \vdots \\ x_n \end{bmatrix} = \Lambda \mathbf{x}$$

where Λ is the diagonal matrix stated in the theorem. $\qquad\qquad\qquad\qquad\square$

In the above theorem, Λ is arrived by some change of basis matrix S: $\Lambda = S^{-1}AS$. This process is called **matrix diagonalisation**. It obviously delivers the eigenvalues but much is less attempted in actual computational methods.

In the theorem, the stipulation of n linearly independent eigenvectors is met when $\lambda_1, \lambda_2, \cdots, \lambda_n$ are *distinct*, but it can be proved that this is not a necessary requirement for the validity of the theorem.

Theorem 9.9 (Gershgorin Inequalities) *If λ is any eigenvalue of a matrix A, then the largest eigenvalue in modulus cannot exceed the largest sum of the moduli of the elements along any row or column, i.e.*

$$|\lambda| \leq \max_{1 \leq i \leq n} \sum_{j=1}^{n} |a_{ij}| \quad and \quad |\lambda| \leq \max_{1 \leq j \leq n} \sum_{i=1}^{n} |a_{ij}|$$

Proof Let the eigenvector corresponding to λ be $\mathbf{x} = [x_1, x_2, \cdots, x_n]^T$. Choose M in such a way that $|x_M| = \max\limits_{1 \le j \le n} |x_j|$. Then according to Eq. (9.1), $\lambda x_M = \sum\limits_{j=1}^{n} a_{Mj} x_j$. Hence,

$$|\lambda| = \left| \sum_{j=1}^{n} a_{Mj} \frac{x_j}{x_M} \right| \le \sum_{j=1}^{n} |a_{Mj}| \cdot \frac{|x_j|}{|x_M|} \le \sum_{j=1}^{n} |a_{Mj}| \le \max_{1 \le i \le n} \sum_{j=1}^{n} |a_{ij}|$$

Since the transposed matrix has the same set of eigenvalues, one also has

$$|\lambda| \le \max_{1 \le j \le n} \sum_{i=1}^{n} |a_{ij}| \qquad \square$$

9.2 Real Symmetric Matrices

Such matrices have real eigenvalues (Theorem 9.3) and are amenable to special treatment. All the methods of computing eigenvalues use *similarity transformation* to reduce a given $n \times n$ matrix A to some special form, from which the eigenvalues are extracted with some ease.

9.2.1 Jacobi's Method

This classical method is important for its conceptual simplicity and is a forerunner of more intricate methods for the general eigenvalue problem that have proved to be highly efficient.

Here, the similarity transformations are based on two-dimensional rotation matrices. In a two-dimensional plane, if the axes are rotated through an angle θ, the change in coordinates of a point is governed by the matrix

$$J = \begin{bmatrix} \cos\theta & \sin\theta \\ -\sin\theta & \cos\theta \end{bmatrix}$$

In n-dimensional space, if we similarly think of rotation in a (p, q)-plane ($1 \le p, q \le n$), the transformation of coordinates will take place according to the matrix

$$
J_{pq} =
\begin{bmatrix}
1 & & & & & & & \\
& 1 & & & & & & \\
& & \ddots & & & & & \\
& & & \cos\theta & \cdots & \sin\theta & & \\
& & & & \ddots & & & \\
& & & -\sin\theta & \cdots & \cos\theta & & \\
& & & & & & \ddots & \\
& & & & & & & 1 \\
& & & & & & & & 1
\end{bmatrix}
\begin{matrix}
\\ \\ \\ \leftarrow \text{row } p \\ \\ \leftarrow \text{row } q \\ \\ \\ \\
\end{matrix}
\tag{9.6}
$$

Evidently, $J_{pq}^{-1} = J_{pq}^{T}$ and the matrix J_{pq} is *orthogonal*. Also $n(n-1)/2$ such matrices can be introduced in the n-dimensional space.

If J_{pq} is used as a similarity transformation of A then we get the similar matrix $A_1 := J_{pq}^{T} A J_{pq}$. To find the structure of A_1, let $A' = A J_{pq}$. It is now easy to verify that $a'_{ij} = a_{ij}$ except when $j = p$ or q, which are given by

$$
a'_{ip} = a_{ip} \cos\theta + a_{iq} \sin\theta, \quad a'_{iq} = -a_{ip} \sin\theta + a_{iq} \cos\theta
\tag{9.7}
$$

for $i = 1, 2, \cdots, n$. If the elements of A_1 are denoted by $a_{ij}^{(1)}$, then it is next easy to verify that $a_{ij}^{(1)} = a'_{ij}$ except when $i = p$ or q, which are given by

$$
a_{pj}^{(1)} = a'_{pj} \cos\theta + a'_{qj} \sin\theta, \quad a_{qj}^{(1)} = -a'_{pj} \sin\theta + a'_{qj} \cos\theta
\tag{9.8}
$$

for $j = 1, 2, \cdots, n$. Thus, the rows and columns of A_1 and A are identical except those for $i = p, q$ and $j = p, q$. The *off-diagonal* elements at the intersection of these two rows and columns are from Eqs. (9.7) and (9.8)

$$
\begin{aligned}
a_{pq}^{(1)} &= (a_{qq} - a_{pp}) \sin\theta \cos\theta + a_{pq} \cos^2\theta - a_{qp} \sin^2\theta \\
a_{qp}^{(1)} &= (a_{qq} - a_{pp}) \sin\theta \cos\theta - a_{pq} \sin^2\theta + a_{qp} \cos^2\theta
\end{aligned}
\tag{9.9}
$$

Both of these elements can be annihilated by invoking the *symmetry* of A for θ satisfying

$$
(a_{pp} - a_{qq}) \sin 2\theta = 2 a_{pq} \cos 2\theta
$$

or

$$
\theta = \frac{1}{2} \tan^{-1} \frac{2 a_{pq}}{a_{pp} - a_{qq}}
\tag{9.10}
$$

where the principal part of the inverse part tangent is taken, that is, $-\pi/4 \le \theta \le \pi/4$.

Suppose that similarity transformations of the type (9.6) through angles of the type (9.1) are repeatedly applied by rotation in the planes $(1, 2)$, $(1, 3)$, $\cdots$, $(1, n)$; $(2, 3), (2, 4), \cdots$, $(2, n)$; etc. Though the annihilated off-diagonal elements may get filled up in succeeding rotations and the iterations may be infinite, it can be shown that the similar matrices converge to a diagonal matrix to yield all the n real eigenvalues. For, from Eqs. (9.7) and (9.8) we note that

$$a_{ip}^{'\,2} + a_{iq}^{'\,2} = a_{ip}^2 + a_{iq}^2, \quad a_{pj}^{(1)^2} + a_{qj}^{(1)^2} = a_{pj}^{'\,2} + a_{qj}^{'\,2}$$

so that the sum of squares of the elements of the matrices A, A', A_1 remain unchanged during the transformations. Now, reverting to Eqs. (9.7) and (9.8), the diagonal elements on the intersection of the rows $i = p, q$ and columns $j = p, q$ are

$$a_{pp}^{(1)} = a_{pp} \cos^2 \theta + 2 a_{pq} \sin \theta \cos \theta + a_{qq} \sin^2 \theta$$
$$a_{qq}^{(1)} = a_{pp} \sin^2 \theta - 2 a_{pq} \sin \theta \cos \theta + a_{qq} \cos^2 \theta$$

(9.11)

where symmetry of the elements $a_{qp} = a_{pq}$ is assumed. The sum of their squares is

$$[a_{pp}^{(1)}]^2 + [a_{qq}^{(1)}]^2 = (a_{pp}^2 + a_{qq}^2)(1 - \tfrac{1}{2} \sin^2 2\theta) + 2 a_{pq}^2 \sin^2 2\theta$$
$$+ a_{pp} a_{qq} \sin^2 2\theta + 2 (a_{pp} - a_{qq}) a_{pq} \sin 2\theta \cos 2\theta$$
$$= a_{pp}^2 + a_{qq}^2 + 2 a_{pq}^2$$

where the equation for θ preceding Eq. (9.10) is used in the reduction of the expression. From the above analysis, we see that annihilation of off-diagonal elements $a_{pq}^{(1)}$ leads to augmentation in the sum of squares of the diagonal elements by $2 a_{pq}^2$. Hence during repeated transformations, with sum of squares of all elements remaining constant, there will be continued augmentation in the diagonal with depletion in the off-diagonal elements. Hence, the assertion on convergence.

For rapid convergence a slightly more elaborate sequence may be adopted, instead of the cyclic order described above, we may scan the off-diagonal elements a_{ij} $(i \neq j)$ and select p, q such that $|a_{pq}|$ is the largest and repeat the procedure till diagonalisation occurs to desired accuracy.

In the sequence of similar matrices say

$$(J_m \cdots J_2 J_1)^T A J_1 J_2 \cdots J_m$$

where the successive rotation matrices are J_1, J_2, $\cdots$, J_m, which tends to a diagonal matrix when $m \to \infty$, the *eigenvectors* of A are the columns of J_1, J_2, $\cdots$, J_m when $m \to \infty$ (see Wilkinson's text for proof).

Remark It can be seen that the matrix A and the sequence of matrices derived from it are symmetric since A is symmetric.

9.2.2   Givens' Transformation

In this more recent method dating 1952, similar matrices based on *finite* number of rotation matrices are obtained. The objective of an ultimate diagonal matrix is abandoned in favour of a *symmetric tridiagonal matrix,* which is treated for eigenvalues.

Consider the rotation matrix J_{pq} given by Eq. (9.6) of the preceding section. The elements of $A_1 = J_{pq}^T A J_{pq}$ are given by Eqs. (9.7) and (9.8), viz. $a_{ij}^{(1)} = a_{ij}$ except for the rows $i = p, q$ and columns $j = p, q$. The elements at the four intersections (p, p), (p, q), (q, p) and (q, q) are given by Eqs. (9.9) and (9.11), and the rest are given by

$$\left. \begin{array}{l} a_{pj}^{(1)} = a_{pj} \cos\theta + a_{qj} \sin\theta \\ a_{qj}^{(1)} = -a_{pj} \cos\theta + a_{qj} \cos\theta \end{array} \right\} \quad j \neq p, q$$

$$\left. \begin{array}{l} a_{ip}^{(1)} = a_{ip} \cos\theta + a_{iq} \sin\theta \\ a_{iq}^{(1)} = -a_{ip} \sin\theta + a_{iq} \cos\theta \end{array} \right\} \quad i \neq p, q$$

$$(9.12)$$

Equations (9.9) and (9.12) show that if A is symmetric, so is A_1. With this observation, we proceed to annihilate one by one the elements below the subdiagonal, so that due to symmetry the ultimate matrix is tridiagonal and symmetric.

To annihilate the first element $(3, 1)$, J_{23} is employed . This element of A_1 from Eq. (9.12) is

$$a_{31}^{(1)} = -a_{21} \sin\theta + a_{31} \cos\theta$$

which vanishes for $\tan\theta = a_{31}/a_{21}$ or,

$$\frac{\sin\theta}{a_{31}} = \frac{\cos\theta}{a_{21}} = \frac{1}{\sqrt{a_{21}^2 + a_{31}^2}} \qquad (9.13)$$

In a similar manner, the elements $(4, 1)$, $(5, 1)$, $\cdots$, $(n, 1)$ of the first column can be annihilated by employing rotations J_{24}, J_{25}, $\cdots$, J_{2n} and the previously annihilated elements remain so during the succeeding rotations. The procedure can be repeated again for elements below the subdiagonal of the second column, leaving out the elements of the first column (and row), where rotations J_{34}, J_{35}, $\cdots$, J_{3n} are to be used. Proceeding in this manner the last element $(n, n - 2)$ can be annihilated.

The symmetric tridiagonal matrix can be generated by subroutine named GIVENS given later, which helps isolation and eventual determination of the eigenvalues.

In order to investigate isolation of the eigenvalues, let the tridiagonal matrix obtained by Given's transformation be represented as

$$B := \begin{bmatrix} d_1 & a_1 & & & & \\ a_1 & d_2 & a_2 & & & \\ & a_2 & d_3 & a_3 & & \\ & & \cdots & \cdots & \cdots & \\ & & & a_{n-2} & d_{n-1} & a_{n-1} \\ & & & & a_{n-1} & d_n \end{bmatrix}$$

Its characteristic polynomial is $\det(B - \lambda I) := P_n(\lambda)$. Expansion of the determinant by row or column easily leads to the following recurrence relations for $P_n(\lambda)$:

$$P_0(\lambda) := 1, \quad P_1(\lambda) = d_1 - \lambda$$
$$P_i(\lambda) = (d_i - \lambda) P_{i-1}(\lambda) - a_{i-1}^2 P_{i-2}(\lambda), \quad i = 2, 3, \cdots, n \tag{9.14}$$

For any given λ, $P_n(\lambda)$ can be computed from Eqs. (9.14) in about $3n$ operations. Hence, the determination of the zeros of $P_n(\lambda)$ can suitably be based on the methods of Chap. 2.

If $a_i = 0$ for some i, then it follows from the characteristic determinant that $P_i(\lambda)$ is a factor of $P_n(\lambda)$. The eigenvalues of B are then the zeros of $P_i(\lambda)$ and $P_n(\lambda)/P_i(\lambda)$. On the other hand, if $a_i \neq 0$ *for all* i, *then the eigenvalues of* B *are distinct*. The localisation of the eigenvalues is helped by the fact that the polynomials $P_0(\lambda)$, $P_1(\lambda)$, $\cdots$, $P_n(\lambda)$ form a *Sturm Sequence*. This allows construction of intervals containing just one eigenvalue. The refinement of an eigenvalue can be accomplished by simple bisection, which may be combined with the secant method or the modified regula falsi method (see Chap. 2).

The subroutine GIVENS incorporating Givens' transformation to tridiagonal matrix and isolation of the eigenvalues follows:

```
SUBROUTINE GIVENS(A,n)
! A=Given n×n symmetric matrix.        (Input)
! The tridiagonal symmetric matrix is returned in A.      (Output)
!*****************************************************************
REAL :: A(n,n), lambda, P(n)
DO i=2,n−1; DO j=i+1,n
IF(A(j,i−1)/=0.0) THEN
asqrt=SQRT(A(i,i−1)**2+A(j,i−1)**2)
asin=A(j,i−1)/asqrt
acos=A(i,i−1)/asqrt
! Change the (i−1)th row
A(i−1,i)=A(i−1,i)*acos+A(i−1,j)*asin
A(i−1,j)=0.0
! Change the ith row
Aii=A(i,i)
DO k=1,n
A(i,k)=A(i,k)*acos+A(j,k)*asin
END DO
A(i,i)=A(i,i)*acos+A(i,j)*asin
```

```
A(i,j)=A(i,j)*acos-(Aii*acos+A(j,i)*asin)*asin
! Change the jth column
DO k=i+1,j
A(k,j)=-A(k,i)*asin+A(k,j)*acos
END DO
A(j,j)=A(j,j)*acos+(Aii*asin-A(j,i)*acos)*asin
! Change the jth row
IF(j/=n) THEN
DO k=j+1,n
A(j,k)=-A(k,i)*asin+A(j,k)*acos
END DO
END IF
! Symmetrise the matrix
DO L=i-1,n; DO K=L,n
A(K,L)=A(L,K)
END DO; END DO
END IF
END DO; END DO
PRINT*, ((A(i,j),j=1,n),i=1,n)
!****************************************************
! Determine the number of variations in sign in the Sturm
! Sequence
! iv=number of variations in sign in Sturm Sequence
! P(n)=value of characteristic polynomial for given lambda
! by Gershgorin inequality
u=MAX(ABS(A(1,1)+ABS(A(1,2)),ABS(A(n,n-1))+ABS(A(n,n)))
DO j=2,n-1
sumrow=ABS(A(j,j-1))+ABS(A(j,j))+ABS(A(j,j+1))
u=MAX(u,sumrow)
END DO
lambda=-u
10 P(1)=A(1,1)-lambda
P(2)=(A(2,2)-lambda)*P(1)-A(1,2)**2
DO i=3,n
P(i)=(A(i,i)-lambda)*P(i-1)-A(i-1,i)**2*P(i-2)
END DO
iv=0; IF(P(1)<0.0) iv=iv+1
DO i=1,n-1
IF(P(i)*P(i+1)<0.0) iv=iv+1
END DO
PRINT*, lambda, '    variations of sign=', iv, P(n)
IF(lambda>=u) RETURN
lambda=lambda+1      ! If necessary reduce the increment step of 1
GOTO 10
END SUBROUTINE GIVENS
```

James Wallace Givens (1910–1993), U.S. mathematician and computer scientist, well known for the Givens rotations. He worked on early generation computers, serving at several laboratories and universities, retiring as professor emeritus at Northwestern University, U.S.A. in 1979.

Example 1. Use subroutine GIVENS to reduce the symmetric matrix

$$A = \begin{bmatrix} 1 & 2 & -1 \\ 2 & 1 & 2 \\ -1 & 2 & 1 \end{bmatrix}$$

to a tridiagonal form. Isolate the eigenvalues of the matrix.

Solution. Writing the main program, which is a simple task, the subroutine gives the output

$$\text{Tridiagonal matrix} = \begin{bmatrix} 1 & 2.236068 & 0 \\ 2.236068 & -0.6 & 1.2 \\ 0 & 1.2 & 2.6 \end{bmatrix}$$

and for the isolation of the eigenvalues the following table is obtained:

$$
\begin{array}{lll}
-4.036068 & \text{variations of sign} = 0 & 74.400050 \\
-3.036068 & \text{variations of sign} = 0 & 21.422310 \\
-2.036068 & \text{variations of sign} = 1 & -7.339021 \\
-1.036068 & \text{variations of sign} = 1 & -17.883940 \\
-0.036068 & \text{variations of sign} = 1 & -16.212460 \\
-0.963932 & \text{variations of sign} = 1 & -8.324564 \\
1.963932 & \text{variations of sign} = 1 & -0.220264 \\
2.963932 & \text{variations of sign} = 2 & 2.100445 \\
3.963932 & \text{variations of sign} = 3 & -7.362439 \\
4.963932 & \text{variations of sign} = 3 & -34.608910 \\
\end{array}
$$

Thus, there are three real roots in $(-3.036, -2.036)$, $(1.964, 2.964)$ and $(2.964, 3.964)$. ☐

Exercises

Isolate the eigenvalues of the following symmetric matrices by using subroutine GIVENS:

1. $A = \begin{bmatrix} 1 & 1/2 & 1/3 \\ 1/2 & 1/3 & 1/4 \\ 1/3 & 1/4 & 1/5 \end{bmatrix}$ [two roots in $(-0.6, +0.4)$ and one root in $(1.4, 2.4)$].

2. $A = \begin{bmatrix} 1 & 2 & -1 \\ 2 & 1 & 2 \\ -1 & 2 & 1 \end{bmatrix}$ $[(-3.04, -2.04),\ (1.96, 2.96),\ (2.96, 3.96)]$.

3. $A = \begin{bmatrix} 4 & -1 & -1 & -1 \\ -1 & 4 & -1 & -1 \\ -1 & -1 & 4 & -1 \\ -1 & -1 & -1 & 4 \end{bmatrix}$ [(0.27, 1.27), three roots in (4.27, 5.27);

    exact roots 1, 5, 5, 5].

4. $A = \begin{bmatrix} 1 & \sqrt{2} & \sqrt{2} & 2 \\ \sqrt{2} & -\sqrt{2} & -1 & \sqrt{2} \\ \sqrt{2} & -1 & \sqrt{2} & \sqrt{2} \\ 2 & \sqrt{2} & \sqrt{2} & -3 \end{bmatrix}$ [(−5, −4), (−3, −2), (1, 2), (3, 4)].

5. $A = \begin{bmatrix} 10 & 7 & 8 & 7 \\ 7 & 5 & 6 & 5 \\ 8 & 6 & 10 & 9 \\ 7 & 5 & 9 & 10 \end{bmatrix}$ [Two roots in (−0.04, 0.96),

    (2.96, 3.96), (29.96, 30.96)].

9.2.3 Householder Transformation: Hessenberg Matrices

A more effective transformation then Given's is the **Householder Transformation**. It is more intricate than the Givens' method that transforms an *arbitrary* matrix to **Hessenberg form**. These latter matrices have zero elements below the subdiagonal of the form

$$\begin{bmatrix} \times & \times & \times & \cdots & \times & \times \\ \times & \times & \times & \cdots & \times & \times \\ & \times & \times & \cdots & \times & \times \\ & & \times & \cdots & \times & \times \\ & & & \cdots & \times & \times \\ & & & & \times & \times \end{bmatrix}$$

where the crosses denote non-zero elements in general. In particular it will be proved that if the original matrix A is symmetric, its transformed Hessenberg form is also symmetric, or the elements above the superdiagonal are also zero. This means that the Hessenberg form is *tridiagonal for symmetric matrices*.

A *Householder transformation* of an arbitrary $n \times n$ matrix A is defined as $S^{-1} A S$, where S is the $n \times n$ matrix

$$S(\mathbf{v}) = I - \alpha \mathbf{v} \mathbf{v}^T, \quad \alpha = 2/\mathbf{v}^T \mathbf{v} \tag{9.15}$$

or

$$S(\mathbf{v}) = \begin{bmatrix} 1 - \alpha\, v_1^2 & -\alpha\, v_1 v_2 & \cdots & -\alpha\, v_1 v_2 \\ -\alpha\, v_2 v_1 & 1 - \alpha\, v_2^2 & \cdots & -\alpha\, v_2 v_n \\ \cdots & \cdots & \cdots & \cdots \\ -\alpha\, v_n v_1 & -\alpha\, v_n v_2 & \cdots & 1 - \alpha\, v_n^2 \end{bmatrix}$$

where $\mathbf{v} = [v_1,\, v_2, \cdots,\, v_n]^T$ is an arbitrary vector. We note that (i) S is symmetric, and (ii) the inverse of S, viz. $S^{-1} = S$; for

$$\begin{aligned} SS &= (I - \alpha\, \mathbf{v}\mathbf{v}^T)\,(I - \alpha\, \mathbf{v}\mathbf{v}^T) \\ &= I - 2\alpha\, \mathbf{v}\mathbf{v}^T + \alpha^2 \mathbf{v}\,(\mathbf{v}^T\mathbf{v})\,\mathbf{v}^T = I \end{aligned} \tag{9.16}$$

by Eq. (9.15). The similarity transformation of A is thus SAS. Also, (iii) when applied over any vector $\mathbf{x}$, the norm of $\mathbf{x}$ does not change, i.e.

$$\|S\mathbf{x}\|_2 = \|\mathbf{x}\|_2 \tag{9.17}$$

For,
$$S\mathbf{x} = (I - \alpha\, \mathbf{v}\mathbf{v}^T)\,\mathbf{x} = \mathbf{x} - \alpha\, \mathbf{v}\,(\mathbf{v}^T\mathbf{x})$$

Hence,

$$\begin{aligned} \|S\mathbf{x}\|_2 &= (S\mathbf{x})^T\,(S\mathbf{x}) \\ &= [\mathbf{x}^T - \alpha\,(\mathbf{v}^T\mathbf{x})\,\mathbf{v}^T]\,[\mathbf{x} - \alpha\,(\mathbf{v}^T\mathbf{x})\,\mathbf{v}] \\ &= \mathbf{x}^T\mathbf{x} - \alpha\,(\mathbf{v}^T\mathbf{x})\,(\mathbf{x}^T\mathbf{v} + \mathbf{v}^T\mathbf{x}) + \alpha^2\,(\mathbf{v}^T\mathbf{x})^2\,(\mathbf{v}^T\mathbf{v}) \\ &= \mathbf{x}^T\mathbf{x} - 2\alpha\,(\mathbf{v}^T\mathbf{x})^2 + \alpha\,(\mathbf{v}^T\mathbf{x})^2\,2 = \|\mathbf{x}\|_2 \end{aligned}$$

Because of property (iii), the transformation is called a *reflection*, just as reflection in a plane (mirror) leaves linear distances unaltered (see $\mathbf{1}^o$).

Of particular interest are the transformations S_k with

$$v_1 = v_2 = \cdots = v_k = 0 \tag{9.18}$$

so that $\mathbf{v} = [0, 0, \cdots, 0, v_{k+1}, \cdots, v_n]^T$. In fact, it can be shown that the similar matrices $A_{k+1} = S_k A_k S_k$, $k = 1, 2, \cdots, n-2$, where $A_1 := A$ can be constructed such that A_{n-1} is of Hessenberg form. In each of the k iterations, the requisite number $n - k - 1$ of zeros are produced in the kth column. To understand the process consider a simple 4×4 matrix

$$A_1 := A = \begin{bmatrix} a_{11} & \times & \times & \times \\ a_{21} & \times & \times & \times \\ a_{31} & \times & \times & \times \\ a_{41} & \times & \times & \times \end{bmatrix} \tag{9.19}$$

with particular attention on the first column in the first iteration $k = 1$. For this step, S_1 is given by Eq. (9.15), with $v_1 = 0$ (Eq. (9.18)) and $n = 4$. It is easy to verify that $A_1 S_1$ has the first column as that of A_1, and

$$A_2 := B_1 = S_1 (A_1 S_1) \tag{9.20}$$

has the form (9.19) with a new first column $\mathbf{b}_1 = [b_{11}, b_{21}, b_{31}, b_{41}]^T$ but an unchanged first row. It follows that $b_{11} = a_{11}$. Because of Eq. (9.20)

$$\mathbf{b}_1 = S_1 \mathbf{a}_1 \tag{9.21}$$

where $\mathbf{a}_1 = [a_{11}, a_{21}, a_{31}, a_{41}]^T$. For getting Hessenberg form, we require that

$$b_{31} = b_{41} = 0 \tag{9.22}$$

To find b_{21}, Eqs. (9.17) and (9.21) yield $\|\mathbf{b}_1\|_2 = \|a_1\|_2$, that is

$$a_{11}^2 + b_{21}^2 = a_{11}^2 + a_{21}^2 + a_{31}^2 + a_{41}^2$$

or

$$b_{21} = \pm\sqrt{a_{21}^2 + a_{31}^2 + a_{41}^2} =: \beta \quad (\text{say}) \tag{9.23}$$

This determines the first column of A_2, except for an ambiguous sign in Eq. (9.23). For the remaining elements, choose

$$\mathbf{v} = \lambda (\mathbf{a}_1 - \mathbf{b}_1) \tag{9.24}$$

(the choice of $\mathbf{v}/\|\mathbf{v}\|_2$ is unique except for a sign, see 2°), (below) for which correctly, $v_1 = \lambda (a_{11} - b_{11}) = 0$. Eq. (9.21), with (9.15) yields (as in the calculation for proving Eq. (9.17))

$$\mathbf{b}_1 = [1 - \alpha\lambda^2 (\mathbf{a}_1^T \mathbf{a}_1 - \mathbf{b}_1^T \mathbf{a}_1)] \mathbf{a}_1 + \alpha\lambda^2 (\mathbf{a}_1^T \mathbf{a}_1 - \mathbf{b}_1^T \mathbf{a}_1) \mathbf{b}_1$$

which is satisfied for $\mathbf{a}_1$ and $\mathbf{b}_1$ when (using Eq. (9.22))

$$\alpha\lambda^2 (\|\mathbf{a}_1\|_2 - a_{11}^2 - b_{21} a_{21}) = 1$$

that is, using Eq. (9.23)

$$\alpha\lambda^2 = 1/[b_{21} (b_{21} - a_{21})] \tag{9.25}$$

Interestingly, by definition of α in Eq. (9.15), viz. $\alpha\lambda^2/\|\mathbf{a}_1 - \mathbf{b}_1\|_2^2$ also reduces to Eq. (9.25). The components of $\mathbf{v}$, from Eq. (9.24) are given by

$$v_1 = 0, \quad v_2 = \lambda (a_{21} - b_{21}), \quad v_3 = \lambda a_{31}, \quad v_4 = \lambda a_{41} \tag{9.26}$$

The selection of sign in Eq. (9.23) can now be done so as to minimise the possible loss of significant digits in the computation of v_2 in Eq. (9.26). Thus we choose

$$b_{21} = -\mathrm{sgn}(a_{21})\,\beta =: -\gamma \quad (\text{say}), \quad v_2 = \lambda\,(a_{21} + \gamma) \qquad (9.27)$$

(In the above sgn(0) may be taken as 1). In the computation of v_2, scaling may be affected by choosing $\lambda = 1/(a_{21} + \gamma)$ to yield

$$v_1 = 0, \quad v_2 = 1, \quad v_3 = \frac{a_{31}}{a_{21} + \gamma}, \quad v_4 = \frac{a_{41}}{a_{21} + \gamma} \qquad (9.28)$$

Finally, by Eq. (9.25) α is given by

$$\alpha = 1 + \frac{a_{21}}{\gamma} \qquad (9.29)$$

This completely determines S_1. The computation of

$$A_1 S_1 = A_1 - \alpha\,(A_1 \mathbf{v})\,\mathbf{v}^T$$

can be carried out in a simple fashion by considering each *row* of the matrices, because a row of A_1 multiplied by $\mathbf{v}$ is simply a scalar. The next product in $S_1 A_1 S_1$ can be written as

$$S_1(A_1 S_1) = A_1 S_1 - \alpha \mathbf{v}\,(\mathbf{v}^T (A_1 S_1))$$

and each *column* can be similarly computed, since $\mathbf{v}^T$ multiplied by a column of $A_1 S_1$ is again a scalar. This completes the generation of A_2.

The succeeding iterations $A_3, \cdots, A_{n-1}$ are based on similar procedure with $k = 2, \cdots, n - 2$. It is easy to verify that in the kth iteration the previous $k - 1$ rows and columns remain unchanged so that A_{n-1} is finally of Hessenberg form.

In particular, if A is *symmetric*, each of the matrices $A_2, \cdots, A_{n-1}$ is symmetric since each of the reflections $S_1, \cdots, S_{n-2}$ is symmetric. Hence, the final matrix is tridiagonal as asserted earlier.

The following subroutine determines a Hessenberg matrix by Householder similarity transformations of any $n \times n$ matrix A. Notations are k for iteration number, i and ii for row number and j, jj for column number:

SUBROUTINE HOUSEHOLDRER(A,n)
```
! A=Given n×n matrix.          (Input)
! Transformed Hessenberg matrix of A, returned in A.    (Output)
!*************************************************************
REAL :: A(n,n), v(n)
DO k=1, n−2
gamma=0.0
DO i=k+1,n
```

```
gamma=gamma+A(i,k)**2
END DO
gamma=SQRT(gamma)
IF(A(k+1,k)<0.0) gamma=−gamma
IF(ABS(gamma)>1.E−10) THEN
scalef=A(k+1,k)+gamma          !scalef is the scaling factor
DO i=1,n
v(i)=0.0
IF(i>k+1) v(i)=A(i,k)/scalef
END DO
v(k+1)=1.; alfa=scalef/gamma
DO i=k,n
scalar=0.0
DO jj=k+1,n
scalar=scalar+A(i,jj)*v(jj)
END DO
DO j=k+1,n
A(i,j)=A(i,j)−alfa*scalar*v(j)
END DO; END DO
DO j=k+1,n
scalar=0.0
DO ii=k+1,n
scalar=scalar+v(ii)*A(ii,j)
END DO
DO i=k+1,n
A(i,j)=A(i,j)−alfa*scalar*v(i)
END DO; END DO
A(k+1,k)=−gamma
DO i=k+2,n
A(i,k)=0.
END DO; END IF; END DO
RETURN
END SUBROUTINE HOUSEHOLDER
```

We now expand on two properties of the Householder transformation in some detail.

1°. *The Householder transformation is a reflection in the hyperplane orthogonal to* **v**.

Proof In three dimensions. let **v** be parallel to the z-axis, with $\mathbf{v} = \mathbf{k}$, then

$$S = \begin{bmatrix} 1 & 0 & 0 \\ 0 & 1 & 0 \\ 0 & 0 & -1 \end{bmatrix}$$

and therefore for any vector $\mathbf{u} = [u_1, u_2, u_3]^T$, $S\mathbf{u} = [u_1, u_2, -u_3]^T$, which is the reflection of $\mathbf{u}$ in the (x, y)-plane. The generalisation to n-dimensions is immediate. □

2°. *If a Householder transformation is written as*

$$S(\mathbf{v}) = I - 2 \frac{\mathbf{v}}{\|\mathbf{v}\|_2} \cdot \frac{\mathbf{v}^T}{\|\mathbf{v}\|_2};$$

then $\dfrac{\mathbf{v}}{\|\mathbf{v}\|_2}$ *determined by* Eq. (9.24) *and satisfying* Eq. (9.21) *is unique except for a sign.*

Proof Let there be another $\mathbf{v}'$ satisfying Eq. (9.21), then

$$\mathbf{b}_1 = (I - \alpha \mathbf{v}\mathbf{v}^T) \mathbf{a}_1 = (I - \alpha' \mathbf{v}'\mathbf{v}'^T) \mathbf{a}_1$$

Thus $\alpha \mathbf{v} (\mathbf{v}^T \mathbf{a}_1) = \alpha' \mathbf{v}' (\mathbf{v}'^T \mathbf{a}_1)$. Now $\mathbf{v}'^T \mathbf{a}_1 \neq 0$, otherwise $\mathbf{b}_1$ is equal to $\mathbf{a}_1$ and no transformation is necessary. Hence, $\mathbf{v}' = \alpha'' \mathbf{v}$, where $\alpha'' = \alpha (\mathbf{v}^T \mathbf{a}_1)/[\alpha' (\mathbf{v}'^T \mathbf{a}_1)]$. The norms satisfy $\|\mathbf{v}'\|_2 = |\alpha''| \|\mathbf{v}\|_2$, yielding $\mathbf{v}'/\|\mathbf{v}'\|_2 = \pm \mathbf{v}/\|\mathbf{v}\|_2$.

Karl Hessenberg (1904–1959), German engineer whose dissertation in the year 1942 (during World War II), investigated the computation of eigenvalues. He was a modest man and not much is known about his life.

Alston Scott Householder (1904–1993), U.S. mathematician. His doctoral dissertation in 1937 was on calculus of variations, but turned to mathematical biology. After the World War in 1946, he shifted to numerical analysis, which retained his interest till retirement in 1974 as Chairman of Mathematics Department at the University of Tennessee, U.S.A.

Example 1. Reduce the matrix

$$A = \begin{bmatrix} 1 & 1 & 1 & 1 \\ 2 & 2 & 2 & 2 \\ 3 & 3 & 3 & 3 \\ 4 & 4 & 4 & 4 \end{bmatrix}$$

to Hessenberg form. Hence, determine the eigenvalues.

Solution. Using subroutine HOUSEHOLDER we obtain the Hessenberg matrix B similar to A given by

$$B = \begin{bmatrix} 1.00000 & -1.67126 & -0.08512 & -0.44682 \\ -5.38510 & 9.00000 & 2.36001 & 0.65601 \\ 0.00000 & 0.00000 & 0.00000 & 0.00000 \\ 0.00000 & 0.00000 & 0.00000 & 0.00000 \end{bmatrix}$$

Hence, the eigenvalues are given by the characteristic polynomial

$$\lambda^2[(1 - \lambda)(9 - \lambda) - 1.67126 \times 5.38516] = 0$$

or
$$\lambda^2[9 - 10\lambda + \lambda^2 - 9] = 0$$

which yields the eigenvalues as $\lambda = 0, 0, 0, 10$.   □

Exercises

Reduce the following matrices to Hessenberg form by using the subroutine HOUSE-HOLDER:

1. $\begin{bmatrix} 2 & -1 & 2 \\ 5 & -3 & 3 \\ -1 & 0 & -2 \end{bmatrix}$ $\quad \left[\begin{array}{rrr} 2.00000 & 1.37281 & 1.76505 \\ -5.09902 & -3.53846 & -2.69231 \\ & 0.30769 & -1.46154 \end{array}\right]$.

2. $\begin{bmatrix} 5 & 6 & -3 \\ -1 & 0 & 1 \\ 1 & 2 & -1 \end{bmatrix}$ $\quad \left[\begin{array}{rrr} 5.00000 & -6.36396 & 2.12132 \\ 1.41421 & -2.00000 & 0 \\ & -1.00000 & 1.00000 \end{array}\right]$.

3. $\begin{bmatrix} 1 & 2 & 3 & 4 \\ 2 & 1 & 2 & 3 \\ 3 & 2 & 1 & 2 \\ 4 & 3 & 2 & 1 \end{bmatrix}$ $\quad \left[\begin{array}{rrrr} 1.00000 & -5.38517 & & \\ -5.38517 & 5.13793 & -1.99524 & \\ & -1.99524 & -1.37449 & 0.28953 \\ & & 0.28953 & -0.76344 \end{array}\right]$.

4. $\begin{bmatrix} 2 & -1 & 3 & -4 \\ 3 & -2 & 4 & -3 \\ 5 & -3 & -2 & 1 \\ 3 & -3 & -1 & 2 \end{bmatrix}$ $\quad \left[\begin{array}{rrrr} 2.00000 & 0 & 3.52315 & -3.68611 \\ -6.55744 & -2.06977 & 4.01933 & -0.77945 \\ & -1.93373 & -1.56518 & -5.10952 \\ & & -0.99206 & 1.63494 \end{array}\right]$.

9.3 General Matrices

In this section, we consider the problem of numerical computation of the eigenvalues of any $n \times n$ matrix, real or complex. This general computational problem received focal attention in the sixth to the eighth decades of the last century, culminating in routines that delivered the eigenvalues in highly efficient manner, even though the methods are somewhat intricate as we shall see. The beginning could be said to be with H. Rutishauser's (1958) development of the *LR Method*.

9.3.1   The LR Method

The reader is well aware of the success of the LU decomposition method for solving
a system of linear algebraic equations (Sect. 3.1.3, Chap. 3). Rutishauser employed
this technique for the evaluation of the eigenvalues of a matrix A, factored as LU.
But instead of calling L and U as lower and upper triangular matrices, he called them
left and *right* triangular matrices, hence the name LR method.

Let the $n \times n$ matrix A be nonsingular, so that employing Gaussian elimination
we have

$$A = LU$$

where L is $n \times n$ unit lower triangular (i.e. lower triangular with *unit diagonal*)
matrix, and U is $n \times n$ upper triangular matrix. For this original decomposition,
attaching a subscript '1' to the above, we have

$$A_1 = L_1 U_1, \quad A_1 := A, \quad L_1 := L, \quad U_1 := U$$

On the matrix A_1, we perform the similarity transformation $L_1^{-1} A_1 L_1$ and obtain

$$A_2 := L_1^{-1} A_1 L_1 = L_1^{-1} (L_1 U_1) L_1 = U_1 L_1$$

Both U_1 and L_1 are nonsingular since A_1 is so ($\det[A_1] \neq 0$). Hence, A_2 is nonsingular. Moreover since $L_1 = A_1 U_1^{-1}$, we have

$$A_2 = L_1^{-1} A_1 L_1 = U_1 A_1 U_1^{-1}$$

Since A_2 is obtained by a similarity transformation of A_1, it has the same eigenvalues
as A_1. Continuing the iteration we can write

$$A_2 = U_1 L_1 =: L_2 U_2$$

$$A_3 = U_2 L_2 = L_2^{-1} A_2 L_2 = U_2 A_2 U_2^{-1} =: L_3 U_3 \quad \text{etc.}$$

$$A_{k+1} = U_k L_k = L_k^{-1} A_k L_k = U_k A_k U_k^{-1} =: L_{k+1} U_{k+1} \quad (k = 1, 2, \cdots)$$

All the iterates have the same eigenvalues as A_1. If A_1 is real, so are the iterates.
The matrices L_k and U_k are, respectively, unit lower triangular and upper triangular
matrices of dimension $n \times n$.

The most important property of sequence $\{A_k\}$ is that under some general restriction, it *converges* to a matrix of upper triangular form, so that the elements of the
principal diagonal of this upper triangular matrix become the eigenvalues of A_1 or
A. We leave the topic here as the method suffers from some inherent difficulties of
practical implementation. The interested reader may see for details in the texts of
Wilkinson 1965 and Ralston and Wilf 1967.

Heinz Rutishauser (1918–1970), Swiss numerical analyst and computer scientist at the Federal Institute of Technology (ETH), Zurich. In spite of heart disease beginning 1955, he made important contributions in numerical analysis and was a team member that developed ALGOL 60—a programming language more advanced than the FORTRAN II language developed a few years earlier. Apart from the LR method, he was the codiscoverer of the so-called QD method for solving polynomial equations and computation of eigenvalues. He passed away prematurely at the age of 52.

9.3.2   The QR Method

This practical method due to J.G.F. Francis (1961–1962) is a powerful alternative to the LR method, covering even singular matrices. Vera N. Kublanovskya (1961) discovered the basics of the method independently during the same period. In this method, according to a theorem of Schur, the matrix A can in general, be factorised into a **Unitary matrix** Q and a right triangular matrix R. A *unitary matrix* Q is defined to be that matrix for which Q^{-1} equals the Hermitian matrix Q^H of Q. When A is *nonsingular*, $A = QR$, in which Q as well as R with non-negative real diagonal elements, are unique. Adopting the iterative procedure of LR, with subscript notations, we get

$$A_1 := A = Q_1 R_1, \quad Q_1 := Q, \quad R_1 := R$$
$$A_2 = Q_1^H A_1 Q_1 = R_1 A_1 R_1^{-1} = R_1 Q_1 =: Q_2 R_2$$
$$\dots\dots\dots\dots\dots\dots\dots\dots\dots\dots\dots\dots\dots\dots\dots\dots\dots\dots\dots\dots\dots\dots$$
$$A_{k+1} = Q_k^H A_k Q_k = R_k A_k R_k^{-1} = R_k Q_k =: Q_{k+1} R_{k+1}$$

Since the transformations are similar, all the iterates have the same eigenvalues as A_1. If A is real so are A_k.

When A is *singular*, the decomposition is no longer unique. However, if the rank of A is r with the first r columns linearly independent, then the last $n - r$ of A_2 will be null and the leading $r \times r$ submatrix will be uniquely defined. Thus isolating the zero eigenvalue, the procedure may be continued on the submatrix.

Therefore reverting back to the case of *nonsingular* A, the main convergence theorem is that under some general conditions, the sequence $\{A_k\}$ tends to a right triangular matrix, so that the eigenvalues are delivered by the diagonal elements.

John G. F. Francis (1934 -) is a British computer scientist. He worked at the National Research Development Corporation and later at Ferranti Corporation Limited, London. Subsequently, he moved to Sussex University and various industrial organisations. He never returned to numerical computation, moving over to artificial intelligence, computer languages and systems. His singular work on the QR method led to explosive growth in researches in numerical matrix algebra by others, which continues to this day.

Vera Nikolaevna Kublanovskya (1920–2012) was a Russian mathematician, noted for her work on computational methods for solving spectral problems of algebra. She worked her entire academic life at the Steklov Institute of Mathematics of Russian Academy of Sciences, St. Petersburg, Russia.

Issai Schur (1875–1941) was a Russian mathematician who worked in Germany. He is famous for many contributions in mathematics, including those in Group Representations, Combinatorics and Number Theory. In late life, during the beginning of World War II, he was forced to leave Germany to die in Tel Aviv, Israel.

9.3.2.1 Convergence of QR Method

The convergence theorem is due to J. H. Wilkinson. It is based on the following two results:

1°. If $\hat{Q}_k = Q_1 Q_2 \cdots Q_k$, so that $\hat{Q}_k$ is $n \times n$ unitary, then

$$
\begin{aligned}
A_{k+1} &= Q_k^H A_k Q_k = Q_k^H \left(Q_{k-1}^H A_{k-1} Q_{k-1} \right) Q_k = \cdots \cdots \\
&= Q_k^H Q_{k-1}^H \cdots Q_1^H A_1 Q_1 Q_2 \cdots Q_k \\
&= \hat{Q}_k^H A_1 \hat{Q}_k
\end{aligned}
\tag{9.30}
$$

2°. If $\hat{R}_k = R_k R_{k-1} \cdots R_1$, so that $\hat{R}_k$ is right triangular with non-negative real diagonal elements, then

$$
\begin{aligned}
\hat{Q}_k \hat{R}_k &= Q_1 Q_2 \cdots (Q_k R_k) R_{k-1} \cdots R_1 = \hat{Q}_{k-1} A_k \hat{R}_{k-1} \\
&= A_1 \hat{Q}_{k-1} \hat{R}_{k-1} = \cdots \\
&= A_1^{k-1} \hat{Q}_1 \hat{R}_1 = A_1^k
\end{aligned}
\tag{9.31}
$$

Theorem 9.10 (Convergence of QR Method) *Let A_1 be diagonalised by a similarity transformation: $A_1 = X \Lambda X^{-1}$, where $\Lambda = diag\{\lambda_1, \lambda_2, \cdots, \lambda_n\}$ is real. If*

(i) $|\lambda_1| > |\lambda_2| > \cdots > \lambda_n > 0$

(ii) $X^{-1} =: Y$ has LU factorisation $L_y U_y$

then $A_k \sim D^k R D^{-k}$ (asymptotic convergence), where R is right triangular and D is diagonal unitary.

Proof Since $A_1 = X \Lambda X^{-1}$, it follows by multiplication that

$$
\begin{aligned}
A_1^k &= X \Lambda^k X^{-1} = X \Lambda^k Y = X \Lambda^k L_y U_y \\
&= X \left(\Lambda^k L_y \Lambda^{-k} \right) \Lambda^k U_y
\end{aligned}
\tag{9.32}
$$

Now focusing on a 3×3 matrix

$$L_y = \begin{bmatrix} 1 & 0 & 0 \\ l_{21} & 1 & 0 \\ l_{31} & l_{32} & 1 \end{bmatrix}$$

we get

$$\Lambda^k L_y \Lambda^{-k} = \begin{bmatrix} 1 & 0 & 0 \\ l_{21}\left(\dfrac{\lambda_2}{\lambda_1}\right)^k & 1 & 0 \\ l_{31}\left(\dfrac{\lambda_3}{\lambda_1}\right)^k & l_{32}\left(\dfrac{\lambda_3}{\lambda_2}\right)^k & 1 \end{bmatrix} =: I + B_k$$

where the element (i, j) of B_k is zero for $i \leq j$ and $l_{ij}(\lambda_i/\lambda_j)^k$ for $i > j$. This form is true for any order n of the matrix. Hence from the given conditions, $B_k \to 0$ as $k \to \infty$. Now, one can always write $X = Q_x R_x$, then Eq. (9.32) yields

$$A_1^k = Q_x R_x (I + B_k) \Lambda^k U_y = Q_x (I + R_x B_k R_x^{-1})(R_x \Lambda^k U_y)$$

In the above equation, $I + R_k B_k R_x^{-1} \to I$ as $k \to \infty$. Hence for $k > K_0$, it becomes nonsingular and we can write it as $\tilde{Q}_k \tilde{R}_k$, where $\tilde{Q}_k \to I$, $\tilde{R}_k \to I$. Hence, the above equation becomes

$$A_1^k = (Q_x \tilde{Q}_k)(\tilde{R}_k R_x \Lambda^k U_y)$$

The first factor is unitary and the second, right triangular with or without non-negative diagonal elements because of Λ^k and U_y. As a remedy let diagonal unitary matrices D and D_1 be introduced such that $D\Lambda$ and $D_1 U_y$ have non-negative diagonal elements. We can then write

$$A_1^k = (Q_x \tilde{Q}_k D_1^{-1} D^{-k})(D^k D_1 \tilde{R}_k R_x \Lambda^k U_y)$$

The second factor in the above equation is right triangular with non-negative diagonal elements (verify). Hence using conditions (ii) and (i), $\hat{Q}_k = Q_x \tilde{Q}_k D_1^{-1} D^{-k}$ and so by Eq. (9.30)

$$\begin{aligned} A_{k+1} &= D^k D_1 \tilde{Q}_k^H Q_x^H A_1 Q_x \tilde{Q}_k D_1^{-1} D^{-k} \\ &\to D^k (D_1 Q_x^H Q_x R_x \Lambda R_x^{-1} Q_x^H Q_x D_1^{-1}) D^{-k} \quad \text{as} \quad k \to \infty \qquad (9.33) \\ &\to D^k (D_1 R_x \Lambda R_x^{-1} D_1^{-1}) D^{-k} \end{aligned}$$

The middle factor is right triangular R and the theorem follows. $\square$

The form of D is $D = \text{diag}\,[e^{i\theta_1}, \cdots, e^{i\theta_n}]$. Hence unless $\lambda_1, \lambda_2, \cdots, \lambda_n$ are all non-negative, D^k and D^{-k} will not tend to definite limits as k tends to ∞. Similarly for D_1. Nevertheless in Eq. (9.33) D and D_1 have no effect on the diagonal elements of $R_x \Lambda R_x^{-1}$, while the off-diagonal elements remain unchanged in modulus only. Hence even though there is no complete convergence of A_{k+1}, the diagonal elements do converge to the eigenvalues. This kind of convergence is called *essential convergence*.

We now state without proof of additional convergence results:

1. If $\lambda_1, \lambda_2, \cdots, \lambda_n$ are real, condition (i) is not required. Thus some of the eigenvalues may be equal or differ by a sign.
2. For a real matrix A, with a pair of complex conjugate eigenvalues, the iterates tend to a right triangular matrix with a 2×2 submatrix along the diagonal. The real diagonal elements yield the $n - 2$ real eigenvalues and the submatrix the complex conjugate pair, though the submatrix itself does not converge to a limiting submatrix.
3. When A is real with multiple complex eigenvalues, the latter are delivered by a string of 2×2 submatrices of the type mentioned in result 2 above, along the diagonal.

James Hardy Wilkinson (1919–1986), British numerical analyst and computer scientist. He began working on ballistics at the beginning of World War II but at the end of it, he turned to a British computer project. Subsequently, he turned to numerical analysis and discovered many significant algorithms. In 1963 he discovered the polynomial $(x - 1)(x - 2) \cdots (x - 20)$ to illustrate the difficulty of finding zeros of a polynomial. He introduced 'backward error analysis' in computational linear algebra and laid theoretical basis for the matrix eigenvalue problem. He led the team to develop NAG (Numerical Algorithms Group) library containing FORTRAN source codes for numerical computation.

9.3.2.2 Preliminary Reduction to Hessenberg Form

The convergence theorem shows that in the limit, the iterations A_k in general tend to an upper Hessenberg matrix. This fact is suggestive that the given matrix be first reduced to Hessenberg form by applying Householder reflections (Sect. 9.2.3) and then the QR method be applied. Several advantages accrue there from. First, if A is in Hessenberg form, so is Q. For instance, if A is a 5×5 matrix:

$$A = [a_{ij}] = \begin{bmatrix} \times & \times & \times & \times & \times \\ \times & \times & \times & \times & \times \\ 0 & \times & \times & \times & \times \\ 0 & 0 & \times & \times & \times \\ 0 & 0 & 0 & \times & \times \end{bmatrix}$$

and $Q = [q_{ij}]$, $R = [r_{ij}]$, $r_{ij} = 0$ for $i > j$ (right triangular), then

$$0 = a_{i1} = q_{i1}r_{11} \Rightarrow q_{i1} = 0 \quad \text{for} \quad i = 3, 4, 5$$
$$0 = a_{i2} = q_{i1}r_{12} + q_{i2}r_{22} \Rightarrow q_{i2} = 0 \quad \text{for} \quad i = 4, 5$$
$$0 = a_{53} = q_{51}r_{13} + q_{52}r_{23} + q_{53}r_{33} \Rightarrow q_{53} = 0$$

Hence the assertion. Secondly, if the product RQ is formed, then it is easily seen that it remains in Hessenberg form. Therefore all the iterations in QR method remain in Hessenberg form. Finally, accounting has shown that the number of additions and multiplications in an iteration of QR drop from an order of n^3 to n^2.

In the sequel, we shall assume that the given matrix is always reduced to this form and such a form will be our A or A_1. In such Hessenberg form, we can assume that *all* subdiagonal elements are *non-zero*. For if one such element vanishes, A can be partitioned into the form

$$\begin{bmatrix} B & | & D \\ -- & -- & -- \\ O & | & C \end{bmatrix}$$

so that the eigenvalues of B and C yield those of A.

9.3.2.3 Shifts of Origin

From the proof of the main theorem, it is obvious that the rate of convergence depends on that of B_k, that is $(\lambda_i/\lambda_j)^k$, tending to zero $(i > j)$. Thus for convergence to the smallest eigenvalue λ_n (assumed real), $a_{n,n-1}^{(k)}$ of Hessenberg form of A_k tends to zero as $(\lambda_n/\lambda_{n-1})^k$ and a_{nn}^k to λ_n. For accelerating convergence, we may consider $A_1 - sI$ whose eigenvalues are $\lambda_i - s$, $i = 1, 2, \cdots, n$. The first of the foregoing elements will now tend to zero as $\{(\lambda_n - s)/(\lambda_{n-1} - s)\}^k$. If s is a close approximation to λ_n, the element will quickly tend to zero yielding the eigenvalue $\lambda_n - s$ in the (n, n) position.

We therefore consider the following modification of the basic QR method:

$$A_k - s_k I = Q_k R_k,$$
$$\qquad\qquad\qquad\qquad k = 1, 2, \cdots$$
$$A_{k+1} = Q_k^H A_k Q_k = R_k Q_k + s_k I,$$

Evidently, A_2, A_3, $\cdots$ have the same eigenvalues as A_1. The basic result $\mathbf{1}^o$ still holds:

$$A_{k+1} = \hat{Q}_k^H A_1 \hat{Q}_k, \quad \hat{Q} = Q_1 Q_2 \cdots Q_k \tag{9.34}$$

and with $\hat{R}_k = R_k R_{k-1} \cdots R_1$, result 2° becomes

$$
\begin{aligned}
\hat{Q}_k \hat{R}_k &= Q_1 Q_2 \cdots (Q_k R_k) R_{k-1} \cdots R_1 \\
&= \hat{Q}_{k-1}(A_k - s_k I)\hat{R}_{k-1} = (A_1 - s_k I)\hat{Q}_{k-1}\hat{R}_{k-1} \\
&\cdots\cdots\cdots\cdots\cdots\cdots\cdots\cdots\cdots\cdots\cdots\cdots\cdots\cdots\cdots\cdots \\
&= (A_1 - s_k I)(A_1 - s_{k-1} I) \cdots (A_1 - s_1 I) =: \phi_k(A_1) \qquad (9.35)
\end{aligned}
$$

In view of these two results, the basic convergence theorem will hold, subject to modifications of condition (i) to

$$
|\phi_k(\lambda_1)| > |\phi_k(\lambda_2)| > \cdots > |\phi_k(\lambda_n)| > 0
$$

For a practical procedure, we consider the general case of a complex matrix A, focusing attention on the eigenvalue of the smallest modulus λ_n delivered by the last element. The choice of shifts may be made in various ways. Initially, in the absence of any knowledge, we assume $s_1 = 0$. In the succeeding iterations, motivated by the particular

$$
\begin{bmatrix}
\times & \times & \times & \times & | & \times \\
\times & \times & \times & \times & | & \times \\
 & \times & \times & \times & | & \times \\
 & & \times & \times & | & \times \\
\hline
 & & & \times & | & \times
\end{bmatrix}
$$

$$
\begin{array}{cc}
\downarrow & \downarrow \\
\epsilon & \lambda_n
\end{array}
$$

case of real matrices with complex conjugate eigenvalues, we determine the complex eigenvalues of the last 2×2 submatrix, α_k and β_k and choose s_{k+1} to be α_k or β_k according as $|\alpha_k - a_{nn}^{(k)}|$ is less than or greater than $|\beta_k - a_{nn}^{(k)}|$. The iterations are continued till $a_{n,n-1}^{(k)}$ becomes sufficiently small, say less than ϵ. Then $a_{nn}^{(k)}$ yields λ_n. The next eigenvalue λ_{n-1} is attacked by deflation, which merely consists of deleting the nth row and column. The procedure is continued till the last deflation yields a 2×2 matrix yielding λ_1 and λ_2.

For factorisation of $A_k - s_k I$ into $Q_k R_k$, we recall that the procedure is to choose Q_k^H such that

$$
Q_k^H (A_k - s_k I) = R_k
$$

Q_k^H may be a series of Householder reflections to triangularise $A_k - s_k I$ into R_k (Sect. 9.2.3).

For real matrices which may have complex conjugate eigenvalues, Francis discovered deeper theory to eliminate complex arithmetic, even though the shifts may be complex and has propounded a highly efficient method called the *double QR method*.

9.3.2.4   The Double QR Method for Real Matrices

In this method two steps of QR are considered together. In the first two

$$A_1 - s_1 I = Q_1 R_1, \quad A_2 = R_1 Q_1 + s_1 I$$
$$A_2 - s_2 I = Q_2 R_2, \quad A_3 = R_2 Q_2 + s_2 I$$

Hence by Eq. (9.35)

$$A_3 = Q_2^H A_1 \hat{Q}_2, \quad \hat{Q}_2 \hat{R}_2 = \phi_2(A_1) \tag{9.36}$$

where

$$\begin{aligned}
\phi_2(A_1) &= (A_1 - s_2 I)(A_1 - s_1 I) \\
&= A_1^2 - (s_1 + s_2) A_1 + s_1 s_2
\end{aligned} \tag{9.37}$$

If s_1 is complex, we choose $s_2 = s_1^*$, making $\phi_2(A_1)$ real in all cases. Hence its Q, R factors $\hat{Q}_2$ and $\hat{R}_2$ also become real and hence A_3 becomes real. In *double QR*, the real sequence A_1, A_3, $\cdots$ is considered, even though s_k may be complex. According to Francis, the iterations are most effectively determined indirectly, based on the following lemma.

Lemma 9.1 *If $H = Q^H A Q$ where A is a given matrix, Q is unitary and H is Hessenberg with positive subdiagonal elements, then H is unique, provided the first column of Q is prescribed.*

Proof We can write $QH = AQ$, that is

$$\sum_{k=1}^{j+1} q_{ik} h_{kj} = AQ$$

or

$$\sum_{k=1}^{j+1} h_{kj} \mathbf{q}_k = A\mathbf{q}_j$$

where $Q = [\mathbf{q}_1, \mathbf{q}_2, \cdots, \mathbf{q}_n]$. Hence

$$h_{j+1,j} \mathbf{q}_{j+1} = A\mathbf{q}_j - \sum_{k=1}^{j} h_{kj} \mathbf{q}_k$$

which yields the subdiagonal elements (since $\mathbf{q}_j$ are orthonormal). Thus

$$h_{j+1,j} = \|A\mathbf{q}_j - \sum_{k=1}^{j} h_{kj}\, \mathbf{q}_k\|_2$$

and

$$\mathbf{q}_{j+1} = \left(A\mathbf{q}_j - \sum_{k=1}^{j} h_{kj}\, \mathbf{q}_k\right) / h_{j+1,j}$$

The above equation yields $\mathbf{q}_j$ *recursively*, provided the elements h_{kj} of the upper triangle are known, for which $H = Q^H A Q$ yields

$$h_{ij} = \mathbf{q}_i^H A\mathbf{q}_j, \quad i \le j$$

which recursively supply the required elements, provided $\mathbf{q}_1$ is prescribed. □

Now from Eqs. (9.36) and (9.37) since $\hat{R}_2$ is *right triangular*, the first column of $\hat{Q}_2$ is

$$\begin{aligned}
\hat{Q}_2\mathbf{e}_1 &= k^{-1}\phi_2(A_1)\,\mathbf{e}_1 = k^{-1}\,(A^2 - \sigma A + \rho I)\,\mathbf{e}_1 \\
&= k^{-1}\,[x_1,\, y_1,\, z_1,\, 0,\, \cdots,\, 0]^T
\end{aligned} \tag{9.38}$$

where k is the first diagonal element ($i = 1,\ j = 1$) of $\hat{R}_2$ and $\sigma = s_1 + s_2$, $\rho = s_1 s_2$. Also, easily

$$\begin{aligned}
x_1 &= a_{11}^2 + a_{12}\,a_{21} - \sigma\,a_{11} + \rho = a_{21}\,[\{a_{11}\,(a_{11} - \sigma) + \rho\}/a_{21} + a_{12}] \\
y_1 &= a_{21}\,(a_{11} + a_{22} - \sigma) \\
z_1 &= a_{32}\,a_{21}
\end{aligned} \tag{9.39}$$

The unitarity of $\hat{Q}_2$, viz. $\hat{Q}_2^T \hat{Q}_2 = I$ requires

$$k^2 = x_1^2 + y_1^2 + z_1^2 \tag{9.40}$$

Now let P_1 be a unitary matrix with the same first column as Eq. (9.38) and of the form

$$P_1 = \left[\begin{array}{ccc|c}
x_1/k & \times & \times & \\
y_1/k & \times & \times & O \\
z_1/k & \times & \times & \\
\hline
 & & & \\
 & O & & I \\
 & & &
\end{array} \right] \tag{9.41}$$

On post and premultiplication of A_1 by P_1 we get the form for $n = 6$:

$$P_1 A_1 P_1 = \begin{bmatrix} \times & \times & \times & \times & \times & \times \\ \times & \times & \times & \times & \times & \times \\ \times & \times & \times & \times & \times & \times \\ \times & \times & \times & \times & \times & \times \\ 0 & 0 & 0 & \times & \times & \times \\ 0 & 0 & 0 & 0 & \times & \times \end{bmatrix} \qquad (9.42)$$

which is not in Hessenberg form. It can however be converted to such a form by $n - 2$ successive Householder transformations $P_2, P_3, \cdots, P_{n-1}$ to yield

$$A' = P_{n-1} \cdots P_2 P_1 A_1 P_1 P_2 \cdots P_{n-1} \qquad (9.43)$$

In this process, the first column of $P_1 P_2 \cdots P_{n-1}$ is that of P_1, that is, of $\hat{Q}_2$. Hence by the uniqueness lemma $A_3 \equiv A'$.

It is natural to select P_1 on the basis of Householder transformation:

$$P_1 = I - \alpha_1 \mathbf{v} \mathbf{v}^T, \quad \alpha_1 \mathbf{v}^T \mathbf{v} = 2$$
$$\mathbf{v} = [u_1, v_1, w_1, 0, \cdots, 0]^T \qquad (9.44)$$

without loss of generality, we may introduce scaling by taking $u_1 = 1$. Hence

$$(P_1 \mathbf{e}_1)^T = [1 - \alpha_1, -\alpha_1 v_2, -\alpha_1 v_3, 0, \cdots, 0]^T$$
$$= \left[\frac{x_1}{k}, \frac{y_1}{k}, \frac{z_1}{k}, 0, \cdots, 0 \right]^T$$

by Eq. (9.41), i.e.

$$-\alpha_1 = \frac{x_1}{k} - 1, \quad -\alpha_1 v_1 = \frac{y_1}{k}, \quad -\alpha_1 w_1 = \frac{z_1}{k}$$

Satisfaction of Eq. (9.43) again yields Eq. (9.40). From the above relations, we get

$$v_1 = \frac{y_1}{x_1 - k}, \quad w_1 = \frac{z_1}{x_1 - k}$$

Writing γ_1 for $-k$, we get

$$u_1 = 1, \quad v_1 = \frac{y_1}{x_1 + \gamma_1}, \quad w_1 = \frac{z_1}{x_1 + \gamma_1}, \quad \alpha_1 = 1 + \frac{x_1}{\gamma_1} \qquad (9.45)$$

where, from Eq. (9.40), we select the sign of the radical so as to minimise possible loss of significant digits in $x_1 + \gamma_1$ by taking

$$\gamma_1 = \text{sgn}(x_1)\sqrt{x_1^2 + y_1^2 + z_1^2} \qquad (9.46)$$

The transformation P_1 (Eqs. (9.44)–(9.46)) is remarkably similar to that of P_2 (and likewise for $P_3, \cdots, P_{n-1}$) (see Sect. 9.2.3) except that it precedes the latter by one stage. The computation of $P_1 A P_1$ may therefore be carried out as described in that section.

Equations (9.28) and (9.29) of Sect. 9.2.3 also show that for Hessenberg reduction of Eq. (9.42), which contains four consecutive non-zero elements in the first column, P_2 requires $\mathbf{v}$ with only three non-zero components starting with the second $u_2 = 1$, v_2 and w_2. Comparison with Eqs. (9.45) and (9.46) shows that for P_2, we can use the procedure for P_1 with a shift, in which x_1, y_1, z_1 are replaced by x_2, y_2, z_2 representing the last three non-zero elements of the first column of Eq. (9.42) for the computation of u_2, w_2, γ_2 and α_2. Significantly, which considerably eases programming, for the computation of P_3, we again require three non-zero components starting with the third, $u_3 = 1$, v_3 and w_3, which may be computed from a P_1 with an additional shift, in the manner just described for P_2 (verify). This kind of procedure holds true for $P_4, P_5, \cdots, P_{n-2}$. At the last stage of reduction P_{n-1}, there is no z term and we require only two non-zero components $u_{n-1} = 1$ and v_{n-1}.

9.3.2.5 Computing the Product for A'

In view of the fact that each P_i is composed of a vector $\mathbf{v}$ with only three non-zero components, the product (9.43) for A' should be computed directly rather than as in Sect. 9.2.3 on Householder transformation. The product with $B_1 = A_1$ can be written as

$$B_{k+1} = P_k B_k P_k, \quad k = 1, 2, \cdots, n-1 \qquad (9.47)$$

$P_k = I - \alpha_k \mathbf{v}_k \mathbf{v}_k^T$, where

$$\mathbf{v}_k^T = [0, 0, \cdots, u_k, v_k, w_k, 0, \cdots, 0]^T$$

$$\uparrow$$
$$k$$

$$u_k = 1, \quad v_k = \frac{y_k}{x_k + \gamma_k}, \quad w_k = \frac{z_k}{x_k + \gamma_k}$$

$$\alpha_k u_k = \frac{x_k + \gamma_k}{\gamma_k}, \quad \alpha_k v_k = \frac{y_k}{\gamma_k}, \quad \alpha_k w_k = \frac{z_k}{\gamma_k}$$

and for $k = n - 1$ there is no v_{k+2}. The vector can be written as $\mathbf{v}_k = u_k \mathbf{e}_k + v_k \mathbf{e}_{k+1} + w_k \mathbf{e}_{k+2}$. The premultiplication in Eq. (9.47) is

$$B'_k = P_k B_k = B_k - (\alpha_k \mathbf{v}_k) \xi_k^T$$
$$\xi_k^T = \mathbf{v}_k^T B_k = (u_k \mathbf{e}_k^T + v_k^T \mathbf{e}_{k+1}^T + w_k \mathbf{e}_{k+2}^T) B_k$$
$$= u_k b_{kj} + v_k b_{k+1,j} + w_k, b_{k+2,j}, \quad j = 1, 2, \cdots, n$$
$$=: \xi_{kj}$$

In any column j, several elements below the subdiagonal are zero, so elements of ξ_k^T for $j < k$ vanish in effect (verify), (also see subroutine HOUSEHOLDER). Also

$$(\alpha_k \mathbf{v}_k) \xi_k^T = \text{matrix of } j\text{th column} = \begin{bmatrix} 0 \\ 0 \\ \vdots \\ \alpha_k u_k \xi_{kj} \\ \alpha_k v_k \xi_{k+1,j} \\ \alpha_k w_k \xi_{k+2,j} \\ \vdots \\ 0 \\ 0 \end{bmatrix}$$

in which columns $j < k$ are null. The computational algorithm for the above is simple:

$$\text{For } j = k, \cdots, n$$
$$\xi \, (= \xi_{kj}) \leftarrow \begin{cases} u_k b_{kj} + v_k b_{k+1,j} + w_k b_{k+2,j}, & k < n - 1 \\ u_k b_{kj} + v_k b_{k+1,j}, & k = n - 1 \end{cases}$$
$$b_{kj} \leftarrow b_{kj} - \alpha_k u_k \xi \qquad\qquad\qquad\qquad (9.48)$$
$$b_{k+1,j} \leftarrow b_{k+1,j} - \alpha_k v_k \xi$$
$$b_{k+2,j} \leftarrow b_{k+2,j} - \alpha_k w_k \xi$$

column number j remaining fixed at a time; these are obviously *row transformations*. The post multiplication in Eq. (9.47) is similarly carried out:

$$B_{k+1} = B'_k P_k = B'_k - \eta_k \mathbf{v}_k^T$$
$$\eta_k = B'_k \alpha_k \mathbf{v}_k = B'_k \alpha_k (u_k \mathbf{e}_k + v_k \mathbf{e}_{k+1} + w_k \mathbf{e}_{k+2})$$
$$= \alpha_k (u_k b'_{ik} + v_k b'_{i,k+1} + w_k b'_{i,k+2}), \quad i = 1, 2, \cdots, n$$
$$=: \eta_{ik}$$

In the above, for which rows η_k vanishes? The first term contributes null for $i > k + 1$ (verify), so the second and the third terms contribute null for $i > k + 2$ and $i > k + 3$, respectively. The remaining product yields

$$\eta_k \, \mathbf{v}_k^T = \text{matrix of the } i\text{th row } [0, \, 0, \cdots, \, \eta_{ik} \, u_k, \, \eta_{ik} \, v_k, \, \eta_{ik} \, w_k, \, 0, \cdots, \, 0]$$

in which rows $i > k + 3$ are null. The algorithm is now:

For $i = 1, \cdots, \min(k + 3, n)$
$$\eta \, (= \eta_{ik}) \leftarrow \begin{cases} \alpha_k \, u_k \, b_{ik} + \alpha_k \, v_k \, b_{i,k+1} + \alpha_k \, w_k \, b_{i,k+2}, & k < n - 1 \\ \alpha_k \, u_k \, b_{ik} + \alpha_k \, v_k \, b_{i,k+1}, & k = n - 1 \end{cases}$$
$$b_{ik} \leftarrow b_{ik} - u_k \, \eta$$
$$b_{i,k+1} \leftarrow b_{i,k+1} - v_k \, \eta$$
$$b_{i,k+2} \leftarrow b_{i,k+2} - w_k \, \eta$$

$$(9.49)$$

Evidently, these are *column transformations*. To top it, a *single cell* is needed for the elements of ξ and η!

9.3.2.6 Vanishingly Small Subdiagonal Elements

The theory of the double QR transformation is based on the hypothesis that the subdiagonal elements are not zero. However, during iteration, several of them may tend to zero as convergence to eigenvalues approaches. Suppose that subdiagonal element in row l tends to zero (machine zero by underflow), viz. $a_{l,l-1} = \epsilon$, say, for example, $l = 3$ in the sth (s odd) iteration of 6×6 matrix:

$$A_s = \begin{bmatrix} \times \times & | & \times \times \times \times \\ \times \times & | & \times \times \times \times \\ \hline \epsilon & | & \times \times \times \times \\ & | & \times \times \times \times \\ & | & \times \times \times \\ & | & \times \times \end{bmatrix} \qquad (9.50)$$

Then the matrix is effectively delinked into two submatrices; one $(l - 1) \times (l - 1)$ at the upper left-hand corner and another $(n - l + 1) \times (n - l + 1)$ at the lower right hand corner. The second submatrix with the first element at (l, l) is picked up for double QR transformation.

If, however, $l = n$ i.e. $a_{n,n-1} = \epsilon \rightarrow$ (machine 0), the smallest eigenvalue λ_n is delivered by a_{nn}. And if $l = n - 1$ i.e. $a_{n-1,n-2} = \epsilon$ (machine 0), two smaller eigenvalues, real or complex are delivered by the submatrix

$$\begin{bmatrix} a_{n-1,n-1} & a_{n-1,n} \\ a_{n,n-1} & a_{nn} \end{bmatrix}$$

These cases are very desirable.

Francis has considered another important possibility, that of consecutive small subdiagonal elements $a_{m,m-1} = \epsilon_1$ and $a_{m+1,m} = \epsilon_2$, such that $\epsilon_1, \epsilon_2 \to 0$ (machine 0). For example, let $m = 3$ in a 6×6 matrix:

$$A_s = \begin{bmatrix} \times & \times & \times & \times & \times & \times \\ \times & \times & \times & \times & \times & \times \\ & \epsilon_1 & \times & \times & \times & \times \\ & & \epsilon_2 & \times & \times & \times \\ & & & & \times & \times & \times \\ & & & & & \times & \times \end{bmatrix}$$

then, we hope to start the product with P_m rather then P_1 (as if we are treating the lower submatrix). In the premultiplication $P_m A_s$, the first $m - 1$ rows remain unaffected (verify) and we obtain the form

$$\begin{bmatrix} \times & \times & | & \times & \times & \times & \times \\ \times & \times & | & \times & \times & \times & \times \\ - & - & - & - & - & - & - \\ \delta_1 & | & \times & \times & \times & \times \\ \delta_2 & | & \times & \times & \times & \times \\ \delta_3 & | & \times & \times & \times & \times \\ & | & & & \times & \times \end{bmatrix}$$

(verify), where by Eq. (9.48) for $j = m - 1$

$$\xi = u_m a_{m,m-1} = \epsilon_1$$
$$\delta_1 = a_{m,m-1} = \epsilon_1 - \alpha_m \xi = \epsilon_1 - 2\epsilon_1/(1 + v_m^2 + w_m^2)$$
$$\delta_2 = a_{m+1,m-1} = -\alpha_m v_m \xi = -2\epsilon_1 v_m/(1 + v_m^2 + w_m^2)$$
$$\delta_3 = a_{m+2,m-1} = -\alpha_m w_m \xi = -2\epsilon_1 w_m/(1 + v_m^2 + w_m^2)$$

Now v_m and w_m containing y_m and z_m have a factor $a_{m+1,m} = \epsilon_2$ and x_m is not small (in general). Hence when $\epsilon_1, \epsilon_2 \to 0$, $\delta_1, \delta_2 \to 0$ and $\delta_1 = -\epsilon_1$. With these stipulations, the row transformations (9.48) can be carried out for $k \geq m$ only. No economy of computation in *column transformations* (9.49) is possible for this case.

9.3.2.7   Implementation

The successive double QR iterations are overwritten on $A = A_1$ in Hessenberg form. The convergence theorem implies that the eigenvalues are delivered in increasing order of magnitude from the bottom upwards. Hence, we consider the bottom-most submatrix

$$\begin{bmatrix} a_{n-1,n-1} & a_{n-1,n} \\ a_{n,n-1} & a_{nn} \end{bmatrix}$$

whose eigenvalues s_1, s_2 satisfy

$$\sigma = s_1 + s_2 = a_{n-1,n-1} + a_{nn}$$
$$\rho = s_1 s_2 = a_{n-1,n-1} \times a_{nn} - a_{n-1,n} \times a_{n,n-1}$$

and use them as shifts in Eq. (9.39). Also, since x_k, y_k, z_k are scaled, we first drop a_{21} in x_1, y_1, z_1 and scale them with respect to $s = |x_k| + |y_k| + |z_k|$ to prevent underflow or overflow in γ. With these choice of σ and ρ, *convergence is usually very fast.*

Since some of the subdiagonal elements tend to zero, those from bottom upwards, these are tested for underflowing to zero. If $a_{l,l-1} = \epsilon \rightarrow$ (machine 0) as in Eq. (9.50), the lower $(n - l + 1) \times (n - l + 1)$ submatrix is considered. If $l = n$, λ_n is delivered by a_{nn} and if $l = n - 1$ two eigenvalues, real or complex, λ_{n-1} and λ_n are delivered. In either case, the matrix is *deflated* by deleting the last row and column or last two rows and columns from A.

For taking advantage of two consecutive subdiagonal elements $a_{m,m-1} = \epsilon_1$ and $a_{m+1,m} = \epsilon_2$, such that ϵ_1, $\epsilon_2 \rightarrow 0$ (machine 0), the criterion

$$|a_{m,m-1}| \frac{(|y_m| + |z_m|)}{|x_m|} << |a_{m-1,m-1}| + |a_{mm}| + |a_{m+1,m+1}|$$

is found satisfactory in experience (theoretically one could have $\|A\|$ on the right-hand side).

The above works in most cases for less than 10 iterations per eigenvalue. In such cases, any shift of the order of $\|A\|$ may be successful. One such is

$$\sigma = 1.5 \times (|a_{n,n-1}| + |a_{n-1,n-2}|)$$
$$\rho = (|a_{n,n-1}| + |a_{n-1,n-2}|)^2$$

With some other minor details, the above scheme is implemented in subroutine named FRANCIS. Though the program is long, execution is usually very fast.

It is important to note the complexity of the algorithm. For reduction to Hessenberg form by Householder transformations, the complexity is $\frac{4}{3} n^3$, while for each QR transformation it is only $5 n^2$. The total complexity for computing n eigenvalues is found to be approximately $8 n^3$.

```
SUBROUTINE FRANCIS(A,n)
! Computes eigenvalues of a real matrix A in Hessenberg form.
! A= The given n× n real matrix in Hessenberg form.        (Input)
! n= Dimension of the matrix.        (Input)
! eigval= n-vector containing the n eigenvalues delivered by the subroutine;
! however the eigenvalues are obtained sequentially backwards from
! eigval(n) to eigval(1).        (Output)
!***********************************************************
REAL :: A(n,n)
COMPLEX :: eigval(n)
! Compute a norm of matrix A for determining L
anorm=ABS(A(1,1))
DO i=2,n; DO j=i−1,n
anorm=anorm+ABS(A(i,j))
END DO; END DO
nn=n; eshift=0.0        ! eshift is for exceptional shift
101 IF(nn==0) RETURN
iter=0
! Compute L for which lowest subdiagonal element A(L,L−1)=0,
! delinking the bottom right hand submatrix of order n−L+1
102 DO LL=nn,2,−1
L=LL; s=ABS(A(L−1,L−1))+ABS(A(L,L))
IF(s==0.0) s=anorm
IF(ABS(A(L,L−1))+s==s) GOTO 103
END DO
L=1
103 A(nn,nn)=A(nn,nn)+eshift
sigma=A(nn−1,nn−1)+A(nn,nn)
rho=A(nn−1,nn−1)*A(nn,nn)−A(nn−1,nn)*A(nn,nn−1)

IF(L==nn) THEN
    eigval(nn)=CMPLX(A(nn,nn),0.0)        ! One eigenvalue obtained.

    PRINT*, 'real eigenvalue=', eigval(nn), ' no. of iterations=', iter
    PAUSE
    nn=nn−1; GOTO 101

ELSE IF(L=nn−1) THEN        ! Pair of eigenvalues obtained:
```

```
        d=0.25*sigma**2−rho;  rtd=SQRT(ABS(d));  hafsig=0.5*sigma
        IF(d>=0.0) THEN
            IF(sigma>=0.0) eigval(nn)=CMPLX(hafsig+rtd,0.0)
            IF(sigma<0.0) eigval(nn)=CMPLX(hafsig−rtd,0.0)
            eigval(nn−1)=rho/eigval(nn)
        ELSE
            eigval(nn)=CMPLX(hafsig,rtd); eigval(nn−1)=CMPLX(hafsig,−rtd)
        END IF
        PRINT*, 'pair of eigenvalues=', eigval(nn), eigval(nn−1)
        PRINT*, 'no. of iterations=', iter
        nn=nn−2
        GOTO 101
ELSE
        IF(L/=1) PRINT*, 'Warning ! Delink at L=', L
        IF(iter==30) THEN
        PRINT*, 'Too many iterations'; RETURN
        END IF

        IF(iter==10 .OR. iter==20) THEN     ! Use exceptional shift
            eshift=eshift+A(nn,nn)
            DO i=1,nn
            A(i,i)=A(i,i)−A(nn,nn)
            END DO
            s=ABS(A(nn,nn−1))+ABS(A(nn−1,nn−2))
            sigma=1.5*s; rho=s**2
        END IF

        iter=iter+1

    ! Search for two small consecutive subdiagonal elements A(m−1,m),
        A(m,m+1):
```

```
DO mm=nn−2,1,−1
    m=mm
    x=(A(m,m)*(A(m,m)−sigma)+rho)/A(m+1,m)+A(m,m+1)
    y=A(m,m)+A(m+1,m+1)−sigma
    z=A(m+2,m+1)
    s=ABS(x)+ABS(y)+ABS(z); x=x/s; y=y/s; z=z/s
    IF(m==1) GOTO 104
    p=ABS(A(m,m−1))*(ABS(y)+ABS(z))
    q=ABS(x)*(ABS(A(m−1,m−1))+ABS(A(m,m))+ABS(A(m+1,m+1)))
    IF(p+q==q) GOTO 104
END DO
104 DO i=m+2,nn
A(i,i−2)=0.0; IF(i/=m+2) A(i,i−3)=0.0
END DO
! Double QR transformation begins:
DO k=m,nn−1
    IF(k/=m) THEN
    x=A(k,k−1); y=A(k+1,k−1); z=0.0
    IF(k<nn−1) z=A(k+2,k−1)
    s=ABS(x)+ABS(y)+ABS(z)
    IF(s==0.0) GOTO 105
    x=x/s; y=y/s; z=z/s
    END IF
gama=SQRT(x**2+y**2+z**2); IF(x<0.0) gama=−gama
xgama=x+gama; yy=y/xgama; zz=z/xgama
x=xgama/gama; y=y/gama; z=z/gama
IF(L/=m) A(m,m−1)=−A(m,m−1)
DO j=k,nn
        v=A(k,j)+yy*A(k+1,j)
        IF(k<nn−1) THEN
        v=v+zz*A(k+2,j); A(k+2,j)=A(k+2,j)−z*v
        END IF
        A(k+1,j)=A(k+1,j)−y*v; A(k,j)=A(k,j)−x*v
    END DO
```

     DO i=1,min(nn,k+3)

       v=x*A(i,k)+y*A(i,k+1)

       IF(k<nn−1) THEN

       v=v+z*A(i,k+2); A(i,k+2)=A(i,k+2)−zz*v

       END IF

       A(i,k+1)=A(i,k+1)−yy*v; A(i,k)=A(i,k)−v

     END DO

     IF(k>m) A(k,k−1)=−gama*s

     105 END DO

     ! End of double QR transformation.

     GOTO 102

END IF

END SUBROUTINE FRANCIS

Remark. We end by stating that there may still be sensitive cases where the above fails. For treatment of such cases see the text of Press et. al. (1996) cited in the Bibliography, who state that the phenomenon is present only in *unsymmetric matrices* and care must be taken in all such cases, before reduction to Hessenberg form is undertaken.

Example 1. Compute the eigenvalues of the Hessenberg matrix

$$A = \begin{bmatrix} 5 & -2 & -5 & -1 \\ 1 & 0 & -3 & 2 \\ & 2 & 2 & -3 \\ & & 1 & 2 \end{bmatrix}$$

using subroutine FRANCIS.

Solution. The main program calling subroutine FRANCIS can be easily written down. On executing the program the output is

$$\lambda_4 \approx -9.999999E - 01$$
$$\lambda_2, \ \lambda_3 \approx 1.000001 \pm 2.000000\,i$$
$$\lambda_1 \approx 3.999999$$

The digits obtained depend on the particular platform used, but the results suggest that the exact values may be $\lambda_4 = -1$, $\lambda_2, \lambda_3 = 1 \pm 2\,i$, $\lambda_1 = 4$. This fact is confirmed in Example 1 of the next subsection. $\qquad\square$

Exercises

Find the eigenvalues of the following matrices using subroutines HOUSEHOLDER and
FRANCIS:

1. $\begin{bmatrix} 3 & 1 & 0 \\ -4 & -1 & 0 \\ 4 & -8 & -2 \end{bmatrix}$ $\qquad$ $[-2,\ 9.999998E - 01 \pm 8.457279E - 04\,i].$

2. $\begin{bmatrix} 4 & 2 & 2 \\ 2 & 5 & 1 \\ 2 & 1 & 6 \end{bmatrix}$ $\qquad$ $[4.486456,\ 8.387617,\ 2.125925].$

3. $\begin{bmatrix} 8 & -1 & -5 \\ -4 & 4 & -2 \\ 18 & -5 & -7 \end{bmatrix}$ $\qquad$ $[9.999978E - 01,\ 2.000002 \pm 4.000001\,i].$

4. $\begin{bmatrix} -2 & 2 & 2 & 2 \\ -3 & 3 & 2 & 2 \\ -2 & 0 & 4 & 2 \\ -1 & 0 & 0 & 5 \end{bmatrix}$ $\qquad$ $[4,\ 3,\ 2,\ 1].$

5. $\begin{bmatrix} 1 & 2 & -2 & 4 \\ 2 & 12 & 3 & 5 \\ 3 & 13 & 0 & 7 \\ 2 & 11 & 2 & 2 \end{bmatrix}$ $\qquad$ $[-2.798580 \pm 5.106368E - 01\,i,$
$\qquad\qquad\qquad\qquad\qquad 18.951210,\ \ 1.645953].$

6. $\begin{bmatrix} 25 & -41 & 10 & -6 \\ -41 & 68 & -17 & 10 \\ 10 & -17 & 5 & -3 \\ -6 & 10 & -3 & 2 \end{bmatrix}$ $\qquad$ $[\text{Delink at } l = 2,\ 2.591981E - 01,$
$\qquad\qquad\qquad\qquad 3.301656E - 02,\ 1.186089,$
$\qquad\qquad\qquad\qquad 98.52168\ \text{(spurious)}].$

9.3.2.8   Check: Characteristic Polynomial by Hyman Algorithm

After λ_i are computed, the best check is to compute $\det[A - \lambda_i I]$. Since we are assuming matrix A to be in Hessenberg form, the determinant can be easily computed by an algorithm due to J.M. Hyman (1957). Consider A in Hessenberg form

$$\det[A] = \begin{vmatrix} a_{11} & a_{12} & a_{13} & \cdots & \cdots & \cdots & a_{1n} \\ a_{21} & a_{22} & a_{23} & \cdots & \cdots & \cdots & a_{2n} \\ & a_{32} & a_{33} & \cdots & \cdots & \cdots & a_{3n} \\ & & \cdots & \cdots & \cdots & \cdots & \cdots \\ & & & a_{n-1,n-2} & a_{n-1,n-1} & a_{n-1,n} \\ & & & & a_{n,n-1} & a_{nn} \end{vmatrix}$$

With arbitrary $k_1, k_2, \cdots, k_{n-1}$ and $k_n = 1$ and column operation $(k_1 C_1 + k_2 C_2 + \cdots + k_{n-1} C_{n-1}) + k C_n$ ($C_1, C_2, \cdots, C_n$ standing for columns) yields

$$\det[A] = \begin{vmatrix} a_{11} & a_{12} & a_{13} & \cdots & \displaystyle\sum_{j=1}^{n} k_j a_{1j} \\ a_{21} & a_{22} & a_{23} & \cdots & \displaystyle\sum_{j=1}^{n} k_j a_{2j} \\ & a_{32} & a_{33} & \cdots & \displaystyle\sum_{j=2}^{n} k_j a_{3j} \\ & \cdots & \cdots & \cdots & \cdots \\ & & & & \displaystyle\sum_{j=n-2}^{n} k_j a_{n-1,j} \\ & & & & \displaystyle\sum_{j=n-1}^{n} k_j a_{nj} \end{vmatrix}$$

Now set the last $n - 1$ elements of the nth column to zero, to determine

$$k_n = 1$$
$$k_i - \frac{1}{a_{i+1,i}} \sum_{j=i+1}^{n} k_j a_{i+1,j}, \quad i = n-1, n-2, \cdots, 1 \tag{9.51}$$

Expanding the determinant by the elements of the nth column,

$$\det[A] = (-1)^n \sum_{j=1}^{n} k_j a_{1j} \times (a_{21} a_{32} \cdots a_{n,n-1}) \tag{9.52}$$

Evidently, the algorithm (51)–(52) also holds for complex A. The following subroutine named HYMAN computes $\det[A]$ for complex A and in general λ_i are complex.

```
SUBROUTINE HYMAN(A,n,det)
! Computes determinant of a complex matrix A of order n
! in Hessenberg form. If A is real, modification is trivial.
! A= given matrix.    (Input)
! n= order of matrix A.        (Input)
! det= determinant of matrix A.        (Output)
!************************************************************
COMPLEX :: A(n,n), det, k(n), sum, prod
k(n)=(1.0,0.0)
DO i=n-1,1,-1
k(i)=(0.0,0.0)
DO j=i+1,n
k(i)=k(i)+k(j)*A(i+1,j)
ENd DO
k(i)=-k(i)/A(i+1,i)
END DO
sum=(0.0,0.0)
DO j=1,n
sum=sum+k(j)*A(1,j)
END DO
prod=(1.0,0.0)
DO j=1,n-1
prod=prod*A(j+1,j)
END DO
det=(-1)**(n-1)*sum*prod
RETURN
END SUBROUTINE HYMAN
```

Example 1. Check the estimates of the computed eigenvalues of Example 1, Sect. 9.3.2.8 by subroutine HYMAN.

Solution. The computed eigenvalues are (i) $\lambda_4 = -9.999999E - 01$, (ii) $\lambda_2, \lambda_3 = 1.000001 \pm 2i$ and (iii) $\lambda_1 = 3.999999$. We write a main program as follows:

```
REAL :: A(4,4)
COMPLEX :: AA(4,4), det
A(1,1)=5; A(1,2)=-2; A(1,3)=-5; A(1,4)=-1
A(2,1)=1; A(2,2)=0; A(2,3)=-3; A(2,4)=2
A(3,1)=0; A(3,2)=2; A(3,3)=2; A(3,4)=-3
A(4,1)=0; A(4,2)=0; A(4,3)=1; A(4,4);=-2
DO i=1,4; DO j=1,4
AA(i,j)=CMPLX(A(i,j),0.)
```

END DO; END DO

    x=−9.999999E−01; y=0.   ! real and imaginary part of a computed eigen
                              ! value

    DO i=1,4
    AA(i,i)=AA(i,i)−cmplx(x,y)
    END DO
    CALL l HYMAN(AA,4,det)
    PRINT*, det
    END

To this program, we have of course to attach subroutine HYMAN. In the above program, the first root λ_4 is tested and the result of execution is

$$\det = -8.583069E - 06$$

In a similar manner writing x=1.000001; y=±2, and x=3.999999; y=0, we get for the cases (ii) and (iii)

$$\det = 9.536743E - 06 \mp 3.814697E - 05$$

and $\det = -5.340576E - 05$

    The digits obtained above depend on the particular platform used. If the eigenvalues are rounded to -1, $1\pm2\,i$ and 4, the value of det is obtained as 0 in all the four cases.   □

Exercises

Use subroutine HYMAN to compute det for the eigenvalues of the Exercises 1–6 of Sect. 9.3.2.7.

[**1.** $-9.440894E-06$, $\pm2.304906E-09\,i$

2. $-4.964799E-06$, $5.678034E-05$, $-7.957226E-06$

3. 0, $7.033955E-05 \pm 3.516977E-05\,i$

4. $2.199042E-06$, $1.272295E-06$, $2.289446E-06$, $3.434170E-07$

5. $-5.303153E-05 \pm 2.305719E-06\,i$, $1.341006E-02$, $9.222874E-05$

6. $5.088465E-05$, $-3.215952E-05$, $-2.372593E-05$, $\underline{-7.77321}$].

9.4   Maximum Modulus Eigenvalue: The Power Method

Let λ_1, λ_2, $\cdots$, λ_n be the eigenvalues of an $n \times n$ matrix A such that they can be arranged in a descending order of magnitude say

$$|\lambda_1| \geq |\lambda_2| \geq \cdots \geq |\lambda_n|$$

so that λ_1 is the largest eigenvalue in modulus. We assume in this section that λ_1 is *real*. If the eigenvectors x_1, x_2, $\cdots$, x_n corresponding to λ_1, λ_2, $\cdots$, λ_n are *linearly independent*, any vector v can be written as

$$v = c_1 x_1 + c_2 x_2 + \cdots + c_n x_n$$

or, absorbing the multiplying scalars c_1, c_2, $\cdots$, c_n into the vectors,

$$v = x_1 + x_2 + \cdots + x_n$$

Hence applying Theorem 5 of Sect. 9.1,

$$\begin{aligned} A^k v &= \lambda_1^k x_1 + \lambda_2^k x_2 + \cdots + \lambda_n^k x_n \\ &= \lambda_1^k \left[x_1 + \left(\frac{\lambda_2}{\lambda_1} \right)^k x_2 + \cdots + \left(\frac{\lambda_n}{\lambda_1} \right)^k x_n \right] \end{aligned} \qquad (9.53)$$

Case (i). Let λ_1 be nonrepeated, i.e. $|\lambda_1| > |\lambda_2| \geq \cdots \geq |\lambda_n|$. Then

$$A^k v \to \lambda_1^k v \quad \text{as} \quad k \to \infty \qquad (9.54)$$

Since a scalar multiplier is inconsequential for an eigenvector, we conclude that $A^k v$ tends to the eigenvector corresponding to λ_1.

To determine λ_1, we obtain a scalar relation by premultiplying Eq. (9.53) by any row vector u^T, obtaining

$$u^T A^k v = \lambda_1^k \left[u^T x_1 + \left(\frac{\lambda_2}{\lambda_1} \right)^k u^T x_2 + \cdots + \left(\frac{\lambda_n}{\lambda_1} \right)^k u^T x_n \right]$$

Hence, it follows that the quotient

$$\frac{u^T A^{k+1} v}{u^T A^k v} = \lambda_1 + O\left[\left(\frac{\lambda_2}{\lambda_1} \right)^k \right] \quad \text{as} \quad k \to \infty$$
$$\to \lambda_1 \quad \text{as} \quad k \to \infty$$

if $u^T x_1 \neq 0$. In particular, one can choose $u = A^k v$ (which tends to the eigenvector corresponding to λ_1 by Eq. (9.54)) and obtain

$$\lambda_1 \rightarrow \frac{\mathbf{u}^T A \mathbf{u}}{\mathbf{u}^T \mathbf{u}} = \left(\frac{\mathbf{u}^T}{\|\mathbf{u}^T\|_2}\right) A \left(\frac{\mathbf{u}}{\|\mathbf{u}\|_2}\right) \tag{9.55}$$

where $\mathbf{u}^T \mathbf{u} = \|\mathbf{u}\|_2^2$. The scalar on the right-hand side of Eq. (9.55) is called the **Rayleigh Quotient** of A and $\mathbf{u}$. With this choice of $\mathbf{u}$ if A is *symmetric* $\mathbf{u}^T = \mathbf{v}^T A^k$, and the Rayleigh quotient becomes

$$\frac{\mathbf{u}^T A \mathbf{u}}{\mathbf{u}^T \mathbf{u}} = \frac{\mathbf{v}^T A^k A A^k \mathbf{v}}{\mathbf{v}^T A^k A^k \mathbf{v}} = \frac{\mathbf{v}^T A^{2k+1} \mathbf{v}}{\mathbf{v}^T A^{2k} \mathbf{v}} = \lambda_1 + O\left[\left(\frac{\lambda_2}{\lambda_1}\right)^{2k}\right] \quad \text{as} \quad k \rightarrow \infty$$

This means that the convergence is twice as fast as in the general case.

Case (ii). Let λ_1 be repeated m times, $\lambda_1 = \lambda_2 = \cdots = \lambda_m$ and $|\lambda_m| > |\lambda_{m+1}| \geq |\lambda_{m+2}| \geq \cdots \geq |\lambda_n|$, then

$$\mathbf{u} = A^k \mathbf{v} = \lambda_1^k \left[\mathbf{x}_1 + \mathbf{x}_2 + \cdots + \mathbf{x}_m + \left(\frac{\lambda_{m+1}}{\lambda_1}\right)^k \mathbf{x}_{m+1} + \cdots + \left(\frac{\lambda_n}{\lambda_1}\right)^k \mathbf{x}_n\right]$$
$$\rightarrow \lambda_1^k (\mathbf{x}_1 + \cdots + \mathbf{x}_m) \quad \text{as} \quad k \rightarrow \infty$$

Since $\mathbf{x}_1 + \cdots + \mathbf{x}_m$ is also an eigenvector of A corresponding to λ_1, we obtain one eigenvector corresponding to λ_1 as the limiting value of $\mathbf{u}$; the other $m - 1$ eigenvectors remain undetermined. However,

$$\mathbf{u}^T A^k \mathbf{v} = \lambda_1^k \left[\mathbf{u}^T (\mathbf{x}_1 + \cdots + \mathbf{x}_m) + \left(\frac{\lambda_{m+1}}{\lambda_1}\right)^k \mathbf{u}^T \mathbf{x}_{m+1} + \cdots + \left(\frac{\lambda_n}{\lambda_1}\right)^k \mathbf{u}^T \mathbf{x}_n\right]$$

Hence
$$\frac{\mathbf{u}^T A^{k+1} \mathbf{v}}{\mathbf{u}^T A^k \mathbf{v}} \rightarrow \lambda_1$$

if $\mathbf{u}^T (\mathbf{x}_1 + \cdots + \mathbf{x}_m) \neq 0$. In particular, the Rayleigh quotient also tends to the eigenvalue λ_1.

The method described above for obtaining the maximum modulus real eigenvalue as the limit of the Rayleigh quotient is called the **Power Method**. The subroutine named POWER that follows, starts with a value of $\mathbf{u}$ as $[1, 1, \cdots, 1]^T$ in the absence of any information regarding λ_1. It is hoped that $\mathbf{u}^T \mathbf{x}_1$ or $\mathbf{u}^T (\mathbf{x}_1 + \cdots + \mathbf{x}_m)$ do not equal zero. In case of failure due to this assumption, the initial value of $\mathbf{u}$ must be changed. However, lack of convergence may be due to the closeness of the next eigenvalue or due to the maximum modulus eigenvalue being complex instead of being real (Exercise 3). Convergence may be poor for case (ii) problems as roundoff errors slightly displace equal eigenvalues (Exercise 5). For this reason although theoretically maximum modulus *complex* eigenvalues should be delivered by using complex arithmetic, practical difficulties appear (see Exercise 6 in this connection). These difficulties have been overcome in the literature and special tricks have been devised to sneak up on a pair of complex conjugate eigenvalues of a real matrix A in real arithmetic. In conclusion, *the power method is not universally applicable*.

```fortran
SUBROUTINE POWER(A,n,eigenval,u)
! A=Given real matrix.    (Input)
! n= Order of the matrix.    (Input)
! eigenval= maximum modulus eigenvalue, assumed real.    (Output)
! u= Corresponding normalised eigenvector.    (Output)
!*************************************************************
REAL :: A(n,n), u(n), unew(n)
maxiter=20    ! Maximum number of iteratiions, may be changed if necessary
DO i=1,n       ! Initialise the vector u
u(i)=1.
END DO
eigenval=0.0
DO k=1,maxiter
! Normalise the vector u
vecmag=0.0
DO i=1,n
vecmag=vecmag+u(i)**2
END DO
vecmag=SQRT(vecmag)
DO i=1,n
u(i)=u(i)/vecmag
END DO
! Compute A*u, the new approximation to u
DO i=1,n
unew(i)=0.0
DO j=1,n
unew(i)=unew(i)+A(i,j)*u(j)
END DO; END DO
! Compute the Rayleigh Quotient
valnew=0.0
DO i=1,n
valnew=valnew+u(i)*unew(i)
END DO
IF(ABS(valnew-eigenval)<1.e-6*ABS(valnew)) RETURN
eigenval=valnew
DO i=1,n
u(i)=unew(i)
END DO
IF(k==maxiter) PRINT*, 'iteration incomplete'
END
RETURN
END SUBROUTINE POWER
```

Example 1. Using subroutine POWER to compute the maximum modulus eigenvalue of the matrix

$$A = \begin{bmatrix} 3 & 1 & 0 \\ 1 & 2 & 2 \\ 0 & 1 & 1 \end{bmatrix}$$

Solution. We write the main program as

REAL :: A(3,3); eigenvec(3)
A(1,1)=3.; A(1,2)=1.; A(1,3)=0.
A(2,1)=3.; A(2,2)=2.; A(2,3)=2.
A(3,1)=0.; A(3,2)=1.; A(3,3)=1.
call POWER(A,3,eigenval,eigenvec)
PRINT*, eigenval
PRINT*, eigenvec
end

Appending subroutine POWER to this program, one obtains

$$\lambda_1 = \text{maximum modulus eigenvalue} = 3.86082$$

with   eigenvector $= [0.73889, \ 0.63608, \ 0.22235]^T$.   □

Remark (**Minimum Modulus Eigenvalue**). By a slight modification of the power method the minimum modulus eigenvalue λ_n and the corresponding eigenvector can be computed. For consider the relation

$$A^{-k}\mathbf{v} = \lambda_1^{-k}\mathbf{x}_1 + \cdots + \lambda_n^{-k}\mathbf{x}_n$$

Then with $\mathbf{u} = A^{-k}\mathbf{v} = (A^{-1})^k\mathbf{v}$, the Rayleigh quotient

$$\frac{\mathbf{u}^T A^{-1}\mathbf{u}}{\mathbf{u}^T\mathbf{u}} \to \lambda_n \quad \text{as} \quad k \to \infty$$

if $\mathbf{u}^T\mathbf{x}_n$ or $\mathbf{u}^T(\mathbf{x}_n + \mathbf{x}_{n-1} + \cdots + \mathbf{x}_{n-m+1}) \neq 0$. Thus we have to first compute A^{-1} by subroutine MATINV (Chap. 3, Sect. 3.1.4) and then use the subroutine POWER.

The idea is further developed in the next subsection to sneak up at *any* real eigenvalue and the corresponding eigenvector.

Exercises
Compute the maximum eigenvalue and the corresponding eigenvector by using the subroutine POWER.

1. $\begin{bmatrix} 1 & -3 & 2 \\ 4 & 4 & -1 \\ 6 & 3 & 5 \end{bmatrix}$

[7.0; $[0.28676, 0.06373, 0.95588]^T$. The other two eigenvalues are complex].

2. $\begin{bmatrix} -15 & 4 & 3 \\ 10 & -12 & 6 \\ 20 & -4 & 2 \end{bmatrix}$

[-20.0; $[-0.66667, 0.33333, 0.66667]^T$].

3. $\begin{bmatrix} 8 & -1 & -5 \\ -4 & 4 & -2 \\ 18 & -5 & -7 \end{bmatrix}$

[Iteration incomplete even for maxitr=100 and relative error 1.E−4. The actual eigenvalues are λ_1, $\lambda_2 = 2 \pm 4i$ and $\lambda_3 = 1$. So the maximum modulus eigenvalues are complex and the routine fails].

4. $\begin{bmatrix} 2 & 1 & 1 & 0 \\ 1 & 1 & 0 & 1 \\ 1 & 0 & 1 & 1 \\ 0 & 1 & 1 & 2 \end{bmatrix}$

[3.56155; $[0.55731, 0.43520, 0.43520, 0.55731]^T$].

5. $\begin{bmatrix} 6 & -3 & 4 & 1 \\ 4 & 2 & 4 & 0 \\ 4 & -2 & 3 & 1 \\ 4 & 2 & 3 & 1 \end{bmatrix}$

[5.2888 with maxitr = 100, relative error 1.E−4 (The actual eigenvalues are $\lambda_1 = \lambda_2 = 3 + \sqrt{5} = 5.236$ and $\lambda_3 = \lambda_4 = 0.764$). Eigenvector = $[0.2677, 0.6148, 0.2377, 0.7027]^T$].

6. Modify subroutine POWER using complex arithmetic for obtaining maximum modulus complex pair of eigenvalues. Show that for Exercise 3 above, there is no convergence.

9.4.1 Any Real Eigenvalue: The Inverse Iteration Method

By a clever manipulation any real eigenvalue and the corresponding eigenvector of a real matrix A can be computed, provided an approximation to the former is known. Let p be an approximation to a real eigenvalue λ_j $(1 \leq j \leq n)$. If as before x_i is an eigenvector corresponding to eigenvalue λ_i, then

$$(A - pI)\mathbf{x}_i = \lambda_i \mathbf{x}_i - p\mathbf{x}_i = (\lambda_i - p)\mathbf{x}_i$$

or,
$$(A - pI)^{-1}\mathbf{x}_i = (\lambda_i - p)^{-1}\mathbf{x}_i$$

Hence, the matrix $\hat{A} := (A - pI)^{-1}$ has eigenvalues $(\lambda_i - p)^{-1}$ and eigenvectors x_i $(i = 1, \cdots, n)$. Of these eigenvalues $(\lambda_j - p)^{-1}$ is the largest in magnitude since $\lambda_j \approx p$. Hence applying the power method, the Rayleigh quotient

$$\frac{\mathbf{u}^T \hat{A}\mathbf{u}}{\mathbf{u}^T\mathbf{u}} \to (\lambda_j - p)^{-1} \quad \text{as} \quad k \to \infty$$

where $\mathbf{u} = \hat{A}^k\mathbf{v}$. Thus if the limit of the Rayleigh quotient is obtained as l_j then

$$\lambda_j = p + \frac{1}{l_j} \tag{9.56}$$

The eigenvector obtained by the method is $\mathbf{x}_j$.

In a Fortran implementation of the program, the matrix $\hat{A}$ can be computed by using the subroutine MATINV (Chap. 3, Sect. 3.1.4). This subroutine of course calls subroutines BANACHIEWICZ and SOLVE (chapter, Sect. 3.1.3) and must be appended to the program. The limit l_j in Eq. (9.56) is delivered by the subroutine POWER.

Example 1. Compute the smallest eigenvalue (in magnitude) of the matrix

$$A = \begin{bmatrix} 1 & 2 & 0 \\ 2 & 1 & 0 \\ 0 & 0 & -1 \end{bmatrix}$$

using inverse iteration.

Solution. To capture the smallest eigenvalue we set $p = 0$ and write the following program:

```
REAL :: A(,3), eigenvec(3)
A(1,1)=1; A(1,2)=2; A(1,3)=0
A(2,1)=2; A(2,2)=1; A(2,3)=0
```

A(3,1)=0; A(3,2)=0; A(3,3)=−1
p=0.
DO i=1,3
A(i,i)=A(i,i)−p
END DO
CALL MATINV(3,A)
CALL POWER(A,3,eigenval,eigenvec)
eigenval=p+1./eigenval
PRINT*, eigenval
PRINT*, eigenvec
END

Appending subroutines MATINV, BANACHIEWICZ, SOLVE and POWER, the eigenvalue and the eigenvector are obtained as −1.0 and [1.524100E−04, 1.524217E−04, 1.0]^T.   □

Exercises

Find the eigenvalues and the eigenvectors near the stated approximation of p to an eigenvalue:

1.
$$A = \begin{bmatrix} 1 & 1/2 & 1/3 \\ 1/2 & 1/3 & 1/4 \\ 1/3 & 1/4 & 1/5 \end{bmatrix} \qquad \text{for } p = 0.5 \text{ and } 1$$

[$p = 0.5$: $\lambda = 0.12233$, eigenvector = $[-0.54719, 0.52687, 0.65038]^T$;

$p = 1$: $\lambda = 1.40832$, eigenvector = $[0.82726, 0.45967, 0.32303]^T$].

2.
$$A = \begin{bmatrix} 1 & 2 & -4 \\ 3 & -1 & 2 \\ 1 & 0 & -1 \end{bmatrix} \qquad \text{for } p = 0, 2 \text{ and } 4$$

[$p = 0$: no result. $p = 2$ and 4: $\lambda = 2.07395$, eigenvector = $[0.63041, 0.74868, 0.20508]^T$].

3.
$$A = \begin{bmatrix} 5 & 7 & 6 & 5 \\ 7 & 10 & 8 & 7 \\ 6 & 8 & 10 & 9 \\ 5 & 7 & 9 & 10 \end{bmatrix} \qquad \text{for } p = 0, 1, 5, 20$$

[$p = 0$: $\lambda = 0.01015$, $[0.83044, -0.50156, -0.20855, 0.12370]^T$

$p = 1$: $\lambda = 0.84311$, $[0.09327, -0.30163, 0.76033, -0.56765]^T$

$p = 5$: $\lambda = 3.85806$, $[-0.39836, -0.61483, 0.27163, 0.62538]^T$

$p = 20$: $\lambda = 30.28870$, $[0.38065, 0.52888, 0.55174, 0.52055]^T$].

9.5 The Characteristic Equation

If one is interested in only the eigenvalues of a real matrix A, then the expanded form of Eq. (9.2) may be considered for solution by methods of Chap. 2. Let the equation be of the form

$$f(\lambda) := \lambda^n + p_1\lambda^{n-1} + \cdots + p_{n-1}\lambda + p_n = 0 \qquad (9.57)$$

In this equation, the direct determination of the coefficients $p_1, p_2, \cdots, p_n$ by expansion of the determinant is extremely laborious for $n > 3$. A systematic method that can be coded in Fortran was given by U.J.J. Leverrier, which was converted into a recursive scheme by Faddeev.

Urbain Jean Joseph Leverrier (1811–1877), was a French astronomer and mathematician who analysed the orbit of Uranus and accurately predicted the existence of another planet nearby. His prediction led directly to the discovery of Neptune in the year 1846.

Leverrier's Method. *The coefficients $p_1, p_2, \cdots, p_n$ satisfy*

$$p_k = -\frac{1}{k}\,\mathrm{tr}\,B_k \quad (k = 1, 2, \cdots, n) \qquad (9.58)$$

where the matrices B_k are given by

$$\begin{aligned} B_1 &= A \\ B_k &= A\,(B_{k-1} + p_{k-1}I), \quad k = 2, 3, \cdots, n \end{aligned} \qquad (9.59)$$

Proof The proof is based on Newton's formulas for the sum of the powers of the roots of a polynomial equation.

Lemma (Newton). *If $\lambda_1, \lambda_2, \cdots, \lambda_n$ are the roots of Eq. (9.57) where each root is repeated as many times as its multiplicity, such that*

$$s_k := \lambda_1^k + \lambda_2^k + \cdots + \lambda_n^k, \quad (k = 1, 2, \cdots, n)$$

then s_k are given by

$$s_k + p_1 s_{k-1} + \cdots + p_{k-1}s_1 = -k\,p_k \qquad (9.60)$$

To prove the formulas, we have identically

$$f(\lambda) = (\lambda - \lambda_1)(\lambda - \lambda_2) \cdots (\lambda - \lambda_n)$$

By logarithmic differentiation

$$\begin{aligned} f'(\lambda) &= \frac{f(\lambda)}{\lambda - \lambda_1} + \frac{f(\lambda)}{\lambda - \lambda_2} + \cdots + \frac{f(\lambda)}{\lambda - \lambda_n} \\ &= n\,\lambda^{n-1} + (n-1)\,p_1\lambda^{n-2} + \cdots + p_{n-1} \end{aligned} \tag{9.61}$$

Now, by *synthetic division* scheme for $f(\lambda)/(\lambda - \lambda_1)$

$$\begin{array}{c|cccc} \lambda_1 & p_1 & p_2 & \cdots & p_n \\ & \lambda_1 & q_1\lambda_1 & \cdots & q_{n-1}\lambda_1 \\ \hline & p_1 + \lambda_1 & p_2 + q_1\lambda_1 & \cdots & p_n + q_{n-1}\lambda_1 \\ & = q_1 & = q_2 & & = q_n \end{array}$$

Hence,
$$\frac{f(\lambda)}{\lambda - \lambda_1} = \lambda^{n-1} + q_1\lambda^{n-2} + \cdots + q_{n-1}$$

In the scheme, we note that $q_n = p_n + q_{n-1}\lambda_1 = p_n + (p_{n-1} + q_{n-2}\lambda_1) = \cdots = p_n + p_{n-1}\lambda_1 + p_{n-2}\lambda_1^2 + \cdots + \lambda_1^n = 0$. This relation is confirmed by the fact that $f(\lambda)$ is exactly divisible by λ_1, so that the remainder $q_n = 0$.

We similarly obtain expressions for $f(\lambda)/(\lambda - \lambda_2), \cdots, f(\lambda)/(\lambda - \lambda_n)$. Hence inserting these expressions in Eq. (9.61), we obtain

$$\begin{aligned} f'(\lambda) &= n\,\lambda^{n-1} + Q_1\lambda^{n-2} + \cdots + Q_{n-1} \\ &= n\,\lambda^{n-1} + (n-1)\,p_1\lambda^{n-2} + \cdots + p_{n-1} \end{aligned}$$

where

$$\begin{aligned} Q_1 &= n\,p_1 + \sum \lambda_1 \\ Q_2 &= n\,p_2 + \sum q_1\lambda_1 \\ &\cdots\cdots\cdots\cdots \\ Q_k &= n\,p_k + \sum q_{k-1}\lambda_1 \\ &\cdots\cdots\cdots\cdots \\ Q_{n-1} &= n\,p_{n-1} + \sum q_{n-2}\lambda_1 \end{aligned}$$

where the summations are taken over all the roots. Comparing the coefficients of $\lambda^{n-(k+1)}$, $(k = 1, 2, \cdots, n-1)$ we obtain

$$Q_k = n\,p_k + \sum q_{k-1}\lambda_1 = (n-k)\,p_k$$

$$\text{or} \qquad \sum q_{k-1}\lambda_1 = -k\,p_k \qquad (9.62)$$

where $q_0 = 1$. Using the definition of $q_1, q_2, \cdots, q_{n-1}$, we obtain the Eq. (9.60). The result is true for $k = n$. For, λ_1 satisfies Eq. (9.57) and we have

$$\lambda_1^n + p_1\lambda_1^{n-1} + \cdots + p_{n-1}\lambda_1 + p_n = 0$$

Similarly for $\lambda_2, \lambda_3, \cdots, \lambda_n$. Adding these equations, we have

$$s_n + p_1 s_{n-1} + \cdots + p_{n-1}s_1 = -n\,p_n$$

$\qquad\qquad\qquad\qquad\qquad\qquad\qquad\qquad\qquad\qquad\qquad\qquad\qquad$ $\square$

Returning to the proof of Eqs. (9.58) and (9.59), Eq. (9.62) can be written as

$$p_k = -\frac{1}{k}\sum q_{k-1}\lambda_1, \quad (k = 1, 2, \cdots, n) \qquad (9.63)$$

Hence in Eq. (9.63), for $k = 1$

$$p_1 = -\sum \lambda_1 = -\operatorname{tr} B_1 \qquad (9.64)$$

Again since $\lambda_1^k, \lambda_2^k, \cdots, \lambda_n^k$ are eigenvalues of the matrix A^k (Theorem 5, Sect. 9.1), we have as before

$$\sum \lambda_1^k = \operatorname{tr} A^k$$

Hence for $k = 2$ in Eq. (9.63)

$$p_2 = -\frac{1}{2}\sum(\lambda_1^2 + p_1\lambda_1) = -\frac{1}{2}\operatorname{tr}(A^2 + p_1 A)$$
$$= -\frac{1}{2}\operatorname{tr}[A(A + p_1 I)] = -\frac{1}{2}\operatorname{tr}[A(B_1 + p_1 I)]$$
$$= -\frac{1}{2}\operatorname{tr} B_2$$

Results for $k = 3, 4, \cdots, n$ similarly follow.

Leverrier's method is coded in Fortran in the following subroutine. The naming of variables and arrays is obvious.

SUBROUTINE LEVERRIER(A,n,p)
! A= Given matrix.   (Input)
! n= Order of the matrix A.   (Input)

! p= n–vector yielding the coefficients p(1), p(2), ⋯ , p(n)
! of the characteristic equation.   (Output)
! The subroutine uses the FORTRAN FUNCTION MATMUL,
! which multiplies two given matrices.
!***
REAL :: A(n,n), p(n), B(n,n)
B=A
p(1)=0.0
DO i=1,n
p(1)=p(1)+B(i,i)
END DO
p(1)=−p(1)

DO k=2,n
    DO i=1,n
    B(i,i)=B(i,i)+p(k−1)
    END DO
    B=MATMUL(A,B)
    p(k)=0.0
    DO i=1,n
    p(k)=p(k)+B(i,i)
    END DO
    p(k)=−p(k)/k

END DO
RETURN
END SUBROUTINE LEVERRIER

The complexity of the algorithm is $O(n^5)$ as there are $O(n^3)$ multiplications and additions involved in the matrix multiplication of A and B. Hence, the method is not efficient for large values of n.

Exercises

Find the characteristic equation for the following matrices using subroutine LEVERRIER:

1.
$$A = \begin{bmatrix} 1 & 2 & 3 \\ 4 & 5 & 6 \\ 7 & 8 & 9 \end{bmatrix}$$
$[\lambda^3 - 15\lambda^2 - 18\lambda = 0]$.

2.
$$A = \begin{bmatrix} 1 & 2 & -4 \\ 3 & -1 & 2 \\ 1 & 0 & -1 \end{bmatrix}$$
$[\lambda^3 + \lambda^2 - 3\lambda - 7 = 0]$.

3.

$$A = \begin{bmatrix} -4 & -3 & 1 & 1 \\ 2 & 0 & 4 & -1 \\ 1 & 1 & 2 & -2 \\ 1 & 1 & -1 & -1 \end{bmatrix}$$

$[\lambda^4 + 3\lambda - 7\lambda^2 - 24\lambda - 15 = 0].$

4.

$$A = \begin{bmatrix} 2 & -4 & 3 & -2 \\ 1 & 1 & -5 & -3 \\ 1 & 1 & -2 & -2 \\ 0 & 1 & -1 & -1 \end{bmatrix}$$

$[\lambda^4 + 2\lambda^2 - 23\lambda - 13 = 0].$

Chapter 10
Fast Fourier Transform

If f is an integrable function of a variable $t \in (-\infty, \infty)$, then its **Fourier integral** or **Fourier transform** may be defined as the complex function F of a variable $\omega \in (-\infty, \infty)$ where

$$F(\omega) := \int_{-\infty}^{\infty} f(t)\, e^{-i\omega t}\, dt, \quad i = \sqrt{-1} \tag{10.1}$$

The most important fact about definition (10.1) is that if it is viewed as an integral equation, its solution under very general conditions is

$$f(t) = \frac{1}{2\pi} \int_{-\infty}^{\infty} F(\omega)\, e^{i\omega t}\, d\omega \tag{10.2}$$

Equation (10.2) is called the **inverse transform** of the **direct transform** of (10.1), and the two constitute the Fourier transform pair.

In practice, one often interprets t as *time* and ω as *angular frequency* equal to $2\pi f$ where f is the frequency. The pair of Eqs. (10.1) and (10.2) assert that one may investigate a function in the frequency domain ω instead of real time t, without any ambiguity. Because of this fact, the Fourier transform finds applications in many diverse subjects such as Antennas, Optics, Acoustics, Geophysics, Linear Systems governed by ODEs, Boundary Value Problems of linear PDEs and Quantum Physics. It is easily recognised that the *characteristic function* of a random variable in Probability Theory is in fact the Fourier transform of the probability density function. The Fourier transform is therefore *ubiquitous* like the computer, present everywhere at the same time.

© Springer Nature Singapore Pte Ltd. 2019
S. K. Bose, *Numerical Methods of Mathematics Implemented in Fortran*, Forum for Interdisciplinary Mathematics,
https://doi.org/10.1007/978-981-13-7114-1_10

If the expression for $F(\omega)$ given by Eq. (10.1) is substituted in Eq. (10.2), one has the *Fourier's Integral Theorem*:

$$\frac{1}{2\pi} \int_{-\infty}^{\infty} d\omega \int_{-\infty}^{\infty} f(t') e^{-i\omega(t'-t)} dt = f(t) \tag{10.3}$$

The proof of this fundamentally important theorem has been much discussed in the theory of Fourier integrals, where a variety of *sufficient conditions* on f have been proposed for the validity of the theorem. For us, it is sufficient to state that the Fourier's integral theorem (10.3) holds *if f is continuous in* $(-\infty, \infty)$ *such that the interval can be broken up into a finite number of partial intervals in each of which f is monotonic and that $\int_{-\infty}^{\infty} |f(t)|\, dt$ exists finitely.* More generally, discontinuities in f may be allowed *provided f is a function of bounded variation.* A function of *bounded variation* is a continuous function except for a finite number of jump discontinuities or even infinite discontinuities that can be excluded by arbitrarily small intervals. In such a case theorem (10.3) also holds at points of continuity if f is piecewise monotonic in a finite number of partial intervals such that $\displaystyle\int_{-\infty}^{\infty} |f(t)|\, dt < \infty$.

Assuming the above facts we are interested in this chapter to compute the integrals appearing in Eqs. (10.1) and (10.2). For this purpose, we have to resort to discretisation. The advanced methods of numerical quadrature are inapplicable in view of rapid oscillation of the exponentials for large values of the integrands.

10.1   Discrete Fourier Transform

Let the infinite range of the integral (10.1) be replaced by a sufficiently large interval beyond which $|f(t)|$ is negligibly small. Let this finite interval be divided into a very large number of small subintervals by a set of N *equally spaced points* $t_0, t_1, \cdots, t_{N-1}$. By the composite rectangle rule (Sect. 5.2.3, Chap. 5), we can write

$$F(\omega) = \Delta t \sum_{k=0}^{N-1} f(t_k) e^{-i\omega t_k}$$

where Δt is the length of a subinterval. If we are interested in computing $F(\omega)$ at N *equidistant points* $\omega_0, \omega_1, \cdots, \omega_{N-1}$, then the above equation takes the discretised form

$$F(\omega_n) = \Delta t \sum_{k=0}^{N-1} f(t_k) e^{-i\omega_n t_k}, \quad n = 0, 1, \cdots, N-1 \tag{10.4}$$

Equation (10.4) is an approximate discretised version of Eq. (10.1).

An exact inversion of Eq. (10.4) is possible. We have, summing Eq. (10.4),

$$\sum_{n=0}^{N-1} F(\omega_n) \, e^{i\omega_n t_l} = \Delta t \sum_{k=0}^{N-1} f(t_k) \sum_{n=0}^{N-1} e^{i\omega_n (t_l - t_k)}$$

If $l = k$,

$$\sum_{n=0}^{N-1} e^{i\omega_n (t_l - t_k)} = \sum_{n=0}^{N-1} 1 = N.$$

If $l \neq k$, the G.P. series yields

$$\sum_{n=0}^{N-1} e^{i\omega_n (t_l - t_k)} = e^{i\omega_0 (t_l - t_k)} \frac{e^{i\Delta\omega(t_l - t_k)N} - 1}{e^{i\Delta\omega(t_l - t_k)} - 1}$$

Now the exponent

$$\Delta\omega(t_l - t_k) \, N = \Delta\omega(l - k)\Delta t \, N$$

$$= 2\pi(l - k)$$

if $\Delta\omega = \dfrac{2\pi}{N\Delta t}$. The G.P. series then vanishes for $l \neq k$ as $e^{2\pi i(l-k)} = 1$ and

$$\sum_{n=0}^{N-1} F(\omega_n) \, e^{i\omega_n t_l} = N \, \Delta t \, f(t_l)$$

or

$$f(t_k) = \frac{\Delta\omega}{2\pi} \sum_{n=0}^{N-1} F(\omega_n) \, e^{i\omega_n t_k}, \quad k = 0, 1, \cdots, N - 1 \quad (10.5)$$

Equation (10.5) is apparently the discretised version of Eq. (10.2) by the composite rectangle rule.

In the formulas (10.4) and (10.5) $\omega_n = n \, \Delta\omega$ and $t_k = k \, \Delta t$. Hence

$$\omega_n t_k = nk \, \Delta\omega \, \Delta t = \frac{2\pi nk}{N}$$

Writing $f(k)$ for $f(t_k)$ and $F(n)$ for $F(\omega_n)$, we obtain the pair

$$F(n) = \Delta t \sum_{k=0}^{N-1} f(k) \, e^{-i \, 2\pi nk/N}, \quad n = 0, 1, \cdots, N - 1 \quad (10.6)$$

$$f(k) = \frac{\Delta\omega}{2\pi} \sum_{n=0}^{N-1} F(n) \, e^{i \, 2\pi nk/N}, \quad k = 0, 1, \cdots, N - 1 \quad (10.7)$$

Equations (10.6) and (10.7) are called **Discrete Fourier Transform** or simply **DFT** pair, where $\Delta\omega = 2\pi/(N\,\Delta t)$.

Equations (10.6) and (10.7) are of the same generic form as far as computation of the two series is concerned. For taking the complex conjugate of Eq. (10.7), we obtain

$$[f(k)]^* = \frac{\Delta\omega}{2\pi} \sum_{n=0}^{N-1} [F(n)]^*\, e^{-i2\pi nk/N}$$

Hence suppressing constant factors Δt or $\Delta\omega/2\pi$, we consider the computation of a series of the form

$$X(n) = \sum_{k=0}^{N-1} x_0(k)\, e^{-i2\pi nk/N}, \quad n = 0,\ 1,\ \cdots,\ N-1 \tag{10.8}$$

where the sequence $\{x_0(k)\}$ and $\{X(n)\}$ are in general complex.

10.2 Fast Fourier Transform

We write the discrete Fourier transform, Eq. (10.8) in the form

$$X_n = \sum_{k=0}^{N-1} x_0(k)\, W^{nk}, \quad n = 0,\ 1,\ \cdots,\ N-1 \tag{10.9}$$

where

$$W = e^{-i2\pi/N} \tag{10.10}$$

Here we note that in the powers of W

$$W^{nk} = e^{-i2\pi nk/N} = W^{[nk \bmod N]} \tag{10.11}$$

where $nk \bmod N$ is the remainder in dividing nk by N. The right-hand side relation in Eq. (10.10) holds because the complex exponential becomes unity for the integer part of nk/N. The system of Eq. (10.9) is of the form

$$\begin{bmatrix} X(0) \\ X(1) \\ \vdots \\ X(N-1) \end{bmatrix} = \begin{bmatrix} W^0 & W^0 & \cdots & W^0 \\ W^0 & W^1 & \cdots & W^{N-1} \\ \cdots & \cdots & \cdots & \cdots \\ W^0 & W^{N-1} & \cdots & W^{2(N-1)} \end{bmatrix} \begin{bmatrix} x_0(0) \\ x_0(1) \\ \vdots \\ x_0(N-1) \end{bmatrix} \tag{10.12}$$

where the high powers of W can be reduced to less than N by using property (10.11). It is observed that to compute an element $X(n)$ we need N complex multiplications and $N-1$ complex additions. Hence in total, we need $N \times N$ complex multiplications and $N \times (N-1)$ complex additions. The number of operations is thus seen to be prohibitively large as N is essentially taken very large because of the oscillatory nature of the exponentials. **Fast Fourier Transform** or **FFT**, in short, are computational techniques to reduce this computational load. These techniques came to prominence by the work of J.W. Cooley and J.W. Tukey in 1965, which is rated amongst the top ten algorithms discovered in the twentieth century. Since then several alternative FFT's have been proposed in the literature. A survey of the history of the subject reveals that in fact simplifying techniques were known even earlier (see the texts of Brigham (1974), Press et al. (1996) and Conte and de Boor (1984) cited in the Bibliography). We, however, restrict in this section to the original method of Cooley and Tukey, because of its relative simplicity.

James W. Cooley (1926–2016), worked at the IBM Corporation U.S.A. Upon retirement from IBM in 1991, he served on the faculty of the computer engineering program at University of Rhode Island, Kingston, U.S.A.

John Wilder Tukey (1915–2000), initially worked in Topology for his Ph.D. degree in 1939. During World War II years he worked with statisticians in the Fire Control Research Office and later when the war ended, he was offered a position in statistics within the mathematics department at Princeton University, U.S.A. He published many papers in statistics. The Cooley–Tukey algorithm was discovered while working as a member of U.S. President's Science Advisory Committee. He coined the term 'software'.

In the Cooley–Tukey framework, N is taken as some integral power of 2 say $N = 2^\nu$. For this choice of N, the $N \times N$ square matrix formed by the powers of W in Eq. (10.11) can be factorised into ν *sparse* matrices, with a permutation of the rows that can be deciphered provided the indices k and n are represented by *binary numbers* 0 and 1! Operations with sparse matrices dramatically reduce the number of arithmetic operations in Eq. (10.12).

In order to unveil the above-stated features consider the simple case $N = 2^2 = 4$. The indices k and n in binary digits 0 and 1 are then

$$k, n = 0, \ 1, \ 2, \ 3 = 00, \ 01, \ 10, \ 11 \ \text{(binary)}$$

The first two representations are obvious. The binary representation of 2 and 3 follow from their polynomial representation as in the case of decimals:

$$2 = 1 \times 2^1 + 0 \times 2^0 = 10 \ \text{(binary)}$$

$$3 = 1 \times 2^1 + 1 \times 2^0 = 11 \ \text{(binary)}$$

In general, therefore, we can write

$$k = 2k_1 + k_0 = (k_1, k_0) \ \text{binary}, \quad n = 2n_1 + n_0 = (n_1, n_0) \ \text{binary}$$

where k_0, k_1, n_0 and n_1 can take on values of 0 and 1 only. Thus, Eq. (10.9) becomes

$$X(n_1, n_0) = \sum_{k_0=0}^{1} \sum_{k_1=0}^{1} x_0(k_1, k_0) \, W^{(2n_1+n_0)(2k_1+k_0)} \tag{10.13}$$

The W^p term ($p = nk$) can be factorised as

$$W^{(2n_1+n_0)(2k_1+k_0)} = \left[W^{4n_1k_1} \right] W^{2n_0k_1} \, W^{(2n_1+n_0)k_0}$$

$$= W^{2n_0k_1} \, W^{(2n_1+n_0)k_0}$$

since $W^4 = 1$ from Eq. (10.10) for the case $N = 4$. Eq. (10.13) therefore becomes

$$X(n_1, n_0) = \sum_{k_0=0}^{1} \left[\sum_{k_1=0}^{1} x_0(k_1, k_0) \, W^{2n_0k_1} \right] W^{(2n_1+n_0)k_0}$$

This equation can be put as the two-stage recursion

$$x_1(n_0, k_0) = \sum_{k_1=0}^{1} x_0(k_1, k_0) \, W^{2n_0k_1}$$

$$x_2(n_0, n_1) = \sum_{k_0=0}^{1} x_1(n_0, k_0) \, W^{(2n_1+n_0)k_0} \tag{10.14}$$

$$X(n_1, n_0) = x_2(n_0, n_1)$$

In the defining equation for $x_1(n_0, k_0)$, k_0 is written as the first bit from the right as it occupies this position in $x_0(k_1, k_0)$ on the right-hand side. Similarly, n_0 retains its position in the definition of $x_2(n_0, n_1)$. The set of Eq. (10.14) represent the Cooley–Tukey FFT for $N = 4$. Written in matrix notation it reads

$$\begin{bmatrix} x_1(0, 0) \\ x_1(0, 1) \\ x_1(1, 0) \\ x_1(1, 1) \end{bmatrix} = \begin{bmatrix} 1 & 0 & W^0 & 0 \\ 0 & 1 & 0 & W^0 \\ 1 & 0 & W^2 & 0 \\ 0 & 1 & 0 & W^2 \end{bmatrix} \begin{bmatrix} x_0(0, 0) \\ x_0(0, 1) \\ x_0(1, 0) \\ x_0(1, 1) \end{bmatrix} \tag{10.15}$$

and

$$\begin{bmatrix} x_2(0, 0) \\ x_2(0, 1) \\ x_2(1, 0) \\ x_2(1, 1) \end{bmatrix} = \begin{bmatrix} 1 & W^0 & 0 & 0 \\ 1 & W^2 & 0 & 0 \\ 0 & 0 & 1 & W^1 \\ 0 & 0 & 1 & W^3 \end{bmatrix} \begin{bmatrix} x_1(0, 0) \\ x_1(0, 1) \\ x_1(1, 0) \\ x_1(1, 1) \end{bmatrix} \tag{10.16}$$

Eliminating the array x_1 from Eqs. (10.16) and (10.15), the solution for x_2 in terms of the x_0 is obtained as a product of the two 4×4 sparse matrices occurring in Eqs. (10.15) and (10.16). The factorisation of the matrix appearing in Eq. (10.12) for the case $N = 4$ into these two sparse matrices with permutation of second and third rows follows from the third equation of (10.14). This fact can be verified independently by actual multiplication of the two sparse matrices. The reader is urged to verify this peculiar factorisation.

The expression for the components of x_1 and x_2 have only two terms, which occur in pairs due to the fact that $W^2 = -W^0$ for $N = 4$. In Eq. (10.15), for instance, we have

$$x_1(0, 0) = x_0(0, 0) + W^0 \times x_0(1, 0), \quad x_1(1, 0) = x_0(0, 0) - W^0 \times x_0(1, 0)$$

$$x_1(0, 1) = x_0(0, 1) + W^0 \times x_0(1, 1), \quad x_1(1, 1) = x_0(0, 1) - W^0 \times x_0(0, 1)$$

Similarly, in Eq. (10.16), $x_2(0, 0)$, $x_2(0, 1)$ and $x_2(1, 0)$, $x_2(1, 1)$ form dual pairs. Dual pair computation further reduces the computational load.

We next investigate the case $N = 2^3 = 8$. The indices k and n now take up the values

$$k, n = \quad 0, \quad 1, \quad 2, \quad 3, \quad 4, \quad 5, \quad 6, \quad 7$$
$$= 000, \ 001, \ 010, \ 011, \ 100, \ 101, \ 110, \ 111 \ (\text{binary})$$

Thus, one can write the indices as

$$k = 2^2 k_2 + 2k_1 + k_0, \quad n = 2^2 n_2 + 2n_1 + n_2$$

where k_i, n_i are 0 or 1. Thus, Eq. (10.9) becomes

$$X(n_2, n_1, n_0) = \sum_{k_0=0}^{1} \sum_{k_1=0}^{1} \sum_{k_2=0}^{1} x_0(k_2, k_1, k_0) \, W^{(4n_2+2n_1+n_0)(4k_2+2k_1+k_0)} \tag{10.17}$$

The W^p, $(p = nk)$ factorises into

$$\left[W^{8(2n_2+n_1)k_2} \right] W^{4n_0 k_2} \left[W^{8n_2 k_1} \right] W^{(2n_1+n_0)2k_1} W^{(4n_2+2n_1+n_0)k_0}$$

Since $W^8 = 1$ for this case, we get for Eq. (10.17)

$$X(n_2, n_1, n_0) = \sum_{k_0=0}^{1} \sum_{k_1=0}^{1} \sum_{k_2=0}^{1} x_0(k_2, k_1, k_0) \, W^{4n_0 k_2} \, W^{(2n_1+n_0)2k_1} \, W^{(4n_2+2n_1+n_0)k_0}$$

which reduces to the three-stage recursion

$$x_1(n_0, k_1, k_0) = \sum_{k_2=0}^{1} x_0(k_2, k_1, k_0)\, W^{4n_0 k_2}$$

$$x_2(n_0, n_1, k_0) = \sum_{k_1=0}^{1} x_1(n_0, k_1, k_0)\, W^{(2n_1+n_0)\, 2k_1}$$

$$(10.18)$$

$$x_3(n_0, n_1, n_2) = \sum_{k_0=0}^{1} x_2(n_0, n_1, k_0)\, W^{(4n_2+2n_1+n_0)\, k_0}$$

$$X(n_2, n_1, n_0) = x_3(n_0, n_1, n_2)$$

(In the above definitions of arrays x_1, x_2, x_3 the bit positions are identified as in the case of $N = 4$.) The set of Eq. (10.18) can be put in the form of matrix factors as in the case of $N = 4$, but this aspect is not important from computational point of view.

The results (10.14) and (10.18) can be generalised for the case $N = 2^\nu$. In this case

$$k = 2^{\nu-1}k_{\nu-1} + 2^{\nu-2}k_{\nu-2} + \cdots + k_0$$

$$n = 2^{\nu-1}n_{\nu-1} + 2^{\nu-2}n_{\nu-2} + \cdots + n_0$$

and Eq. (10.9) can then be put in the recursive form

$$x_1(n_0, k_{\nu-2}, \cdots, k_0) = \sum_{k_{\nu-1}=0}^{1} x_0(k_{\nu-1}, k_{\nu-2}, \cdots, k_0)\, W^{2^{\nu-1}n_0 k_{\nu-1}}$$

$$x_2(n_0, n_1, k_{\nu-3}, \cdots, k_0) = \sum_{k_{\nu-2}=0}^{1} x_1(n_0, k_{\nu-2}, \cdots, k_0)\, W^{(2n_1+n_0)\, 2^{\nu-2}k_{\nu-2}}$$

$$\cdots\cdots\cdots\cdots\cdots\cdots\cdots\cdots\cdots\cdots\cdots\cdots\cdots\cdots\cdots\cdots$$

$$x_\nu(n_0, n_1, \cdots, n_{\nu-1}) = \sum_{k_0=0}^{1} x_{\nu-1}(n_0, n_1, \cdots, k_0)\, W^{(2^{\nu-1}n_{\nu-1}+2^{\nu-2}n_{\nu-2}+\cdots+n_0)\, k_0}$$

$$X(n_{\nu-1}, n_{\nu-2}, \cdots, n_0) = x_\nu(n_0, n_1, \cdots, n_{\nu-1})$$

$$(10.19)$$

The proof is a simple generalisation of the cases for $\nu = 2$ and 3. It uses the property that $W^{2^\nu} = 1$.

10.2.1   Signal Flow Graph of FFT

A signal flow graph is a network of connected arrows, each arrow joining two *nodes* representing some kind of *flow* of some amount or *weight*. For the case $N = 2^2 = 4$, the two-stage iterations (10.14) or Eqs. (10.15) and (10.16) can be represented by

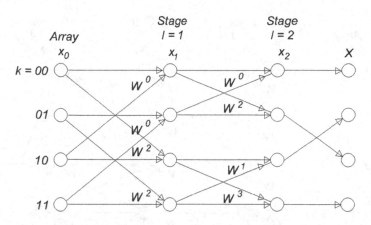

Fig. 10.1 FFT signal flow graph for $N = 4$

the signal flow graph shown in Fig. 10.1. The trivial weight factor 1 of an arrow is not shown in the figure.

In a similar manner Eq. (10.18) can also be represented by a graph for the case $N = 2^3 = 8$. It is shown in Fig. 10.2.

In these figures, we easily identify that the network consists of *dual nodes* that always originate from the same pair of nodes. In Fig. 10.1 for stage $l = 1$, the pairs are $(0,2)$ and $(1,3)$ while for stage $l = 2$ these are $(0,1)$ and $(2,3)$. These pairs are, respectively, separated by 2 and 1. In Fig. 10.2, on the other hand, these are $(0,4)$, $(1,5)$, $(2,6)$, $(3,7)$ for $l = 1$, $(0,2)$, $(1,3)$, $(4,6)$, $(5,7)$ for $l = 2$ and $(0,1)$, $(2,3)$, $(4,5)$, $(6,7)$ for $l = 3$. The separation of the nodes is therefore 4, 2 and 1, respectively. We therefore conclude that in general the separation of the dual nodes is given by the formula $N/2^l$.

The weighting factor involved in the dual node computation is equal in magnitude but opposite in sign. In Fig. 10.1 for $N = 4$, we have in fact $W^2 = -W^0$ and $W^3 = -W^1$. Similarly, in Fig. 10.2 for $N = 8$, $W^4 = -W^0$, $W^6 = -W^2$, $W^5 = -W^1$, and $W^7 = -W^3$. In general if the weighting factor is W^p, that at the dual node is $W^{p+N/2} = -W^p$. The computation at a node is therefore given by

$$x_l(k) = x_{l-1} + W^p x_{l-1}(k + N/2^l) \tag{10.20}$$

and that at its dual by

$$x_l(k + N/2^l) = x_{l-1} - W^p x_{l-1}(k + N/2^L) \tag{10.21}$$

The last term in Eq. (10.21) having been already computed in Eq. (10.20), only one operation of addition is required in the dual node computation. In the computation along an array, the dual nodes that are down by $N/2^l$ places can therefore be *skipped*,

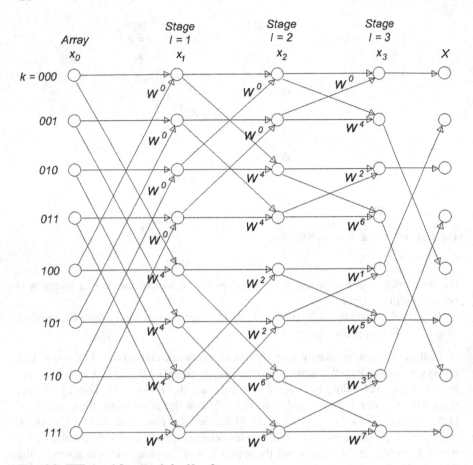

Fig. 10.2 FFT signal flow graph for $N = 8$

if use is made of Eq. (10.21). Evidently, the skipping of nodes is to be stopped when a radix index $k > N - 1$ is reached.

The power p in the weighting factor W^p appearing in Eqs. (10.20) and (10.21) are related to the node k in a peculiar manner. It follows the following rule: (a) Represent k in binary digits, ν bits long as in Figs. 10.1 and 10.2. (b) Slide or scale this binary number $\nu - l$ bits to the right, filling in the newly opened bit positions on the left by zeros (keeping the number ν bits long). And (c) Reverse the order of the bits. This *bit-reversed number* is the number p. As an illustration, consider Fig. 10.1 ($\nu = 2$) with $k = 2 = 10$ (binary). If $l = 1$, $\nu - l = 1$, so following the operation (b) we get the number 01 and bit-reversing following (c) we get 10=2 (decimal). Hence $p = 2$ as in the figure. If $l = 2$, $\nu - l = 0$, so by (b) we get 10 and by (c) 01=1 (decimal) or $p = 1$. Similarly, in Fig. 10.2 ($\nu = 3$) consider $k = 5 = 101$ (binary). For $l = 1$, $\nu - l = 2$; $101 \rightarrow 001$ [by (b)] $\rightarrow 100$ [by (c)] or

$p = 4$. For $l = 2$, $\nu - l = 1$; $101 \to 010$ [by(b)] $\to 010$ [by(c)] or $p = 2$. And for $l = 3$, $\nu - l = 0$; $101 \to 101$ [by (b)] $\to 101$ [by (c)] or $p = 5$.

Finally, we recall that bit-reversal is also required in identifying the elements of the array X. This is implied in the last equations of (10.14) and (10.18) as well as of Eq. (10.19).

10.2.2 Bit-Reversal

For bit-reversal of a binary integer k, we require a method for identifying the bits of k. If k is a three-bit integer as in the case of $\nu = 3$, $k = b_3 b_2 b_1$ (binary) $= b_3 \times 2^2 + b_2 \times 2^1 + b_1$ (decimal). Then $[k/2]$ (integer part of $k/2$) $= b_3 \times 2 + b_2$, and therefore $k - 2[k/2] = b_1$. This identifies the bit b_1, whether it is 0 or 1. Similarly, the difference $b_3 b_2 - 2[b_3 b_2/2] = b_2$ identifies the bit b_2. Finally, we have $b_3 - 2[b_3/2] = b_3$ identifying b_3. The bit-reversed number is $b_1 b_2 b_3 = b_1 \times 2^2 + b_2 \times 2 + b_3$ (decimal). This latter polynomial can be computed as a nested multiplication $b_3 + 2 \times (b_2 + 2 \times b_1)$.

10.2.3 FFT Subroutine

With the preparation of Sects. 10.2, 10.2.1 and 10.2.2, we are in a position to write down the subroutine FFT for the computation of (10.9), which represents the direct DFT (10.6). The inverse DFT (10.7) can be computed from this very subroutine by using the trivial procedure of complex conjugation described in the paragraph following this equation. The input of the program is the complex array x_0 and the output complex array is overwritten on x_0. The subroutine uses an *integer function subprogram* BITREV which reverses the bits of a given integer k.

```
SUBROUTINE FFT(X,N,nu)
! X= Complex input array x.₀              (Input)
! The output array is returned in X.      (Output)
! N= Dimension of array X.                (Input)
! nu= Exponent such that N=2 raised to the power nu.    (Input)
!*************************************************************
COMPLEX :: X(N), z
INTEGER :: BITREV
k=0; n2=N/2; nu1=nu−1
DO L=1,nu
10 DO i=1,n2
```

```
p=BITREV(k/2**nu1,nu)
k1=k+1; k1n2=k1+n2
z=CEXP(CMPLX(0.0,-6.283185)*p/N)*X(k1n2)
X(k1n2)=X(k1)-z; X(k1)=X(k1)+z
k=k+1
END DO
k=k+n2
IF(k<N) GOTO 10
k=0; nu1=nu1-1; n2=n2/2
END DO
DO k=1,N
i=BITREV(k-1,nu)+1
IF(i>k) THEN
z=X(k); X(k)=X(i); X(i)=z
END IF
END DO
RETURN
END SUBROUTINE FFT
!************************

INTEGER FUNCTION BITREV(k,nu)
k1=k
BITREV=0
DO i=1,nu
k2=k1/2
BITREV=(k1-2*k2)+2*BITREV
k1=k2
END DO
RETURN
END FUNCTION BITREV
```

One can easily find the complexity of the Cooley–Tukey algorithm from the above subroutine. If the operations involved in the complex exponential and the indices are ignored, there is just one complex multiplication involved in the computation of z. Since it is looped in $\nu \times N/2$ times (over L and i), the number of complex multiplications is $\frac{1}{2} N \log_2 N$. In the next step of the program, two subtractions/additions are encountered. In total therefore, there are $2 \times \nu \times N/2 = \nu N = N \log_2 N$ addition/subtraction in the algorithm. Thus, there is a dramatic reduction from N^2 complex multiplications and $N \times (N-1)$ complex additions involved in straightforward computation of Eq. (10.12).

Example 1. Compute the Fourier cosine transform

$$\int_0^\infty \frac{\cos \omega t \, dt}{t^2 + \beta^2}, \quad (\beta > 0)$$

for $\beta = 1$ (say), and $\omega \approx 0, 0.5, 1.0, \cdots, 4.5, 5.0$ by using subroutines FFT and BITREV. Compare with the exact value $\frac{\pi}{2\beta} e^{-\beta\omega}$.

Solution. The function $1/(t^2 + \beta^2)$ does not decay very fast and so we take the upper limit of integration up to 100. In order to take the steps sufficiently small we take $\nu = 14$, and $N = 2^\nu = 16384$. The integral is equivalent to

$$Re \int_0^\infty \frac{e^{-i\omega t}}{t^2 + \beta^2} dt$$

which can be computed by writing a program as follows:

```
COMPLEX :: X(16400)
beta=1.; pi=3.141593; nu=14; N=2**nu
upper_limit=100.; deltat=upper_limit/N
DO k=1,N
tk=(k−1)*deltat
X(k)=CMPLX(1./(tk**2+beta**2),0.0)
PRINT*, k, X(k)
END DO
CALL FFT(X,N,nu)
DO m=0,10; om=0.5*m
DO nn=1,N
omn=2*pi*(nn−1)/(N*deltat)
fomn=pi/(2*beta)*EXP(−beta*omn)          !exact value of the integral
tol=0.03
IF(abs(omn−om)<tol) THEN
PRINT*, omn, REAL(deltat*X(nn)), fomn; EXIT
END IF
END DO; END DO
END
```

Appending the two subroutines FFT and BITREV, the computed and exact values of the integral are obtained as

ω_n	Computed	Exact
0	1.56385	1.57080
0.50265	0.95325	0.95021
1.00531	0.57785	0.57480
1.50797	0.35076	0.34771
2.01062	0.21339	0.21034
2.51327	0.13029	0.12724
3.01593	0.08002	0.07697
3.51858	0.04961	0.04656
4.02124	0.03122	0.02817
4.52389	0.02009	0.01704
5.02655	0.01336	0.01031

tol is set 0.03 in order to obtain all of the required values in the table; some are missed if it is set 0.01 or 0.02.

Note the degree of agreement of the two results. For better agreement the upper limit of integration needs to be raised and the value of ν increased. In order to avoid roundoff errors, double precision programming may be adopted.   □

Exercises

1. Compute the following Fourier integrals using subroutines FFT and BITREV:

(i) $\displaystyle\int_0^\infty e^{-\beta t} \cos\omega t\, dt$ $\left(= \dfrac{\beta}{\omega^2 + \beta^2} \right)$

(ii) $\displaystyle\int_0^\infty \frac{\cos\omega t\, dt}{(t^2 + \beta^2)^2}$ $\left(= \dfrac{\pi}{4\beta^3} (1 + \beta\omega)\, e^{-\beta\omega} \right)$

(iii) $\displaystyle\int_0^\infty \frac{\cos\omega t\, dt}{t^4 + \beta^4}$ $\left(= \dfrac{\pi}{2\beta^3} e^{-\frac{\beta\omega}{\sqrt{2}}} \sin\left(\dfrac{\pi}{4} + \dfrac{\beta\omega}{\sqrt{2}} \right) \right)$

(iv) $\displaystyle\int_0^\infty \frac{\sin\omega t\, dt}{t(t^2 + \beta^2)}$ $\left(= \dfrac{\pi}{2\beta^2} (1 - e^{-\beta\omega}) \right)$

(v) $\displaystyle\int_0^\infty \frac{\sin\omega t\, dt}{\sinh\beta t}$ $\left(= \dfrac{\pi}{2\beta} \tanh\dfrac{\pi\omega}{2\beta} \right)$

for $\beta = 1$ (say) and $\omega \approx 0, 1, 2, 3, 4, 5$. Satisfy yourselves by also computing the exact values given in parentheses. Take $\nu = 14$ and unless otherwise stated, upper_limit $= 10$, and tol $= 0.03$.

[The computed values are:
(i) (0, 1.00026), (1.25664, 0.38801), (1.88496, 0.21993), (3.14159, 0.92301), (3.76991, 0.06604), (5.02655, 0.03838);

(ii) (0, 0.78537), (1.25664, 0.50471), (1.88496, 0.34433), (3.14159, 0.14087), (3.76991, 0.08667), (5.02655, 0.03136);

(iii) (0, 1.11069), (1.25664, 0.64282), (1.88496, 0.35400), (3.14159, 0.02319), (3.76991, −0.03298), (5.02655, −0.04154);

(iv) (upper_limit = 50, tol=0.1, write tk=(k−1)*deltat+1.e−10), (0, 0), (1.00531, 0.99445), (2.01062, 1.35739), (3.01593, 1.48922), (4.02124, 1.53649), (4.90089), 1.55163);

(v) (0, 0), (1.25664, 1.51090), (1.88496, 1.56179), (3.14159, 1.56965), (3.76991, 1.56960), (5.02655, 1.56925)].

2. (Fourier Coefficients). If as in Chap. 8, Sect. 8.6, a periodic function in $[0, 2\pi]$, is approximated by a trigonometric polynomial written in complex form

$$f(t) \approx \sum_{n=-N}^{N} c_n e^{int}$$

then as in the case of Eq. (10.2), prove that

$$c_n = \frac{1}{2\pi} \int_0^\infty f(t) e^{-int} dt \approx \frac{1}{N} \sum_{k=0}^{N-1} f(t_k) e^{-int_k}, \quad c_{-n} = c_n^*$$

where star denotes complex conjugate. Modify the program of Example 1, for computing the Fourier coefficients c_n for $n \geq 0$ in the case of:

(i) $f(t) = t$, (ii) $f(t) = \begin{cases} t, & 0 \leq t \leq \pi \\ 2\pi - t, & \pi \leq t \leq 2\pi \end{cases}$, (iii) $f(t) = e^t$

Satisfy yourselves by also computing the exact values of:

(i) $c_n = i/n$, $(n \neq 0)$, $c_0 = \pi$, (ii) $c_n = [(-1)^n - 1]/(\pi n^2)$, $(n \neq 0)$, $c_0 = \pi/2$, and (iii) $c_n = [e^\pi \sinh \pi (1 + in)]/[\pi(1 + n^2)]$.

[In the program write upper_limit = 2*pi, declare fomn as complex, set tol=0.1 and take the output as print*, omn, X(nn)/N, fomn.

(i) 3.14140, −1.92E−04 + i, −1.92E−04 + 0.5i, −1.92E−04 + 0.33333i, −1.92E−04 + 0.25i, −1.92E−04 + 0.2i (the real parts −1.92E−04 are due to accumulated errors),

(ii) 1.57080, −0.63662, 0, −0.70736, 0, −0.02546 (complete agreement with exact values),

(iii) 85.05073, 42.51721+42.53352i, 16.99710+34.02682i, 8.49039+25.52011i, 4.98763+20.01577i, 3.25549+16.35905i].

10.2.4   Canonic Forms of FFT

The Cooley–Tukey algorithm described in the preceding subsections can be arranged in other ways in order to perform the computation of Eq. (10.9). Such variants of the algorithm are, in a sense, *canonic*. One canonic variation is bit-reversal of the index k in the input data $x_0(k)$. It then transpires that in the corresponding signal flow graph, the output $X(k)$ is in the natural order. The powers p of W also turn out to be in natural order.

A distinct form of FFT is known as the **Sande–Tukey algorithm** developed in 1969. Here in contrast to the Cooley–Tukey approach, the components of n are separated instead of the components of k. For the case $\nu = 2$, for example, we have in Eq. (10.9)

$$W^{(2n_1+n_0)(2k_1+k_0)} = \left[W^{4n_1k_1} \right] W^{2n_1k_0} W^{n_0(2k_1+k_0)}$$

$$= W^{2n_1k_0} W^{n_0(2k_1+k_0)}$$

since $W^4 = 1$. Thus, Eq. (10.9) can be written as

$$X(n_1, n_0) = \sum_{k_0=0}^{1} \left[\sum_{k_1=0}^{1} x_0(k_1, k_0) W^{2n_0k_1} W^{n_0k_0} \right] W^{2n_1k_0}$$

If we introduce intermediate computational steps, then the above equation can be written as

Fig. 10.3 Sande–Tukey signal flow graph for $N = 2^2$

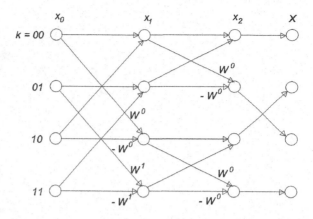

$$x_1(n_0, k_0) = \sum_{k_1=0}^{1} x_0(k_1, k_0) \, W^{2n_0 k_1} \, W^{n_0 k_0}$$

$$x_2(n_0, n_1) = \sum_{k_0=0}^{1} x_1(n_0, k_0) \, W^{2n_1 k_0} \qquad (10.22)$$

$$X(n_1, n_0) = x_2(n_0, n_1)$$

The signal flow graph describing Eq. (10.22) is shown in Fig. 10.3.

We note that in this figure, the input data is in natural order, but the output data is in bit-reversed order. The powers of W however occur in natural order. The Sande–Tukey algorithm can evidently be extended to the case of $N = 2^3$ and in general to the case $N = 2^\nu$.

A canonic variant of the above algorithm can be obtained by bit-reversing k in the input data. The output in this case is obtained in natural order. The powers of W however appear in bit-reversed order.

All the above variants have been converted into computer codes in the literature. For references see Brigham (1974).

To conclude, the assumption of $N = 2^\nu$ was generalised later by other authors to the case $N = r_1 r_2 \cdots r_m$ where $r_1, r_2, \cdots, r_m$ are certain integers, not necessarily prime numbers. Corresponding computer codes have also been written that are far more complicated without being significantly more efficient than the original one of Cooley and Tukey.

Bibliography

1. N.I. Achieser, *Theory of Approximation* (Ungar, New York, 1956)
2. E. Akin, *Object-Oriented Programming via FORTRAN 90/95* (Cambridge University Press, Cambridge, 2003)
3. K.E. Atkinson, *An Introduction to Numerical Analysis* (Wiley, New York, 1989)
4. C. de Boor, *An algorithm for numerical quadrature, in Mathematical Software* (Academic Press, New York, 1971), pp. 201–209
5. W.S. Brainerd, C.H. Goldberg, J.C. Adams, *Programmer's Guide to Fortran 90* (McGraw-Hill, New York, 1990)
6. E.O. Brigham, *The Fast Fourier Transform* (Prentice-Hall, New Jersey, 1974)
7. S.J. Chapman, *Fortran 95/2003 for Scientists and Engineers* (McGraw-Hill, New York, 2008)
8. E.W. Cheney, *Introduction to Approximation Theory* (McGraw-Hill, New York, 1966)
9. L. Collatz, *The Numerical Treatment of Differential Equations* (Springer, Berlin, 1960)
10. S.D. Conte, C. de Boor, *Elementary Numerical Analysis: An Algorithmic Approach* (McGraw-Hill, New Delhi, 1980)
11. N.I. Danilina, N.S. Dubrovskaya, O.P. Kvasha, G.L. Smirnov, *Computational Mathematics* (Mir Publishers, Moscow, 1988)
12. B.P. Demidovich, I.A. Maron, *Computational Mathematics* (Mir Publishers, Moscow, 1973)
13. C.T. Fike, *Computer Evaluation of Mathematical Functions* (Prentice-Hall, New Jersey, 1968)
14. G.E. Forsythe, M.A. Malcolm, C.B. Moler, *Computer Methods for Mathematical Computations* (Prentice-Hall, New Jersey, 1977)
15. C.E. Fröberg, *Introduction to Numerical Analysis* (Addison–Wesley, Reading, Massachusetts, 1965)
16. C.W. Gear, *Numerical Initial Value Problems in Ordinary Differential Equations* (Prentice-Hall, New Jersey, 1973)
17. S.K. Gupta, *Numerical Methods for Engineers* (New Age International, New Delhi, 1995)
18. E. Hairer, S.P. Nørsett, G. Wanner, *Solving Ordinary Differential Equations I: Nonstiff Problems* (Springer, 2008)
19. G. Hämmerlein, K.H. Hoffmann, *Numerical Mathematics* (Springer, New York, 1991)
20. F.B. Hildebrand, *Introduction to Numerical Analysis* (Dover, New York, 1987)
21. M.K. Jain, S.R.K. Iyengar, R.K. Jain, *Numerical Methods for Scientific and Engineering Computation* (New Age International, New Delhi, 2007)
22. M.K. Jain, *Numerical Solution of Differential Equations* (Wiley Eastern, New Delhi, 1984)

© Springer Nature Singapore Pte Ltd. 2019
S. K. Bose, *Numerical Methods of Mathematics Implemented in Fortran*, Forum for Interdisciplinary Mathematics,
https://doi.org/10.1007/978-981-13-7114-1

23. L.V. Kantorovich, V.I. Krylov, *Approximate Methods of Higher Analysis* (Interscience, New York, 1958)
24. D. Kincaid, W. Cheney, *Numerical Analysis: Mathematics of Scientific Computing* (American Mathematical Society, 2010)
25. G. Meinardus, *Approximation of Functions: Theory and Numerical Methods* (Springer, New York, 1967)
26. K.W. Morton, D.F. Mayers, *Numerical Solution of Partial Differential Equations* (Cambridge University Press, Cambridge U.K., 2005)
27. J. Muller, *Elementary Functions, Algorithms and Implementation* (Birkhäuser, 2003)
28. R. Piessens, E. de Doncker-Kaperga, C. Überhuber, D. Kachner, *QUADPACK: A Subroutine Package for Automatic Integration* (Springer, New York, 1983)
29. W.H. Press, B.P. Flannery, S.A. Teukolsky, W.T. Vetterling, *Numerical Recipes, The Art of Scientific Computing* (Cambridge University Press, Cambridge, 1986)
30. W.H. Press, S.A. Teukolsky, W.T. Vetterling, B.P. Flannery, *Numerical Recipes in Fortran 90* (Cambridge University Press, Cambridge, 1996)
31. A. Ralston, H.S. Wilf, *Mathematical Methods for Digital Computers, vols* (I and II) (Wiley, New York, 1967)
32. S. Ray, *A Textbook of Fortran/2003* (Narosa, New Delhi, 2009)
33. J.R. Rice, *Approximation Theory* (Addison-Wesley, London, 1964)
34. T.J. Rivlin, *Introduction to the Approximation of Functions* (Blaisdell, Walthem (Massachusetts, 1969)
35. T.J. Rivlin, *The Chebyshev Polynomials* (Wiley, New York, 1970)
36. S.S. Sastry, *Introductory Methods of Numerical Analysis* (Prentice-Hall of India, New Delhi, 2006)
37. J.B. Scarborough, *Numerical Mathematical Analysis* (Oxford-IBH, New Delhi, 1966)
38. F. Scheid, *Numerical Analysis, Schaum's Outline Series* (McGraw-Hill, New York, 1984)
39. K.G. Steffens, *The History of Approximation Theory from Euler to Bernstein* (Birkhäuser, Berlin, 2006)
40. J. Stoer, R. Bulirsch, *Introduction the Numerical Analysis* (Springer, New York, 2002)
41. A.H. Stroud, D. Secrest, *Gaussian Quadrature Formulas* (Prentice-Hall, New Jersey, 1966)
42. T. Veerarajan, T. Ramchandran, *Theory and Problems in Numerical Methods with Programs in C and C++* (Tata McGraw-Hill, New Delhi, 2004)
43. E.A. Volkov, *Numerical Methods* (Mir Publishers, Moscow, 1986)
44. B. Wendroff, *Theoretical Numerical Analysis* (Academic Press, New York, 1966)
45. J.H. Wilkinson, *Rounding Errors in Algebraic Processes* (Prentice-Hall, New Jersey, 1963)
46. J.H. Wilkinson, *The Algebraic Eigenvalue Problem* (Clarendon Press, Oxford, 1965)

Index

© Springer Nature Singapore Pte Ltd. 2019
S. K. Bose, *Numerical Methods of Mathematics Implemented in Fortran*, Forum for Interdisciplinary Mathematics,
https://doi.org/10.1007/978-981-13-7114-1